The Tradition

of Technology

LANDMARKS OF WESTERN TECHNOLOGY IN THE COLLECTIONS OF THE LIBRARY OF CONGRESS

Leonard C. Bruno

LIBRARY OF CONGRESS

WASHINGTON

1995

Endpapers: This busy fifteenth-century workshop scene celebrates—in a sixteenth-century work—the "new art" of engraving on copper. The bespectacled engraver sits at the right, cutting lines into the metal surface with a burin. The two men at the low center table are warming and polishing the engraved plates, while on the left, the paper is being rolled over the inked plate. Youthful apprentices are seen assisting their masters, trying their hands at engraving, and simply watching in awe and admiration. This busy image captures the essence of the tradition of technology as an active, purposeful, tangible endeavor that is uniquely human. Drawn by Jan van der Straet; engraved by Phillip Galle. For the series *Nova Reperta*. Prints and Photographs Division, Library of Congress, LC-USZ62-16701.

Designed by Stephen Kraft

Library of Congress Cataloging-in-Publication Data

Bruno, Leonard C.
The tradition of technology: landmarks of Western technology in the collections of the Library of Congress / Leonard C. Bruno
p. cm.
Includes bibliographical references (p.) and index.
ISBN 0-8444-0888-3
—— ——— Copy 3 Z663.7 .T73 1993
1. Technology—History. 2. Technology—Library resources.
3. Library of Congress. I. Library of Congress II. Title.
T15.B685 1993
016.6'094—dc20 93-16406
CIP

For sale by the Superintendent of Documents
U.S. Government Printing Office
Washington, D.C. 20402

To Anna, my mother, and the memory of my father, Pasquale

Contents

Preface

This book is very much like its predecessor, *The Tradition of Science: Landmarks of Western Science in the Collections of the Library of Congress*. As a companion to that earlier work, it is guided by the same principles and methods. It is similarly, therefore, neither a qualitative survey of those collections nor a merely descriptive guide. Finally, it is not real history either. Rather, this book is a highly selective and subjective effort and is perhaps best described, like the first, as one person's trip through the general and special collections of a great collecting institution. Its overall purpose is to identify the major published works in technology available in the Library of Congress and to offer the reader some information of interest or usefulness about them.

Many of the works on technology chosen for inclusion are well known and often recognizable as being on anyone's list of great books. Unlike science, however, the history of technology embraces a far greater number of what might be called unknown classics, and many of these are included here also. All of these selected works are first organized in an essentially chronological manner and then treated somewhat thematically. Thus each chapter encompasses a broad period of time as the book progresses from Greek and Roman times to the twentieth century. Because of these great expanses of time and the wide sweep of technological accomplishment, both of which were necessarily compressed in this brief treatment of a large subject, it was often necessary to impose simplistic patterns on certain eras or epochs. This made them more tractable for the writer but sometimes less rich for the reader, lacking the subtleties embedded in the often complex and contradictory reality. This taking refuge in the house of conventional history was as unavoidable here as it was in the writing of *The Tradition of Science*.

Similarly, this treatment is by no means a critical or analytical review, nor is it inclusive. It deals only with works contained in the Library's collections and discusses them in an admittedly friendly manner. For the most part, when a book is discussed, it is as a primary source in the history of technology. No secondary works are mentioned. Finally, this book is written for the general reader by a generalist, with no authority claimed.

Although a particular work may be located in one of the Library's many special collections, no reference to that collection is made in the text, since the emphasis here is always on the individual work itself. The bibliography provides information about particular Library collections to which a book or manuscript may belong, as well as specific bibliographical information. Title translations and Library call numbers will also be found there.

This book was begun with the support and encouragement of Joseph W. Price, Chief of the Science and Technology Division, and completed under the Acting Chief, John F. Price. I thank the former for seeing the worth of this companion volume, and the latter for his steadfast belief in both me and the idea of this book. Little could have been done without the cooperation of the Library's many special collection divisions and their extremely helpful, cooperative, and professional curators. Specifically, the entire staff of the Rare Book and Special Collections Division were always graciously willing to help.

Dana J. Pratt, Director of the Library's Publishing Office, was the first to suggest that this book be written. He wisely recommended that the single chapter on technology originally written for *The Tradition of Science* not be included there for two reasons: its subject was inappropriate, since the rest of that book was entirely science; and its large, important subject demanded separate treatment. I hope I have proven him correct. As with our first book, Mr. Pratt reviewed each chapter as it was written and gave his usual constructive comments and criticisms. His unflagging support and quiet encouragement were a steady source of confidence to the author, who is grateful and very appreciative.

Johanna T. Craig, Production Manager, saw to it that this volume had the same designer as its predecessor, and real continuity was therefore splendidly achieved. Once again, the insight, talent, and empathy of Stephen Kraft manifested itself in his sensitive, knowledgeable, and artistic book design. He is a craftsman in the best and most complete sense of the word. The editor was Anne Rollins, an exacting, conscientious, and extremely hard-working individual. She not only edited my work in a very rigorous manner, but kept the reader in mind throughout, regularly requesting from me the needed detail or clarifying explanation. Evelyn E. Sinclair of the Library's Publishing Office oversaw the editing of the work. Her wise counsel and continuity again proved indispensable. I consider myself a lucky author to have come under her scrutiny. Finally, Ruth S. Freitag, my colleague in the Science and Technology Division, again agreed to provide the book with a bibliography, the likes of which no one else but she could do. I thank her for her extreme diligence and very hard work.

This book was earlier described as one person's journey through the Library's collections in technology, and indeed, it probably reflects all that is both good and bad about a one-man job. It is therefore fitting that I accept responsibility for any and all errors of fact and omission, and I gladly, even enthusiastically, embrace this task. This is so because I have had the privilege, satisfaction, and even joy of daily dwelling with the great works of great individuals housed in a great national institution. It is a distinct honor to associate with the first two and to represent the third.

Introduction

Most discussions on almost any aspect of technology seem inevitably to consider the issue of its relation to science. Some identify the two, although most make adequate distinctions which reveal something of the singular essence of both fields. Science and technology are certainly distinguishable and ought to be so. Although the effort to distinguish could be considered merely academic or conceptual, the understanding that derives from such a logical exercise is worth the effort.

Science could be described as mankind's intellectual pursuit of truth, or more specifically as its attempt to gain a deeper understanding of the natural world or the nature of things. As science works ultimately within the realm of ideas, technology is concerned rather with the tangible world of invention or the construction of things and processes which are useful. Historically, both science and technology have been subjected to warping stereotypes or unthinking notions that distort the true nature of both. One of these opinions raises science to a higher plane and regards technology as existing on a significantly lower level, saying it is merely the application or elaboration of scientific principles. According to this simplistic model, because science pursues knowledge for its own sake, it deserves all of the positive implications and connotations that go with the term "pure science." In this, the scientist is totally objective and disinterested, with no biases, no ulterior motives, no hopes or suspicions as to where the truth might lie, neither ego involvement nor any distracting or demeaning ambition save the glowing satisfaction of learning one of nature's secrets.

The stereotype of technology is, not surprisingly, the opposite of that of science, and equally simplistic. In this view, technology pursues only useful or practical knowledge, caring not for the truth, beauty, symmetry, or simplicity or the underlying principles (which were discovered by science). Unlike pure science and its abstract domain of ideas, technology operates in a world of the senses—a tangible, concrete, and therefore decidedly tainted place. The technologist is thus but a talented mechanic who applies the principles of science, caring only for the technical solution to his grimy, limited problem.

Obviously, both extremes are exaggerated and certainly neither is correct, both being essentially caricatures. The scientist is not a dreamy metaphysician, nor is the technologist just a mechanic with dirty hands. Science does indeed have as its paramount goal the pursuit of knowledge for its own sake, while technology seeks to employ knowledge to some positive advantage. Yet the two are related by a contingent rather than a

Opposite page:
Since earliest times, mankind has used technology to make its mark upon the earth—from the prehistoric cave paintings at Lascaux to this fourth-dynasty Sphinx and the even-older Pyramid in the background, to the soaring medieval cathedrals, the ugly and brutal mines and factories of the Industrial Revolution, and the twentieth-century ribbons of concrete highways that can be seen even from outer space. Throughout its fairly short history, mankind has used this seemingly innate technical inclination not only to imagine new things and new ways but actually to make them come to be. In many ways, the tradition of technology can be viewed as the universal human impulse first to tinker with the natural state of things and then to change it to suit our wants, needs, or even whims. For good as well as for ill, mankind has left its mark with its technology wherever it has tread, even saying on the moon, "I was here." Prints and Photographs Division, Library of Congress, LC-USZ62-104407

necessary relationship, with neither inherently superior to the other. It is safe to say that during the greater part of mankind's history, technological inventions and improvements were made with no help from science. Until relatively recently in the history of civilization, most technological advances were made by individuals blithely unaware of any theoretical principles. Until the scientific revolution that began sometime in the late sixteenth century, this was most often the case. Technology advanced by knowing how but not necessarily why. It was only later that technical advancement came as a result of the direct application of scientific knowledge. This might be described as knowing how based on knowing why. In the former case (knowing only how), technology was most properly described as an art, and bore all of its characteristics. It progressed by trial and error and had all the aspects of what has been called a "craft mystery." Here, its essential processes were closely held by its secretive practitioners, and the techniques or the art were both based on and passed on by experience or actual doing. Today's technological experience is a more open, predictable, and even deliberate process that is quite often linked intimately to science, which, in turn, also has benefited from the relationship.

Innumerable examples can be given of both science-driven technology and technology-pushed science, and few today would deny technology its own separate standing. Thus it might be more productive to examine the nature of technology itself to try to discern what makes it separate and unique. One of the first observations easily made is that technology is an essentially human endeavor. Mankind is often referred to as "toolmakers," and despite the many examples of the use of tools by animals, it is safe to characterize only the human race as such. Toolmakers we have been from the beginning, it seems. In fact, mankind might be considered as much a technological animal as a social animal. Our whole history seems to attest to that description.

Another obvious observation is that technology is a purposeful, fairly rational activity that usually results in some sort of tangible object. Simply, it involves the making of things that had no prior natural existence. This is very different from science, whose principles exist irrespective of human discovery, for the fact that human intervention or involvement is so necessary to technology makes it an activity that tells us a great deal about the intervening humans. In one way, the object made could be said to be a physical manifestation of the mind that created it. Without belaboring the point to the extent of saying "you are what you make," there would seem to be some validity and usefulness in searching for the man or woman within the made object itself. Just as the material remains of ancient civilizations enable today's archaeologists to compile retroactively a profile of a particular culture, so will our own technology become an eloquent and telling spokesman to future investigators. To a large extent, technology contains the silent and sometimes coded imprint of the values of its creators.

Stated another way, technology is decidedly culture-sensitive. Who we are, how we live, and what we believe conditions and affects not only the technology we produce but also the ends we pursue with it and the way in which we use it. There are reasons why the ancient Greeks put technology in the service of warfare, leisure, and amusement but seldom used it to advance production or profit. One of those reasons was their outlook on life—their inner-oriented philosophy and essentially nonmaterial values—and their love of the marvelous. The example of the Greeks is repeated throughout the history of technology—that the made object is but the visible expression of the mind that created it. Henry Adams says essentially this very thing when he discusses the different cultures and values responsible for the medieval cathedrals and the modern dynamos.

Another useful generalization about technology is that its domain is almost totally that of the particular. Unlike science, which strives for ever more inclusive and general theories, technology glories in the details. Embedded as it is firmly in the minutiae of a particular solution to a particular problem, technology must be detail-specific or it will not be successful—that is, it will not work. And working or functioning properly is another characteristic of technology. Its values are those of the everyday world—efficiency, economy, expediency—and not those of the realm of ideas. A technical design or solution may be more attractive and pleasing if it has some sense of aesthetics about it, but such notions are essentially peripheral to its basic purpose (unless the design is somehow a factor in its function). Startling exceptions to the functional specificity of technology are found occasionally throughout history, and there are several cases when a certain technology has gone well beyond its specific limited purpose and has effected some broad, nontechnical, change. The role of gunpowder in the demise of feudalism is one well-known example, as is the part played by the automobile in the transformation of American life. This irony of the toolmaker being remade by his own tools is even more relevant to today's technology-saturated world.

A final observation on the nature of technology goes against the often widely held opinion that technology progresses via major leaps. In fact, its history shows that technological accomplishment is much more evolutionary than most would think, with the real, unprecedented breakthrough being a truly rare phenomenon. One of the best examples of this thesis is the experience of Leonardo da Vinci, regarded by most as one of history's most creative and technologically inspired individuals. Few have bothered to inquire as to his background and training, preferring to hold him aloft as a solitary genius who owes little to anyone. In fact, however, it takes nothing away from his true technical genius to state that he was a member of the *second* generation of Italian Renaissance engineers, and that much of his engineering work followed on already established practice and tradition.

Tradition plays a large role in the history of technology, as it does in the life story of any worthwhile human pursuit. For technology, this tradition is most directly and obviously expressed as "toolmaking." Whether it is the early craft technology which owed little or nothing to science (knowing how but not why) or the more applied science of a recent technology (which knows and uses the why), the tradition of technology is active, purposive, even tangible [illegible] itself possessing all the characteristics

Opposite page: This busy fifteenth-century workshop scene celebrates—in a sixteenth-century work—the "new art" of engraving on copper. The bespectacled engraver sits at the right, cutting lines into the metal surface with a burin. The two men at the low center table are warming and polishing the engraved plates, while on the left, the paper is being rolled over the inked plate. Youthful apprentices are seen assisting their masters, trying their hands at engraving, and simply watching in awe and admiration. This busy image captures the essence of the tradition of technology as an active, purposeful, tangible endeavor that is uniquely human. Drawn by Jan van der Straet; engraved by Phillip Galle. For the series *Nova Reperta*. Prints and Photographs Division, Library of Congress, LC-USZ62-16701.

of a tool. And whether the tool produced is a thing or a process, it stands as both a symbol and a product of the individual mind and culture that produced it. As a uniquely human endeavor whose origins and motivations are as diverse as the individual minds which created it, a given technology can be as telling and evocative as the most complex and beautiful work of art. What we choose to make and what we are able to make not only say a great deal about what kind of people we are, but in turn, affect and influence us in a very fundamental way. Mankind and tools are inseparable—making the tradition of technology not only the oldest but perhaps the truest.

A final word must be said about what might be called the dangers of technology. In our modern society, the pace and level of technological accomplishment have reached such a high level that they have begun to impinge on the most intimate and essential aspects of the natural world. By now, most of us are aware that negative secondary effects of technological exploitation may be endangering the health and physical future of the earth itself. Although there is disagreement about long-term effects, few could argue that mankind has not immensely altered the entire ecosystem, often in very bad ways. At this critical point in the history of technology, it appears we have reached a threshold of sorts, for now technological man must, for probably the first time in his history, ponder all the consequences of a technology *before* he fully develops it. Further compounding this already complex situation is the fact that those making decisions must make these judgments not solely on technical grounds or merit, but from the broader terrain of diverse and often conflicting systems of interests and values. No longer can we afford to invent and implement haphazardly. Our technology is too powerful, too effective, and too far-reaching.

The technological competence possessed by today's world undoubtedly surpasses in every way the dreams of even the wildest Renaissance speculator. Surely the learned person who thought on such things centuries ago would have believed that once mankind was able to learn and master nature's secrets—and thus be able to fly, to walk on the moon, or to send invisible messages and images through the air—a truly golden age would have arrived, and that mankind would certainly be privy to all there was to know about nature and the universe. And yet how far we seem from knowing it all or even from possessing greater wisdom or prudence than our forebears. When technological sophistication becomes an end in itself and is divorced from the larger, human context of values and purpose, it becomes at best a dazzling diversion, and at worst, a dangerous dominator. There seems no escape from the conclusion that unless technology is created and used to serve what might be called noble and just ends, it has little chance itself of possessing either virtue. For as with any tool, it is judged by the end it serves.

19.
SCVLPTVRA IN ÆS.
Sculptor noua arte, bracteata in lamina
Sculpit figuras, atque prælis imprimit.

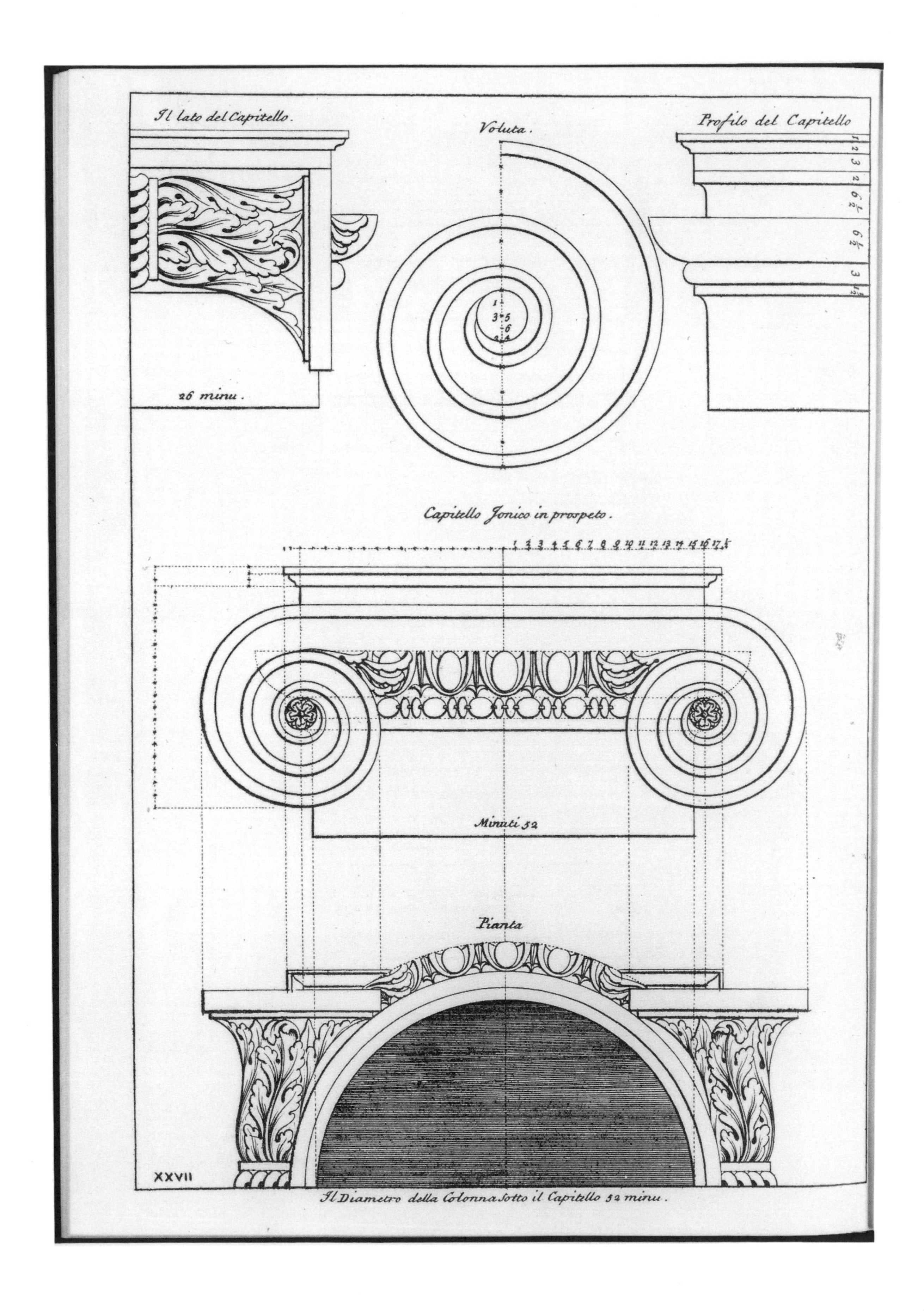

Il Diametro della Colonna sotto il Capitello 52 minu.

1. *The Technology of Greece and Rome*

By the time ancient Greece was just beginning to flourish, around 600 B.C., several millennia of technological accomplishment had already passed. Mankind had long mastered the rudimentary aspects of everyday life and industry, and its technical skills and understanding had reached levels of surprising sophistication. The treasures found in the pharaoh Tutankhamen's tomb bear witness to this.

Although the Greeks are most often acknowledged for their seminal contributions to science—that is, by their seeking knowledge for its own sake—they were of necessity also a very practical people, their economy being almost totally based on commerce. Traditional wisdom has the Greeks despising physical labor and its result, or indeed anything practical, yet their siege engines, drydocks, ship building, waterclocks, and architecture seem to portray a busy, vigorous, earthy people.

Attempting to resolve this apparent contradiction underscores a revealing aspect of ancient Greek culture, temperament, and values. Apparently this marvelously talented and unique race could have been technological virtuosi had they so chosen or been so inclined. Yet technology was relegated to something of a secondary status, a pursuit whose derivative nature made it unworthy of intense or prolonged attention. Perhaps if they were less individualistic, more worldly, or less aristocratic they would not have felt so keenly the common, homely taint characteristic of the manual arts. More likely, if they had had a more modern sense of technology as, in part, applied science or as the concrete application or tangible embodiment of a principle or an idea, they may have acted differently.

The cultural and intellectual bounty bequeathed to the world by ancient Greece is so rich that its technological heritage seems rather poor in comparison. So it is that Greek technology, when compared to Greek accomplishments in nearly any other field—science, politics, art, literature, or even the art of being human—is regarded as not having realized anything near its full potential. Here the Greeks seem to fall short by the standards that history has learned from the Greeks themselves. When a people or a nation becomes so identified with or defined by its superlative achievements that the regular, accepted standards of excellence no longer apply, history's expectations are inevitably raised.

It has become a historical cliché to say that the Greeks despised anything smacking of the technical, practical, or manual arts. Indeed, the intelligentsia did feel this way. As concerned and content as they were

Opposite page:
The classical tradition of architecture, with its regular proportion and symmetry, is seen in this Renaissance depiction of an Ionic capital. The three stone orders of architecture—Doric, Ionic, and Corinthian—defined the pattern of the columnar facades of most Greek temples and gave the West a standard of harmonic proportion. To the Greeks, good design was synonymous with beauty. *Ten Books on Architecture* (1955). Copyright Da Capo Press, Inc., 1955. 1485. Leone Battista Alberti. LC-USZ62-110076.

with their pursuit of the intellectual aspects of physical nature and human society, they felt little cultural impulse to flex their technological muscles and let theory serve practice. The seeds for a real industrial revolution were ready for planting, and many a historian has pondered why they were not sown. The reasons can most likely to be found in Greek culture and its general system of values. As a society, its highest values granted precedence to things intangible—to honor, duty, loyalty, to the intellect, the spirit, and the gods. Since Aristotle's time at least, Greeks were always more intrigued with the metaphysical "why" than with the practical "how." At any rate, Greek contributions to science are sufficiently secure so as not to be threatened by criticism that they could have and therefore should have gone even further than they did.

There is little doubt, however, that had they wished, they could have produced technological marvels. Their knowledge of mechanics, as evidenced by the devices and systems they did produce, was of a kind and degree not seen again for another fifteen hundred years. Seemingly, the Greeks regarded technology in only two ways: either it was a plaything to amuse and divert, or it was a lowly craft to produce the tools or techniques necessary for a civilized and secure life. It appears that it was simply not a part of the Greek cultural make-up to become a real technological society.

Whatever the Greeks may have lacked in this regard—if one could call it a deficiency—the Romans possessed in force. Not surprisingly, the scientific reputation of Rome is the antithesis of that of Greece. Where the Greek tended to disdain the vulgar application of a principle or the careful elaboration of a technique, the Roman took to both with vigor and even excelled. Roman aqueducts and bridges, war engines and road systems are all tangible products of the realistic and pragmatic Roman. Unlike the Greek, whose culture, values, and style inclined him toward what we today would call pure science, the Roman's culture and values placed greater emphasis on the more active and worldly aspects of life, values which esteemed technology more than science. A consideration of whether these values were determined by Rome's imperial ambitions or whether the opposite was more likely does little to explain the Roman's essential proclivity toward technology. As with the Greeks, whose very "Greekness" was the sum total of a unique human experience and history which left them more disposed to things scientific than technological, so "Roman-ness" might suffice in a tautological way to explain the Roman's disinclination to science and disposition toward technology.

The technological legacy of this Greco-Roman civilization is substantial and significant even if the number of extant technical writings is small. Aside from the Hippocratic corpus of medical writings and some writings on mechanical problems from the Peripatetic school, both of which could be called Hellenic or the product of pure Greek culture, all of the writings discussed here are the product of Hellenistic (the post-Alexander blend of East and West) or Greco-Roman culture (Hellenistic as adopted by the Romans). This entire period of ten centuries (ca. 600 B.C.–ca. A.D. 400), which spans the time from when the Greeks

The Romans were vigorous, pragmatic builders with an imperial sense of grandeur. This eighteenth-century drawing of the Castel Sant'Angelo shows it to be a circle surrounded by a square. Built in A.D. 135–39 by the emperor Hadrian as his tomb, it was converted into a fortress in the fifth century and served throughout the Middle Ages as a refuge for the popes in times of trouble. Today the statue of the Archangel Michael can still be seen atop the castle. It was placed there in A.D. 590 by Pope Gregory the Great, who had a vision of the angel above the fort which signified to him the end of the plague. The bridge in the foreground is the Pons Aelius (A.D. 134), to which Bernini added angels. It is now known as the Ponte Sant'Angelo. *Le Antichita romane,* 1756. Giovanni Battista Piranesi. LC-USZ62-110277.

first began to flourish to the fall of the Roman Empire, is generally called classical antiquity.

In the field of what might be called applied mechanics, perhaps the best-known name of classical antiquity is Archimedes, the famed mathematician and mechanical genius. He is credited with the invention of the compound pulley and the hydraulic screw among others and his name is linked to the explanation of the principles of the lever and of the buoyancy of liquids. Ironically, the collected works of one of the most ingenious inventors of all time contain little on such practical subjects. For the most part, his extant works could be characterized as wholly theoretical. The Library of Congress has the first complete edition of his works, a 1544 collection called *Opera omnia*, printed in Basel. The work was first made available to the West in 1269 when the Flemish Dominican monk William of Moerbeke completed his translations from Greek into Latin. It was from Moerbeke's manuscript translations that the first printed version of the partial works of Archimedes was made in Venice in 1503. Apart from this small tract and an imperfect Latin *Opera* published in 1543 in Venice, the 1544 edition in both Greek and Latin is the definitive Archimedean work. Containing many mathematical diagrams, this

That the Greeks knew of the motive power of steam is evidenced in the work of Hero of Alexandria, himself an inheritor of Ctesibius who lived two centuries before. In this sixteenth-century book, the inner workings of Hero's miraculous altar are shown. The temple doors open when a fire is lit and close shut when it is extinguished. The fire heats compressed air in the base which then expands in the sphere H, compressing its water and forcing it into receptacle M. This sinks and pulls the cord running over the pulley P, which moves concealed hinges on their pivots and opens the doors. Cooling down slackens the rope and the doors close. Hero's approach to mechanics was typically Greek, as he ignored the practical aspects of this transfer of energy and applied it solely to toys and wondrous devices. *Gli artifitiosi et curiosi moti spiritali di Herrone,* 1589. Giovanni Battista Aleotti. LC-USZ62-110278.

work inspired the so-called Archimedean renaissance and influenced directly the mathematical work of Galileo, Descartes, and Newton.

It is not to Archimedes himself, however, but rather to Plutarch that we must look for information concerning the technological accomplishments of Archimedes. In his *Life of Marcellus*, Plutarch recounts how Archimedes applied his knowledge to help defend the city of Syracuse against the Roman general Marcellus during the second Punic War. At the request of his friend and possible relation King Hieron of Syracuse, Archimedes is said to have devised catapults, iron hooks to capture ships, and other devices including the famous "burning glasses"—composite mirrors used to set the Roman fleet on fire. Plutarch also recounts Archimedes's general dislike for such technical displays, saying that "he considered practical mechanics and especially any art governed by a particular need as manual and vulgar." In this attitude and in his preference for the contemplative life, Archimedes characterized the Greek ideal.

Nevertheless, writings of a decidedly practical nature began to appear more frequently after Archimedes. Of the three great post-Archimedean inventors—Ctesibius, Philo of Byzantium, and Hero of Alexandria—only the writings of Hero have survived. What we know of Ctesibius comes mainly from Vitruvius, who referred to a book written by Ctesibius, and from Hero, who also wrote of his inventions. The most significant accomplishment attributed to Ctesibius is his improvement of the Egyptian clepsydra or water clock. Philo of Byzantium also invented many pneumatic machines, mostly for use in battle, and his most famous device is a sort of compressed-air gun. His major work was an encyclopedia or compendium of applied mechanics in nine books, only fragments of which still exist.

Many describe Hero of Alexandria as the greatest technician of classical antiquity, and his fame as a clever inventor certainly rivals that of Archimedes. The fact that many of his writings survived antiquity could be a factor, but he was an ingenious and versatile inventor. His most famous works deal with air and mechanics and are well represented in the Library's collections. The Library of Congress has the earliest Latin translation of his *Pneumatica*, titled *Spiritalium liber*, published in Urbino in 1575. The translator, Frederico Commandino, died before it was published. The work contains numerous woodcuts showing double-action pumps, siphons, and the famous aeolipile. Hero's work on mechanics, *Automata*, is represented in the Library by Giovanni Battista Aleotti's 1589 Italian translation titled *Gli artifitiosi et curiosi moti spiritali di herrone*, published in Ferrara. The Library also has the 1589 Italian translation by Bernardino Baldi called *Di gli automati*, published in Venice. All three of these translations contain beautiful engravings of Hero's mechanical devices. Among those depicted are his whirling aeolipile or aeolian wheel, a hollow sphere which spun like a top when propelled by steam escaping through two tube outlets. This was a true steam engine, which would be reinvented seventeen centuries later.

It was Hero who first demonstrated the motive power of steam. Also, his knowledge of the properties of air was far ahead of his time and he employed his insight and skill to produce a treasure house of working toys which would whistle, pivot, siphon water, move gears, and flap wings. These devices were mostly employed in recreational pursuits such as the theater. It is intriguing that the Greeks used these potential power sources only to amuse or entertain—to operate a penny-in-the-slot machine for holy water or to open temple doors when a fire was made on the altar. Hero's inventions were not translated into the practical realm of everyday life, not even by Hero himself. Rather, despite his insights and inventions, Hellenistic technology advanced traditionally, by the regular but slow accretion of the craftsman's techniques and skills. Hero's work was not all diversion, however, for his writings served as textbooks for the school he founded at Alexandria.

As Hero's work stands nearly alone as our major source on the practical mechanics of classical antiquity, so the work of Marcus Vitruvius Pollio constitutes the only surviving treatise on the architecture and engineering of that same period. His famous work, *De architectura*, not only contains details of Roman architecture and building methods but is the only Roman work extant that is based on the classical architecture of Greece, and as such assumes an even greater significance. Written in ten books, each with a separate preface, the treatise makes no distinction between architecture and mechanical engineering, the learned Roman architect being both theoretician and practitioner. Vitruvius served Julius Caesar in some capacity, probably as a military engineer, and later served the emperor Augustus. Because of the terrible Latin prose of *De architectura*, many historians believe that its author was not highly educated. But as one authority put it, "He writes atrocious Latin, but he knows his business." Vitruvius did know his business, to the extent that he claimed his ten books to be the first real comprehensive work on architectural theory and practice. While that may or may not be so, since he had his favorites among his sources and did not survey all the masters, his admiration for Greek architecture and contempt for the work of many of his contemporaries resulted in a work which informs us more about the architecture of classical Greece than of Roman imperial architecture.

Although it had little or no influence or applicability in medieval times, at least fifty-five manuscript copies of *De architectura* were made. After its rediscovery in 1414 by Gian Francesco Poggio Bracciolini, it exerted an enormous influence on Renaissance architecture. It was first printed sometime between the years 1483 and 1490 in Rome. The Library does not have this first edition or the first illustrated edition (Venice, 1511). Rather, it has the Como edition of 1521, the first in Italian and perhaps the most beautiful. This magnificent edition was translated and edited by Cesare Cesariano, a pupil of Donato Bramante and Leonardo da Vinci. Some of its illustrations have been attributed to Leonardo. One of them is a full-page plate showing the plan and elevation of the Milan Cathedral, said to be the first measured architectural drawing to appear in a printed book. As such, it is also one of the earliest illustrations of Gothic architecture. Altogether, *De architectura* has rightly been called the bible of architectural theory, dealing as it does with subjects

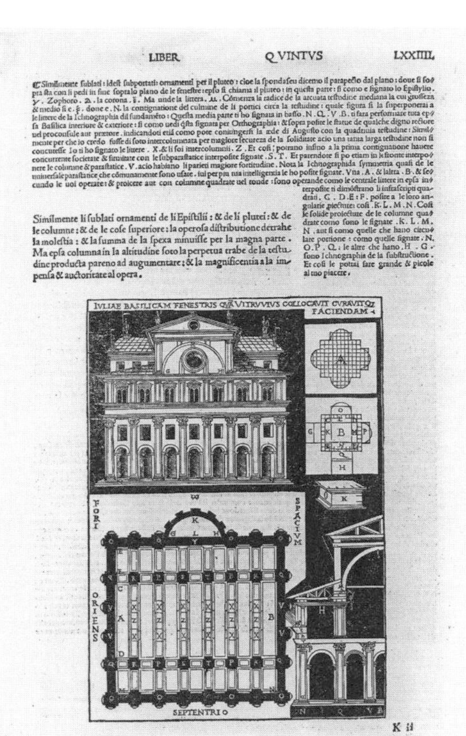

LIBER QVINTVS LXXIIII

Similmente sublati : idest subportati: ornamenti per il pluteo : cioe la spondaseu dicemo il parapecto dal plano : doue si sopra sta con li pedi in fine sopra lo plano de le fenestre : epso si chiama il pluteo : in questa parte : si como e signato lo Epistylio . γ . Zophoro . la corona . Ma unde la littera . Comenza la radice de la arcuata testudine mediana la cui grossezza & medio si e . doue e . N . la contignatione del culmine de li portici circa la testudine : quale figura si la superponerai a le littere de la Ichnographia dil fundaméto : Questa media parte ti ho signata in basso . N . Q . V . B . ti fara performare tuta epsa Basilica interiore & exteriore : si como uedi q̃sta signata per Orthographia : & sopra posite le statue de qualche digno rectore uel proconsule aut praetore . indicandoti etiá como pote coniungersi la aede di Augusto con la quadriuia testudine : Similmente per che io credo fusse di soto intercolumnata per magiore secureza de la soliditate acio una tanta larga testudine non si concutesse Io ti ho signato le littere . X . & li soi intercolumnii . Z . Et cosi ; potrano insino a la prima contignatione hauere concurrente societate & firmitate con le subparastatice interposite signate . S . T . Et parendote si po etiam in le fronte interponere le columne & parastatice . V . acio habiano li parieti magiore fortitudine . Nota la Ichnographida symmetria quasi de le uniuersale parastatice che cōmunamente sono usate . sul per piu tua intelligentia le ho posite signate . Vna . A . & laltra . B . & secundo le uoi operare : & proicere aut con columne quadrate uel tonde : sono operande como le centrale littere in epsa interposite ti dimóstrano li infrascripti quadrati . G . D . E : F . posite a le loro angulare piectute : cosi . K . L . M . N . Cosi le solide proiecture de le columne quadrate como sono le signate . K . L . M . N . aut si como quelle che hano circulare portione : como quelle signate . N . O . P . Q . : le altre che hano . H . G . sono Ichnographia de la substructione . Et cosi le potrai fare grande & picole al tuo piacere .

Similmente li sublati ornamenti de li Epistilii : & de li plutei : & de le columne : & de le cose superiore : la operosa distributione detrahe la molestia : & la summa de la spexa minuisse per la magna parte . Ma epsa columna in la altitudine soto la perpetua trabe de la testudine producta pareno ad augumentare : & la magnificentia a la impensa & auctoritate al opera .

A handbook on classical architecture, this treatise is the only Roman work inspired by Greek architecture that has survived. Written by Vitruvius, a Roman contemporary of Julius Caesar and Augustus, it deals with the classic principles of building and discusses all aspects of mechanical engineering as well. Vitruvius was largely unknown through the Middle Ages, and it is with the Renaissance that his influence began. This ancient floor plan obviously served as a model for many a Renaissance project. *De architectura,* 1521. Cesare Cesariano. LC-USZ62-110279.

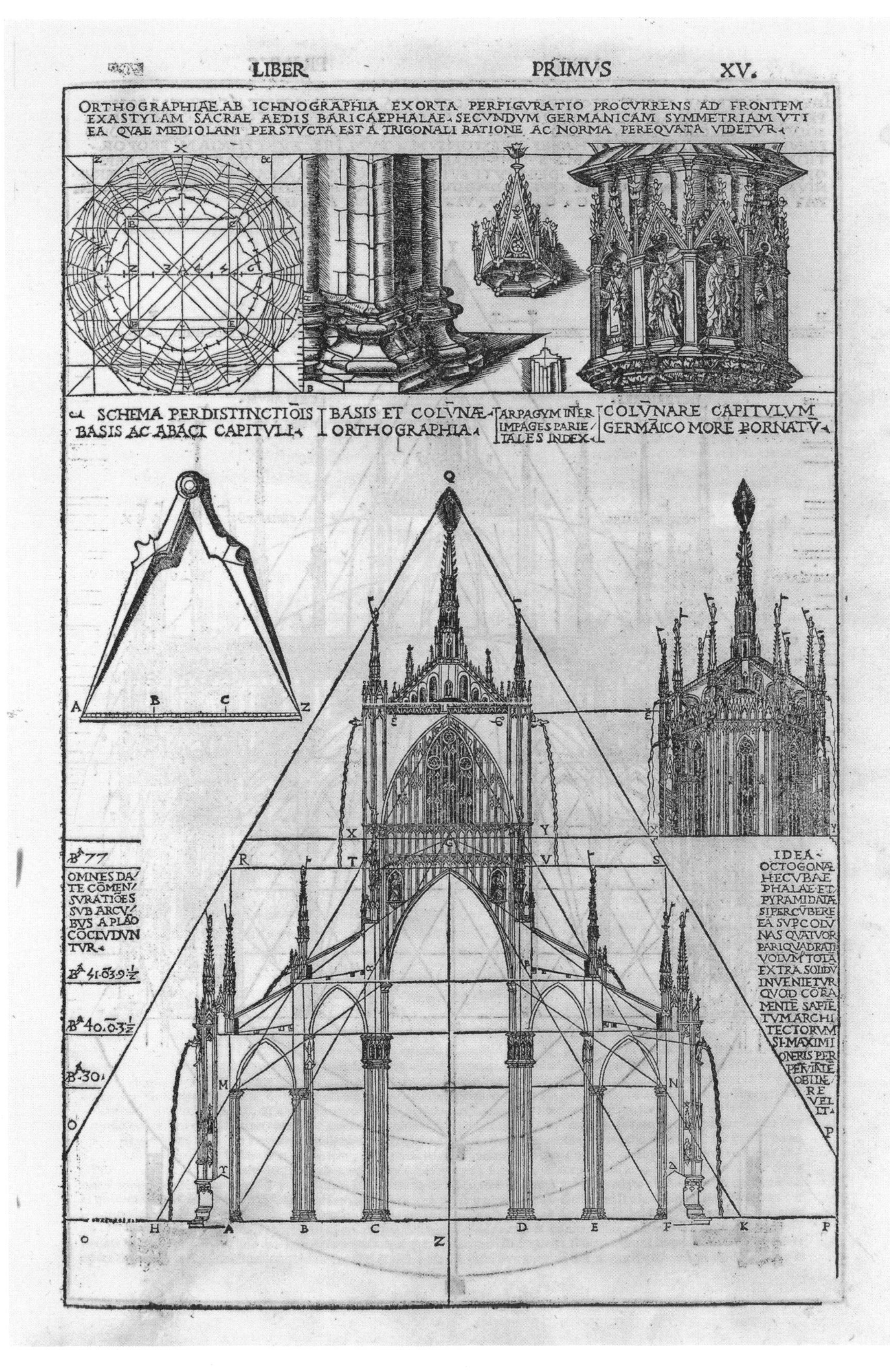
LIBER PRIMVS XV.
ORTHOGRAPHIAE AB ICHNOGRAPHIA EXORTA PERFIGVRATIO PROCVRRENS AD FRONTEM EXASTYLAM SACRAE AEDIS BARICAEPHALAE. SECVNDVM GERMANICAM SYMMETRIAM VTI EA QVAE MEDIOLANI PERSTVCTA EST A TRIGONALI RATIONE AC NORMA PEREQVATA VIDETVR.
SCHEMA PERDISTINCTIOIS BASIS AC ABACI CAPITVLI.
BASIS ET COLVNAE ORTHOGRAPHIA.
ARPAGVM INER IMPAGES PARIETALES INDEX.
COLVNARE CAPITVLVM GERMAICO MORE ЭORNATV.
OMNES DATE COMENSVRATIOES SVB ARCVBVS A PLAO COCLVDVNTVR.
IDEA OCTOGONAE HECVBAE PHALAE ET PYRAMIDATAE SIPERCVBERE EA SVPCOLVNAS QVATVOR PARIQVADRATI VOLVM TOTA EXTRA SOLIDV INVENIETVR QVOD CORAMENTE SAPIETVM ARCHITECTORVM SI MAXIMI ONERIS PERPETVITATE OBTINERE VELIT.

ranging from the specifics of building materials and religious, domestic, and public architecture, to the nature of architecture and the temperament and training of an architect. The later books provide valuable details on water transport, clocks and dials, and military technology and engineering.

Fifteen hundred years after Vitruvius had served his two Caesars, the classical principles of architecture were re-echoed and reformulated by the learned humanist and great architectural theorist Leone Battista Alberti. Alberti defined beauty, whether in art or in architecture, as a kind of objective mathematical harmony knowable to all mankind. His major work on architecture, *De re aedificatoria*, was based on Vitruvius and

Opposite page:
This full-page plate showing the plan and elevation of the Milan Cathedral is considered to be the first measured architectural drawing to have appeared in a printed book. It was one of the many new illustrations included in the first Italian edition of Vitruvius, translated and edited by Cesare Cesariano. It is also one of the earliest illustrations of Gothic architecture. *De architectura,* 1521. Cesare Cesariano. LC-USZ62-110280.

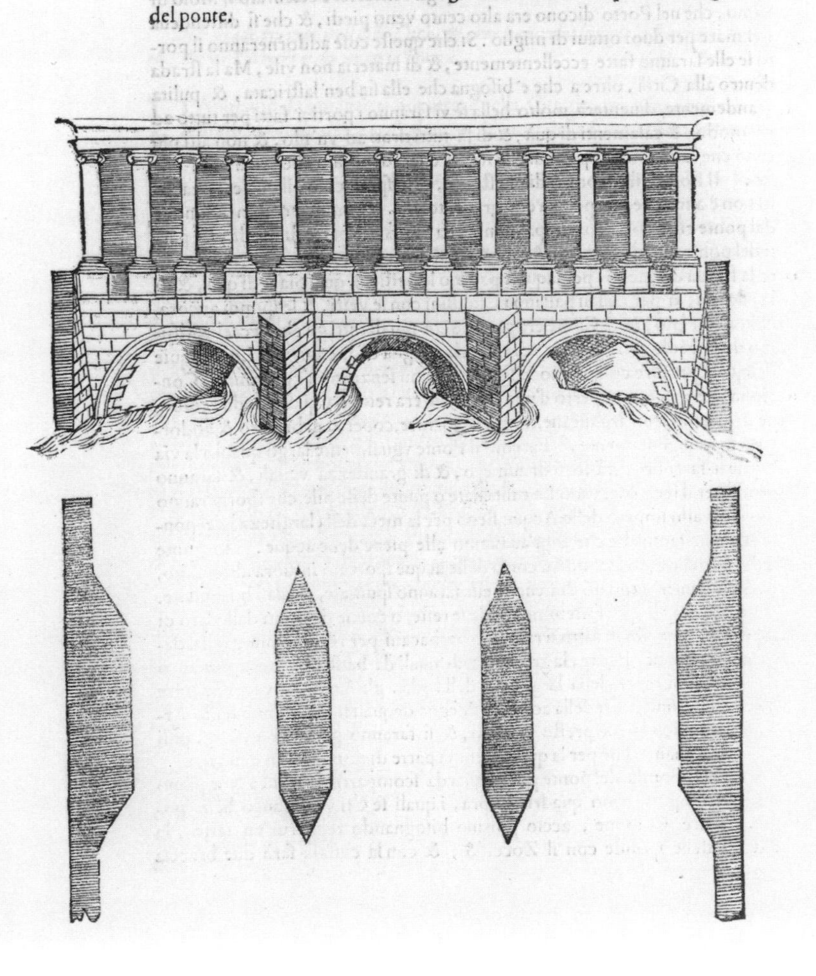

288 DELLA ARCHITETTVRA

& infra l'uno zoccolo, & l'altro, o vero fra l'un piediſtallo & l'altro metterai laſtroni per ritto, o ver muro, l'uno, & l'altro di queſti habbia per cimaſa vna goletta, o piu toſto vna ondetta, tirata per tutta la lunghezza della ſponda, il zoccolo da piede corriſponderà parimente alla cimaſa. Saranno quei duoi andari di quà, & di là, che mettono in mezo la ſtrada di mezo del Ponte fatti, perche vi vadino le donne, & i pedoni, duoi ſcaglioni piu alti, che queſta via del mezo; laquale per amore delle caualcature ſi laſtricherà di Selici l'altezza delle colonne con gli ornamenti ſarà quanto la larghezza del ponte.

In this plan of a bridge, Alberti is ever the humanist, stressing the aesthetics of design. He suggests that an odd number of arches is pleasing to the eye. He also notes that the shape of the piers is critical and that they must not impede the water passing through. They should be angled at both ends, he says, like the shape of a ship, "that the force of the water may be broken by the angle." *L'architettura di Leonbatista Alberti,* 1550. Leone Battista Alberti. LC-USZ62-110281.

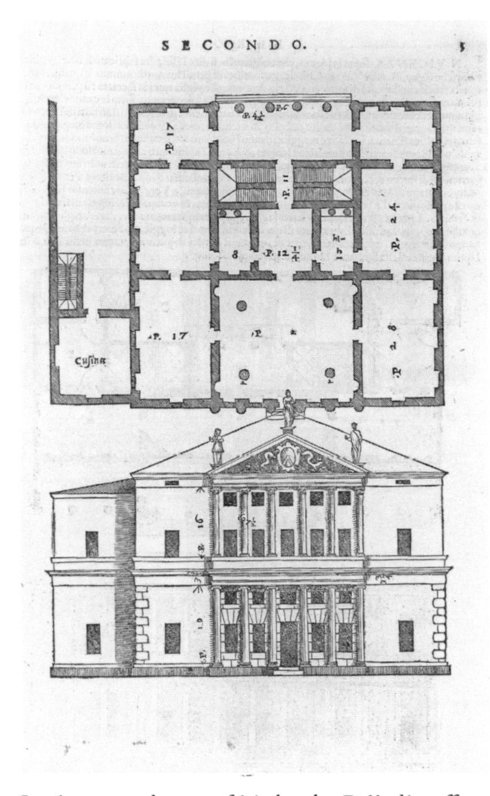

In the second part of his book, Palladio offers his designs for town and country houses built in the classical manner. As the Renaissance interpreter of the grandeur of ancient Greece and Rome, he inspired, with his book and his actual buildings, such works as Chiswick, near London, and Thomas Jefferson's home Monticello, near Charlottesville, Virginia. *I quattro libri dell'architettura,* 1570. Andrea Palladio. LC-USZ62-110283.

Opposite page:
Possibly the most influential architectural book ever printed, this work reached the height of its influence in the architecture of eighteenth-century England, Ireland, and the United States. Known as Palladianism, after its author, Andrea Palladio (1508–1580), this architectural style is best characterized by the Greco-Roman temple front or roofed porch supported by columns. *I quattro libri dell'architettura,* 1570. Andrea Palladio. LC-USZ62-110282.

was similarly titled and organized into ten books. The Library has the first edition printed in Florence in 1485. Alberti may have finished the work as early as 1450. Published posthumously, it became the first great Renaissance treatise on architecture. As an architect, Alberti designed churches in Florence, Rimini, and Mantua. As a writer, he discussed not only the principles of architecture and engineering but their specifics as well, including the construction of vaults, the design of towers, and even the workings of canal locks. The Library has in its collections a later version of this work, edited by Geofroy Tory and printed in Paris in 1512, as well as a 1550 edition titled *L'Archittetura di Leonbatista Alberti*, printed in Florence.

One hundred years after Alberti, Andrea Palladio, regarded as one of the greatest architects of the sixteenth century, was still moved by the classic architectural models of Greece and Rome (via Vitruvius and Alberti) and became one of the most influential figures in Western architecture. Palladio's influence was spread more through his writings than through his building, and his major treatise was titled *I quattro libri dell' architettura*. The Library has the first edition, published in Venice in 1570, as well as two later editions (1581 and 1601). The Library also recently acquired the first English translation of part of this work, *The First Book of Architecture*, translated by Godfrey Richards and published in London in 1663. Palladio pursued and perfected the classical ideals of harmony, order, and a mathematical proportionality of part to part. In fact, he had taken his surname from the name of an angel who represented classical architecture in an epic poem he admired. Born at Padua, his real name was Andrea di Pietro. The Palladian style has been especially dominant in American public architecture, with the United States Supreme Court building, completed in 1935, a recent example.

The final apostle of classical architecture and engineering mentioned here is Giovanni Battista Piranesi. As a Venetian architect and engraver, Piranesi produced an extraordinary number of works illustrating the architecture and antiquities of ancient Rome. The Library has his four-volume folio, *Le antichita Romane*, published in Rome in 1756. The Library also has his magnificent twenty-volume *Opera*, published in Rome between 1761 and 1799. Piranesi was a prolific engraver whose huge prints captured the grandeur of Roman architecture in both their drama and their unparalleled accuracy.

The Vitruvian apostles—Alberti, Palladio, and Piranesi—admirably fostered the architectural ideas of that early Roman who, in fact, was but a conduit of Greek ideas. Where Vitruvius was both a theorist and a practitioner, another Roman, Sextus Julius Frontinus, commends himself to us because of his exactitude in compiling a wealth of practical technical information about every aspect of the Roman water supply system. Although the Greeks brought water to their larger towns via aqueducts, tunnels, and pipelines, nothing they constructed compared in scale and complexity to the Roman system of aqueducts. The system was begun by the blind censor Appius Claudius and continued to be expanded and repaired up to the days of the Emperor Justinian. Aqueducts were built

REGINA VIRTUS
I QVATTRO LIBRI
DELL'ARCHITETTVRA
Di Andrea Palladio.
Ne' quali, dopo un breue trattato de' cinque
ordini, & di quelli auertimenti, che sono
piu necessarij nel fabricare;
SI TRATTA DELLE CASE PRIVATE,
delle Vie, de i Ponti, delle Piazze, de i Xisti, et de' Tempij.
CON PRIVILEGI.
IN VENETIA,
Appresso Dominico de'
Franceschi.
IAC. AVG. 1570. THVANI.

Although tunnel aqueducts were built in Persia and Assyria centuries before Rome, the Romans are considered the greatest aqueduct builders. In the time of Frontinus (A.D. 97), the combined length of all aqueducts entering Rome totaled about 260 miles, of which only 30 miles were on above-ground arches. During a span of over five hundred years, from 312 B.C. to A.D. 226, Rome built eleven major aqueducts across the plains of the Campagna to supply itself with water. Empire-wide, over two hundred arched aqueducts were built. This sixteenth-century depiction compresses a typically lengthy and elaborate water supply system into one frame. *De architectura,* 1521. Cesare Cesariano. LC-USZ62-110284.

Opposite page:
Roman warfare commonly evokes its massed and disciplined legions, yet Rome had a significant and varied arsenal of hardware at its disposal. The siege machine being raised in this sixteenth-century image was known by Vegetius, the Roman military authority, who offered medieval and Renaissance Europe all manner of destructive siege craft. *Scriptores rei militaris,* 1532. LC-USZ62-110285.

throughout the empire, in Egypt and Asia Minor, Spain, France, Germany, and Britain.

Frontinus was both a Roman soldier, three times consul, imperial legate in Britain, and an engineer who in A.D. 97 was appointed *curator aquarum*, described variously as commissioner, administrator, or superintendent of the water supply of Rome. He wrote a treatise in two books about that city's water supply titled *De aquaeductibus urbis Romae*, which provides considerable information about Rome's nine aqueducts and water system. This work somehow survived the ages and was first printed in Rome sometime between 1483 and 1490. The Library does not have this edition but it does have his work as part of a collection of Greek and Roman works titled *Harmonicum introductorium*, published in Venice in 1497. This edition of *De aquaeductibus* contains no illustrations and totals nineteen folio pages. The earliest separate edition of Frontinus in the Library is a 1792 edition published in Altona, Germany.

Frontinus's book is the result of his attempt to establish correct data with which he could administer the system. Consequently, it provides valuable specifics of Roman engineering as well as details on the output, capacity, and dimensions of the aqueducts. Fontinus measured the quantity of water delivered at the sources, reservoirs, and even to the consumers. Unlike Vitruvius, he was one Roman who declared his independence of and technical superiority to the Greeks: "With such an array of indispensable structures carrying so many waters, compare, if you will, the idle Pyramids or the useless, though famous, works of the Greeks."

If there was one area of both Greek and Roman life in which technology played an active, regular, and deliberate role, it was in warfare. The military tradition was an integral part of the history and culture of Greece and Rome, the Greeks being more fratricidal warriors than conquerors, whereas the Romans were mainly imperialists. At various times for each, military success was the necessary difference between expansion or survival, prosperity or annihilation. Each linked much of its status and fate to that of the military. It is not surprising then that both Greece and Rome should attempt to advance the technology of war to a greater degree than any other field of endeavor. And although the Romans may have been the most successful in practicing the art of warfare, it was the Greeks of the Hellenistic period who made most of the major innovations. These weapons usually took the varied forms of siege engines—battering rams and movable towers with drawbridges to surmount walls—and artillery—all manner of catapults and ballistae (devices similar to huge crossbows). Other aggressive devices had names like onager, mangonel, and mantelet. Altogether these machines demonstrated how skill, ingenuity, and motivation could embellish a basic understanding of the principles of the lever, wedge, and screw to a sometimes frightening degree.

Perhaps the best source on the technology of classical warfare is a compilation of four works titled *Scriptores rei militaris*. Published first in Rome in 1487, this collection contains the *Strategematicon* of Frontinus, *De re militari* by Vegetius, *De instruendis aciebus* by Aelianus, and *De vocabulis rei*

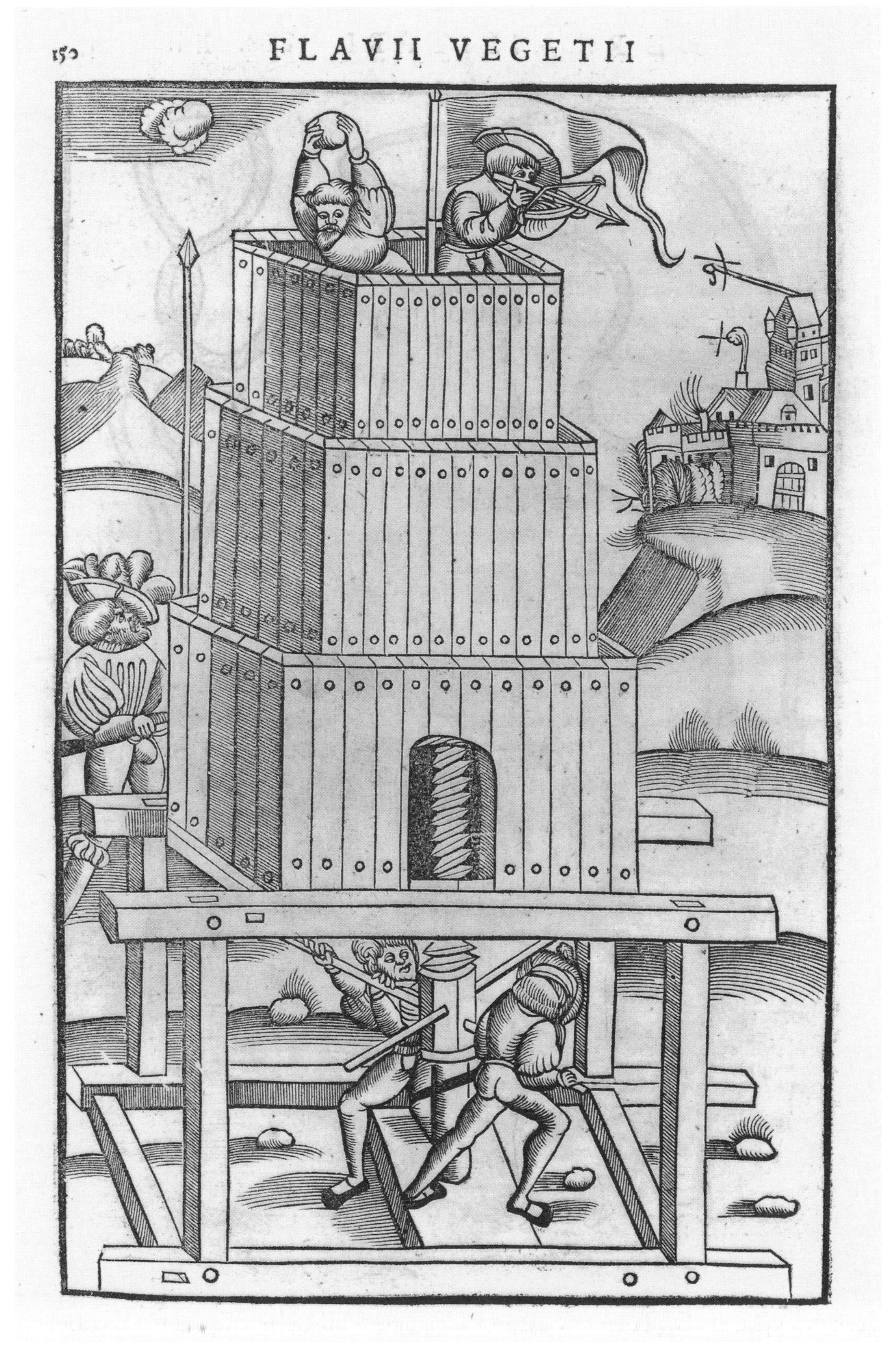
150
FLAVII VEGETII

An indication of the resourcefulness and creativity of Greece and Rome are these sixteenth-century versions of their ideas. The first shows a type of air mattress made of inflated animal skins for resting soldiers. The other suggests how a diver might use an underwater breathing device made of leather to walk on a river bottom. *Scriptores rei militaris,* 1532. LC-USZ62-110286 (air mattress); LC-USZ62-110287 (diver).

militaris, ascribed to Modestus. The Library has the *Scriptores rei militaris* in this 1487 first edition as well as two other editions published before 1500 (1494 and 1495). Of the five editions in the Library's collections that were printed during the sixteenth century, by far the most interesting is the 1532 Paris folio edition. In this, only the Vegetius work is illustrated, but its illustrations are so numerous and varied that they make up for any lack in the others. This work is packed with scores of engravings which depict all manner of weaponry and siege machines, some in full-page illustrations.

Each of the four books contained in the *Scriptores rei militaris* was written during the first four centuries A.D. The earliest was the *Strategematicon* of Sextus Julius Frontinus, the same Frontinus who wrote with such detail about Rome's water supply system. Here he has written a treatise on the art of war, which he begins by offering a collection of the sayings and deeds of the most renowned military leaders of antiquity. Then more specifically, he describes the various contingencies which may precede a battle, the battle itself and its results, the forming and raising of sieges, and various aspects of the internal discipline of the army and duties of the commander.

The next oldest work in this collection was written sometime during the second century A.D. by the Greek tactician Aelianus Tacticus, or Aelian. Dedicated to the emperor Hadrian, *De instruendis aciebus* is a treatise on the military tactics of the Greeks and is said to be derived from the writings of Asclepiodotos, who flourished over two hundred years before. The work also gives military formations, illustrated by diagrams of group and type. The third treatise, *De vocabulis rei militaris*, was written during the following century and is addressed to the emperor Tacitus. It is ascribed to Modestus, possibly a pseudonym, about whom nothing is known. This eleven-page work is valuable in that it contains an outline of the system used during that period to classify and discipline soldiers.

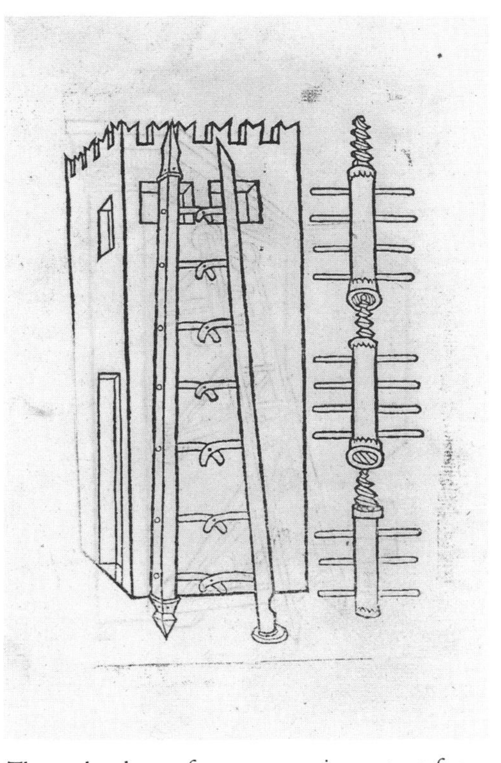

The technology of war was an important factor to the Roman military authority Vegetius, who argued that an army's effectiveness could be multiplied "by mechanical ingenuity." Among the many devices he suggests is a portable siege ladder that can be assembled on the spot. *De re militari*, ca. 1475. Flavius Renatus Vegetius. LC-USZ62-110288.

The final and largest treatise in this collection is *De re militari*, by Flavius Renatus Vegetius, considered by many to be the single most influential military author in the West. Ironically, his writings had little impact on his own times, coming as they did near the end of the western Roman Empire. His work discusses the levying and training of troops and includes instructions for camp fortifications, organization of the legion, operations of an army in the field, attack and defense of fortresses, and even marine warfare. Although Vegetius wrote specifically for his times in the hope of arresting the decay of the Roman army, his work does not tell us how things actually were in his own era but rather how they were during Rome's prime. Vegetius states frankly that his aim was to collect from ancient sources the military wisdom that made Rome great. He accomplished this very well and his popular work was reproduced throughout medieval times. Richard the Lion-Hearted is said to have carried Vegetius's *De re militari* with him throughout all his military campaigns. In addition to its inclusion in the collected edition, this work is also represented separately in the collections of the Library of Congress. The Library has two editions of *De re militari*, both published around 1475. The first is an Augsburg edition published by Johann Wiener, said to be the first technical book printed in Germany. The second is a Cologne edition published by Nicolaus Goetz.

Besides Vegetius, the rapidly declining Roman world summoned a small, last burst of mechanical inventiveness and produced *De rebus bellicis*. Some say it was Rome's only serious attempt to advance the technology of its day. Its author is unknown, although he lived around A.D. 370. He dedicated the work to the emperor, telling him he could double the force of his army, "not by sheer strength alone, but also by mechanical ingenuity." The *De rebus bellicis* is represented in the Library's collections as a very small part of *Notitia dignitatum*, a sort of imperial

Expoſitio Liburnæ.

Roman technical ingenuity is best evidenced in the military tools Romans devised. Whether or not this oxen-powered, paddle-wheel ship was ever built, it nonetheless suggests the combining of known technologies in an imaginative and doable way. *Notitia dignitatum,* 1552. LC-USZ62-110289.

Expoſitio Thoracomachi.

Roman soldiers wore into battle a breastplate or cuirass such as those pictured here. They were usually made of leather that was overlapping or layered for a protective, plated effect. It was not until the second century A.D. that the Roman army adopted armor made of metal scales. Not surprisingly, the army was the largest consumer of leather, and a remarkable degree of specialization existed among the army's leatherworkers. There were thong-makers, saddlers, tent-makers, shield-makers, breast-plate-makers, wineskin makers, and the waterskin makers. *Notitia dignitatum,* 1552. LC-USZ62-110290.

Exemplum baliſtæ quadrirotis.

To both the Greeks and Romans, the main artillery machines were the "ballistae," which shot arrows or bolts using a taut rope like a bow-and-arrow. This Roman ballista is a type of catapult designed to shoot huge arrows against enemy fortifications. Note the armor worn by the horses. *Notitia dignitatum,* 1552. LC-USZ62-110291.

record book which indicated the order of precedence of those who held office throughout the Roman Empire. The Library's well-illustrated *Notitia* was edited by Sigmund Gelen and was published in 1552. *De rebus bellicis* takes up only eighteen folio pages but it has twelve illustrations depicting, among other things, two new ballistae on wheels and pulled by horses, a ship with a series of paddle wheels on its sides which are driven by oxen on the ship's deck, a two-horse chariot with scythes on its wheels, a foot soldier's armor and hardware in some detail, and even a sort of small bridge made of inflated skins. Interestingly, its anonymous author paid homage to the technical ingenuity of the advancing barbarians, saying "they are by no means considered strangers to mechanical inventiveness . . . a quality which we see granted without respect of persons."

One of the most famous as well as most popular works to document the military technology of classical antiquity emerged from Renaissance Italy's fascination with all things Greek. It was commissioned by Sigismondo Malatesta, one of the more dynamic members of a family that ruled Rimini for nearly two centuries. As the prototypical Italian Renaissance prince, Sigismondo was both a vigorous militarist and a patron of the arts. It was in this context that he requested his private secretary, Roberto Valturio, to write a treatise on the art of war. Valturio wrote his own *De re militari* sometime between 1455 and 1460 and approached the subject as a humanist scholar, for he was not an engineer or an architect. The Library of Congress has two undated manuscript editions as well as the first printed edition (Verona, 1472). The two manuscripts contain extensive markings and corrections, most of which were incorporated in the 1472 first edition. This suggests that the manuscripts could have served as the pre-printing originals and the corrections may well have been made by Valturio himself.

From *De re militari,* 1472. Roberto Valturio. See p. 103.

In its book form, the folio contains eighty-two woodcuts, many of which depict the military technology of the quattrocento as well as some fantastic devices. While the illustrations show the writer's interest in the technology of his own time, the majority of the book's text attests to his passion for antiquity. Valturio's text treats the art of war mainly from a historical point of view and provides a sort of compendium of military engineering up to his time, focusing on that paradigm, the Roman war machine. Valturio's book was the first printed book to contain technical or scientific illustrations, and it became extremely popular, going through four more editions before the turn of the century. It is known that Leonardo da Vinci not only possessed a copy but copied passages of the text and commented on them.

One of the more interesting sources of information concerning classical warfare was written by a fourteenth-century poetess of courtly love, Christine de Pisan. Widowed with three children at twenty-five after ten years of marriage, she wrote love poems and romantic ballads to support herself. Achieving immediate success, she turned to longer and more philosophical themes, all the while embellishing the sound education she received at the French court. By 1410 her reputation as a scholar was such that Philip, duke of Burgundy, requested that she prepare a manual of warfare to be used in the young dauphin's military education. The work she produced has four parts. The first deals with a discussion of what constitutes a "just" war and mentions logistics; the second considers methods of defending castles and towns and discusses the operation of assault machines; the third is devoted to legal questions of warfare; and the fourth discusses rules applying to civilian and military populations.

Like any good humanist scholar, Christine de Pisan made substantial reference to the classic works on warfare, such as Vegetius and Frontinus, and her work was much more of a summary than an original piece. In manuscript form, the work was called *Le livre des fais d'armes et de chevalerie.* It was first printed as a book in 1488 by the Parisian publisher and bookseller to King Charles VIII, Antoine Vérard, under the title *Faits d'armes et de chevalerie.* The Library has this first printed edition in its collections. Written in French in double columns with rubricated initials, it contains four half-page woodcuts, two of which depict battle scenes. The Library also has the first English version of this work, published by

William Caxton at his Westminster Abbey press in London in 1489. Al

❧ Cy apres ſẽſuyt la tierce
partie de ce preſẽt traictie La
q̃lle parle des drois darmes
ſelon les loix et droit eſcript.
Et deuiſe le premier chappi/
tre par quel moyẽ lacteur ad
iouſta a ce liure ce qui eſt dit
en droit des faitz darmes
❧ Premier chappitre.

A Inſi que ie pretendoye
a entrer en ceſte iiie.
partie de ce preſent liſ-
ure mon entendement aſſez tra/
uaille de la peſãteur de la matiere
ou labeur des precedẽtes parties
Adonc ſurprins de ſomme en mõ
lit couche: me apparut en dormãt
par ſemblance vne creature treſ/
ſolẽnel/dabit de chiere et de main
tien ancien ſage et auctoriſe iuge
qui me diſt ainſy. ❧ Chier amy
duquel en fait ou en pẽſee labeur
en nulle heure ne ceſſe dexcercite
deſtude que tu as aux choſes que
lettres peuent demonſtrer p eſpe-

though the title of this version is the same as Vérard's, the text is in English, translated by Caxton himself.

Where Caxton's version rightly cited Christine de Pisan as the author, the earlier French edition by Vérard made no mention of her. Further, Vérard implied that the author was a man by changing the original feminine pronouns to masculine ones. He also suppressed altogether Christine's introductory invocation to Minerva, the goddess of arms and chivalry, whose help she requests, "for I am like you, an Italian woman." Despite Vérard's posthumous snub, Christine de Pisan's place in history was already secure. As the first professional woman writer in France, she was of course a champion of women's causes. Also, it was she who introduced the writings of Dante to France. It may only be coincidence that the last poem she wrote was dedicated to another woman of greatness, Joan of Arc, to commemorate her victory at Orléans. Christine died shortly before Joan was martyred.

The final work mentioned here deals with the war engines of classical antiquity. It is a very late sixteenth-century book titled *Poliorecticon sive de machinis tormentis*. Written by the Belgian scholar and professor at Leiden and Louvain, Justus Lipsius, it was first published in 1596. The Library has the 1599 edition, published by Plantin. This treatise not only describes the machines used by the ancients both for attack and for defense of fortified places, but it also contains forty-four engravings (sixteen full-page and twenty-eight half-page) attributed to Otto Venius or Otto van Veen, teacher to Peter Paul Rubens. The engravings provide heavily detailed pictures of all sorts of battering rams, catapults, ballistae, and sling devices.

Opposite page:
This fifteenth-century printed version of Christine de Pisan's treatment of the military classics shows a battle scene in which the combatants are dressed and equipped as knights of that century would be. *Faits d'armes et de chevalerie,* 1488. Christine de Pisan. LC-USZ62-110293.

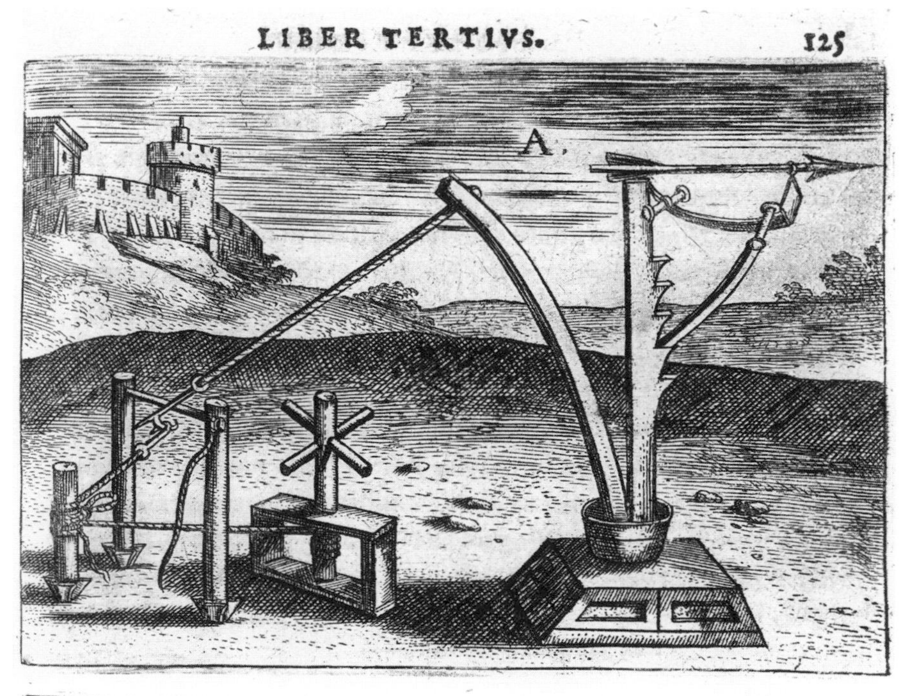

A rudimentary gear is formed by a rope and pulley that is tightened by torsion and then released to snap the bent plank against the upright one, firing the arrow. Notches in the upright support can change the arrow's elevation. Such machines were used by the Romans to shoot both arrows and rocks (up to 175 pounds) more than fifteen hundred feet. *Poliorecticon, sive De machinis tormentis,* 1599. Justus Lipsius. LC-USZ62-110294.

It is perhaps significant that the technology of early warfare is so well documented, relative to other fields. Whether one subscribes to the idea that conflict and aggression are intrinsic human instincts or to the notion that war is but an organized form of theft, it is obvious from the earliest written records that warfare has been with mankind from the beginning. The military culture was an integral part of both Greece and Rome and indeed was often responsible for their maintenance and growth. It is not surprising therefore that both societies devoted a great deal of their physical and intellectual efforts to advancing and refining the art of war. Just as nations today compete earnestly and secretly for a technological breakthrough in modern warfare, so both the Greeks and the Romans dedicated a substantial portion of their resources to the search for better and more lethal weaponry. The books already discussed are but a small, surviving part of these ancient efforts documenting the technology of war.

In extreme contrast to the lethal machines of Greco-Roman times were those invented to sustain life—devices and techniques that made the everyday practice of agriculture a bit easier, more efficient and productive. Although the Greeks used such devices as the water mill and beam press, they were not a people overly smitten with the joys and tasks of producing food. Their mountainous topography probably had as much to do with this attitude as did their natural aversion to the practical arts. In contrast to this Greek disdain for the vulgar application of a principle, the Romans excelled and even seemed to enjoy it. Indeed, the Roman inclination toward the soil and toward the idea of the farm and farming was characteristically deep and lasting. Even when the jaded urban populace of Rome had reached the height of its decadence, it still clung to the conceit that Romans were basically a simple, almost peasant, people. Rome jealously guarded the tradition of its rustic origins, and no better evidence of the Roman's romantic attachment to the earth is available than the *Scriptores rei rusticae*. This is a fifteenth-century compilation of four separate works relating to the agriculture of ancient times. Each was written by a different author and together they span five centuries of Roman agricultural practice from the second century B.C. to the fourth century A.D.

Each of the four texts in the collection originally bore the title *De re rustica* and were written respectively by Cato, Varro, Columella, and Palladius. These authors were greatly separated in time, the earliest being Marcus Portius Cato. Better known as Cato the Censor or Cato the Elder (to distinguish him from his illustrious great-grandson who opposed Caesar and who committed suicide rather than be captured), he wrote in his old age the first work on agriculture in the Latin language. It was a haphazardly organized work that was intended to serve as a practical handbook and thus covered all aspects of farming. It also contains several chapters detailing the construction of an olive press and a crushing mill.

During the next century, Marcus Terentius Varro wrote his treatise on agriculture and husbandry. Varro has been called the most learned of the Romans, and was considered so by both Quintilian and Petrarch. Almost ninety when he died, Varro was a man of vast learning who wrote an estimated 630 books under seventy-four different titles. Of these, only two remain, one of which is his *De re rustica*, which he wrote at age eighty-one. Written in the form of an easy-going dialogue, it treats agriculture in general, the rearing of cattle and sheep, and the breeding of smaller stock. Compared to Cato's work, it is much more logical and fluent but less exhaustive. Although a scholar, Varro devoted much of his life to public service, and more than once narrowly escaped death during the time of Rome's civil war.

The next two writers both drew from Varro. The first, Lucius Junius Moderatus Columella, wrote during the reign of Tiberius and spoke of Varro as a contemporary of his grandfather. His work is the most voluminous of the four writers and provides comprehensive treatment of farming and animal husbandry. His work is both practical and scientific, and it is from him that we learn of such devices as a "hot-bed," heated from below by fermenting manure. Using this method, he tells us, Tiberius raised cucumbers under glass during the winter.

The last writer, Rutilius Taurus Palladius, flourished during the first half of the fourth century. Palladius drew freely from both Varro and Columella, but unlike them, he wrote as a proprietor giving instructions to subordinates and not as one farmer speaking to his equal. Although there is much repetition, he often mentions a variation in procedure or a notable technological development, such as his lengthy and detailed description of a mechanical grain harvester.

The works of these four Roman writers were first published together in book form in Venice in 1472 by Nicolaus Jenson under the title *Scriptores rei rusticae*. There was, however, a single earlier printing (1471) of one of Columella's twelve books. The Library of Congress has several editions of the *Scriptores rei rusticae* compilation. Although it does have the first edition (1472), its copy is imperfect, containing only the work of Palladius. Of the later incunabula editions, it has two copies of the 1482 Reggio Emilia edition as well as the 1496 and 1499 editions from that same city. Later editions include the 1514 Venice edition and one published in Paris in 1533.

From the standpoint of technology, the writings of these four Romans contain glimpses and sometimes details of many agricultural techniques and tools. For example, Cato mentions ploughshares and a flour mill turned by donkeys, and describes a beam press for squeezing grapes and olives. Varro and Columella offer grafting techniques and details of viticulture, and Palladius mentions water mills. However, when Cato the Censor and Varro, Rome's greatest scholar, devote their energies to tracts on farming, one suspects a larger purpose is being served. Such books were written not only to be valuable handbooks of practical instruction but also in the hope of maintaining the traditional customs and virtues of rural family life that had served Rome so well in the past.

Nearly a thousand years after the last of these Roman writers flourished, their works were revived and adapted to the needs of a new time by a lawyer from Bologna. In 1299, Petrus de Crescentiis (or Pietro Crescenzi) retired from his active legal and political career and withdrew to

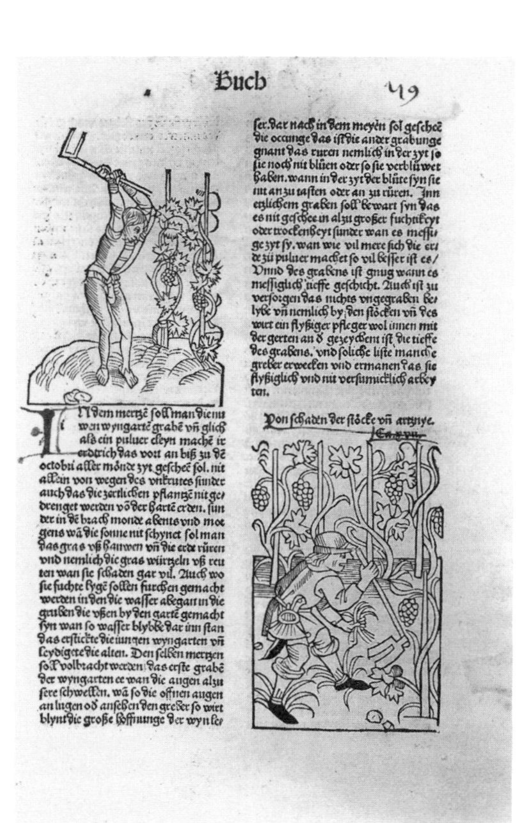

Buch 49

By the time of the Romans Varro and Columella, the cultivation of grapes was already centuries old. Both discuss the necessary techniques of grafting, training, and pruning. This fifteenth-century page tells of the importance of training and cultivating the vine. *Ruralia commoda,* 1493. Pietro de Crescenzi. LC-USZ62-110295.

Nothing was more important in medieval times than successful husbandry and agriculture. A failed harvest could mean disaster. In 1299, Pietro Crescenzi harkened back to Roman agricultural wisdom to compile a practical encyclopedia on how to run a successful estate. This image depicts an obviously bounteous harvest. *Ruralia commoda,* 1486. Pietro de Crescenzi. LC-USZ62-110296.

his country estate, Villa d'Olmo, in Urbizzano. Nearly seventy years of age when he retired, he was to live another twenty years. During that time, he devoted himself to the composition of the great book on husbandry and agriculture that was to make him famous. Completed sometime during the first decade of the fourteenth century, his *Ruralia commoda* was a comprehensive work composed of twelve books or chapters. His aim was to be encyclopedic, and he drew upon all available sources of information. His work is mentioned here because his primary sources were the Latin authors Cato, Varro, and Palladius, as well as Pliny's *Historia naturalis*. He did not slavishly reproduce these writers' words, however, but critically revised and updated the Roman writers to meet the needs of his time. Because of this, his work became one of the most important and popular books of the fourteenth century and remained a standard for nearly three more centuries. It was thus primarily owing to Crescenzi's work that the Latin agricultural classics became known again. The *Ruralia commoda* was spectacularly popular and was issued in Latin, French, Italian, German, and Polish printed editions, with thirteen separate editions issued during the fifteenth century alone. The Library of Congress has several of these, the most significant being the first edition (Augsburg, 1471) and two illustrated versions, one in French (Paris, 1486), and one in German (Speier, 1493).

The final work noted here is perhaps the world's oldest cookbook. Entitled *De re coquinaria*, which means simply "On cookery," the work is a compilation of Greek and Roman recipes intended for the Epicurean. It is attributed to M. Gabius Apicius, who may have been any one of three Romans named Apicius who lived at different times between the rise of Julius Caesar and the rule of Trajan, a period of about 150 years. Adding to the confusion over the identity of Apicius is the fact that an anonymous Roman transcribed into Latin 470 of Apicius's recipes during the fourth century A.D., and another Roman writer, Vinidarius, contributed some of his own to the text during the next century. The issue of authorship aside, the book remains a fascinating glimpse into the cooking and eating habits of Rome's elite. The Library's copy is one of the very rare first editions, published in Milan in 1498.

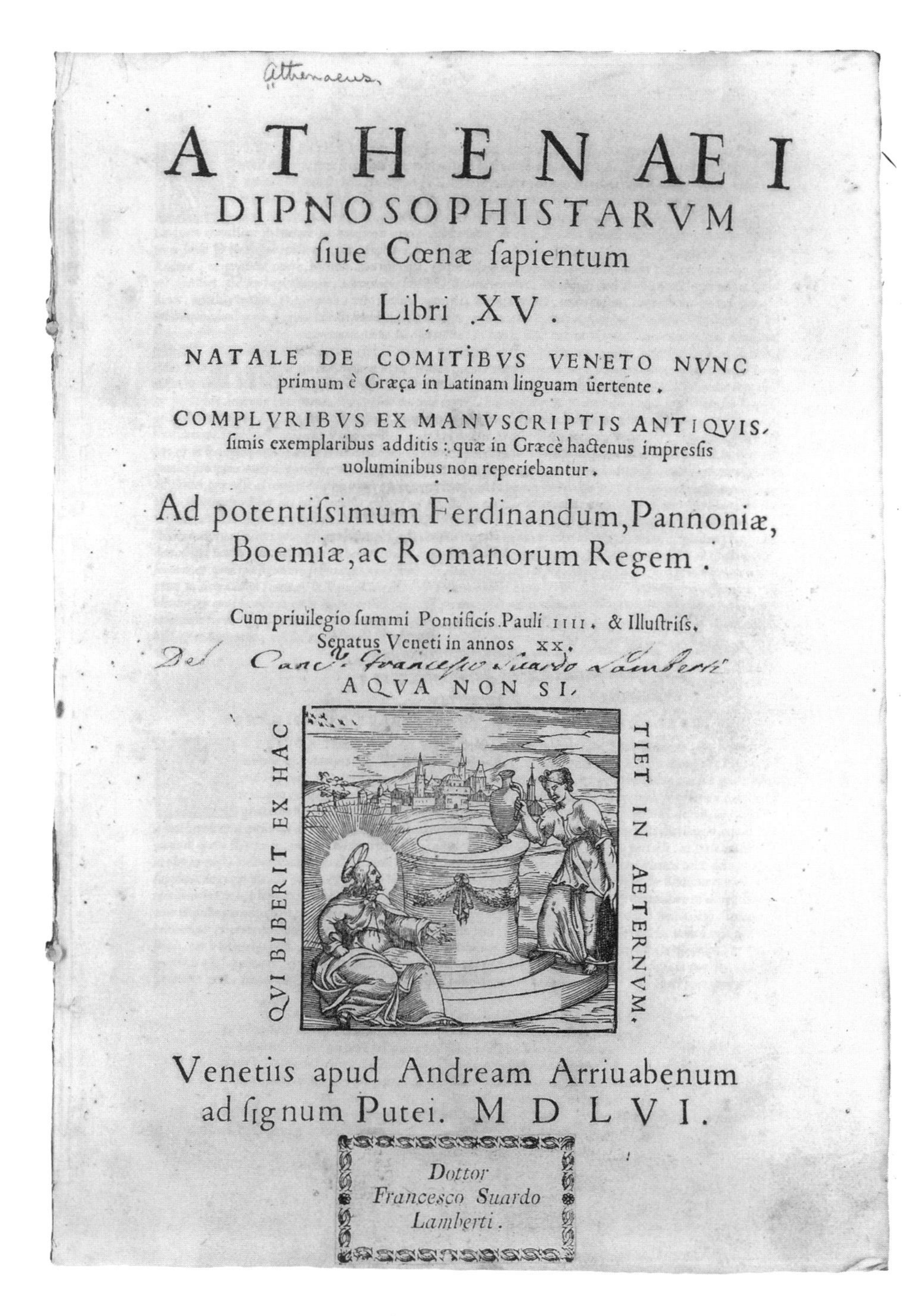

Athenaeus

ATHENAEI
DIPNOSOPHISTARVM
siue Cœnæ sapientum
Libri XV.

NATALE DE COMITIBVS VENETO NVNC primum è Græça in Latinam linguam uertente.

COMPLVRIBVS EX MANVSCRIPTIS ANTIQVISsimis exemplaribus additis: quæ in Græcè hactenus impressis uoluminibus non reperiebantur.

Ad potentissimum Ferdinandum, Pannoniæ, Boemiæ, ac Romanorum Regem.

Cum priuilegio summi Pontificis Pauli IIII. & Illustriss. Senatus Veneti in annos XX.

AQVA NON SI

QVI BIBERIT EX HAC

TIET IN AETERNVM.

Venetiis apud Andream Arriuabenum ad signum Putei. MDLVI.

Dottor Francesco Suardo Lamberti.

A major source for information about ancient gastronomy is the *Deipnosophistae* of Athenaeus, the Greek grammarian who flourished around A.D. 200. Translated variously as "The Gastronomers" and "The Philosophers at Table," it offers a large number of quotations from ancient, lost works and provides in the form of a dialogue between poets, doctors, artists, and philosophers, a great deal of extremely varied information about all aspects of food and cookery. This copy belonged to Thomas Jefferson and was translated into Latin by Natale Conti from the Aldine Greek text. *Dipnosophistarum, sive Coenae sapientum,* 1556. Athenaeus. LC-USZ62-110297.

As this brief survey indicates, there are relatively few written sources from ancient times, and the history of ancient technology is found not so much in the written record as in the physical remains of the tools and devices that have somehow survived the centuries. Thus much more is learned when an ancient grain mill is unearthed and studied than is gained from all the books about such devices. Some ancient remains are so large or so bold and obvious that they need no specialists to interpret them. Besides the architectural survivors that easily come to mind—the great pyramids of Egypt, the Colosseum, the preserved city of Pompeii—there are scores of others that stand as testaments to the technological ambition, ingenuity, and determination of the ancients.

Among the most remarkable of these physical testaments were the thousands of miles of roads built by the Persians and the Romans to speed the movement of their armies. The Romans alone had over 87,500 miles of well-kept and well-supervised roads stretching from Spain to Asia Minor. Caesar once traveled 800 miles in ten days. Many of these roads still exist today, including an old Roman provincial highway near Nimes, France, that leads to the Pont du Gard, a Roman bridge that one can visit and cross even now. The bridge is formed by three tiers of arches, none of which was laid with cement. The river channel beneath the bridge was, however, lined with cement to a width of four feet. Also in Nimes, which became Roman in 120 B.C., is an amphitheater, Les Arenes, that seats 24,000 and is still regularly used. The two-tiered aqueduct at Segovia, Spain, known as the Devil's Bridge is over a half-mile long and 119 feet high, and rests on tiers only 8 feet wide.

Perhaps the most infamous of these extant architectural and engineering wonders is the mighty siege ramp built at Masada by the Roman governor of Judea, Flavius Silva. During the gruesome Roman capture of Jerusalem in A.D. 70, about one thousand Zealots took refuge atop the great mesa called Masada. It took a Roman force of almost fifteen thousand nearly two years to take the fortress, but only after they had constructed an enormous earth and masonry ramp up the side of the mesa and crowned it with a stone causeway that would support the weight of the siege towers. The Jewish Zealots chose to commit suicide rather than be taken and enslaved by the Romans. The mighty siege ramp stands today, a technological monument and a symbol of Jewish heroism.

Given the major and lasting contributions of Greece and Rome to the fields of philosophy, politics, and art, their contributions to technology may seem relatively limited and insubstantial. We tend to think of a genius like Archimedes, whose intellect could match that of any person of any age, typically living in near Stone-Age primitiveness when it came to the meaner functions of everyday life. But this was certainly not the case. Although no age has witnessed anything near our own in terms of sheer technological accomplishment, the average well-to-do individual of classical antiquity was a person of relative technological sophistication.

The books already discussed make this point very well. In them, we discover details of everyday life that, upon consideration, indicate a surprising degree of technological awareness as well as a real inclination to

Opposite page:
The Romans were the first scientific road builders, as they approached the job rationally and systematically. They constructed a network of carefully aligned and well-laid-out roads throughout the empire, over which their legions could march rapidly. A major highway like the Appian Way was constructed of four main layers whose thickness varied from three to five feet. Its total width was about thirty-five feet with a two-way, crowned and curbed central lane paralleled by two one-way lanes. By the peak of the empire, the Romans had built nearly fifty-three thousand miles of road. *Le Antichita romane,* 1756. Giovanni Battista Piranesi. LC-USZ62-110298.

Veduta dell'antica Via Appia, che paſsa ſotto le mura già descritte nelle paſsate tavole dell'Ustrino, oggi ricoperta nelle rovine del medesimo. A. Letto del terreno ben ſodato, e battuto con pali, prima di stendere la groſsa riempitura alta palmo uno a ſimiglianza di lastrico composto di calce pozzolana, e scaglie di ſelci, e ſopra di eſso piantati a forza i ſelci B. tagliati nel roverscio a punta di diamante. C. Altri ſelci posti a guisa di Cunei, i quali stringono e gagliardamente rinserrano i ſelci sudetti, che lastricano la Via gia detta, fra quali ogni 30. palmi evvene uno D. piu eminente e ſuperiore degli altri di tal fatta, quale dovea servire forſe a quelli, che montavano, e ſmontavano da cavallo, e di riposo a Viandanti. Questo e gli altri inferiori ſono piantati sopra un grosso muro di riempitura di ſimili ſcaglie di ſelci, ma piu grandi delle mentovate di ſopra

Piraneſi Archit. dis. inc.

manipulate the natural state of things. Thus it is not surprising to find in Vegetius a recipe for a flaming arrow fuel made of sulfur, resin, and pitch, wrapped in oakum and soaked in oil. Nor is it remarkable to learn that Greek women of wealth would wear a face mask of meal all night and wash it off with milk in the morning. The perfume trade was one of the largest business concerns in the city of Alexandria, with both Greeks and Romans (even Roman soldiers, says Pliny) using scents and hair oils. Romans often wore hairpieces and Greeks would dye their hair. Leaves of Malabar were often chewed as a breath freshener. Such cosmetic indulgences seem to mirror our own fascination with the luxuries that technology can provide.

On a larger scale, Empedocles drained the swamps in Greece in 450 B.C. and Appius Claudius brought pure water into Rome via an underground aqueduct system. In Hellenistic times, portable braziers or charcoal fires were in general use as home heaters, with the Romans eventually developing the hypocaust system that heated buildings through the floor. The Greeks raised viticulture to a science and the Romans applied their talents for organization to it, resulting in a wine label that told not only the vintage date and origin but the date the wine was tapped and who tapped it, its purity, the vintner's name, the color, what type of grapes, the number of the jar and the name of the cellarer. In some cases, we are behind the Romans.

Although considered together within the useful time frame known as classical antiquity, the technological traditions of our two Western forebears are essentially separate and radically different. While the Greeks usually did excel at whatever technology they attempted, their technological goals were both modest and limited. Culturally, they seemed to feel little need to flex their technological muscles by letting theory serve practice. Thus we see a Greek technological tradition that is creative but deliberately limited and restrained.

In contrast, the Romans sought after and excelled at anything technological. As an ambitious and pragmatic people, they esteemed the many practical advantages and conveniences of technological accomplishment. Evidence of their technological zeal is with us today in the remnants of their road systems, aqueducts, bridges, and buildings. Roman technological tradition is thus decidedly one which was in no way ambivalent but rather embraced wholeheartedly the notion of *homo faber*.

It could be argued that the present Western love-hate relationship with technology has its origins in this Greek-Roman duality, with our technological antipathy reflecting the Greek tradition and our technological zeal drawn from the Roman tradition. Such a perspective might be a beginning to further understanding our own attitudes toward technology.

Opposite page:
As one of the major technological achievements of the ancient past still with us, the Roman Colosseum stands as a noble example of the technical abilities of Greece and Rome. Called more correctly the Flavian Amphitheater, it was begun by Vespasian between A.D. 70 and 72 and was inaugurated by Titus in A.D. 80. As a freestanding structure of stone and concrete, it measures one-third of a mile around and has a 160-foot facade. Able to seat fifty thousand, it was adorned with Doric columns on the ground floor, Ionic columns on the next, and Corinthian columns on the top. It was in this grand symbol of all that is best in the classical tradition that brutal gladiatorial matches and contests between humans and beasts took place. *Le Antichita romane,* 1756. Giovanni Battista Piranesi. LC-USZ62-110299.

Tav. XXXVII
Fig. I
Veduta dell'Anfiteatro Flavio detto il Colosseo. A Arco mancante del rispettivo numero ordinale con cui son notati tutti gli altri per rincontranza delle persone che anda
vano agli spettacoli. sopr'a quest'arco appoggiavasi il ponte, che dal portico di Claudio portava all' Anfiteatro, come dimostra la mancanza fra lo stesso arco e'l su
periore della ricorrenza dell'architrave, e delle altre membra, nella quale s' incastrava il ponte. B Avanzi de' parapetti fra gli archi. C Forami nella cornice per
dove passavano le antenne alle quali era raccomandata la tenda che copriva l'anfiteatro. D Mensole sulle quali posavano le antenne. E Muri moderni.
Piranesi Architett. dis. inc.

Das.viiii.cap. Wie nütze sy ackern vnnd graben.

2. *The Technology of the Middle Ages*

The temporal division of human history into periods called ancient, medieval, and modern are convenient generalities that should serve only as a starting point. When used rigidly without qualifiers, they distort the complex reality of the past, forcing out both subtlety and diversity with the power of their universal explanations.

Of these three arbitrary periods, the medieval stereotype is perhaps most capable of distorting and repressing the reality behind the idea. Unlike its two partners in this convenient trinity, the medieval era evokes many negative feelings and images. Possessing neither the grandeur of ancient times nor the familiarity of the modern period, medieval times have come to connote ignorance and stagnation. When referred to as the Dark Ages—a name heavy with images of dullness, gloom, and even fear—the negative attributes of this time stand in even greater relief between the implied light of the periods both preceding and following it. The word *medieval*, meaning literally "middle age," has now entered our jargon to imply in a derogatory way backwardness and extreme credulity.

Upon examination, however, medieval times are revealed to be, as common sense would seem to dictate, less monolithic than was first thought, since the time encompassed is so large. As one historian has noted, the differences between the Europe of 800 and that of 1300 are greater than their similarities. Further, it was a time characterized as much by change as by continuity. Different parts of medieval Europe did not develop at a uniform pace but progressed unequally. Some areas were more receptive to change, some less; some were more inventive, some not so. As with most of history, closer scrutiny of medieval times reveals them to be more complex and less susceptible to simple explanations than was thought at first glance.

So it is with technology's development during the Middle Ages. Examination of its history reveals an uneven, undirected, trial-and-error process that has been described as an "empirical groping." Further, medieval technology pursued a road entirely separate from the development of science. If anything, medieval technological advances often provided science with many new, vexing problems to consider.

These problems emerged as the result of Europe's gradual transformation from an agrarian, subsistence economy to one that could eventually sustain both trade and industry. The early medieval West achieved a number of technological innovations in its agricultural methods, such that by the end of the ninth century it was employing the heavy plow,

Opposite page:
The Middle Ages was several ages in one, but all were characterized by the gradual transformation from an agrarian, subsistence economy to one that could eventually sustain trade and industry. It was during this supposedly static time that the seeds of modern technology were sown and a static society began to evolve slowly into a more dynamic one. *Ruralia commoda,* 1471. Pietro de Crescenzi. LC-USZ62-110300.

rotating its crops triennially, and maximizing its horsepower with nailed horseshoes and the padded shoulder harness. By the end of the twelfth century, the waterwheel and windmill were harnessing natural forces whose energy was used to greatly increase productivity.

The entire medieval span witnessed many other significant technological innovations that cumulatively place in doubt the blanket characterization of a contemplative, other-worldly society totally obsessed with symbol, allegory, and other modes of Platonic thought. Architecturally, both Romanesque and Gothic construction required technological breakthroughs to achieve their effects. Militarily, there are the disparate developments exemplified by gunpowder and cast iron which produced artillery, and the stirrup which revolutionized mounted warfare. Although little progress if any was made in land transportation, sea transport saw major advances when the traditional square sail was combined with the triangular lateen sail of the Arabs, in addition to the adoption of the sternpost rudder and the magnetic compass. These and many other technological achievements indicate that the Middle Ages was several ages in one, all the time evolving slowly from a static society to one characterized by an emerging intellectual curiosity and dynamic development. By the middle of the fourteenth century, the seeds of modern technology had been sown and the intellectual climate prepared.

A glance at medieval ways, mores, and institutions offers some beginning insight into the attitudes and methods of those individuals who lived and worked so long ago. Without overemphasizing the well-known portrait of a medieval individual as one whose religious or spiritual view of life dominated most aspects of living, it is easy to see how a strongly ideological tradition would make a difficult partnership with the rigorous

Left:
Nailing a U-shaped metal plate to a horse's hoof improved the efficiency of the medieval horse and created one of the great staple crafts—that of blacksmith. This in turn contributed to the development of metallurgy. *Ruralia commoda,* 1471. Pietro de Crescenzi. LC-USZ62-110301.

Right:
The lateen or triangular sail seen at this ship's bow peaking above its two square sails was an Arab import to medieval Europe. Affixed to a crossbar, it was able to take the wind on either side, unlike the traditional square sail, which allowed sailing only before the wind. With ships now able to tack into the wind, medieval trading ships could explore their fullest range. The lateen contributed not only to the resurgence of commerce but to the later voyages of ocean exploration. *Instauratio magna,* 1620. Francis Bacon.

With the fall of Rome, the physical deterioration of its empire-sustaining infrastructure began. Roman buildings, roads, bridges, and aqueduct systems started their inevitable decline once maintenance stopped. No one captured the fallen grandeur of Rome better than the eighteenth-century artist Giovanni Battista Piranesi, whose romantic etchings of the eternal city evoke the drama of the past. This is the remains of a sepulchre he found on the Appian Way. *Le Antichita romane,* 1756. Giovanni Battista Piranesi. LC-USZ62-110302.

pursuit of material advancement. Although this invisible world of religion often led to a passive, contemplative state altogether opposite to the typically active, dissatisfied, and energetic fervor one associates with inventiveness, it will be noted that, in fact, religion eventually provided many of the necessary stimulants and nurturing conditions essential to the development of technology.

In addition to religion itself, there are a surprisingly large number of factors, conditions, or influences that stimulated the progress of medieval technology. First among these are the obvious remnants of the old Roman order that persevered after the collapse of Rome itself. Terrible social and political disorder occurred when Rome fell to the invading hordes, but, as we know from both past and recent wartime and disaster situations, the simple, most powerful requirements of everyday living have a way of ordering, in their own fashion, even the most chaotic of situations. People must eat, sleep, and have shelter. In fulfilling these and other very fundamental needs, the survivors and their barbarian conquerors had at their disposal a fairly impressive array of physical and institutional support systems. Most obvious was what remained of the physical infrastructure, such as the Roman road system, buildings, aqueducts, and grinding mills of all kinds. Less obvious, although as important, were Roman systems of law and administration, in addition to newer institutions like the Christian church. The collapse of the western empire of Rome, although catastrophic to Rome's existence and certainly to its imperial goals, did not mean that daily life in the former empire stopped altogether or that people forgot old ways. Crafts of all kinds continued to be passed on from one generation to another, and farming and manufacturing techniques were maintained and even improved.

A second stimulant to the development of medieval technology was a literal infusion of new blood into a creaking, overripe old order. The invading Germanic tribes—barbarians to the Romans—did not suddenly swoop down in 476, the traditional date of Rome's fall, and conquer the seat of power. The empire had been divided earlier, east and west, by the Emperor Diocletian (284–305), and Rome itself had been replaced during that time by Milan as the imperial residence. In the late fourth century, Rome did not resist the Visigoths' settlement in the heart of its western empire, and in 410 Alaric led the Visigoths in the sack of Rome. Further invasions of the city followed and, by the end of the fifth century, the western empire was overrun by barbarians and severed from its eastern half.

This wrenching political chaos and social upheaval was to be a stimulant to the advance of technology. For all the disruption of trade, industry, and town life the tribes caused, which could be considered retrograde in terms of technological development, the barbarians more than compensated, not only with a technology of their own but also with a vitality that made change possible and an inventiveness that came in as fresh air. Unlike the more rigid, traditional ways and techniques of the people they conquered, the barbarians had a vigorous, pragmatic approach that allowed for innovation, so long as it worked for the better. Some argue that in certain respects their technology was superior to that of the Romans, and among their specific contributions to medieval life are the heavy plow, trousers and the habit of wearing furs, the ski, cloisonné jewelry, barrel making, soap for cleaning, butter in place of olive oil, and the cultivation of rye, oats, and hops.

But if the invaders were superior to Rome in military strength, vitality, and motivation, theirs was the weaker force in the clash of cultures that would ensue. In the fusion of peoples and cultures that followed, the Goths and the Franks became more romanized than the Romans were germanized. History documents their eventual embrace of the Latin language, Nicene Christianity (as opposed to their own Arian Christianity), and Roman law and government. It was from this amalgam of cultures that Europe was born.

The third stimulant to medieval technology also came from without—this time not from the north like the Germanic invaders, but from the east. Roughly one hundred years after Rome fell, the prophet Muhammad was born in 570. After receiving his call to prophesy at forty years of age, he founded the religion of Islam (meaning submission to the

will of Allah) and preached to his followers, called Muslims ("true believers"). Offering a spiritual idea that proved to be a unifying cultural and political force to Arabs, Islam spread rapidly. By 750, Europe and the eastern Roman Empire, or Byzantine Empire, found itself severely threatened by Muslim conquest. Indeed, the Muslims controlled all of Spain as well as all areas south and east of the Mediterranean, which itself became a "Muslim lake." Pressed by Islam on three sides, it is not surprising that Europe adopted technical novelties and innovations from the Muslim Near East as well as from the Byzantine Empire itself. More important, the conquering Muslims had already assimilated and then embellished the technological achievements of Hellenic civilization. So it was through them that this technical legacy of Greece became available to the West, first through the sporadic contacts of conquest and commerce, and later through the entrenched presence of the Moors in Spain and the Arabs in Sicily. The Muslims also acted as a conduit to the West for some of the technological achievements of the Far East, especially from China and India. Gunpowder, kites, and papermaking are among several of the oriental exports introduced to Europe in this manner. As Western technology had benefited from the positive hybridizing effect of the barbarians from the north, so it became even richer because of this Eastern influence.

The fourth major stimulant to the development of medieval technology—religion—may be the most powerful and pervasive. As a force in the daily life of the medieval individual, its influence cannot be over-

Gunpowder or black powder is one of the defining technologies of the Middle Ages in Europe, although it arrived there from China via the Arabs. Its immediate use in Europe was not for peaceful purposes such as mining but almost exclusively for warfare. Here, a battery of guns discharges a volley of fire tubes used to frighten cavalry and set ships and buildings aflame. The author of this treatise cannot overemphasize the violence of which gunpowder is capable, saying, "this thing made by art is more harmful to the life of man than the deadly poisons in numerous animals, herbs, and in so many other things produced by Nature, or than the very thunderbolts of the sky." *De la pirotechnia,* 1540. Vannuccio Biringucci. LC-USZ62-110303.

IL PADRE SAN BENEDETTO CON L'ESPOSITIONE D'IL R. PADRE FRATE ROGIERO DI BARLETTA, MONACHO CELESTINO.

S.BENEDETTO ABB.

F. ROGIERO.

F. Iacobus Lytiensis Baccalaureus ad Authorem.

Ia. Siste gradũ Rogeri. Ro. quid vis tibi? Ia. cãdida portas?
Ro. Sensa patris, quicquid pagina sacra parit.
Ia. Grande sophos, quo? Ro. lucis in auras. Ia. ergo iuuẽtus
Docta eme, Dispeream ni tibi grata dabit.

Saint Benedict of Nursia is seen here presenting his system of monastic regulations, called the *Regula,* to a kneeling monk. Devised by him in 529 for the monastery of Monte Cassino in Italy, depicted in the background, it was a guide to communal religious life based on moderation and cooperation. The Cistercians, founded in 1098, were a later, reform branch of the Benedictines. Their initial zeal for hard manual labor eventually transformed their abbeys into efficient, money-making factories. *Il padre san Benedetto con l'espositione d'il r. padre frate Rogiero de Bartletta,* 1539. LC-USZ62-110304.

Opposite page:
This sixteenth-century waterwheel is much more sophisticated than those used by the medieval abbeys, but it was this device that gave even the remotest of European villages the ability to reap the economies of mechanization. Unlike the familiar vertical wheel, this one has horizontally set paddles and was popular because it could generate excellent grinding power even with a small water flow. The gearing and power-transmission principles demonstrated by the watermill were familiar to every medieval builder and carpenter, and it was the direct ancestor of the turbine engine. *Le diverse et artificiose machine,* 1588. Agostino Ramelli. LC-USZ62-110305.

emphasized. Above all, it was a source of unity, even in the wildest times following the Germanic invasions. These invaders came as Christianized barbarians, albeit heretical Christians. Their Arian Christianity, which denied that Jesus Christ was both fully human and equal with God the Father, had been condemned by a council of Christian bishops who met at Nicaea, near Constantinople, in 325. Although this action introduced major new divisions, the Nicene or Catholic interpretation won out in the long run. The victory was inevitable after Clovis, the king of the Franks, converted his people about the year 500. From then on the many centrifugal forces that sometimes worked in favor of disorder were countered by the unity and continuity of the Catholic church.

The church as a force for order and stability in medieval times is also exemplified by the early Latin monks whose near monopoly on literacy made them the sole guardians of learning in the West. The foundations of Western monasticism were laid by St. Benedict, who was born shortly after Rome's fall. His *Regula*, or system of regulations, written for his new monastery at Monte Cassino, Italy, incorporated study as well as work and prayer into the regime of the monks' lives. During the next five centuries, Benedictine monasteries were the chief repositories not only of Christian literature but also of secular learning. As these monasteries multiplied in size and wealth, they necessarily became more bureaucratic and worldly, prompting reform movements that emphasized the original, simple values of their founder. One group of Benedictines founded the Cistercian order in 1098, emphasizing a simpler, more solitary and austere life. These Cistercians became a force under the guidance of St. Bernard de Clairvaux who, by the time of his death in 1153, had established 338 Cistercian abbeys.

The Cistercians were perhaps the most concrete example of religion spurring on technological development. First, they incorporated physical labor as an essential part of the group's daily regimen. This was not a simple requirement to do a bit of work, but rather was derived from an institutionalized prescription that each abbey be self-supporting in its material needs. The Cistercians embraced the notion that hard labor yielded a real, spiritual reward, and this in turn fostered a change in the way work was regarded and a keen receptivity for new techniques and methods. The eager Cistercians thus sought out, sampled, and usually adopted anything that could make their toil more productive. Because they were educated, organized, and numerous, and because they sought to mechanize whenever possible, the Cistercians contributed greatly to the spread of new technologies throughout Europe.

The abbeys were always located near a river, and the order became expert in hydraulics, using water power to drive various crushing mills, fulling mills, and tanneries. They also became very skilled in the related mining technologies of forging and smelting. In addition to these agricultural and industrial techniques, they used the newest construction methods available and introduced them to the rest of Europe. Thus exploiting any and all forms of natural power—water, wind, and animal—and making use of new technologies, the industrious Cistercians met

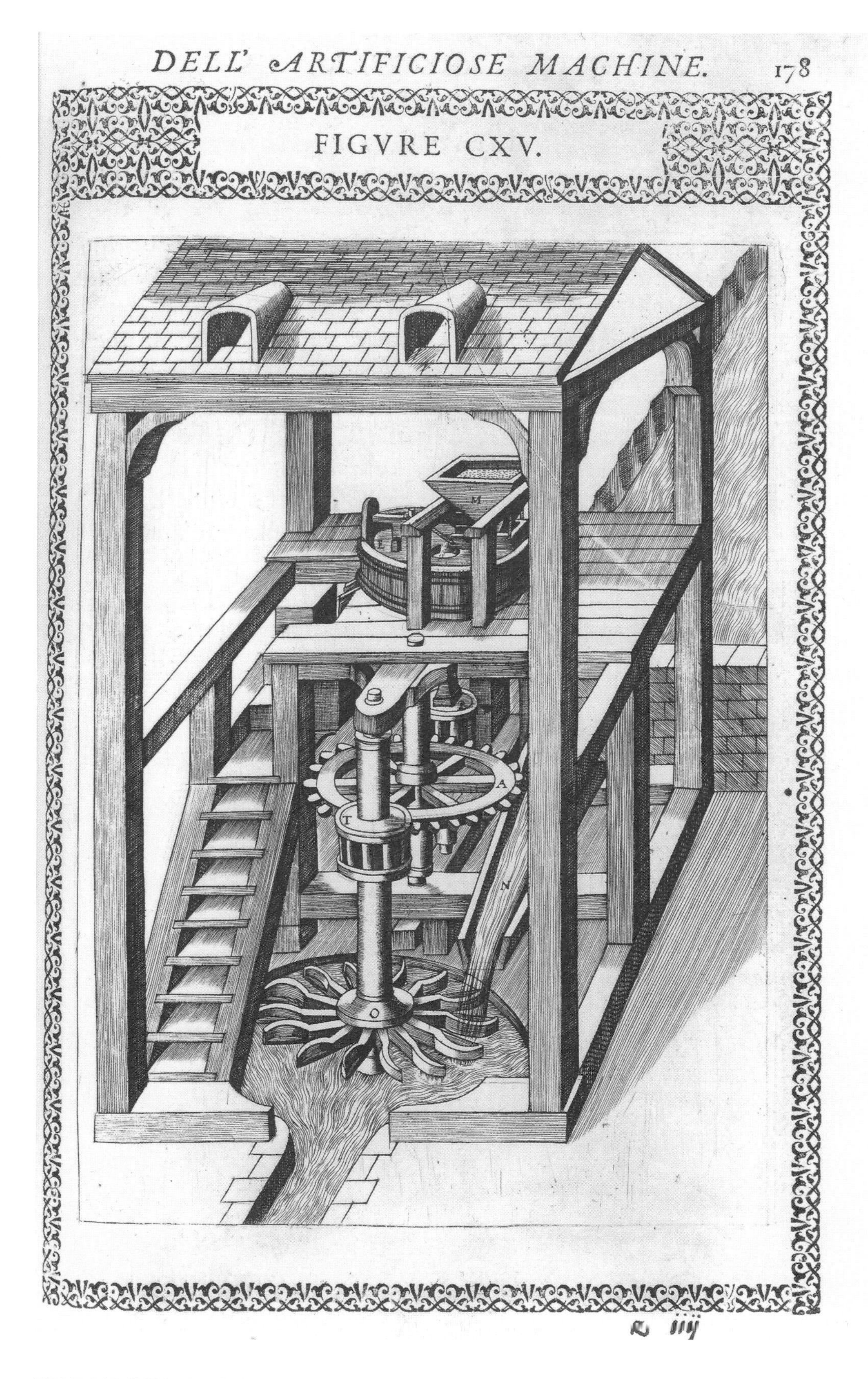
DELL' ARTIFICIOSE MACHINE. 178
FIGVRE CXV.

ter mouet: ac scd'm dispositionem diuersam
materie apparet diuersimode. Cum enim
cometa in medio est grossioris substantie et
in extremis ptib' subtilioris: tunc i medio ap
paret lume mag' opacu: et i extremis magis
claru: et vocat stella comata: sed qn illa materia extendit i longu: vocat caudata: et qn h3
ptes subtiles inferi' vocat barbata. ¶Et
sic p3 q cometa e exhalatio terrestri calida et
sicca: grossa piguis et viscosa: cui' ptes multu adinuice circuiacet: eleuata ab isti' iferio
rib' vsq ad supma regione3 aeri, in q pp mo-
tu illi' regiois inflamat, et circularit mouet.
¶Etia p3 q cometa si multu gnat i hyeme
ppter multa frigiditatez. Nec i estate ppter
nimia caliditate, s3 bn in Autuno et in vere.
Min' tn in ve: ppter nimiam frigiditatez
¶P3 et ppter qd cometa ta diu durat et vn
puenit diuersitas coloru. Durat em diu ppt
grossitie materie: et q3 otinue noua attrahit
materia: q qn est bn inflamabilis apparet co
meta alb' vel pallid': qn vo e mediocris ap
paret cometa rube': et qn e croce' appet vt
carbo ardes. ¶Un cometa e signu multoru
Prio ei est signu magne caliditat' et multoru
vetoru. Scdo est signu traditionu: mortalita
tu: belloru: et pestiu: q3 tpe gnatiois comete
viget caliditas ignea: et calor i hoib' multu
auget, et etia colera: et sic icitant hoies ad iur
gia et bella. Tertio e signu mortalitat' pnci-
pu: q3 pncipes delicati' viuut: et io citi' infla
mantur. ¶Et si qrat cui' nature e cometa.
¶Dr q est nature calide et sicce. Qd p3. Tu
prio: q3 tuc e magna siccitas qn cometa apparet. Tum scdo: q3 tuc e multitudo magno
ru vetoru. Ite est exptu q cu appuit cometa
sicca: fuit hyes et vet' borealis. Ite etia tpe
comete i egyptiis fluuiis cecidit lapis ex ae
re a veto eleuat'. ¶Ite sub pncipe Nicoma
cho factus fuit cometes circa eqnoctialem
circulu3: q a vespa facies ortu paucis dieb'
durauit: et tuc vehemes vent'. Corinthu ve-
xauit igitur cometa est natura calide et sicce

¶De galacia seu via lactea.

DE galacia dixerut atiq q sol semel du-
ct' a phetote exiuit via sua: et non re-
cte duxit curru suu: et io cobussit illa p
tez q nuc videt alba: et q dr via lactea Et hec
opio est pithagoricoru. ¶Ipa tn est falsa et
friuola. Tu3 prio: q3 zodiac' sub quo mouet
sol iam ess3 combust' qd est falsum. ¶Sed
anaxagoras et democrit' dixerut galacia ee
lumen quoruda astroru aggregatoru. Et non
videt tale lumen de die: q3 obfuscatur a so
le. ¶S3 hec opio est falsa: q3 tuc galacia ali
qn deberet appere circa aliquas stellas: et
aliqn circa alias: sicut sol: nuc istas: nuc illas
stellas obfuscat. ¶Item celum est maius
magnitudie terre: ergo terra no impedit qn
interpoit inter sole et illa astra qn illuminet
illa astra. ¶Alij vo dixerut q lume astroru
immiscet vapori huido et reflectit ad celu: et
sic galacia e lume reflexu ab aere vel corpe
huido vsq ad celu. ¶S3 hec opio falsa est:
q3 tuc galacia sp deberet apparere in alio et
alio loco: qd est otra experientia. ¶Ite galacia pt appere denocte in aq vel in speculo.
et tn tuc no fit reflexio vsq ad celu igr. ¶Io
dicendu est q galacia est lumen astroru rece
ptu in exhalationib' calidis et siccis eleuat'
p calorem solis ab istis inferiorib' vsq ad
supmam regionez aeri. ¶Probat. Tu3 pri
mo: q3 p virtutem vni' astri eleuat sufficies
materia p cometa: igit per virtutem pluriu
astroru eleuatur sufficiens materia pro gala
tia. Tu3 scdo q3 galacia non fit inter tropi-
cos: et hoc non est nisi q3 ibi est nimia caliditas: igitur signu est q illo mo fit galacia.
¶Sed oppositum nunc tenetur: cum dica-
tur galacia esse circulus celestis qui parum
recipit de lumine.

Stella coata. Stel. caudata. Stel. barbata. Diffinitio comete. Signata comete. De natura comete. Experimeta. Opio pithagoricoru. Confutatio. Opio Anaxagoras et Demo. Confutatio. Opio alioru. Confutatio. Opio ppria quid galatia.

with great success. They sold wine from Burgundy and wool from England, and they ran the most modern factories in Europe. Although the ruling members eventually became more managers and businessmen than monks, they had nonetheless provided all of Europe with a living example of how technology can raise one's standard of living.

Besides the dramatic example of the Cistercian order, religion provided another, much less tangible condition that was conducive to technology's advance. This was in the realm of doctrine or ideology. The fundamental Christian teaching that all men are created free and are equal before God made slavery, as practiced in antiquity, theologically impermissable. This meant that medieval Europe could not advance solely on the backs of its chattel slaves. Although European serfdom was created to produce a large underclass of virtually servile individuals, Europe did not consider itself a society based on slavery. It was thus amenable to deriving power from machinery rather than from serfs when it became in its economic interest to do so. Further, it has been pointed out by many historians that the Western church always emphasized the great importance of correct action to an individual's salvation. In contrast to its Eastern counterpart, which placed more importance on correct thought, the Western church favored positive action over quiet contemplation. Such an activist attitude served to encourage a medieval craftsman to try something new and strange. Unlike the era's dogmatic approach to many other matters, this particular emphasis on the voluntaristic aspects of human life encouraged individual attempts at innovation.

Perhaps the most important contribution of religion to the development of medieval technology came during the late Middle Ages. When the whole work of Aristotle became available to the West at the beginning of the thirteenth century, Christianity found itself confronted with a rival, pagan way of explaining the entire cosmos. This new view was threatening since its idea of a natural universe held no place for the notion of God or the role of a deity and led to such heretical conclusions as the existence of other worlds. The church reacted quickly and strongly to this new, pagan threat, and in 1210 it first condemned all of Aristotle's works on natural philosophy and prohibited their teaching in the universities, on penalty of excommunication.

Despite this, by 1255 the works of Aristotle constituted the core of a university education because the power of Aristotelian ideas and the philosophical tools of analysis they offered proved irresistible. Acceptance of Aristotle would not have occurred so quickly (although official annulment of the condemnation had to wait until 1325) had it not been for the work of Albertus Magnus and of his pupil, Thomas Aquinas. The great achievement of Aquinas was to reconcile these two apparently conflicting ways of understanding man and the cosmos. The form of Christian Aristotelianism that Aquinas produced gave scientific truth its due, and said that it in no way contradicted religious or revealed truth. By devising a system that embraced both scientific knowledge and theology while also keeping them separate, he offered a rationale whereby the Christian educational system could accept and pursue secular knowledge

Opposite page:
The medieval embrace of Aristotle's natural philosophy and metaphysics was essential if the age was to come to regard the natural world as knowable. Once Thomas Aquinas had reconciled religious faith and secular knowledge, making the riches in Aristotle's natural books available to be taught in the universities, the tools of logic and analysis could benefit technology as well as science. This is an illustrated page from a late fifteenth-century edition of Aristotle. *Textus abbreviatus in cursum totius Physices et Metaphysicorum Aristotelis,* 1496. LC-USZ62-110306.

without endangering its faith. Aquinas argued quite simply that natural philosophy is a self-sufficient discipline since it seeks only to understand the natural world. Theology deals with faith and revealed truth. Thus, just as reason fails in explaining revealed truths, dogma is equally useless in understanding the workings of nature. This prudent and pragmatic rationale eventually became the official position of the church, thus preparing the way for the great intellectual expansion and application of knowledge that was soon to follow.

The extant literature that documents this story of medieval technology is certainly a weak and very thin corpus. It is ironic that more written documentation has survived of Greek and Roman technological development than from the Middle Ages. This becomes somewhat understandable upon examination.

First, whenever a medieval technique was described, a methodology detailed, or a mechanism explained in writing, it was always done in manuscript form, the printing press being a mid-fifteenth-century invention. This writing and copying by hand necessarily limited the number of manuscripts that were made, limiting further the number that would eventually survive the ravages of environment, neglect, and caprice. Copying a manuscript was also extremely labor intensive and was done mostly in the monasteries. It must surely have been the rare monk who would give up valuable time transcribing religious writings to copy a base "how-to" work.

Second, the long-standing attitude that anything related to labor was somehow tainted and less noble certainly mitigated against raising it to a higher plane than it was thought to deserve. One of the more pervasive cultural antipathies shared by pre-twelfth-century religious leaders, members of the ruling families, and intellectuals alike was a disdain for work. Manual labor of any kind was a common act done by commoners. This attitude was by no means new, reflecting as it did opinions inherited from the Greeks.

Third, where the Greeks possessed a formal logical system that enabled them to treat the study of practical things in an almost abstract manner, it was the rare medieval writer on technology who attained anything near even the most general of conclusions. This Greek bias for the abstract or the transcendent necessarily gave their work a more universal appeal, as evidenced by the Arab enthusiasm to preserve it. Thus by the time the mechanical printing press was invented, printers turned not to the scarce and often anonymous medieval manuscripts of relatively recent production but rather to the elegantly ordered, exotic, and recently rediscovered technological knowledge of the Greeks and Romans.

Other reasons why medieval techniques and technology are less documented than their counterparts from classical antiquity reflect both the nature of technology itself and the nature of medieval life in general. The world of technology has been described as an essentially nonverbal one—one in which the emphasis is decidedly on the doing, the active, the tangible. The exigencies of life during the Middle Ages, however, surely exaggerated its nonverbal aspects. If, as Cyril Stanley Smith said,

One of the better-known medieval scribes was the Burgundian editor-copyist Jean Miélot. In this well-appointed room, he works at a typical sloping desk and copies from the open book above. Also seen are three inkpots, several large block books kept on shelves and in wall cabinets, a rolled manuscript, and a trunk below the desk, which probably also held books and manuscripts. The reproduction of a book by hand was a necessarily slow and costly process. *Miracles de Nostre Dame,* 1885. Jean Miélot. LC-USZ62-110307.

"men . . . whose hands were accustomed to both the hammer and the pen, have always been rare," they indeed were exceedingly few during the Middle Ages. The steep decline of learning during that time made such individuals all the more exceptional.

Further compounding the factors that mitigated against the documentation of medieval technology was the pervasive habit of secrecy ingrained into even the youngest apprentice. Obviously, a tradition that regards the methods and means of a craft as a closely guarded secret does not adapt easily to the idea that those methods should be written down and made available to one and all. On the contrary, the transmission of technical knowledge during the Middle Ages was accomplished personally and by demonstration. The system was so successfully closed that, to an outsider, the details of a particular technology would take on the aura of what has been called a "craft mystery."

Such a mind-set was hardly conducive to documenting and disseminating information, and it is no surprise that many individuals took measures that would obfuscate or even mislead when they were required to write down formulas or techniques. No craftsmen exemplified this deliberate lack of openness more than the medieval alchemists. Nearly every aspect of their craft was carefully protected by symbolism, codes, or equivocation, making their art a real craft mystery.

In summary, many factors conspired to minimize the number of medieval technological manuscripts that exist today, not the least of which was the relatively low number of copies produced. A system characterized by

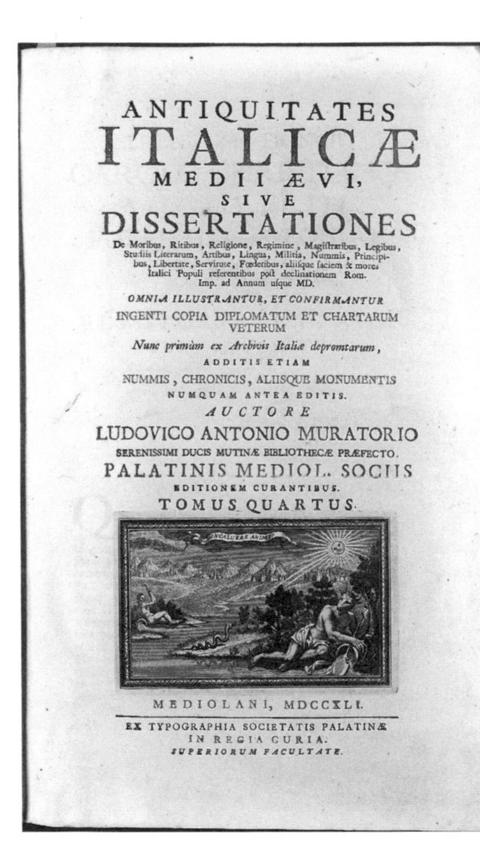

ANTIQUITATES
ITALICÆ
MEDII ÆVI,
SIVE
DISSERTATIONES
De Moribus, Ritibus, Religione, Regimine, Magistratibus, Legibus, Studiis Literarum, Artibus, Lingua, Militia, Nummis, Principibus, Libertate, Servitute, Fœderibus, aliisque faciem & mores Italici Populi referentibus post declinationem Rom. Imp. ad Annum usque MD.
OMNIA ILLUSTRANTUR, ET CONFIRMANTUR
INGENTI COPIA DIPLOMATUM ET CHARTARUM VETERUM
Nunc primùm ex Archivis Italiæ depromtarum,
ADDITIS ETIAM
NUMMIS, CHRONICIS, ALIISQUE MONUMENTIS
NUMQUAM ANTEA EDITIS.
AUCTORE
LUDOVICO ANTONIO MURATORIO
SERENISSIMI DUCIS MUTINÆ BIBLIOTHECÆ PRÆFECTO.
PALATINIS MEDIOL. SOCIIS
EDITIONEM CURANTIBUS.
TOMUS QUARTUS.

MEDIOLANI, MDCCXLI.
EX TYPOGRAPHIA SOCIETATIS PALATINÆ
IN REGIA CURIA.
SUPERIORUM FACULTATE.

The Italian scholar Ludovico Antonio Muratori was a priest who studied original documents of the Italian Middle Ages. As a man of letters, he was the first to give many unknown or ignored medieval manuscripts any serious consideration. His six-volume opus *Antiquities of the Italian Middle Ages* contains several penetrating studies of the techniques and methods of Italian artists. *Antiquitates Italicae medii aevi*, 1738–42. Ludovico A. Muratori. LC-USZ62-110308.

a traditional disdain for labor, which was both secretive and nonverbal and which emphasized the particular and the pragmatic as opposed to the general, cannot be expected to have left a written record that even begins to document the nature or diversity of its technological accomplishments.

By far the oldest medieval source of written technical information is *Compositiones ad tingenda musiva*, better known as *Compositiones variae*. It ranks first in a body of literature that might be called the technology of the decorative arts. Dating from the time of Charlemagne, the *Compositiones variae* represents an even older, Hellenistic tradition and may have been first written some two hundred years before, around A.D. 600. These "recipes for coloring" were translated from Greek into Latin around 800 and today are found in Codex 490 of the Biblioteca Capitolare in Lucca, Italy. This anonymous medieval manuscript received its first serious attention a millenium later when the Italian scholar Ludovico Antonio Muratori included it in his sumptuous six-volume study of Italian art after the fall of Rome, *Antiquitates Italicae medii aevi*, published in Milan between 1738 and 1742. The Library of Congress has the first edition of this work. It was Muratori who gave the untitled manuscript the descriptive caption that became its name.

The recipes included in the *Compositiones variae* represent a practical or technical tradition, rather than an alchemical one. It contains a wide range of specific recipes for making pigments and dyes, for calligraphy and glass making, and for using colored metals in their related trades. The fragmentary and abbreviated nature of most of the recipes indicates they were intended as reminders for the already skilled practitioner. They include no alchemical symbols or any illustrations, nor is there any systematic method or arrangement. Besides being the earliest source extant, the *Compositiones variae* stands as a link with classical antiquity, having preserved some of the technical knowledge of a lost time.

Contemporary with the *Compositiones variae* or perhaps only slightly later is a larger collection of the same kind, called *Mappae clavicula*. The earliest dated evidence for this work is in the library catalog of the Reichenau monastery on Lake Constance, dated 821–22. That original manuscript no longer exists. The *Clavicula*, or "Little Key to Painting" as it has come to be called, includes virtually all of the recipes of the *Compositiones* plus many more. Like its predecessor, it is a randomly assembled list of practical recipes that stands basically as a compilation of compilations. Although its main focus is on pigments and painting, it deals with a wide range of substances and techniques such as metallurgy, glass making, and gilding. There is even a section that mentions incendiary devices. Like *Compositiones variae*, the *Mappae clavicula* is much more informative about classical techniques than it is about those of its own times. The two works appear to have had their own spheres of influence, the *Compositiones* being used south of the Alps and the *Mappae* to their north.

Despite the existence of at least fifteen manuscripts of the *Mappae clavicula*, all in Europe, the work was not published until 1847, in the British journal *Archaeologia*. The Library of Congress has this journal in its collections. The sixty-two page article consists of a four-page introductory letter from the manuscript's owner and fifty-eight pages of untranslated Latin text. This text of the *Mappae* was taken from a twelfth-century copy. In 1974, a facsimile of an earlier *Mappae* manuscript was published, with substantial introductory and historical text by Cyril Stanley Smith and John G. Hawthorne. This special publication of the American Philosophical Society is in the Library's collections. In addition to the facsimile reproduction of the *Mappae*, it includes a full and complete translation of each chapter or recipe, totaling well over three hundred. The casual reader of this English translation may be surprised to learn how often such ingredients as urine ("take clean urine from good wine and healthy men and . . . clarify it at once") and animal dung ("take the dung of a dog, a dove, and a cock: dissolve it into a broth, and put de-haired skins into it") are recommended, especially for dyeing. The *Mappae clavicula* contains much practical knowledge of materials and techniques useful to medieval decorative arts and, despite the inclusion of what today might be regarded as questionable ingredients or eccentric methodology, neither the *Compositiones variae* nor the *Mappae clavicula* suggest a reliance on superstition or alchemy.

Somewhat of a companion to these two works is *De coloribus et artibus Romanorum*, a craftsman's manual attributed in part to the Roman Eraclius, also called Heraclius. This compilation of recipes, "On the Paints and Arts of the Romans," is composed of three books. The first two are in verse and are believed to have been compiled by Eraclius in the tenth century. The third book is in prose and was added in France during the twelfth century. Few manuscript copies of Eraclius have survived, and the work was first published and translated into English by Mary Philadelphia Merrifield in her *Original Treatises Dating from the XIIth to the XVIIIth Centuries on the Arts of Painting* (London, 1849). The Library of Congress has this two-volume work in first edition, as well as a later one published by Albert Ilg, *Heraclius, Von den Farben und Kunsten der Romer* (Wien, 1873). Ilg adapted Merrifield's text in his edition. It was Merrifield who posited that the first two books in verse were written by an Italian from the Lombard dukedom of Benevento, whereas the third prose book was done much later in France. Sometime after the twelfth century, all three books were united and given the title of the first two books.

Examining Merrifield's English translation that accompanies the original Latin text, we note that the first two books detail such techniques as painting earthen vases, sculpting glass, decorating with gold leaf, polishing gems, and cutting crystal. The third book, which is much longer, contains similar instructions as well as descriptions of refining gold and silver and recipes for metal alloys. It also includes techniques of amalgamation and soldering.

Unlike the works discussed so far, the next is an original piece composed by a craftsman and not a compilation prepared by a list-maker. Variously titled *Diversarum artium schedula* or *De diversis artibus*, it was written in the twelfth century by a man who called himself Theophilus. The importance of this work in learning about the material crafts of the

Aurifaber. Der Goltſchmit.

AVrifaber rutilo nimium ſpectabilis auro,
Aureus auratis omnia fingo modis.
Ignibus argentum ſpectamus, & ignibus aurum,
Arte mea gemmas terq; quaterq; proba.

Arte mea Reges utuntur ubiq; potentes,
Vtitur & coniunx Cæſaris arte mea.
Arte mea ſiquidem conflatur, amabile quicquid,
Aut oculus pulchrum iudicat eſſe bonus.
Sed quia ſæpe latent incommoda mille ſub auro,
Auro te cupiam ſæpè carere faber.

Fuſor

Gold-working, gilding, and the decorative arts declined during the Middle Ages. Compared to the work of such ancient master gilders as the Egyptians or the magnificent jewelry produced during Hellenistic times, the arts of the medieval era suggest a fallow period. It was followed, however, by the truly high levels of Renaissance achievement by artists whose rediscovery of classical techniques was made possible by the existence of such humble medieval sources as *Compositiones variae* and *Mappae clavicula*. Here a smith beats gold by hand into leaves sometimes as thin as 1/280,000 inch (0.00009 mm). *Panoplia*, 1568. Hartmann Schopper. LC-USZ62-110309.

This sixteenth-century image illustrates the several steps and processes involved in cupellation, or separating silver (or other noble metals) from impurities. As recommended by the medieval metalworker Theophilus, it is best accomplished over a fire on a "test," which is a shallow, earthenware dish containing a porous bed of ashes that will absorb the oxidized products. By the fifteenth century, this was being done on a large scale in furnaces similar to the one depicted here. *Beschreibung aller furnemisten mineralischen Ertzt vnnd Bergwercks Arten,* 1598. Lazarus Ercker. LC-USZ62-110310.

Middle Ages cannot be overemphasized, and it has long been recognized by serious art history scholars as a seminal work. Unlike its literary predecessors, it is an integral, unified work that offers much detail about complex technical processes. As such, it not only reflects the technology of its time but can be considered the first written technological history since before the fall of Rome.

What distinguishes this work of Theophilus, besides its unity, logical order, and originality, is the high level of very intricate and useful detail it provides about medieval painting, glass making, and metalwork. In addition we have, almost as bonus, an insightful personal account of how a medieval artist conceived of and related to his chosen work. To the art historian, its description of certain Romanesque art techniques, of which there are no known examples extant, makes it an invaluable source. To the historian of technology, its reference to "Byzantine parchment" is probably the first direct reference to paper in the West. Further, it provides the earliest medieval description of bell founding, as well as the most complete medieval account of how to build an organ.

In this Latin text and its accompanying translation, the words of the twelfth-century monk Theophilus became available in English in 1847. Here he describes how to achieve a fleshlike tone when painting. Although using oil as a painting medium was known as early as the eleventh-century, Theophilus was one of the first to write about it. The white and cinnabar pigments mentioned here were ground in drying oils. *Diversarum artium schedula,* 1847. Theophilus. LC-USZ62-110311.

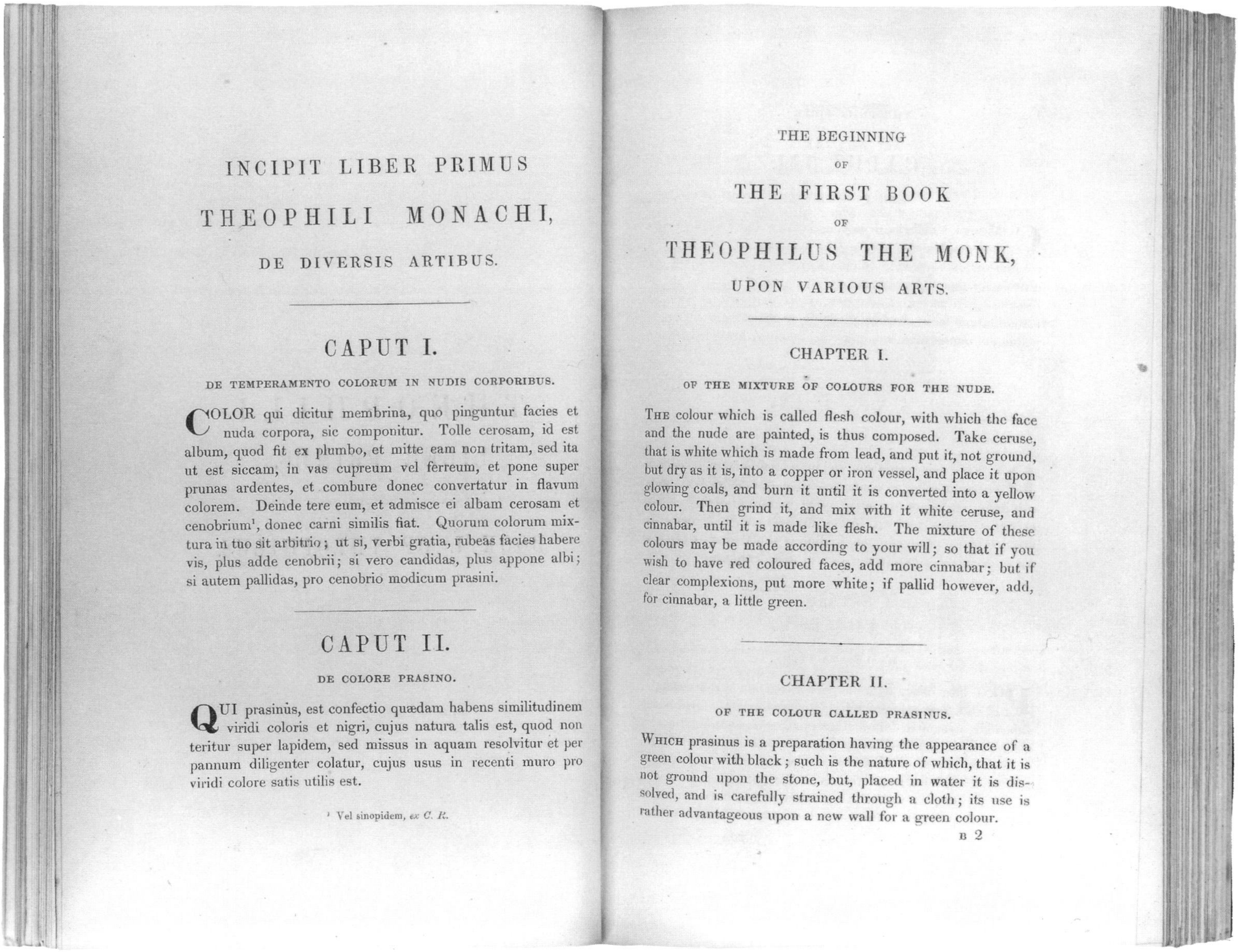

INCIPIT LIBER PRIMUS

THEOPHILI MONACHI,

DE DIVERSIS ARTIBUS.

CAPUT I.

DE TEMPERAMENTO COLORUM IN NUDIS CORPORIBUS.

COLOR qui dicitur membrina, quo pinguntur facies et nuda corpora, sic componitur. Tolle cerosam, id est album, quod fit ex plumbo, et mitte eam non tritam, sed ita ut est siccam, in vas cupreum vel ferreum, et pone super prunas ardentes, et combure donec convertatur in flavum colorem. Deinde tere eum, et admisce ei albam cerosam et cenobrium[1], donec carni similis fiat. Quorum colorum mixtura in tuo sit arbitrio; ut si, verbi gratia, rubeas facies habere vis, plus adde cenobrii; si vero candidas, plus appone albi; si autem pallidas, pro cenobrio modicum prasini.

CAPUT II.

DE COLORE PRASINO.

QUI prasinus, est confectio quædam habens similitudinem viridi coloris et nigri, cujus natura talis est, quod non teritur super lapidem, sed missus in aquam resolvitur et per pannum diligenter colatur, cujus usus in recenti muro pro viridi colore satis utilis est.

[1] Vel sinopidem, *ex C. R.*

THE BEGINNING

OF

THE FIRST BOOK

OF

THEOPHILUS THE MONK,

UPON VARIOUS ARTS.

CHAPTER I.

OF THE MIXTURE OF COLOURS FOR THE NUDE.

THE colour which is called flesh colour, with which the face and the nude are painted, is thus composed. Take ceruse, that is white which is made from lead, and put it, not ground, but dry as it is, into a copper or iron vessel, and place it upon glowing coals, and burn it until it is converted into a yellow colour. Then grind it, and mix with it white ceruse, and cinnabar, until it is made like flesh. The mixture of these colours may be made according to your will; so that if you wish to have red coloured faces, add more cinnabar; but if clear complexions, put more white; if pallid however, add, for cinnabar, a little green.

CHAPTER II.

OF THE COLOUR CALLED PRASINUS.

WHICH prasinus is a preparation having the appearance of a green colour with black; such is the nature of which, that it is not ground upon the stone, but, placed in water it is dissolved, and is carefully strained through a cloth; its use is rather advantageous upon a new wall for a green colour.

B 2

Despite some controversy about who Theophilus really was, most now agree he was the German Benedictine monk and metalworker, Roger of Helmarshausen, who wrote his treatise between the years 1110 and 1140. The original was copied often and there are many manuscripts scattered about the major libraries of Europe. Gotthold Ephraim Lessing, first printed the work in the eighteenth century, giving it the title *Diversarum artium schedula*, from a phrase in the prologue. The Library of Congress has part of that work as included in a thirteen-volume edition of Lessing's collected works published in Berlin from 1838 to 1840. The first English translation was published by Robert Hendrie in London in 1847 under the all-encompassing title of *Theophili, qui et Rugerus, presbyteri et monachi, libri III. de diversis artibus: seu, Diversarum artium schedula*. The Library has a copy of this work. Since the appearance in 1961 of C. R. Dodwell's English translation and newly collated Latin text, which the Library also has in its collections, his new title, *De diversis artium*, has taken hold.

The final work in this small body of literature on the technology of the decorative arts is a late-thirteenth-century work titled *De coloribus faciendis*. Written by the French cleric Peter of St. Omer (also called Petrus de S. Audemar), this work "on making colors" contains a list of substances available for painting, along with recipes for the preparation of various pigments and their specific uses. He describes different ways to prepare greens from copper and vegetables, white from lead, black from charcoal, and blue from silver, copper, and flowers. He also teaches what vehicle is most useful for each color, and so we learn that colors were applied to walls *in secco*; in books and in miniatures, tempered with egg or gum; and on wood, with oil. The mention of oil painting—in this case linseed oil mixed with some sort of resin called vermix or glassa—is not surprising, since oil as a painting medium is recorded as early as the tenth century in Eraclius. The use of oil paints in easel painting is, however, a much later technique. Like Eraclius's work, *De coloribus faciendis* was of course first written in manuscript form and was not reproduced in book form until Merrifield's *Original Treatises . . . on the Arts of Painting* (London, 1849), noted earlier. This landmark compilation has become one of the standard reference texts in the technology of painting from the Middle Ages to the seventeenth century. Intended originally to establish the historical techniques of painting, in particular Italian oil painting, the work eventually was broadened in scope to include texts outside the area of oil painting as well as those that predated its appearance. Merrifield's work published each treatise in full, in both its original language and in English translation.

It is instructive to note that of all the various fields of medieval technical pursuits that could have generated a body of literature, the decorative arts are the best represented. This may be significant, providing a possible key to understanding one aspect of medieval life. Viewed thus, it reveals a steady passion for ornamentation and embellishment, of even the most ordinary or mundane objects of everyday life. This love of color and decoration can be seen in the material objects that remain from that time—

Opposite page:
The medieval passion for ornamentation and embellishment is seen in this twelfth-century manuscript whose binding is the oldest in the collections of the Library of Congress. Written about 1150 in an early Gothic writing style, the 104 leaves of biblical manuscript are held in a well-preserved, stamped leather binding. Unlike most manuscripts whose bindings were plain leather over a wooden board, this one is bound in leather showing the use of deeply engraved metal dies very similar to those used on coins and medals of this period. The use of seven different dies—including dies for a lion and a griffin—to decorate the outer covers indicates that the medieval impulse for ornamentation sometimes went beyond illumination of the individual manuscript page. *Expositio mistica super exodum*, ca. 1150. LC-USZ62-110312.

THE TECHNOLOGY OF THE MIDDLE AGES

the tapestries, furniture, clothing, armor, and even weapons. The medieval craftsmen skillfully and lovingly put into practice the belief that the end product of their craft should be not only well made but aesthetically pleasing. This zeal for beauty was cultivated not only in the monasteries, which were often centers of fine craftsmanship, but also later in the towns, where craft guilds maintained a high standard of workmanship. Because of this, we have the curious situation in which a twentieth-century discussion of the extant literature of medieval technology focuses primarily and necessarily on the techniques of achieving beauty. This alliance of labor and art is characteristic of the medieval decorative arts and was personified by the medieval artist, who was always a craftsman as well.

Two additional sources of information about medieval technology are also somewhat unexpected. First are the twelfth- and thirteenth-century classifiers of knowledge who were, in turn, forerunners of the second group, the encyclopedists. Both considered the subject matter called "mechanica" to some degree in their work, although seldom at length. Nonetheless, even the most incidental inclusion of almost any medieval technical subject is informative and often revealing, given our relative lack of knowledge.

First among these classifiers or philosophers is Hugo de S. Victore, also called Hugh of St. Victor, an Augustinian cleric born at the very end of the eleventh century. An educator as well as a theologian—he was master of the Paris Abbey School of St. Victor—Hugh realized the need for an orderly and coordinated approach to knowledge and wrote his most important nontheological work, *Didascalicon: De studio legendi*, to that end. Written in the very old tradition of didascalic or didactic literature, this work sought to select and define all the areas of knowledge important to man and to demonstrate that they not only possessed an integrity among themselves, but were essential if man was to achieve his divine destiny. Hugh may have been the first to raise technology to the dignity of a science and, indeed, his inclusion of "mechanica" as one of the four parts of his quaternary or master categories of knowledge represents a significant step toward the eventual embrace of secular or profane knowledge as worthwhile. He divided the mechanical sciences into seven categories, all described as "adulterate" since they are concerned "with the artifer's product, which borrows its form from nature." His seven mechanical sciences are fabric-making, armament, commerce, agriculture, hunting, medicine, and theatrics. Each term is intended to be very general and to include a great variety of technical activities. They are not perhaps the best terms available, since by *commerce* he meant not only all aspects of an economic transaction but navigation as well. Further, he includes the preparation of food and the related duties of baker, butcher, cook, and tavern keeper in the hunting category. Nonetheless, it was very popular and influential, coming as it did at the beginning of the twelfth-century reawakening, and can be said to have prepared the way for technology to be eventually regarded as a serious and worthy area of human endeavor and study. The Library of Congress has the *Didascalicon* bound with a fifteenth-century Latin and German dictionary and glossary by Wenceslaus Brack entitled

Opposite page:
On this title page from the seventeenth-century publication of the collected works of the great classifier Hugo of Saint Victor, chapter 6 is seen to include his treatment of "mechanica." This late twelfth-century addition of some aspects of technology to the master categories of knowledge was a major step in the eventual official embrace of what might have been called profane, adulterate, or secular knowledge. *Opera omnia tribvs tomis digesta,* 1617. Hugo of Saint Victor. LC-USZ62-110313.

M. HVGONIS DE SANCTO VICTORE AD REGVLAM S. AVGVSTINI CANONICI, DIDASCALI LIBRI VII.

De ſtudio legendi: quorum primi hæc ſunt Capita:

Sequitur Liber primus de ſtudio legendi.

CAPVT I.

Duo eſſe præcipua, quibus ad ſcientiam quisque inſtruitur: Et quæ ſit operis huius materia.

DVæ præcipuè res ſunt, quibus quisque ad ſcientiam inſtruitur: videlicet lectio & meditatio, è quibus lectio priorem in doctrina obtinet locum: & de hac tractat liber iſte, dando præcepta legendi. Tria autem ſunt præcepta magis lectioni neceſſaria. Primùm vt ſciat quisque quid legere debeat. Secundùm, quo ordine legere debeat, id eſt, quid prius, quid poſterius. Tertiũ, quomodo legere debeat. De his tribus per ſingula agitur in hoc libro. Inſtruit autem tam ſecularium quàm diuinarum ſcripturarum lectorem: vnde & in duas partes diuiditur, quarum vnaquæque tres habet diſtinctiones. In prima parte docet lectorem artium, in ſecunda diuinum lectorem. Docet autẽ hoc modo, oſtendendo primum, quid legendum ſit: deinde quo ordine, & quomodo legendum ſit. Vt autem ſciri poſſit, quid legendum ſit, in prima parte enumerat originem omnium artium: deinde deſcriptionem, & partitionem earum, id eſt, quomodo vnaquæque contineat aliam, vel contineatur ab alia, ſecans philoſophiam à ſummo vſque ad vltima membra. Deinde enumerat authores artium: & poſtea oſtendit, quæ ex his videlicet artibus præcipuè legendæ ſint. Deinde etiam quo ordine, & quomodo legendæ ſint, aperit. Poſtremò legentibus vitæ ſuæ diſciplinam præſcribit: & ſic finitur prima pars. In ſecunda parte determinat, quæ ſcripturæ diuinæ appellandæ ſint: deinde numerum & ordinem diuinorum librorum & authores eorum, & interpretationes nominum. Poſtea agit de quibusdam proprietatibus diuinæ Scripturæ, quæ magis ſunt neceſſariæ. Deinde docet, qualiter legere debeat ſacram ſcripturam is, qui in ea correctionem morum ſuorum, & formam viuendi quærit. Ad vltimum docet illum, qui propter amorem ſcientiæ eam legit: & ſic ſecunda quoque pars finem accipit.

CAPVT II.

De origine artium, & animæ perfectione.

OMnium expetendorum prima eſt ſapientia, in qua perfecti boni forma conſiſtit. Sapientia illuminat hominem, vt ſeipſum agnoſcat, qui cæteris ſimilis fuit, cũ ſe præ cæteris factum eſſe non intellexit. Immortalis quippe animus ſapientia illuſtratus reſpicit principium ſuum, & quàm ſit indecorum, agnoſcit, vt extra ſe quicquam quærat: cui quod ipſe eſt, ſatis eſſe poterat. Scriptum legitur in tripode Apollinis: Γνῶθι σεαυτόν, id eſt, *noſce teipſum*: quia nimirum homo ſi non originis ſuæ immemor eſſet, omne quod mutabilitati obnoxium eſt, quàm ſit nihil, agnoſceret. Probata apud philoſophos ſententia, *animam ex cunctis naturæ partibus aſſerit eſſe compactam.* Et Timæus Platonicus *ex diuidua & indiuidua, mixtaque ſubſtantia: itemque eadem & diuerſa, & ex vtroque commixta natura, quo vniuerſitas deſignatur*, ἐντελέχειαν *formauit.* Ipſa namque & initia, & quæ initia conſequuntur, capit: quia & inuiſibiles per intelligentiam rerum cauſas comprehendit: & viſibiles actualium formas per ſenſuum paſſiones colligit: ſectaque in orbes geminos motum glomerat: quia ſiue per ſenſus ad ſenſibilia exeat, ſiue per intelligentiam ad inuiſibilia aſcendat, ad ſeipſam rerum ſimilitudines trahens regyrat: & hoc eſt, quod eadem mens, quæ vniuerſorum capax eſt, ex omni ſubſtantia atque natura, quo ſimilitudinis repræſentet figuram coaptatur. Pythagoricum namque dogma erat, *ſimilia ſimilibus comprehendi*: vt ſcilicet anima rationalis niſi ex omnibus compoſita foret, nullatenus omnia comprehendere poſſet, ſecundum quod dicit quidam:

Terram terreno comprendimus, æthera flammis,
Humorem liquido, noſtro ſpirabile flatu.

Nec tamen exiſtimare debemus viros in omni rerum natura peritiſſimos, hoc de ſimplici eſſentia ſenſiſſe, quod vlla ſe partium quantitate diſtenderet, ſed vt apertius mirabilem eius demonſtraret potentiam, dicebant ex omnibus naturis conſtare: non ſecundum compoſitionem, ſed ſecundum compoſitionis rationem. Neque enim hæc rerum omnium ſimilitudo aliunde aut extrinſecus animæ aduenire credenda eſt, ſed ipſa potius eam in ſe, & ex ſe

Hugonis de S. Victore Tom. 3. A natiua

This illustration from a fifteenth-century text is used to depict the ages or stages of man, from infancy to old age. It also can be viewed as giving us an unusual picture of medieval life, for it includes such revealing particulars as the tightly wrapped baby's cradle, children's toys, the schoolboy's uniform, the young man's hair style and dagger, the grown man's footwear, and the old man's crutch and rosary or worry beads. *De proprietatibus rerum,* 1486. Bartholomaeus Anglicus. LC-USZ62-110314.

Vocabularius rerum, first published in Basel in 1483.

Although Hugh of Saint Victor relied on purely Western sources in Latin in his classification of knowledge, the twelfth-century Spanish translator and philosopher Domincus Gundissalinus was the first to provide the West with a classification that was based on Arabic sources, most notably from Alfarabi. Although his *De divisione philosophiae* contains little if any discussion of technical subjects, it exerted a considerable influence on later classifiers and encyclopedists such as Robert Kilwardby, Michael Scot, and Vincent de Beauvais. It remained in manuscript form until 1903, when it appeared in the journal *Beitrage zur Geschichte der Philosophie des Mittelalters*. The 142 pages of the Latin text of *De divisione philosophiae* are followed by Ludwig Baur's 253 pages of commentary and analysis. The Library of Congress has a complete set of this journal.

We know a good deal about Alexander Neckam, a late-twelfth-century Augustinian abbot in England. In addition to a popular encyclopedia, he wrote *De nominibus utensilibus*, a vocabulary intended to illustrate in descriptive form the meanings of the "names of instruments." Neckam's Latin manuscript first appeared in book form in 1857 in a two-volume work entitled *A Volume of Vocabularies*, with commentary by Thomas Wright, which the Library has in first edition. Although Neckam's text gives no technical details on the structure or operation of the instruments described, his work is valuable as a sort of inventory, albeit selective, of what devices were available and being used in the twelfth century. Perhaps most notable is his mention of the magnetic compass being used as a navigational aid, especially since he does not treat it as a novelty. A brief look at Wright's glosses indicates the range of Neckam's interests, including many taken from daily life—a horse comb, leather thimble, weathervane, writing instruments, and weaving apparatus. Further, he mentions details and techniques related to architecture and sculpture. The date of Neckam's birth—September 8, 1157—is known with certainty, for on that date Richard the Lion-Hearted, son of Henry II and Eleanor of Aquitaine and later king of England, also was born. Eleanor bore Richard, the third of her eight children with Henry, at Beaumont Palace, just within the city gates of Oxford. That same night, a woman of St. Albans named Hodierna bore her own son, Alexander. When Hodierna was chosen as wet nurse and eventual foster mother for Richard, she "gave suck to Richard with her right breast, and to Alexander with her left breast." Alexander's mother was given an annuity and he no doubt benefited from this early princely alliance.

Like Hugh in France and Gundissalinus in Spain, Neckam in England exemplifies a kind of twelfth-century renaissance in which inquiry into nature was regarded as an important and necessary step in human intellectual development. Unlike much earlier times when such inquiry was considered an exercise in human pride (an Augustinian concept), this new attitude was a foreshadowing of things to come. This fledgling twelfth-century reassessment of man and the natural world also extended to the mechanical arts and that growing interest is evident in the work of Hugh and Neckam. Such classifiers of knowledge were mostly documenting things as they were rather than innovating, but by incorporating "mechanica" in their classification systems, they were bringing theory more into line with practice. This incorporation of technical subjects into what might be called doctrinal or official knowledge was a significant new idea. It marked a growing trend that was to flower in the Renaissance praise for devices and techniques and preoccupation with them.

In the next century, Michael Scot was one of the first writers whose work reflected the influence of his predecessors—he adapted the classification schemes of Gundissalinus. Little is known of his origins, although, because of his name, it is assumed that he was born in either Scotland or Ireland. As a translator of several of Aristotle's works, he spent most of his life in either Toledo or Sicily. Because of his closeness to Emperor Frederick II—he was court astrologer and overall scientific companion to the "baptized sultan of Sicily"—Scot came to be regarded posthumously as an evil but wise wizard. A century after his death his notoriety was sufficient for Dante to place him in Hell. Scot's classification of science, called *Divisio philosophica*, exists only in fragments and is preserved in the work of Vincent de Beauvais.

Vincent de Beauvais's primary accomplishment was a massive three-part encyclopedia whose overall title was *Speculum majus*, or the *Great Mirror*. Although the Library's copy of this huge work bears no publication date, it is known that it was printed in Strasbourg by Johannes Mentelin between 1473 and 1476. Of its three parts—*Speculum naturale*, which is an encyclopedia of nature; *Speculum historale*, a history of man-

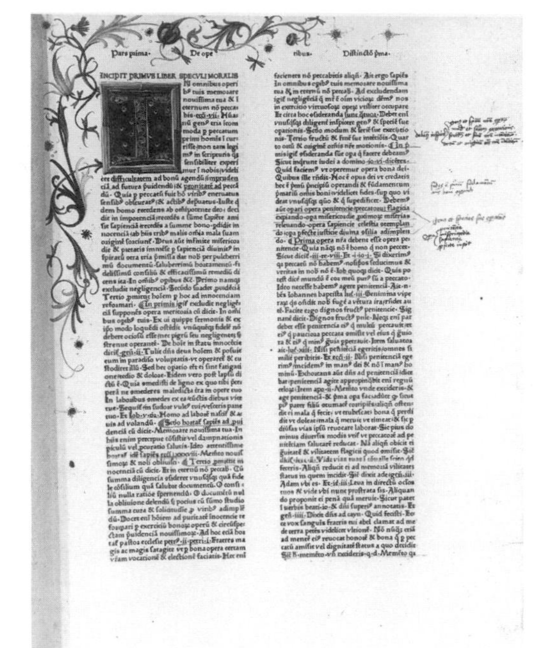

The late thirteenth-century French scholar and encyclopedist Vincent de Beauvais included what he calls "mechanical arts" in his massive work. In this, he continued a trend begun a century earlier by Hugo of Saint Victor in France and carried on by Gundissalinus in Spain, Neckam in England, and Scot in Sicily. This colorful, rubricated page is the first in what is considered to be the largest incunabulum known. *Speculum morale,* 1476. Vincent de Beauvais.

kind beginning with creation; and *Speculum doctrinale*, a summary of all the learned arts—it is the last that is of interest here. Vincent includes the mechanical arts in the *Speculum doctrinale*, or *Mirror of Teaching*, together with such subjects as husbandry, trades, medicine, surveying, and weights and measures. In particular, he has much to say about architecture, especially as it spurred the development of specific knowledge by its use of arithmetic for estimates, of music for acoustics, and of astronomy for sundials. Vincent's entire *Speculum* is the largest incunabulum known, making him the greatest encyclopedist of the Middle Ages. Although none of it was original or scarcely even reflective or comparative, Vincent was scrupulous in attributing each work he excerpted to its respective author.

Whereas Hugh of St. Victor's work was based entirely on Latin sources and that of Gundissalinus relied on Arabic sources, a mid-thirteenth-century Dominican, Robert Kilwardby, used both Latin and Arabic sources in his classification of the sciences. A philosopher and grammarian, Kilwardby was archbishop of Canterbury from 1272 to 1278. In *De ortu scientiarum* he discusses the nature and division of scientific knowledge. Kilwardby divided the mechanical arts into a trivium (agriculture, dietetics, medicine) and quadrivium (costuming, armor-making, architecture, commerce) that reflected the traditional seven-part organization of knowledge in the liberal arts as formalized by the emerging university system. This large treatise was very influential in its time, but remained in manuscript form until it was first printed with a substantial introduction by Albert G. Judy in 1976. The Library has this recent work, published jointly by the British Academy and the Pontifical Institute of Mediaeval Studies of Toronto. Kilwardby was the last of the great classifiers. His fourteenth-century counterparts were little concerned with continuing research on the nature, scope, and classification of the sciences, and preferred rather to confront the actual problems of those sciences.

A contemporary of Kilwardby's and a rival Franciscan (although only a tertiary or lay member), the Spanish scholar Ramon Lull generated one of the better and more interesting medieval examples of classifying knowledge in a concrete manner. His use of the diagrammatic symbol of a tree or "arbor" to make his dichotomies explicit reflected a popular medieval illustrative trend. Drawing a tree enabled the author to show the branching or subdividing of knowledge and its hierarchy while still demonstrating its essential interrelatedness and unity. His great encyclopedia, *Arbor scientiae*, is relevant here for in his review of the sciences he cited, apart from the trivium and quadrivium, metallurgy, building, clothing, agriculture, commerce, navigation, and the military arts. In this book of sixteen trees, written during 1295 and 1296 while he was in Rome, Lull consistently emphasized not only the technicians' need for theoretical knowledge, but also the fact that sometimes a technique itself can become a science. The Library's copy of Lull's *Arbor scientiae* was published in Lyons in 1515 and is in Latin. It was the second longest of Lull's works originally written in his native tongue, with the Catalan title *Arbre de sciencia*.

As a Catalan, Lull wrote most of his treatises in that language first and then later translated them into Latin. His writing was so prolific that he single-handedly transformed the Catalan language into both a literary medium and a scholarly one, nearly a century before any other Romance vernacular was so developed.

By his own admission, Lull was a worldly, passionate man in his youth who, after a series of visions of the crucified Christ, transferred the strength of his passions to God. Following his visions, his goals became to attain martyrdom and to combat the influence of the infidel. As would be expected, the Muslim infidels dominated his thinking, and nearly all of his writings can be seen as an attempt to refute Muslim arguments by reason alone. Lull's attempt to provide a rational basis in his disputes with heretics, and not appeal to faith alone, is a forerunner of the Renaissance deemphasis on authority.

Roger Bacon, the famed "Doctor mirabilis," was by far the best known figure of this high medieval period who was in any way related to technology. Bacon was an English Franciscan, albeit a dissatisfied one, and therefore tended to think in terms of the Neo-Platonism of Augustine. As such, he had an obvious relish for mathematics, in distinction to the Franciscan's rivals in Paris, the Dominicans. His Aristotelianism was therefore somewhat different from theirs, although both emphasized the importance of verification by experience in the study of nature. Unlike most classifiers of his time, Bacon ranked mathematics above metaphysics in the hierarchy of human knowledge, saying that "it is impossible to know the things of this world, unless one knows mathematics."

This strong disposition toward mathematics is of more than simple academic interest, for Bacon has come to be seen as an intellectual bridge between antiquity and the Renaissance. In his belief that mathematics is essential to what he called "scientia experimentalis" in understanding the natural world, Bacon prefigured the practical methods of modern science. In this experimental science Bacon reveals his desire not only to discover nature's hidden secrets but to harness and use its powers as well. In the letter known as *Epistola de secretis operibus artis et naturae et de nullitate magiae*, or *Letter Concerning the Secret Works of Arts and of Nature*, written to either William of Auvergne (died 1248) or John of Basingstoke (died 1252), Bacon offers examples of the potential usefulness of scientific experiment. His famous descriptions of what he calls "natural marvels" include such mechanical devices as flying machines and driving carriages, ships, and submarines that move swiftly "without the help of any living creature." Among other marvels in which "there is no magic whatsoever," he includes artificial fire, a self-activated working model of the heavens, spectacles, and the camera obscura, and he provides directions on how to make gunpowder. Bacon states that these marvels were made in antiquity (except for the flying machine) and that he is "familiar" with them. What is significant in these thirteenth-century writings by Bacon is not so much the technological details or actual hard information they contain, for they offer virtually none. Rather, their importance to the history of technology resides in the attitude they exemplify. His concluding statement—made in the middle of the thirteenth century—that "infinite

The image of a tree was a useful and common medieval teaching device symbolizing interrelatedness and essential unity, and here the Catalan mystic and poet Ramon Lull offers a literal depiction of what he calls the tree of knowledge. Although he reacted severely against the Church's reconciliation with Aristotelianism and argued that theological and philosophical truth were inseparable, he did not dispute the trend that technical knowledge had a proper place somewhere in his tree. *Arbor scientiae,* 1515. Ramon Lull. LC-USZ62-110315.

other such things can be made"—is a remarkable expression of optimism, hope, and a belief in material progress.

This switch from the older medieval notion of the cyclical nature of human history to something approaching a premodern idea that human events proceed in a linear fashion did not begin with Bacon, but in his time he was its best promoter. What details are known of his life indicate that he was not only ahead of his time but a man of unsettling contrasts. Educated at Oxford and Paris, he chose to enter the Franciscan order, a restrictive sect that made his life's work and interests subject to innumerable difficulties and, eventually, to condemnation. In his writings he espoused the primacy of mathematics and the necessity of experimentation, when he knew little of the former and barely practiced the latter. Throughout his life he relied upon authority for a proper hearing of his views, yet he held it in disdain and took every opportunity to write against it. His fortunes rose with Pope Clement's election in 1265—the pope commissioning what would become Bacon's major contributions, the *Opus majus*, *Opus minus*, and *Opus tertium*—and began to decline with the pope's death in 1268. He was eventually imprisoned, although his offenses are unknown.

Bacon's legacy to technology is essentially the vision he communicated in his *Epistola*. Instead of details, recipes, or specifc plans, he offered an idea, a benign picture of nature mastered, of natural power both understood and controlled. Perhaps because of his censure and condemnation by the church, his fame for centuries afterward resembled legend more than fact, and it is significant that there are no incunabular editions of his work. One of the first of his works to be printed was his *Epistola*, in Paris in 1542. It is found in Lazarus Zetzer's *Theatrum chemicium* (Strasbourg, 1659–61) and in Jean Jacques Manget's *Bibliotheca chemica curiosa* (Geneva, 1702). The Library has both of these works in its collections.

In contrast to the power and romance of Roger Bacon's bold technological forecasting is the workmanlike listing of the actual contemporary mechanical arts and inventions done by an Italian Dominican named Giovanni da San Gimignano (Johannes de Sancto Geminiano, in Latin). A popular lecturer, or "contionator," at the beginning of the fourteenth century, he prepared a large encyclopedia for his fellow preachers entitled *Liber de exemplis similitudinibus rerum*, in which he attempted to share the benefit of his oratorical experiences. As a kind of reference book of natural, moral, and legal knowledge that preachers could use to prepare their lectures, the book contained many concrete illustrations from both history and daily life to enliven sermons and make them relevant. From a modern perspective, however, this book serves as a real history of everyday technology, offering us discussions of scores of technical subjects, from weights and measures, agriculture, glassmaking, fortifications, windmills and watermills, to painting, limning, and smithing. His book had five separate incunabular printings, the earliest of which is 1477. The Library of Congress has the 1497 and 1499 Venice editions, as well as a 1499 edition, published in Basel.

Besides these chroniclers, encyclopedists, and classifiers whose writings were not directly concerned with technology, but whose work nonetheless

Opposite page:
This seventeenth-century portrait of the English Franciscan friar Roger Bacon shows him performing an experiment—something he did only infrequently. His restless intellect, energy, and zeal for learning made him, however, a sort of bridge to the Renaissance. His inclination toward alchemy and astrology seemed to dispose him to dangerous "novelties" of the mechanical and marvelous type, and he wrote about all manner of technical wonders he could envision. Although many discoveries have been improperly attributed to him (such as spectacles and gunpowder), his major contribution to technology was his precocious and apparently unshakable optimism, hope, and belief in material progress. *Symbola aureae mensae dvodecim nationum,* 1617. Michael Maier. LC-USZ62-110316.

Bacon ob eruditionē Magiæ ſuſpicionem incurrit apud vulgū. Liuor, vt hedera aſcēdit, aliorum ope.

Ocherius Bacon Anglus monachus minoris ordinis agens, ſua eruditione, variarumque artium cognitione in id celebritatis faſtigium aſcendit, vt non ſolum in ore omnium eſſet, ſed apud inuidos quoque eum ſupernaturali quodam modo, hoc eſt, magico, ſapere, ſuſpicionem, aut potius calumniā incurrerit: Sic liuor ſumma ingenia, vt *Hedera* arbores, petit: Hæc per ſe aſcendere nequit, vtpote fragilis & inſtabilis, ſed ſemper ſuſtentaculo eget, nempe arbore vireſcente & co impinged on the field in some way, there were a few individuals who made actual contributions that were also documented by technical treatises. Not surprisingly, the earliest lived in the late twelfth century and the last died just after the beginning of the fourteenth century. Their original writings were all, of course, in manuscript form, recopied by hand, and eventually printed during the fifteenth and sixteenth centuries.

One of the first to discuss and experiment with the magnifying power of lenses was an Oxford University scholar and later bishop of Lincoln, Robert Grosseteste. Best known as the teacher of Roger Bacon, Grosseteste exhibited not only a strong curiosity about the facts of the natural world but also, more important, an appreciation for the need for experimentation in understanding those facts (*propter quid*). A scholar of many interests and an experimenter before he took on the duties of bishop, Grosseteste wrote two treatises relevant to the history of technology, one on optics and the other on the calendar. He believed that light was the first of all forms of matter to be created and therefore that the science of optics or "perspectiva" was the most fundamental of all the sciences. In *De iride*, he discusses his experiments with lenses and gives practical accounts of magnification and diminution using refracting mediums. Although it is not known for certain when spectacles were invented, it is documented that they were being used in Italy by the time of Bacon's death just before the end of the thirteenth century. Just as Grosseteste was the first medieval European to draw attention to the feasibility of spectacles, so his original influence was also felt in the technical domain of calendar reform. His authoritative treatise on the *Compotus* offers intelligent and useful proposals for reforming the inaccurate Julian calendar.

Although the Library of Congress has neither of these treatises on lenses nor the calendar, it does have his important *Commentaria in Posteriora Aristotelis*, which details and exemplifies the formal structure of his mature scientific method. The *Commentaria* was first published in Venice in 1494, and the Library's copy is the 1497 Venice edition. Many of Roger Bacon's ideas and methods have been attributed to Grosseteste, and Bacon later eulogized his teacher in appreciation.

A French contemporary of Grosseteste about whom very little is known is the engineer called Villard de Honnecourt. Because of the chance survival of his sketchbook, done between 1225 and 1235, Villard's name and work are recognized by every student of medieval art. Despite his fame, most of the "facts" we have about Villard himself and the origins of the manuscript he left behind are not held with any real certainty. Villard is unrecorded anywhere in history, except by himself in his own manuscript. It is not known if he was an artist, engineer, or sculptor, or whether the book of sketches, drawings, and notes was the result of random efforts or was a more purposeful product intended as a textbook or shop manual for instruction. Whatever his reasons for compiling the sketchbook, Villard produced a highly valuable and unique glance into the architecture and related technology of the early thirteenth century.

The manuscript itself, or *Album* as the French call it, is in the Bibliothèque Nationale in Paris and consists of thirty-three parchment folios folded once and bound in pigskin. The *Album* was published in facsimile for the first time in 1858 by J. B. A. Lassus and A. Darcel. The Library's copy is the 1859 translation of the Lassus work by R. Willis, entitled *Facsimile of the Sketch-Book of Wilars de Honecort*, published in London. In these modern editions one can study the 207 pen-and-ink drawings, many of which illustrate the architecture and machinery of the Gothic age, the

In this illustration of a medieval bibliophile, the obsessive book collector is seen wearing a pair of pince-nez eyeglasses. These early spectacles were convex lenses that helped only far-sighted people, and it was believed wrongly that they had to be placed at the center of the face where eyesight converges rather than at the exact center of each eye. The availability of spectacles certainly extended the productive life of medieval scholars and copyists and also gave some indication that many a mechanical solution to the limitations of physical nature was available to the inventive mind. *Das Narrenschiff,* 1494. Sebastian Brant. LC-USZ62-110317.

This thirteenth-century drawing by Villard de Honnecourt of a cross-section of the buttress-system of Reims cathedral makes clear what this newest development of the Gothic style accomplished. Basically, the flying buttress is an arch built against a wall with another support a short distance away. Introduced first in 1163 at Notre Dame cathedral in Paris, the flying buttresses of Gothic architecture not only added to the external beauty of a building but allowed its interior columns to soar to new heights. *Facsimile of the Sketch-book of Wilars de Honecort,* 1859. Robert Willis. LC-USZ62-110318.

last part of the Middle Ages. Nearly every type of activity pursued by a medieval builder's guild is documented, including details on mechanical devices, carpentry, masonry, surveying, architectural motifs, and even church furnishings. Most drawings are necessarily concerned with churches, since building cathedrals was the primary outlet for both civic and religious pride at this time.

Villard's drawings capture Gothic architecture in full flower, showing the west rose window at Chartres, the west cathedral tower at Laon, the floor plan of Cambrai, and both the completed and planned structural members at Reims. In these and other drawings, the essential architectural features of the Gothic style are documented—the flying buttress, the pointed arch, and the ribbed vault—as well as its characteristic effects of great height, lacey walls of colored glass, and overall openness. This splendid emphasis on the vertical is best seen today in the Strasbourg spire, whose height of 466 feet (the equivalent of a forty-story building) was not exceeded until the iron and steel constructions of the late eighteenth century. Whether built as symbols of civic pride, individual audacity, or spiritual aspirations, these overwhelming masonry structures did eventually reach their physical limits. No doubt, the spectacular collapse of the incredibly high choir vault of the Beauvais cathedral in 1284 sent such a message throughout the guilds of Europe. In fact, by the time the over-spaced piers and inadequate columns and buttresses gave way at Beauvais, the trend was already toward smaller buildings.

Villard traveled widely, as did many of the architects and engineers of his time who hired themselves out to other cities, and he sketched many of the most important structures of his age. He was fascinated by the tower at Laon—"such as I have never seen anywhere else"—and said he drew the classic window at Reims simply "because I loved it." As a man of his time, he was also preoccupied with harnessing natural sources of energy. Among the several devices he sketched and described in his work are a water-powered saw and a catapult. He also offered several gadgets or automatons as well as a perpetual motion wheel that he explained worked by means of an uneven number of mallets and quicksilver. The sketchbook left by this unknown technician reveals that he was not a first-rate draftsman. However, its worth lies not in the skill of its drawings but in the unique and obviously fresh approach of a medieval man in full possession of what would later be described as Renaissance attributes—intellectual curiosity, independence of judgment, and an appreciation for both the artistic and the practical. Many see Villard and his sketchbook as a precursor of Leonardo da Vinci.

Pierre de Maricourt lived in the same century as Villard and hailed from the same region of northern France—Picardy. Like Robert Grosseteste, he may have been a teacher of Roger Bacon. Bacon's writings established Maricourt's reputation as a great mathematician and sagacious experimentalist, Bacon calling him "Magister Petrus." He is better known as Petrus Peregrinus, the cognomen Peregrinus, meaning "pilgrim," being an honorific title given to those who either participated in the crusades or went on a pilgrimage to the Holy Land.

Aside from what Bacon has to say about him and his work, most of what we know about Peregrinus comes from a letter he wrote in 1269, the only work attributed to him with certainty. Called the *Epistola Petri Peregrini de Maricourt ad Sygerum de Foucaucourt, militem, de magnete*, the letter was written to Syger or Sygerus of Foucaucourt, a soldier whom Peregrinus calls his "dearest of friends," during the siege of Lucera. It is

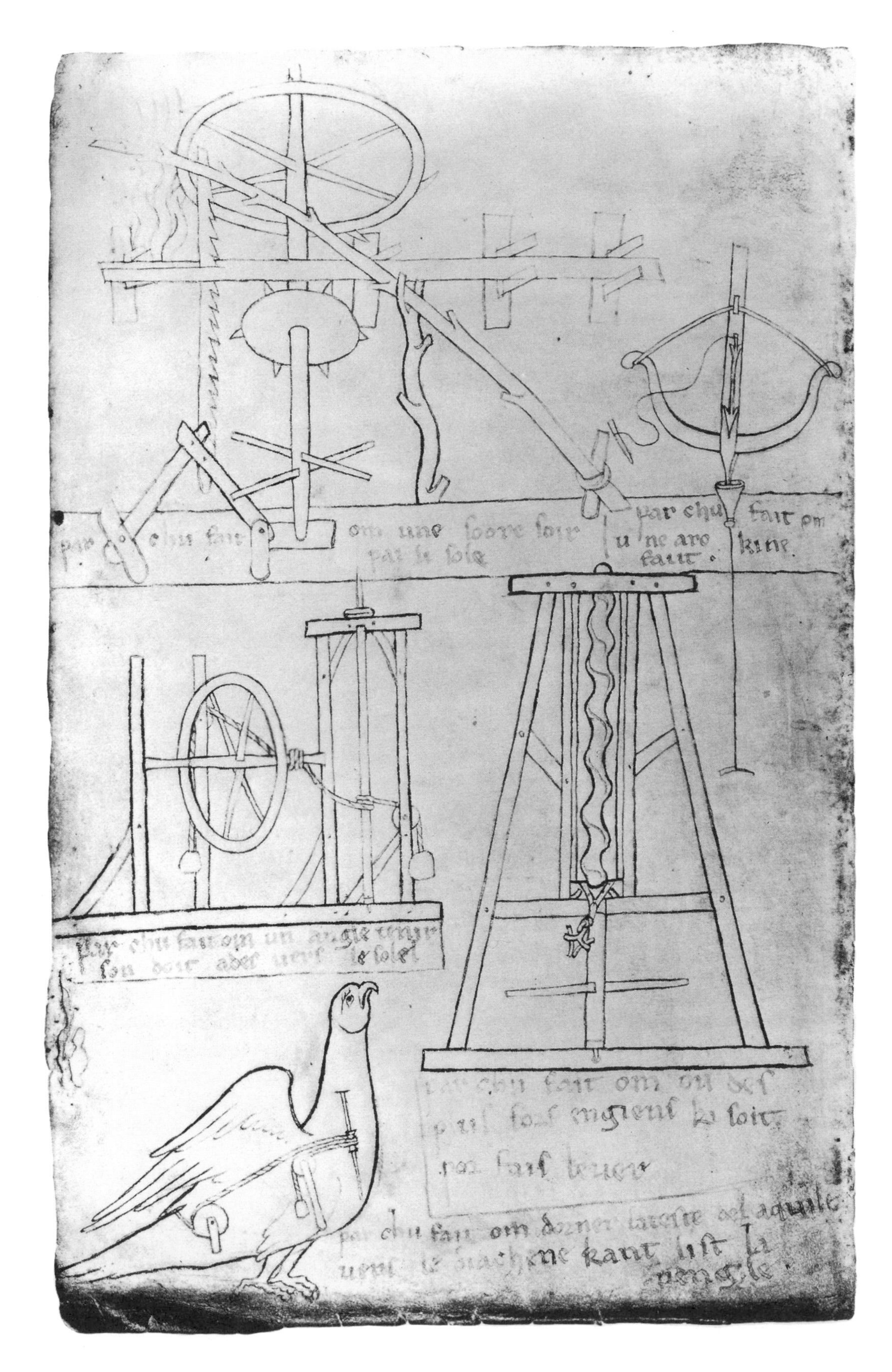

believed that Peregrinus was part of the military force of Charles of Anjou, king of Sicily, who was then personally directing the assault on Lucera, which had rebelled against its French master. Why Peregrinus should write a 3,500-word battlefield letter to his friend back home in which he describes methodically the properties of a lodestone and the discoveries he had made, can only be conjectured. However, as part of Charles's engineering corps, he may have been stimulated by the mechanical problems he encountered or perhaps he was simply taking advantage of the sometimes slow pace of a siege to do some creative work.

His letter is divided into two parts and treats both the theory and the application of lodestones or magnets. It is the second part of the letter that is most relevant here, in that it discusses and illustrates three instruments that put magnetism to use. First, Peregrinus described an improved version of the wet mariner's compass, which consisted of a magnetic needle floating in water in a tight wooden case. He divided the case's circumference into 360 degrees, creating for the first time a mariner's compass with divisions that could determine not only a ship's direction but also the azimuths of the stars. This was a valuable improvement over the early magnetic compasses, which presumably came to the West from China sometime during the last decade of the twelfth century.

Opposite page:
The varied nature, subject, and style of the sketches in Villard de Honnecourt's notebook seem to presage the genius displayed in Leonardo da Vinci's notebooks two centuries later. At the top and to the left of his rudimentary crossbow is seen his version of a water-powered saw. Below that is a device for mounting a wheel on an axle, and to its right is a screw jack that operates much like today's automobile jack. Honnecourt's automated eagle is reminiscent of the marvelous machines of the Greeks and was intended for use in church, where it would turn its head toward the reader of the Gospel. *Facsimile of the Sketch-book of Wilars de Honecort,* 1859. Robert Willis. LC-USZ62-110319.

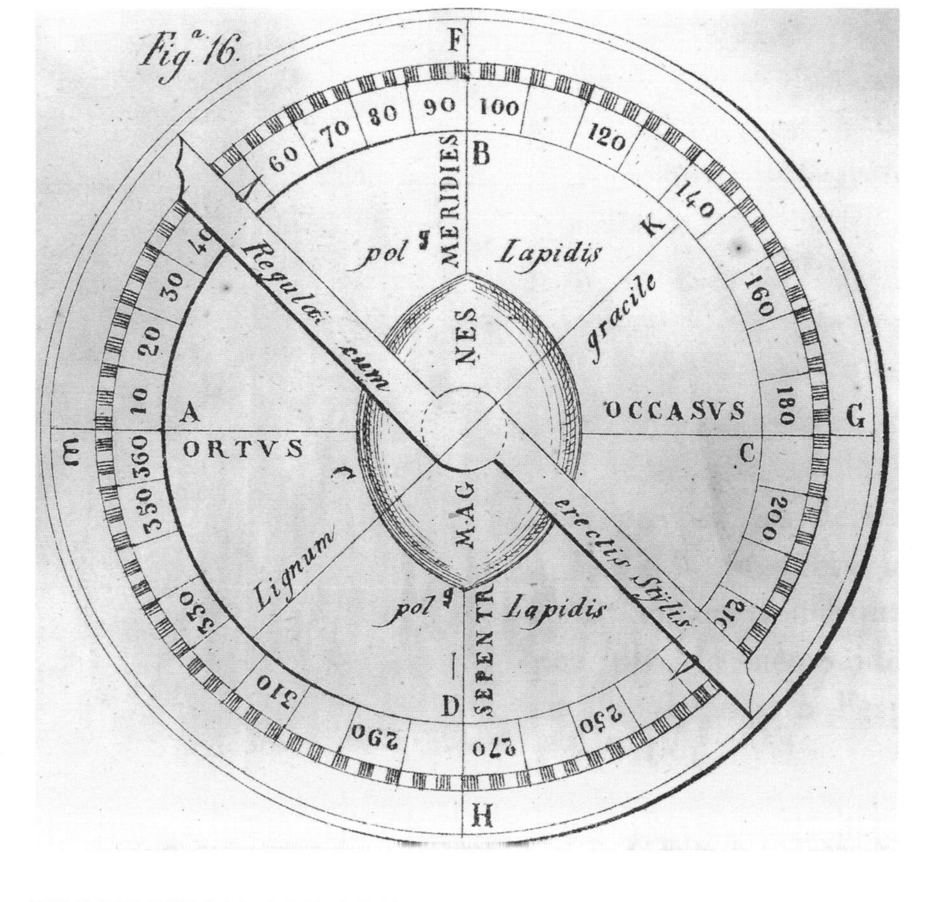

The first detailed description of the compass as an instrument of navigation was rendered by the French crusader Petrus Peregrinus de Maricourt. Here is his version of a mariner's compass that could be used to determine a ship's direction as well as the azimuths of the stars. Made of brass with its circumference divided into 360 degrees, it had a lodestone in the middle pivoting on a wooden float. The gradual development of the compass would eventually transform navigation from an art to a science and make voyages of ocean exploration possible. *Bulletino di Bibliografia e di Storia delle Scienze Matematiche e Fisiche,* 1868. Timoteo Bertelli. LC-USZ62-110320.

His second device, the dry and pivoted compass, marked the major breakthrough in the application of magnetism, for it offered the sailor real portability. In his detailed drawing, Peregrinus showed a round, glass-covered box, again surrounded by degree markings, within which was a magnetic needle turning on a metallic pivot. This is the modern compass in all its essentials. The effects of such a simple, easily produced device upon navigation, and therefore upon commerce and exploration, were pronounced and dramatic. Once a navigator could accurately plot his course, sea voyages became safer and more predictable, improving the returns on capital investment as well as encouraging more frequent and ambitious trips. With the development of the sternpost rudder, the use of the lateen or triangular sail with the square sail, and the invention of the portolan chart and Peregrinus's compass, the technology was emerging that would soon permit Europeans to explore the limits of the seas.

Although his sketches and descriptions of the two compasses resulted in real, working devices that did what Peregrinus claimed they would, his third creation, for which he also provided a drawing, did not work. Since he was describing how a perpetual motion machine might use magnetic power to keep it running continuously, it is not surprising that his machine failed to work. His device was essentially a gear powered by an oval magnet. Its importance lies not in its proposed functions nor in the cleverness of the idea itself. Rather, it shows that Peregrinus possessed certain traits that were more prevalent then than we tend to believe today—namely, an intellectual inclination to experiment rather than accept authority, an incipient belief in man's ability to manipulate his environment for his own practical benefit, and a willingness to believe in essential change. If two thirteenth-century individuals as varied in background and status as Peregrinus and Villard de Honnecourt could both propose seriously the construction of perpetual motion wheels—an idea strictly opposed by the prevailing Scholastic theory—it would appear that a pliant, creative mind was not such an exceptional phenomenon in the late Middle Ages.

The *Epistola* of Peregrinus became well known in its time, and its influence was considerable. It was the first known attempt to treat exclusively all aspects of the magnetic phenomenon. Its popularity throughout the thirteenth and fourteenth centuries is evidenced by the existence of thirty-one separate manuscripts. The treatise did not appear in print form, however, until 1558, when it was prepared by a Bavarian physician, Achilles Glasser. The Library of Congress does not have this very rare edition. The earliest copy of the *Epistola* in the Library's collections is the version by Guillaume Libri published in 1838 in the journal *Histoire des sciences mathematiques en Italie*. Libri did not comment on the famous letter, but merely reproduced its Latin text. A more valuable, if later, edition of the *Epistola* was published in 1868 by the Barnabite monk Timoteo Bertelli, who collated and compared the readings of two Vatican manuscript copies with other texts and added notes, comments, and references in his scholarly appraisal written in Italian. This appeared in the inaugural volume of the *Bulletino di Bibliografia e di Storia delle Scienze Matemataiche e Fisiche* published in 1868, which the Library of Congress has in its collections.

Contrary to the stereotypical image of medieval times as dull, closed, and unchanging, a close look at what that maligned age produced indicates that technologically there was movement and real progress. Since its arbitrary time frame is so long, it was also several ages in one, with each contributing something to its slow evolution from a relatively static beginning after Rome's fall to a dynamic conclusion and the beginning of the Renaissance. While not a paragon of technical progress, neither was it a do-nothing, dead-end period of no advancement.

The medieval technological tradition was a way of doing things that was characterized by much unstructured "empirical groping," which often resulted in surprising accomplishment. There certainly was little or no science or theory to inform its practitioners, and what technological progress there was occurred, as it were, in the dark by trial and error. Following the accomplished but lost technological achievements of Rome, medieval empiricists were compelled in many cases literally to start again and reinvent or rediscover many processes and methodologies which their Mediterranean predecessors had taken for granted. There were many good reasons why early medieval times are called the "dark ages," but from a technological perspective, the entire age was not a wasted time, and many real, substantive accomplishments were made.

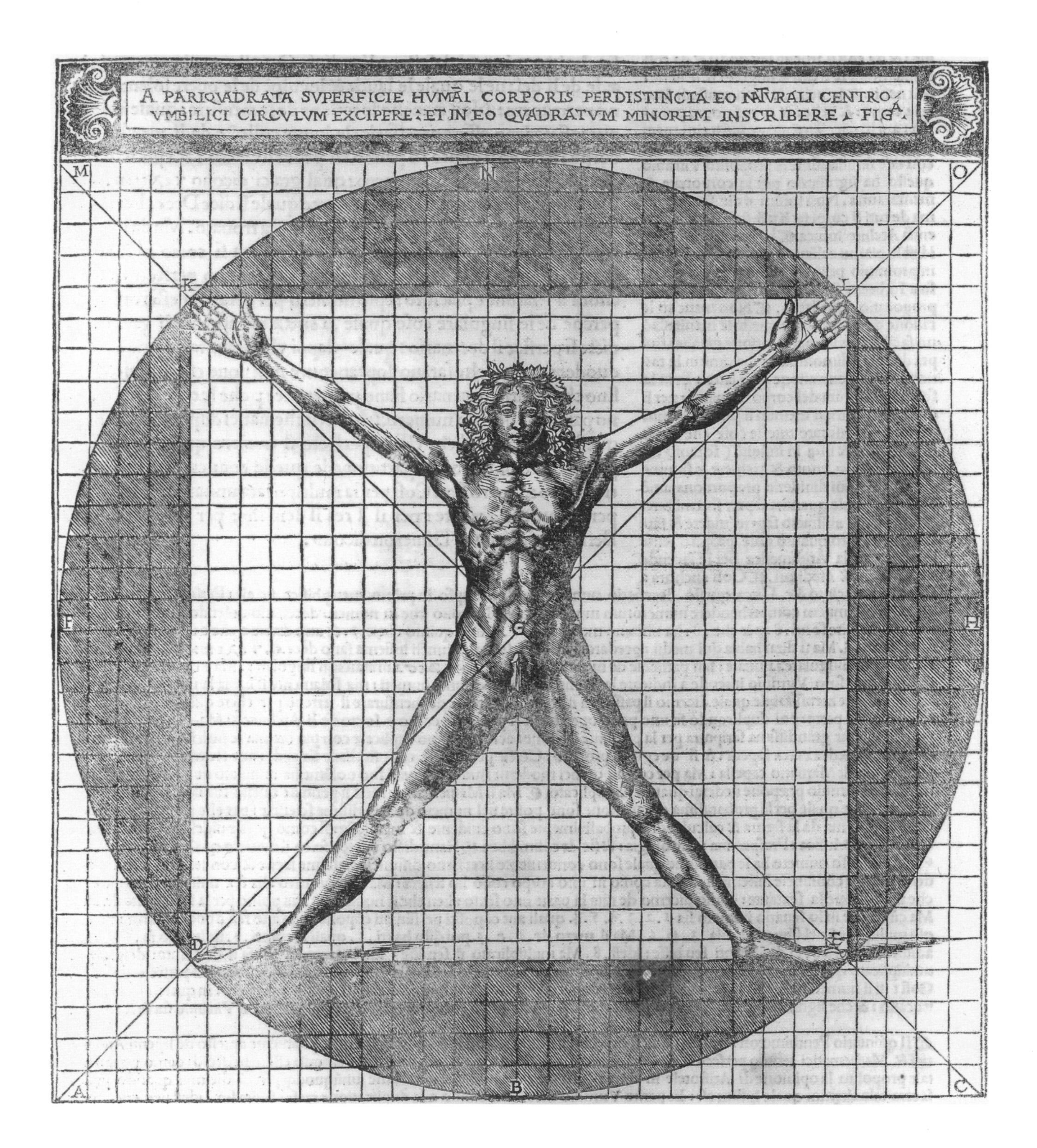
A PARIQVADRATA SVPERFICIE HVMĀI CORPORIS PERDISTINCTĀ EO NĀTVRALI CENTRO
VMBILICI CIRCVLVM EXCIPERE: ET IN EO QVADRATVM MINOREM INSCRIBERE. FIG. A.
M
N
O
K
L
F
G
H
D
E
A
B
C

3. *Renaissance Technology*

The previous chapter considered the Middle Ages as generally an extension of earlier times and chose to emphasize those aspects favoring continuity rather than those suggestive of essential change. However, if there was ever a time in the history of the West when a sudden, dramatic, and portentous break with the past occurred, it was the horrific time that began in 1347 with the first outbreak of the Black Death. Not even with the fall of Rome nearly a thousand years before had organized Western society been so struck by the forces of discontinuity and chaos. Where the actual sack of Rome was but an anticlimax to a progressively slow decline and inevitable collapse, the desolation and disorder wreaked by the plague happened with a dizzying suddenness, sometimes literally overnight.

It has been variously argued elsewhere that the social, economic, and political changes wrought by the plague were such that the basis was laid for everything momentous that followed, from the Reformation to the creation of modern Europe. While such claims of direct causation seem exaggerated, they nonetheless properly reflect the underlying reality that something truly calamitous had happened after which nothing would ever again be the same.

Focusing on the great plague that swept over Europe at mid-century as the trigger mechanism that loosed the forces of change is a convenient and useful explanatory device if it is not overplayed. The Renaissance was many things to many people and places, but it was in essence a time of basic changes—changes of attitude and perception as well as the more obvious outward changes of form and design.

During the two-and-a-half centuries from the mid-fourteenth century to the beginning of the seventeenth, momentous and multitudinous events occurred in nearly every aspect of Western social, political, economic, and religious life. Along with the rebirth of interest in classical antiquity came a humanistic spirit that leavened the medieval respect for authority by applying a new, practical, more earthly standard. Petrarch characterized much of this when he taught that learning was not to remain abstract and impractical. "It is better to will the good than to know the truth," he taught. Such a human-centered idea has wide and significant implications.

This emphasis on the present undoubtedly had much to do with the decline of spiritual authority and the concomitant rise of dynastic rivalries and national monarchies that this period witnessed. These attitudinal or intangible factors were accompanied by the very tangible technical advances that did much to alter the foundations of the medieval way of life.

Opposite page:
The Renaissance notion that ideal human proportions could be represented geometrically was taken directly from the Roman Vitruvius. This striking image was no doubt inspired by the famous squared circle of Leonardo da Vinci, especially since the artist, Cesare Cesariano, was a Milanese painter and architect who once had studied in Leonardo's own atelier. In this first edition of Vitruvius published in a modern language, Cesariano takes license with proper human proportions. To make the centers of the circle, square, and human torso coincide, the artist drew the man's arms much too long (they would reach his knees), and his hands and feet are also too large. *De architectura libri dece*, 1521. Vitruvius. LC-USZ62-110321.

For example, the full-rigged ship, or three-masted carrack, was built in response to practical military needs in the middle of the fifteenth century, but it also made possible the great transoceanic voyages of discovery and trade. Domestic life changed with the development of better tools and instruments, like the spinning wheel and the mechanical clock, and the use of gunpowder and the development of the cannon altered the nature of warfare altogether. Where the cannon thus threatened the status quo, and the carrack actually enlarged European horizons, the invention of mechanical printing revolutionized the use and spread of information.

All of this and more took place in a historical and cultural context in which upheaval was endemic. A glance at some of the major events dominating the fourteenth century alone begins with the famine of 1315–17, jumps to the beginning of the disastrous Hundred Years' War (1337), then to the plague years of 1347–50, and on to the Great Schism (1378–1417) which split the papacy, and ends in the peasant revolts and uprisings in France and later in England in 1381.

Except for the schism of the papacy, it could be argued that technology, in the broadest sense, played some role in each of the above major events. Both the famine and the plague were the culmination of a series of interconnected and sequential developments that were directly related to technological progress. Medieval innovations such as the heavy plow and the horse harness, coupled with an open-field system and three-field crop rotation, resulted in a real revolution in the production of food. By the thirteenth century, Europe was better fed and therefore healthier than it ever had been, and its population increased accordingly. The heightened prosperity, growth of towns, and general economic and population boom were largely founded on regular agricultural surpluses. These surpluses eventually were absorbed as the population surged, especially in the increasingly overcrowded towns. When a series of crop failures upset this precarious balance, real and widespread famine struck Europe. Famine meant not only considerable loss of life but also financial disaster for many of the great mercantile and financial centers of Europe, since everyone's fate was tied to agricultural success. As technology was unable to provide any further agricultural innovations, the living conditions of the peasantry worsened and the surviving population was left in a physically weakened condition. Such was the impoverished living condition of Europe as it was about to face infection from abroad.

Like a rapacious and knowing predator, the Black Death, so called because it discolored its victims' bodies, came to Europe from the East, following the major trade routes. Entering Sicily in late 1347, it then came into Italy via the ports of Venice, Genoa, and Pisa, to be spread rapidly westward into Spain, southern France, and then up to northern Europe. As it raged and finally waned in a span of two years, its aftermath left no doubt that something catastrophic had occurred, for between one-third and two-fifths of the population of western Europe had been killed, horribly and quickly.

To famine and pestilence, the fourteenth century added war. The Hundred Years' War (1337–1453) was begun by the English king, Ed-

A preoccupation with death and dying was not only a holdover from medieval times but was made very real by the horrors of the plague. In this German block book, a dying man is surrounded by the central images of his Christian faith, as well as by angels and devils who vie for his soul. In reality, the plague rarely took its victims in so calm and orderly a fashion. *Ars moriendi,* 1466. LC-USZ62-110322.

ward III, with his claim to the French throne. Before it was through, the war was to ravage every aspect of life in a significant part of the Western world. It was particularly ruinous to the economies of both England and France, which had to finance the war, but it also blocked the traditional exchange channels, disrupting trade for all of Europe and beyond. Not surprisingly, technology was to play a major role in the exercise and eventual outcome of this war. When it began, combatants employed methods and tactics dating from antiquity and used strictly medieval weapons. By its conclusion when the French artillery defeated the English at the Battle of Castillon, both sides were using not only advanced tactics and strategy but the newest artillery. While the first powder cannons in Europe appeared in Flanders about 1314, in England about 1321, and in France five years later, it was not until the second phase of the Hundred Years' War, after the Battle of Agincourt, that they were improved and mobile enough to be a decisive factor. From 1450 on, artillery rather than mounted knights played the key role in every battle, and in 1453, a long line of French culverins—small, long cannons—brought the war to an end with their withering, point-blank fire. In the same year, Constantinople fell before the cannons of Islam and the surging Ottoman dynasty took control.

The devastations of war, want, and disease took their toll in human misery and always settled most firmly and finally on the peasantry. As food prices first soared and then plummeted during the agricultural crises, and labor costs rose with the sudden loss of so much manpower to the plague, landlords tried to cut their losses, imposing direct taxes, tithes, and other unpopular measures to shore up their weakening positions. Inevitably they pushed the peasants too far, and local uprisings began to appear after the middle of the fourteenth century. As estates changed hands or were divided up, the attachments that held peasant to lord became weaker. Finally the Peasants' Revolt of 1381—the first great popular rebellion in English history—broke out. Although it lasted only one month and met with no real success other than eliminating the dreaded poll tax, it represented a major landmark in a changing time. By mid-fifteenth century, feudal society had lost its economic and political base, as well as its justification. The modern world was beginning to take shape as political power became more centralized in larger units and people felt a growing national consciousness. During this crucial first century of the Renaissance, the development and use of technology was an important factor in many of its shaping events and determining forces. For the next 150 years, to 1600, technology would become an increasingly significant part of the foundation upon which a new order would be built.

The field of architecture—designing and building structures—is probably the most conspicuous and permanent example of the changes brought about by Renaissance technology. Most historians regard the construction methods Filippo Brunelleschi used to build the brick dome for the church of Santa Maria dei Fiori in Florence as the starting point of Renaissance architecture. To the astonishment of his contemporaries, Brunelleschi raised the cupola without any false work, that is, he built the huge double dome without using scaffolding, abutments, or flying buttresses. When he first proposed this idea, it seemed to contradict not only existing practice (and the professional opinions of his peers) but common sense as well. Since no one shared Brunelleschi's daring, new conception of a dome as a series of rings, one atop another, the idea that it could be self-supporting during its erection seemed impossible. It was only through his temerity and stubbornness that a real break with the medieval past was accom-

Opposite page:
The long, thin cannon on the right, called a culverin, evolved from the early fire-stick or hand cannon and was adapted by the French. Its projectiles had little trajectory and could be aimed to fly straight to the mark, even at long range. From the middle of the fifteenth century on, artillery would play a central role in warfare and would increase considerably the devastation wrought by armed conflict. *Military Antiquities,* 1812. Francis Grose. LC-USZ62-110323.

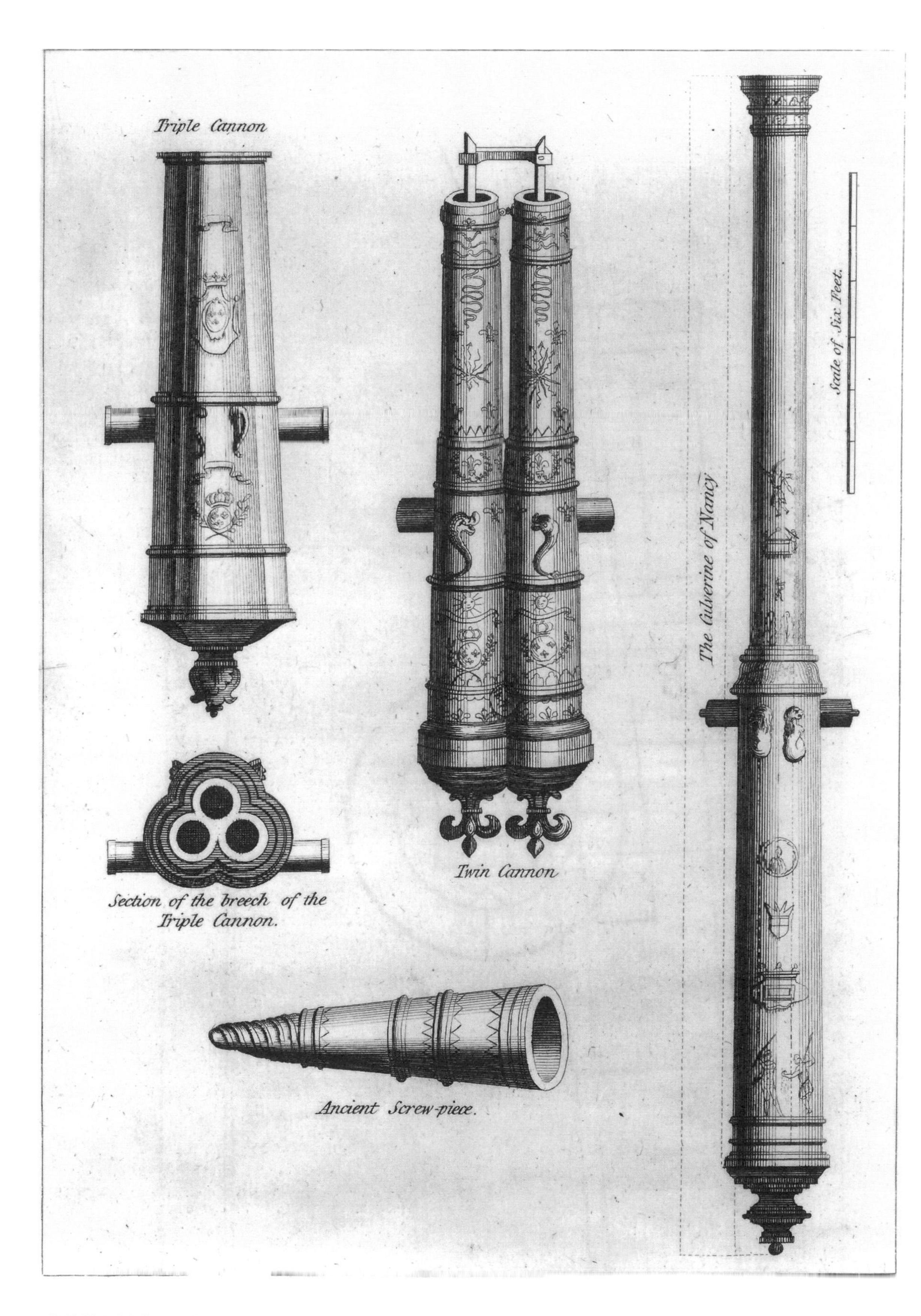
Triple Cannon
Twin Cannon
Scale of Six Feet.
The Culverine of Nancy
Section of the breech of the
Triple Cannon.
Ancient Screw-piece.

plished. Brunelleschi's dome is a great and beautiful triumph.

Filippo Brunelleschi was what we would now call a Renaissance man, and it is significant that he preceded the most famous of all, Leonardo da Vinci. In fact, along with his peers and sometimes rivals—Ghiberti the sculptor, Taccola the military engineer, and the noble and inventive Alberti—he belonged to the first generation of truly Renaissance engineers, with Leonardo a member of the second generation. Born seventy-five years before Leonardo, Brunelleschi began his career as a goldsmith and sculptor. After a journey to Rome with the sculptor Donatello to study architecture, he returned to Florence to further hone not only his architectural skills but also those he applied to mechanics, hydraulics, and the invention of machines and other mechanical devices.

Not surprisingly, Brunelleschi turns up in Giorgio Vasari's biographical history of art, *Vite de piu eccelenti architteti, pittori, et scultori Italiani*, published first in 1550 and then in an enlarged edition in 1568, both in Florence. The Library of Congress has both editions of this most influential book. Vasari treats Brunelleschi very reverently and never offers the slightest criticism. When he states that "no one was ever more kind and lovable than he," we must ask if this is the same Brunelleschi who relentlessly taunted his rival Ghiberti, who went to prison rather than pay a fee that he disputed, and who stayed away from the dome construction site just to prove how much he was needed? In fact, Vasari based much of his work on an admiring biography of Brunelleschi written in the 1480s by Antonio de Tuccio Manetti. The Library has a modern version of Manetti's work published first in Florence in 1812 as part of a larger work (*Vita*, edited by Domenici Moreni) and later separately as *Vita di Filippo di ser Brunellesco*, by Antonio Manetti (Florence, 1927).

Vasari's treatment of Brunelleschi does not focus solely on personal details but discusses his professional work and achievements as well. He examines some early experiments Brunelleschi conducted on building without armature (or temporary supports) and describes in great detail an elaborate machine Brunelleschi invented for use as a stage set for an annual religious play in Florence. Knowledge of Brunelleschi's technological skill was given the greatest boost during the nineteenth century with the publication of two volumes of invaluable information on the details of the construction of both the church and its dome. This rare information was extracted from the archives of the Cathedral of Florence by Cesare Guasti, superintendent of the Archives of Tuscany. The first volume, *La Cupola di Santa Maria del Fiore*: was published in Florence in 1857, and the second, *Santa Maria del Fiore: La Costruzione della Chiesa e del Campanile*, was published in the same city in 1887. The Library of Congress has both of these volumes in their original editions.

A revival of ancient classical culture and forms, Renaissance architecture gradually replaced the Gothic style of the late Middle Ages. Engineers and architects of the first half of the fifteenth century could learn from and be inspired by the classical style in two ways: they could visit the actual remains of these buildings in France and Spain as well as in Rome itself—as Brunelleschi did when he visited Rome with his friend

Opposite page:
The ingenious Tuscan architect Filippo Brunelleschi, designer and builder of the dome of the Santa Maria del Fiore cathedral (shown here in cross-section), was a link between Gothic and Renaissance architecture. Not only did he build this eight-sided vault using no external supports, but he did it without any mathematical formulas, guiding data, or quantitative notions of any kind. He had, instead, what can only be described as a qualitative or intuitive understanding of such critical factors as stress and loads. Brunelleschi also built many spectacular cranes and hoisting machines to construct his dome, and can be considered the pioneer Renaissance architect and engineer. *Descrizione e studi dell'insigne fabbrica di S. Maria del Fiore,* 1733. Bernardo S. Sgrilli. LC-USZ62-110324.

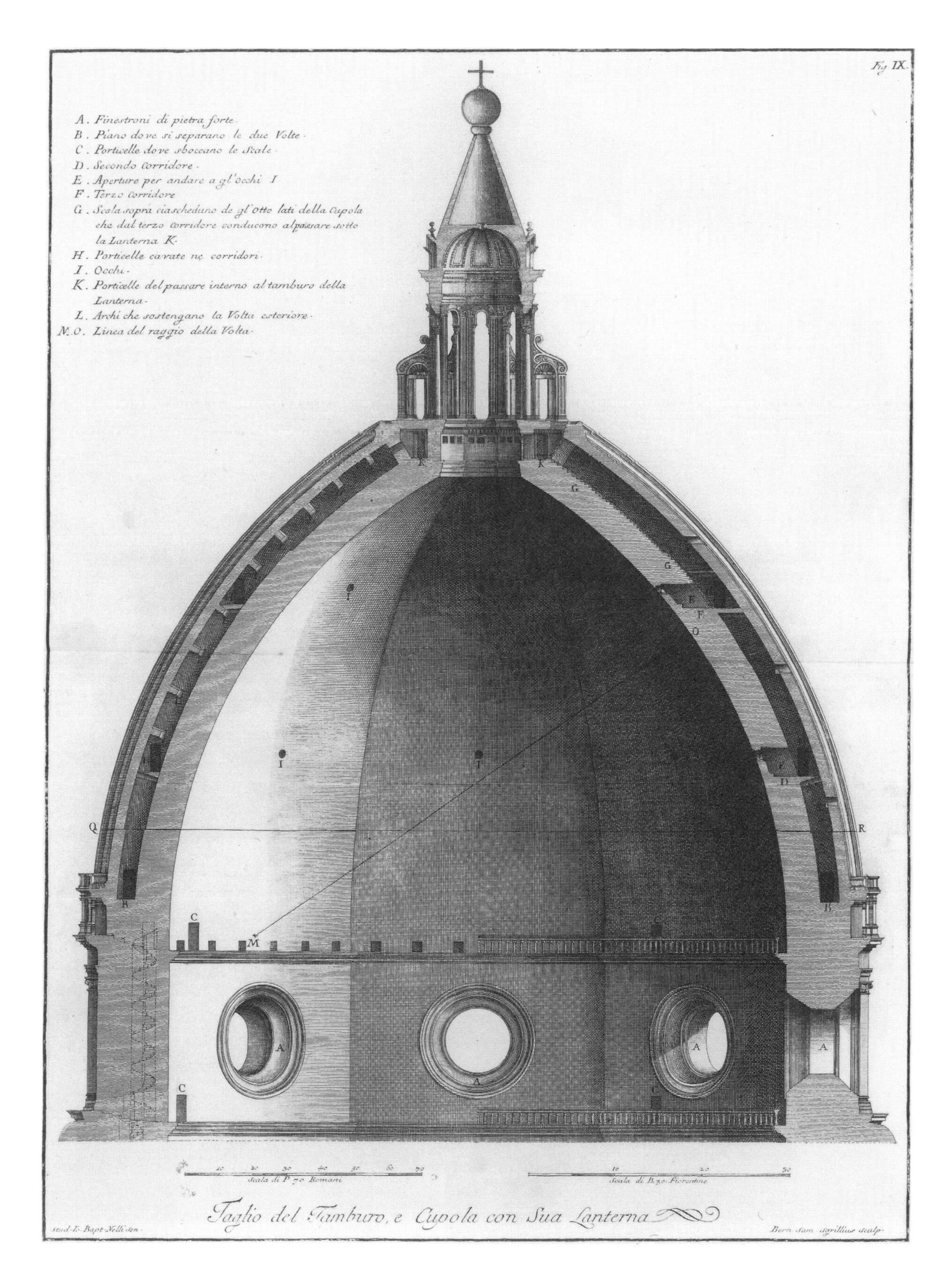
Fig. IX.
A. Finestroni di pietra forte.
B. Piano dove si separano le due Volte.
C. Porticelle dove sboccano le Scale.
D. Secondo Corridore.
E. Aperture per andare a gl'occhi I
F. Terzo Corridore
G. Scala sopra ciascheduno de gl'Otto lati della Cupola che dal terzo Corridore conducono al passare sotto la Lanterna K.
H. Porticelle cavate ne corridori.
I. Occhi.
K. Porticelle del passare interno al tamburo della Lanterna.
L. Archi che sostengano la Volta esteriore.
N.O. Linea del raggio della Volta.
Scala di P. 70 Romani
Scala di B. 30 Fiorentine
Taglio del Tamburo, e Cupola con Sua Lanterna
Bern. Sam. Sgrillius sculp.

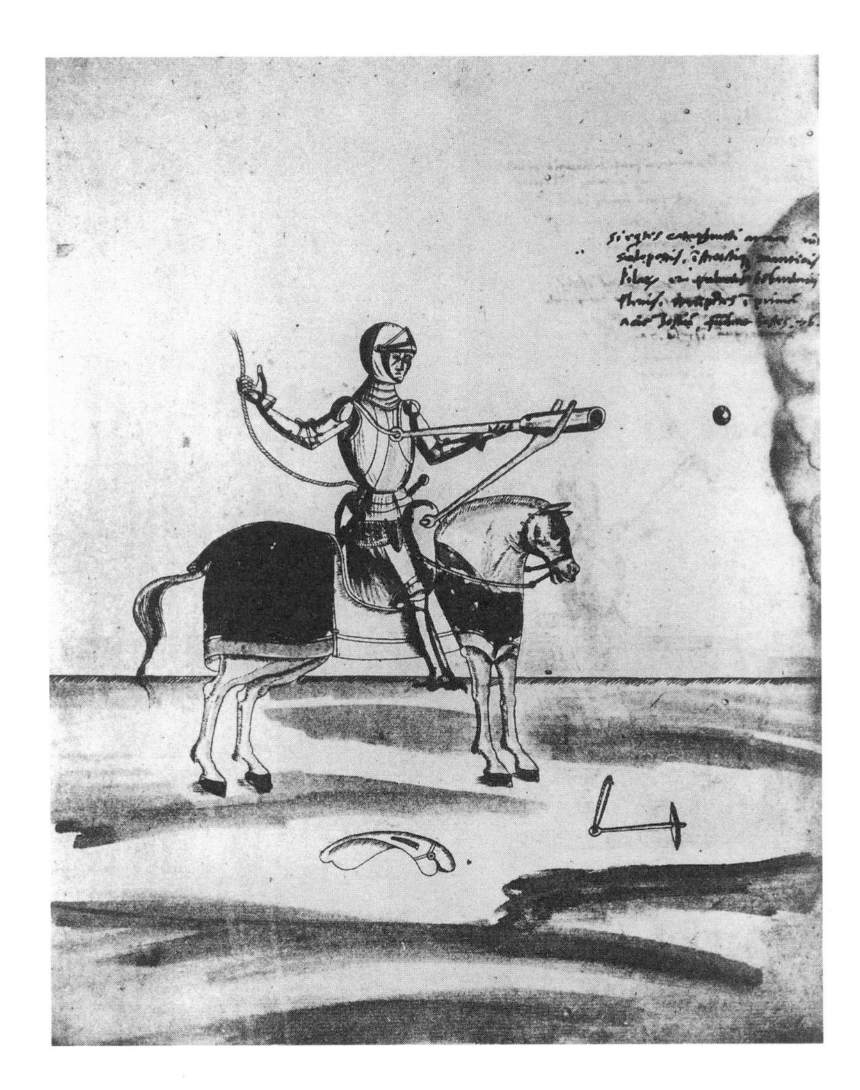

Like any Renaissance engineer, Mariano Taccola was expert in military matters. Here he shows an emerging military technology of the mid-fourteenth century, the fire-stick. A precursor of both the rifle and the hand cannon and first used in siege warfare, it was a simple iron tube with a hole. Taccola shows an armored knight on horseback who has discharged the ball using a long, lighted wick. *Arte militare meccanica medievale,* 1946. Giovanni Canestrini. LC-USZ62-110325.

Donatello—and they could study the ancient architecture and building methods detailed in the lately rediscovered treatise, *De architectura*, written by the Roman engineer Vitruvius (see chapter 1). No one abandoned Gothic forms to embrace the classical style more quickly or better than Leone Battista Alberti. The posthumously published *De re aedificatoria* (see chapter 1) of this multitalented man—artist, scholar, architect, musician, mathematician—became the first original Renaissance treatise on architecture. Alberti, although much younger than Brunelleschi, was nonetheless a close associate and was an integral member of the first generation of Renaissance architects and engineers.

Where Renaissance architecture resulted in grand structures like Brunelleschi's church in Florence that may be visited and studied even today, little remains of the ubiquitous machines whose ingenuity and usefulness reflected the many talents of their curious creators. Along with other first-generation leaders like Brunelleschi and Alberti in architecture and Lorenzo Ghiberti and Donatello in art, Mariano di Jacopo, called Taccola, was the first of many famous Renaissance military engineers to produce fascinating and complicated machines. Although it is not known if Taccola ever left his native Siena, south of Florence, he did know Donatello well and met Brunelleschi when the famous architect visited Siena. Taccola was less educated than most of his peers and was more of a specialist, working exclusively with machines or devices. But in his lifetime, his stature was great and he was known as the "Sienese Archimedes." He is recognized today as influencing later Italian engineers more with his technological erudition and his encyclopedic knowledge of technical capabilities than with his originality. Taccola's main writings, *De ingeneis* (1433) and *De machinis* (1449), remained in manuscript until this century. In 1946, Giovanni Canestrini reproduced 84 drawings from a later manuscript copy of Taccola's works in the Biblioteca Marciana in Venice. The Library of Congress has this sumptuous modern edition, titled *Arte militare meccanica medievale*. In 1972, Frank D. Prager and Gustina Scaglia published *Mariano Taccola and His Book De Ingeneis*, which contains 129 selected drawings. The Library also has this volume.

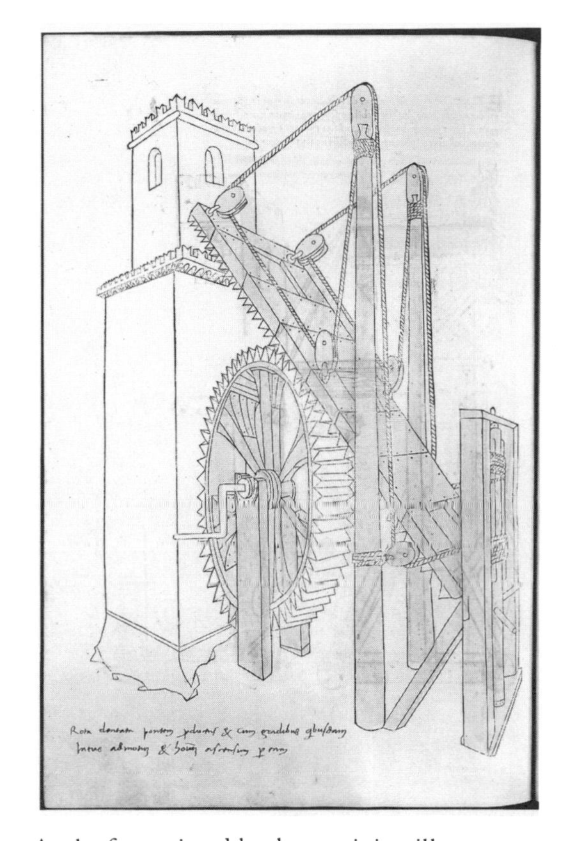

As the first printed book containing illustrations of a technical nature, Valturio's work reconstructs many ancient instruments of warfare. This, however, is one of his more contemporary illustrations, showing how a scaling ladder or bridge might be used to climb a tower. Valturio was employed by the prototype of Italian Renaissance princes, Sigismondo Malatesta (1417–1468), whose family had ruled Rimini since the thirteenth century. *De re militari*, 1472. Roberto Valturio. LC-USZ62-110326.

Ironically, the first technical treatise to be printed mechanically—that is, using the new technology of Gutenberg and not hand printing—was compiled by a man who was probably neither an engineer nor an architect. Roberto Valturio is mentioned in the earlier chapter on Greek and Roman technology because his famous treatise *De re militari* focuses on the war machines of ancient Rome. Completed by Valturio in 1455 and first published in Verona in 1472, the work treats the historical technology of war and is by no means original to Valturio, who was more a man of letters. In fact, many feel that the book may actually be the work of Valturio's prince, Sigismondo Malatesta of Rimini. It is known, however, that the illustrations were done by engraver and medallist Matteo dei Pasti who, in turn, borrowed a significant amount from the earlier drawings by Taccola. Because of this, Valturio's work is considered part of the second generation of Italian Renaissance engineers. As mentioned earlier, the Library of Congress has Valturio's work in first edition.

Before considering the work of the most famous member of this second generation, Leonardo da Vinci, another of its representatives, Francesco di Giorgio, must receive his due. Known also as Francesco di Giorgio Martini, he was in his time so famous that governments competed for his services. During his career he was employed by no less powerful and famous Renaissance leaders than Federico (Urbino), Ferdinand I and Alfonso (Naples), and Gian Galeazzo Sforza (Milan), among others. Until recently, Francesco's fame rested on his achievements as a painter, sculptor, and architect, but recent discoveries of his many writings on technology indicate that he was also a skilled engineer. Having studied painting and sculpture, he then learned architecture from Lorenzo di Pietro, friend of Donatello. When he was thirty, he shared a commission from the city of Siena to oversee the city's water supply system. From there, he went on to be steadily employed by many a *condottiere*, in building fortresses, supervising military operations, designing and building war machines, and even working on the construction of a cathedral. It was on such a job in Milan that Francesco met Leonardo in 1490, when both

THE TRADITION OF TECHNOLOGY

were inspecting the cathedral's construction. Francesco's documented writings reflect the full range of his interests and abilities and, of what survives in manuscript form, those sections on hydraulics and mechanics are considered the most original. However, it is not only the actual devices themselves that impress us, but his mastery in the art of drawing machines. The Library of Congress has a copy of his only work ever printed, *Trattato di archittetura civile e militare.* Written by Francesco between 1477 and 1480, this work on architecture, city planning, fortification, and mechanics was very popular in its own day, but was not printed until the nineteenth century. Edited by Carlo Promis, it was published in two volumes in Torino in 1841. This edition is incomplete, however, and is surpassed in usefulness by a newer version edited by Corrado Maltese and titled *Trattati di architettura ingegniera e arte militare.* The Library also has this recent work, published in Milan in 1967.

In many areas Francesco was Leonardo's equal, but the latter possessed a depth of natural curiosity and insight and an intellectual and philosophical dimension unavailable to Francesco, as it is to most of us. To these intangible qualities was added a crowning touch—the aesthetics and talents of a natural artist. Altogether, they made for an extraordinary individual whose ability to master whatever he took up depended only on his will to do so. Thus we can describe Leonardo as a painter, sculptor, engineer, architect, inventor, writer, musician, philosopher, anatomist, botanist, geologist, paleontologist, optician, and astronomer. The remarkable aspect of his activities in each of these very different fields is that in many of them he was not a mere dabbler but the equal of anyone of his time.

Typically unique and enigmatic in everything he did, in 1519 Leonardo bequeathed to his pupil, Francesco Melzi, a rich and confusing treasure of his writings. These were Leonardo's notebooks—odd-sized pages in his own hand that included rough sketches, detailed instructions, quickly noted observations, descriptions, reflections, and even notes to himself to "buy some tablecloths and towels" or to "get Messer Fazio to show his *De proportione.*" Written by the left-handed genius in his peculiar right-to-left mirror style, the notes and text are crammed into any blank spaces left around his sketches. Lacking a formal education, Leonardo used language and grammar that were understandably irregular. This tendency, combined with his stylistic eccentricities like simplified spelling, hieroglyphs, and abbreviations, makes reading and understanding his notes extremely difficult. But Leonardo wrote for himself and not for others. It is believed that Leonardo kept all of his notebooks with him and that they always traveled wherever he went. The only known reference to them made during his lifetime was by the cardinal of Aragon's secretary who, in 1517, described "an endless number of volumes as seen with our own eyes." Of these notebooks, about six thousand manuscript sheets and pages are known to exist—roughly one-third the total Melzi inherited.

The history and provenance of these notebooks is a long and interesting story, but those of particular interest to technology were obtained

Opposite page:
Although the Middle Ages is known as the age of fortresses, it was during the Renaissance that real creative energy went into developing new, permanent defenses against heavy cannons. Francesco di Giorgio was among the earliest to exhibit this Renaissance talent to marry engineering to fortifications. Here, his four pointed bastions extend outward to defend all approaches. *Trattato di architettura civile e militare,* 1841. Carlo Promis, editor. LC-USZ62-110327.

near the end of the sixteenth century by a sculptor, Pompeo Leoni, who cut up the portions he received and rearranged them roughly according to subject. Thus Leonardo's anatomical drawings and notes were all put together and, as a unit, eventually made their way to England's Royal Library, where they are known as the Windsor Castle collection. Most of the material relating to machines and inventions was done on very large sheets, and after Leoni assembled these, they became known as the Codice Atlantico or Atlantic Codex, so named because their large size resembled an atlas folio. This codex today resides in the Ambrosian Library in Milan, the bulk of it presented to that library in 1625 by Count Galeazzo Arconati. The codex was away from Italy for nineteen years, however, for when Napoleon entered Milan in 1796, he sent it to the Bibliothèque Nationale. After Napoleon's fall, the codex was returned to Milan in 1815.

It was not until 1894, however, that the first facsimile copies of the Atlantic Codex were published. In a massive, ten-year publishing effort led by editor Giovanni Piumati that resulted in the eventual publication of nine separate volumes (four two-part volumes plus an index), the Atlantic Codex first became available to the public. Four volumes of text are accompanied by four volumes of tables, which contain a total of 698 plates. It is a physically impressive edition, each volume measuring twenty by fourteen inches (51 by 38 cm). Titled *Il Codice atlantico di Leonardo da Vinci nella Biblioteca ambrosiana di Milano*, only 280 copies of this edition were printed. The Library of Congress has number 192. The Library also has the more recent twelve-volume edition published in 1975 by Giunti-Barbera of Florence.

In the Atlantic Codex, as in all of Leonardo's work, we see the convergence of science and art. A glance through its beautiful and remarkable plates reveals highly realistic drawings for a range of diverse projects. Included are plans for diverting the Arno River, sketches of military machines and weapons, drawings of water-powered mills and spool-winding machines, and details of bridges, bells, and fortifications, as well as plans for singular items like the perspectograph, a device invented by Leonardo that permitted the seated artist to see and to draw from a fixed position in true perspective, essentially by tracing the outlines of a model onto a pane of glass.

Ever since Pompeo Leoni divided up Leonardo's notebooks according to his own scheme and broke the original integrity of the manuscript, an artificial order has prevailed. Thus many other examples of Leonardo's technological work and musings exist outside of the Atlantic Codex. In 1965, through a fortuitous sequence of events, two more of Leonardo's notebooks were rediscovered in the Biblioteca Nacional, Madrid, after being officially lost since 1830. Called the Madrid Codices, these notebooks were never actually lost but simply could not be found because of a cataloging mistake. Two years after their rediscovery, the Spanish Ministry of Education and Science authorized Taurus Ediciones and the McGraw-Hill Book Company to publish these notebooks in facsimile for the first time. In 1974, a five-volume facsimile edition called *Madrid Codices* was pub-

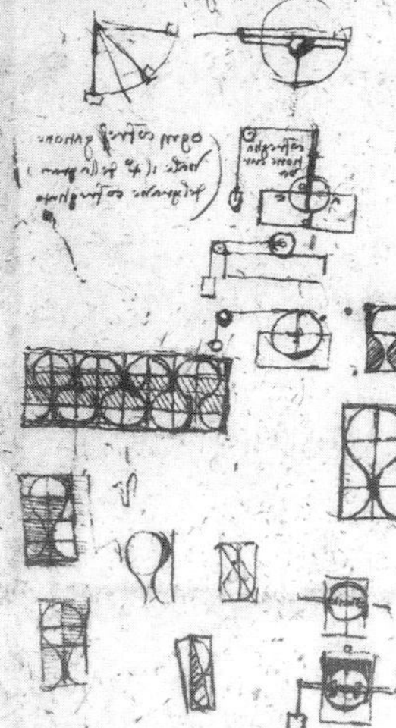

A typical page from Leonardo's notebook demonstrates the innate curiosity that must have driven this multitalented genius. The breadth and range of his interests are easily apparent to us, without even attempting to translate his right-to-left mirror writing or analyzing his geometric designs. *Il Codice atlantico di Leonardo da Vinci nella Biblioteca ambrosiana di Milano,* 1894. LC-USZ62-110328.

This audacious giant crossbow drawn by Leonardo is over seventy-six feet long. The entire mechanism rests on three pairs of wheels that are canted to provide a stable firing platform. The bow arms are built separately in laminated sections. To the left are two drawings showing alternative release mechanisms. This is but one of the many new and terrifying weapons he designed (and probably never built) for his belligerent employers. From 1490 to 1507, central Italy saw incessant warfare, and Leonardo served five different masters as military engineer. *Il Codice atlantico di Leonardo da Vinci nella Biblioteca ambrosiana di Milano,* 1894. LC-USZ62-110329.

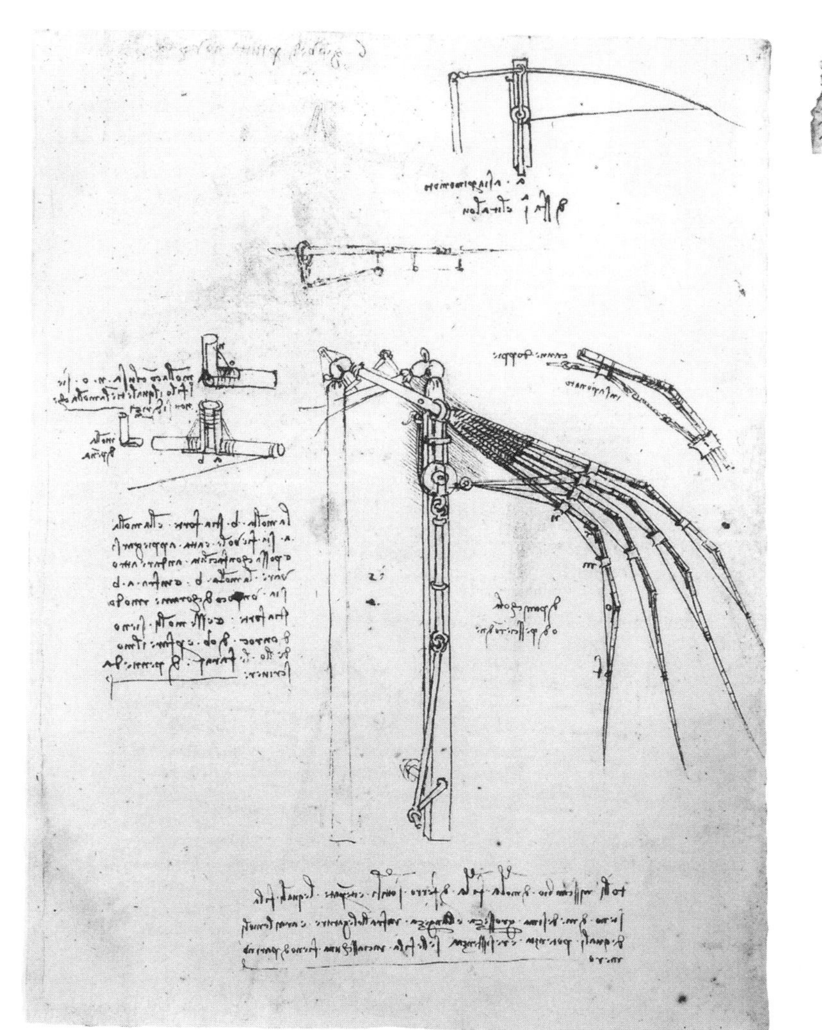

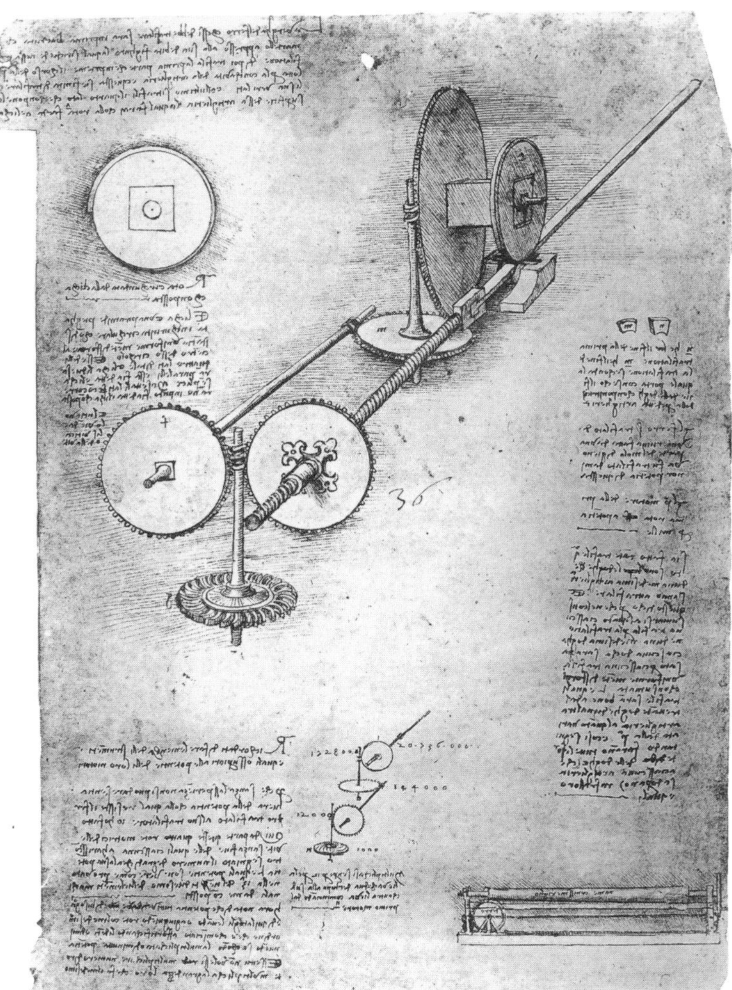

The delicacy and beauty of this flapping mechanism shows Leonardo drawing at his exacting best. The complex designs for these articulated "fingers" worked by pulley arrangements are ingenious. His investigations of flight were based, however, on two misconceptions: that man had sufficient muscle power to flap like a bird and that birds beat their wings downwards and backwards. From this complex design, he progressed to several different but simpler designs for ornithopters (flapping-wing aircraft). He also sketched designs for gliders, a parachute, and a helical-screw helicopter. *Il Codice atlantico di Leonardo da Vinci nella Biblioteca ambrosiana di Milano,* 1894. LC-USZ62-110330.

lished in English and Italian with transcription, translation, and commentary by Vincian scholar Ladislao Reti. The Library of Congress has this edition, the publication of which increases by about 20 percent the aggregate information about Leonardo's work. Moreover, the original codices are by far the best preserved of all that has survived, making them even more valuable.

Reti's study of the Madrid Codices reveals their intimate connection with the Atlantic Codex. Reti discovered that many of the elaborate and beautiful drawings found in the Madrid Codices had their rough precursors in the sketches of the Atlantic Codex. The two works are becoming increasingly valuable to researchers, who have found that the preliminary drawings of the Atlantic Codex cannot be interpreted correctly without their finished counterparts in the Madrid version. Further, when an elaborated drawing has no explanatory text in the Madrid Codices, it often is found in the Atlantic Codex. Reti argues that the first part of the Madrid Codices, Codex Madrid I, "is not a notebook, but a carefully

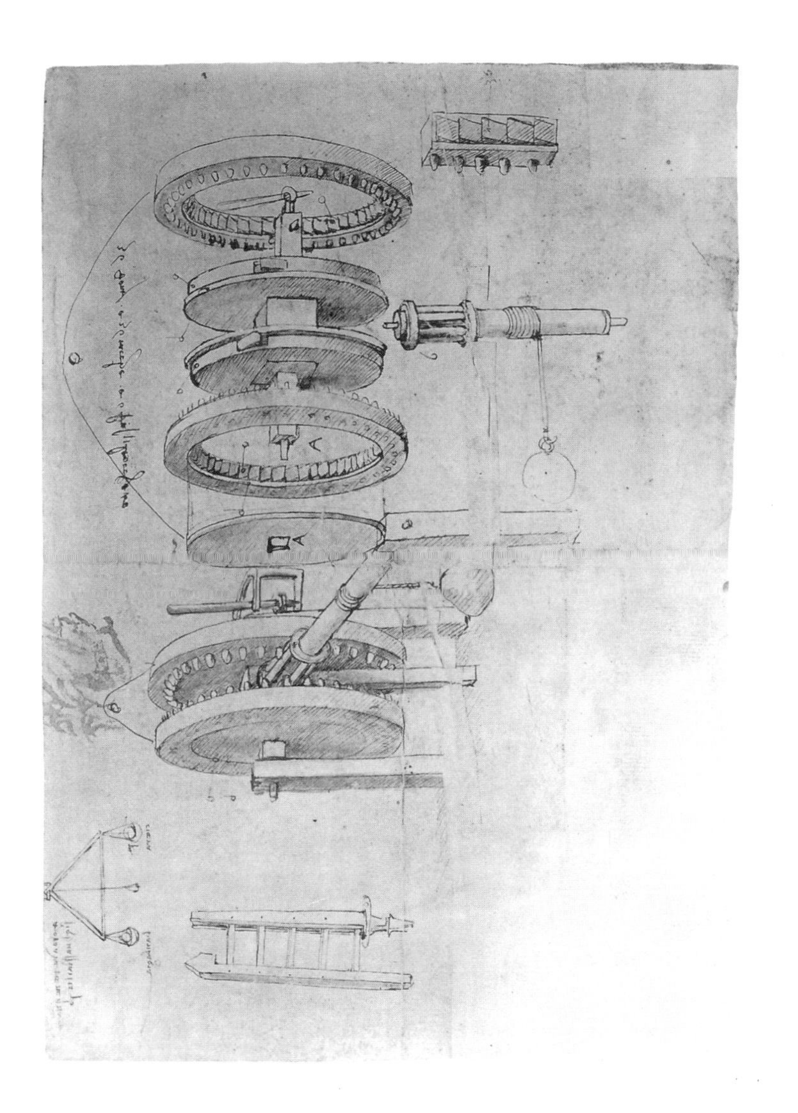

Center:
The precision of Leonardo's hand and mind as well as his complete mastery of the machinery of his time are apparent in this drawing of a water-powered machine for making large cannon barrels. Two sets of worm drives were used to advance the sheet iron and roll it at the same time. The small schematic at the center bottom shows the gearing reduction of the four gear wheels involved. This drawing indicates how clearly Leonardo saw the potential of automation, since his machine would provide a degree of precision and uniformity not obtainable by hand-forging. By the middle of the fifteenth century, Leonardo had apparently understood the essentials of mechanization that were to be central to an industrial revolution that did not come for another three centuries. *Il Codice atlantico di Leonardo da Vinci nella Biblioteca ambrosiana di Milano,* 1894. LC-USZ62-110331.

Right:
This startlingly modern "exploded" image of a disassembled mechanism underscores the creative convergence of ideas that must take place in the mind of a genius. Technically, it illustrates the inner workings of a device that changes the rocking motion of an upright lever into the rotary motion of a shaft for the purposes of lifting a great weight. What is so significant, however, is the nearly anatomical manner in which Leonardo lays open the inner workings of a machine. It is almost as if he applied the lessons learned at the dissecting table (and doing his cross-section sketches) to his technical experiments. The impulse behind this image is as creative and original as any work of his artistic efforts. *Il Codice atlantico di Leonardo da Vinci nella Biblioteca ambrosiana di Milano,* 1894. LC-USZ62-110332.

planned treatise on machine elements and mechanisms." Whether this is indeed the case, and the notebooks are yet another example of Leonardo's penchant for not finishing things, remains to be seen.

Leonardo trained no students, nor did he leave any flourishing school of acolytes behind. Further, no one has demonstrated that the notebooks exerted any real influence on his contemporaries or on the immediate subsequent generations. Judging by his work alone, it is difficult to say whether Leonardo was more engineer than artist. Most would agree that the artistic talents displayed in such an accomplishment as *The Last Supper* qualified him to stand with any of the great painters of his or any other time. Yet Leonardo showed an obvious disinclination to paint. Why was this? And why was the preponderance of his time spent with pen in hand devoted almost exclusively to drawing machines and not Madonnas?

Leonardo's own words may offer partial insight into his seemingly deliberate emphasis on the technological. In the Atlantic Codex, he contrasts himself to the educated men of his time:

Although I may not, like them, be able to quote other authors, I rely on that which is much greater and more worthy, on experience, the mistress of their masters. They go out puffed up and pompously dressed and decorated with [the fruits] not of their own labors but of those of others. They will scorn me as an inventor, but how much more might they be blamed who are not inventors but vaunters and declaimers of the works of others.

Here Leonardo may be seen as a man keenly aware of his own genius for technical inventiveness and mechanical originality, and he argues that although the subject matter to which he devotes his creativity is not highly regarded, the essence of that thought is pure and true. Later, he writes with what sounds like false modesty when he says that he chose to become an inventor

since the men who have come before me have taken for their own every useful or necessary theme—I must do like one who, being poor, comes last to the fair, and can find no other way of providing for himself than by taking all the things already seen by others and not taken, but refused by reason of their lesser value.

This statement could be interpreted as Leonardo subtly saying to posterity that only he can see the true value of these supposedly "lesser" things. It might be argued that in such a remarkable age for art—Leonardo's artistic contemporaries were very impressive people, like Botticelli, Dürer, Michelangelo, Titian, and Raphael—Leonardo as a painter would have been one of many. But by choosing to be what he called an inventor, he selected a field where he could truly distinguish himself as an innovator and indeed as a real creator.

Finally, it should be restated that Leonardo's inspired originality notwithstanding, neither the man nor his work sprang fully grown from out of nowhere. Leonardo has been described earlier as a member of the second generation of Italian Renaissance engineers, and indeed his work followed an already established and traditional practice. Thus when the young Leonardo was brought by his father to the studio of Andrea del Verrochio—a versatile painter, sculptor, and founder—he was exposed to and trained in everything from preparing paints and trying perspective drawing to casting statues, bells, and guns. When Leonardo's father brought him to Verrochio's *bottega*—asking "Do you think he will make his way as a painter?"—the master was then working on the final piece of the late Brunelleschi's masterpiece, the lantern or top of the dome of the Florence cathedral. It is possible that Leonardo was thus exposed to the engineering tradition begun by Brunelleschi, and that he sketched and studied closely the actual machines that Brunelleschi had designed and built. Here Leonardo entered a tradition in which the apprentice painstakingly copies the masters' works before evolving his own original style. Thus viewing Leonardo in no way diminishes his creative genius. In fact, this very contextuality of his work, as emerging from a certain tradition, makes his relentless curiosity and inspired originality all the more remarkable.

Probably the most obvious and best known example of dramatic technological advances in the Renaissance is the invention of mechanical printing. Although included here as a Renaissance invention, printing from movable type was accomplished first in the West in Gothic Germany in mid-fifteenth century, before the Italian or Mediterranean Renaissance reached that northern country. Although printing with movable type was used in China as early as the eleventh century and its related techniques and processes gradually developed in other parts of the Far East, only the production of paper made its way from that source to the West, via the Arabs, and then on to Spain.

Knowledge of the typographic process itself had to be reinvented in the West. Despite its high medieval culture and ways, the southwestern German city of Mainz—known as "golden Mainz" by its neighbors because of its prosperity—fostered all the requisite elements to make it the first home of mechanical printing. Technically, the city's many goldsmiths were expert in their use of punches for decorations; the metal manufacturers were familiar with spherical shot molds and other such casting methods; wine presses were naturally in plentiful supply in the Rhine district; ink had been introduced in the early 1400s; and rag paper had long since replaced parchment as the preferred medium. While this combination of techniques and industries existed in other cities, Mainz also had Johann Gensfleisch zur Laden, known as Gutenberg (although he carried out his experiments in Strasbourg), and it was he who provided the guiding focus and necessary creative synthesis of other technologies. The crucial breakthrough in this synthesis was Gutenberg's individually cut pieces of metal type, creating letterforms that were movable and therefore reusable.

It appears that Gutenberg had solved all of the basic mechanical problems involved in printing by 1448, for in that year he produced an astronomical calendar. This is the first known dated piece of European printing. By about 1455, Gutenberg finally completed an entire book. He had chosen, appropriately, the most influential book of all time—the Bible. Known now as the Gutenberg Bible, it was first called the Mazarin Bible (since one was owned by the powerful and influential statesman and collector Cardinal Mazarin) and then called the Forty-Two-Line Bible (to distinguish it from a later, thirty-six-line rival edition). Of the more than two hundred copies originally printed, only forty-eight are known to exist, thirty-six of which are printed on paper and twelve printed on vellum. The Library of Congress acquired its Bible in 1930. Printed on vellum, it has 641 leaves (1,282 pages). It is printed in Gothic type in Latin, with capital letters and headings ornamented by hand in color. Each page has forty-two lines, double columns throughout, and typically no colophon. The Library's Bible was for nearly five hundred years in the possession of a European Benedictine order.

Since Gutenberg's printed Bible had to compete with the best that hand-production could offer, it sought and achieved the highest standards in all technical aspects, from design to the quality of paper and ink. Many regard it as the greatest of all printed books. At nearly the same moment

The first book printed mechanically or with movable type is also considered the greatest of all printed books. Printing so large a book—641 leaves in one volume—took over a year, but the spectacular result set standards in all its technical aspects, from book design to the quality of paper and ink, that have seldom been matched. Letterpress printing or typography is therefore unlike most technologies in that its technical apex was achieved at its inception. The Library's copy of the Gutenberg Bible is one of three perfect vellum copies. Bible, in Latin, 1454-55. Johann Gutenberg. LC-USZ62-110333.

that Gutenberg was achieving this technical success, however, he was suffering a financial defeat. He had borrowed 800 guilders from Johann Fust, a Mainz lawyer, who took about half of Gutenberg's plant and equipment when the printer was unable to repay his debt. The Gutenberg-Fust partnership ended and Fust hired Peter Schoeffer, thus founding the first major printing house. Schoeffer had been Gutenberg's assistant and later married Fust's daughter, so it is thought that Fust dissolved the partnership with Gutenberg when he saw the project would be a success, and thus gave the completion of the Bible over to Schoeffer to maximize his profits. The earliest book from the Fust-Schoeffer press in the collections of the Library of Congress is the *Rationale divinorum officiorum*, written by a thirteenth-century bishop, Gulielmus Durantis. Printed on vellum in Mainz in 1459, the volume's numerous initials and headlines in red and blue are characteristic of the Fust-Schoeffer house, which was the first to use three-color printing.

The Library of Congress collections contain not only a perfect example of Gutenberg's initial and spectacular effort, but other outstanding examples as well of the best of what was being produced immediately preceding the invention of mechanical printing. The late medieval or early

Renaissance bookmaker produced a "book" that was essentially a bound collection of hand-written manuscript pages. One of the greatest items in the Library's collections is just such a hand-produced book, the *Biblia Latina*, known as the Giant Bible of Mainz. The huge (57.5 by 40.5 cm) two-volume book is even larger than Gutenberg's Forty-Two-Line Bible and was produced about the same time. What is known about its origin and production indicates that it was regarded and conducted as a project of exceptional importance upon which no expense was to be spared. The best of both materials and human skills were employed. At the end of its 459 white vellum leaves, some of which contain strikingly beautiful illuminated borders, is an inscription saying that the "faithful Pen" had begun his hand-copying on April 4, 1452, and finished fifteen months later on July 9, 1453. Few manuscripts were the work of a single master scribe as this was, since economics dictated the splitting up of pages among several scribes, thus getting the job done more quickly.

The most immediate precursor to the mechanically printed book was the block book. Appearing in Europe around 1400, block books were in a way the first printed books, although they did not involve the use of

From *Apocalypsis Sancti Johannis,* ca. 1470. See p. 104.

movable type. Rather, they were books in which the illustration and the text were all cut on woodblocks. These individual woodcut prints were then sewn together to form one book, usually devotional in nature. The Library of Congress has ten of these very rare block books, the earliest being the 1465 *Biblia pauperum* which contains woodcuts of many well-known biblical scenes. Called the "poor man's Bible," this edition is characteristic of the typical block book, since it was simple and inexpensive to produce after the woodcuts were made and did not even require that a press be used. The Library of Congress also has four different copies of *Ars moriendi*, all printed in German, which tell their Christian readers how to die a proper death. Perhaps the best-preserved and most brilliantly illustrated block book in the Library's collection is the *Apocalypsis Sancti Johannis*. Yet by the time all these block books were produced, between 1465 and 1475, they were already obsolete. Typographic printing, or printing with movable type and a press, had already begun a phenomenally rapid spread. Cologne had its first press in 1465; Rome and Venice in 1469; Paris in 1470; Westminster in 1476.

It has been estimated that during the first fifty years of printing, approximately forty thousand titles or recorded editions of books were produced, totaling some eight million individual books. Other estimates give a lower number of separate editions (between thirty and thirty-five thousand) but a considerably higher number of total copies (from fifteen to twenty million). Although two-thirds of these volumes came from presses in Germany and Italy, twelve other European countries had active presses, the new technology spreading east to west, from Budapest to Valencia. Besides Gutenberg and Fust and Schoeffer, there were individuals in each country whose publishing work was both distinctive and pioneering. The Library of Congress collections contain examples of the work of many of these early printers, including: the prolific publisher and excellent typecutter Nicolaus Jenson, a Frenchman whose press was in Venice; Erhard Ratdolt, who published out of Augsburg and then Venice; Anton Koberger in Nuremberg; the first great Italian printer, Aldo Manuzio, a scholar-printer in Venice who is better known by his Latin name, Aldus Manutius; the Frenchmen Antoine Vérard (Paris) and Jean DuPré (Lyons and Paris); and finally, the first great English printer, William Caxton, whose press was in Westminster. All of these individuals published before 1500, the arbitrary cut-off date generally acknowledged as the end of the incunabula or infancy period of the art.

The work of these and other printers also serves to document the essential differences between Gothic and Renaissance styles of typeface or lettering and illustration. The first printed books of German origin were of the rich black-letter, Gothic type then in common use. Since most of these books were of a theological nature, the heavy, serious black lettering was entirely appropriate and thus predominated in the north. In the south, however, the spirit of humanism carried over into the printing trade and affected not only the type of books that were printed (Cicero's works, for example, instead of the Bible) but also the way the printed books looked. To Italian eyes, the black, thick lettering of the north exemplified all that was archaic, and they developed the open, round roman typeface to symbolize the new freedom of Renaissance thought. Typically, book illustration followed suit, and the open, light, and airy designs of the early Italian books contrast starkly to the often masterful but still pre-Renaissance look of the German books.

Examples of both styles abound in the Library's collections. Hartmann Schedel's *Liber chronicarum* or *Nuremberg Chronicle* (Nuremberg, 1493) contains the Gothic illustrations of Michael Wolgemut and Wilhelm Pleydenwurff. This can be contrasted to the *Hyperonotomachia Poliphili* or *Strife or Love in a Dream of Poliphilus*, a mixture of religious symbolism and erotic allegory written by Francesco Colonna. It was first published in Venice in 1499 by Aldus Manutius and its illustrator is unknown. One of the best examples in the Library's collections of this new look of lightness was produced in Paris in 1529 by Geofroy Tory, whose *Champ fleury* or *The Field of Flowers* introduced a modified, lighter roman typeface, as well as more open line illustrations. Brimming with illustrations by Tory and his pupil Claude Garamond, the book is considered one of the more tangible (and perhaps extreme) humanist gestures against the archaic, medieval style. Moreover, by following the lead of Luca Paccioli and Albrecht Dürer in the more geometrical construction of letters, Tory influenced fine printing for two centuries to come.

The invention and proliferation of typographical printing technology was a unique event central to the history of mankind. Indeed, some regard it alone as signaling not only the end of the Middle Ages but the real beginning of the modern world. In fact, to science itself, the notion of any cumulative progress at all is unthinkable without the preservation and dissemination of knowledge made possible by the printing press. In many ways, printing is the quintessential technology, having been some part of nearly every subsequent technological advance. With the invention of printing, perhaps more than with any other device or technique, the machine and all its implications had come into the world to stay.

For technology as well as for science, the Renaissance can be regarded as a time of transition. The relative stagnation of the earlier age (however one regards medieval times) was breaking apart and fostering expansive ideas about the possibilities and potential of the human intellect. One aspect of this incipient change was that technology was advancing in part—albeit in the smallest of ways—because of a scientific stimulus. For centuries technology had trod the same path of trial and error, accomplishing often prodigious things but understanding only the why of them, never the how. As the Renaissance laid the foundation for the coming scientific revolution of the seventeenth century, it also moved the practical world of technology closer to its theoretical counterpart, science. This is best seen in the Renaissance development of scientific instruments. As science and its practitioners gradually became more experimental in their method, the demand arose for more and better measuring devices. This in turn spurred the invention of some instruments and the improvement, especially in terms of accuracy, of others.

Among the incunabula in the collections of the Library of Congress,

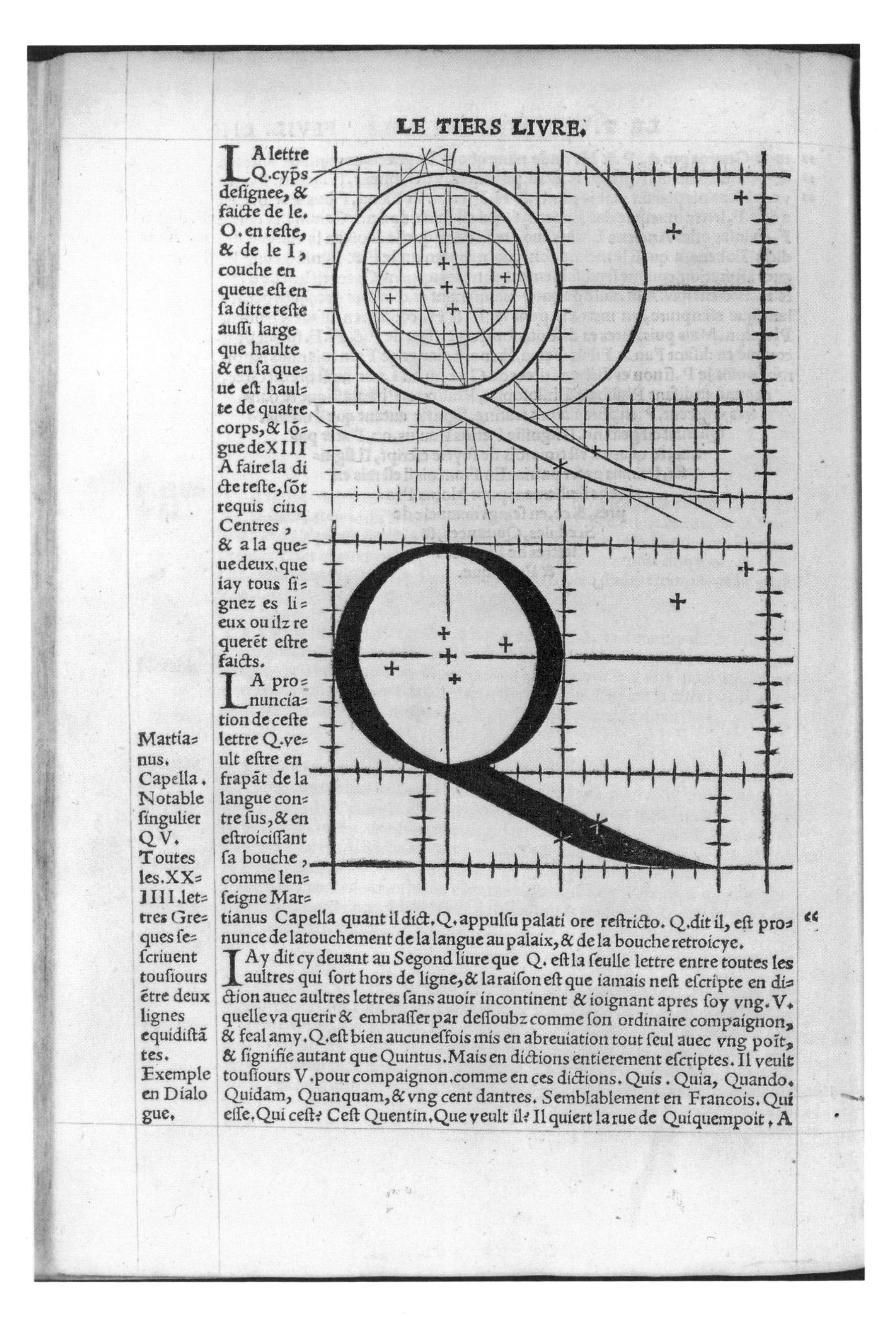

LE TIERS LIVRE.

LA lettre Q. cy pŝ deſignee, & faicte de le. O. en teſte, & de le I, couche en queue eſt en ſa ditte teſte auſſi large que haulte & en ſa queue eſt haulte de quatre corps, & lõgue de XIII A faire la dicte teſte, ſõt requis cinq Centres, & a la queue deux, que iay tous ſignez es lieux ou ilz requerẽt eſtre faicts.

LA pronunciation de ceſte lettre Q. veult eſtre en frapãt de la langue contre ſus, & en eſtroiciſſant ſa bouche, comme lenſeigne Martianus Capella quant il dict, Q. appulſu palati ore reſtricto. Q. dit il, eſt pronunce de latouchement de la langue au palaix, & de la bouche retroicye.

IAy dit cy deuant au Segond liure que Q. eſt la ſeulle lettre entre toutes les aultres qui ſort hors de ligne, & la raiſon eſt que iamais neſt eſcripte en diction auec aultres lettres ſans auoir incontinent & ioignant apres ſoy vng. V. quelle va querir & embraſſer par deſſoubz comme ſon ordinaire compaignon, & feal amy. Q. eſt bien aucuneſfois mis en abreuiation tout ſeul auec vng poĩt, & ſignifie autant que Quintus. Mais en dictions entierement eſcriptes. Il veult touſiours V. pour compaignon. comme en ces dictions. Quis. Quia, Quando. Quidam, Quanquam, & vng cent dantres. Semblablement en Francois. Qui eſſe. Qui ceſt? Ceſt Quentin. Que veult il? Il quiert la rue de Quiquempoit, A

Martianus. Capella. Notable ſingulier QV. Toutes les. XXIIII. lettres Greques ſe ſcriuent touſiours ẽtre deux lignes equidiſtãtes. Exemple en Dialogue.

one of the earliest to contain information on what could be called scientific instruments is a 1470 publication of the collected works of Priscian or Priscianus of Caesarea, the best known of all the Latin grammarians. In his *Opera*, published in Venice nearly a thousand years after he lived, is a book titled *Carmen de ponderibus et mensuris*. In this work on weights and measures, the medieval grammarian describes two measuring instruments—a trutine (*trutina* in Latin), which is a balance used to weigh

Opposite page:
For books, the break from Gothic to Renaissance style was dramatic. Before 1520 or so, books had heavy, black letter type and mannered, not very natural illustrations. One of those most responsible for the sudden switch to what became characteristic Renaissance lightness and naturalness was Geoffroy Tory, a French publisher and engraver. Tory had visited Italy and returned to France imbued with its new, humanist values. The hallmark of his books was a mixture of roman and italic types, delicate borders, and illustrations with cool, pure, light lines. In this book advising lettering design based on human proportions, Tory also offered the notion of accents and punctuation marks like the apostrophe and cedilla. *Champ fleury,* 1529. Geoffroy Tory.

IVLIANO CONSVLI AC PATRICIO PRI-
SCIANVS SALVTEM.

CVm omnis eloquẽtiæ doctrinam et omne ſtudiorũ
genus ſapientiæ luce præfulgens a græcorũ fontibus
deriuatũ latinos proprio ſermone ĩgenio celebraſſe:
et in omnibus illorum ueſtigia liberalibus conſecu-
tos artibus uideo: nec [illegible] ab
illis ſunt ſcripta: ſed etiam quoſdam errores eorum
amore græcorum doctorũ deceptos imitari: in qbus
maxime uetuſtiſſimi grãmatica in arte arguũtur peccaſſe: cuius auctores quanto iuniores tanto pſpicaciores: et ingeniis floruiſſe: et dilıgẽtia
ualuiſſe omniũ iudicio confirmatur eruditiſſimorum. Quid enim Herodiani artibus certius? Quid Apollonii ſcrũpuloſis quæſtionibus enu-
cleatius poſſit ĩueniri? Cum igitur eos omĩa fere uitia quæcũq; ãtiquorum
græcorum cõmentariis ſunt relicta artis grãmaticæ expurgaſſe cõperio:
certiſq; rationis legibus emendaſſe: noſtrorũ autem neminem poſt
illos imitatorem eorum extitiſſe. quippe in negligentiam cadentibus
ſtudiis litterarum ppter inopiam ſcriptorum: quis audacter ſed non ĩ-
pudenter ut puto conatus ſum pro uiribus meis rem arduam quidẽ:
ſed officio profeſſionis non indebitam ſupra nominatorum præcepta ui-
rorum: quæ congrua ſunt uiſa: in latinũ transferre ſermonem: collectis
etiam omnibus fere quæcunq; neceſſaria noſtrorũ quoq; inueniuntur
artium cõmentariis grãmaticorum: quod gratum fore credidi tempera-
mentum: ſi ex utriuſq; liguæ moderatoribus elegantiora in unũ coeant
corpus meo labore faciẽte. Quia nec uituperãdum me eſſe credo ſi eos
imitor qui principatũ inter graios ſcriptores artis grãmaticæ poſſidẽt:
cum ueteres noſtri in erroribus etiam, ut dictũ eſt græcos æquipantes
maximam tamen laudem ſunt cõſecuti. Exemplum etiam proponere
placuit: ne pigeat alios etiam a me uel ignorãtia forte prætermiſſa: uel
uitioſe dicta. Nihil enim ex omni parte perfectũ in humanis inuenti-
onibus eſſe poſſe credo: ſua quoq; Induſtria ad cõmunem litteratoriæ
pfeſſionis utilitatẽ congrua rationis pportione uel addere uel mutare
tractantes. Nanq; feſtinãtius q̃ uolui hos edere me libros compulerũt:
qui alienis laboribus inſidiãtes: furtimq;: et quaſi per latrocinia ſcripta
ab aliis ſurripientes unius nois ad titulũ pertinentis infanda mutatioe
totius operis gloriam in ſe transferre conant̃. Sed quoniã ĩ tanta opis
materia ĩpoſſibile eſt aliquid perfectũ breuiter exponi ſpacii quoq; ue-
niam peto: quãuis ad herodiani ſcriptorum pelagus: et ad eius patris
Apollonii ſpacioſa uolumia meorũ cõpendioſa ſunt exiſtimãda ſcripta
librorũ: huius tamen operis te hortatorẽ ſortitus Iudicem quoq; facio
Iuliane conſul ac patricie: cui ſũmos dignitatis gradus ſũma acquiſiuit

From *Carmen de ponderibus et mensuris,* 1470. Priscianus of Caesarea. See p. 105.

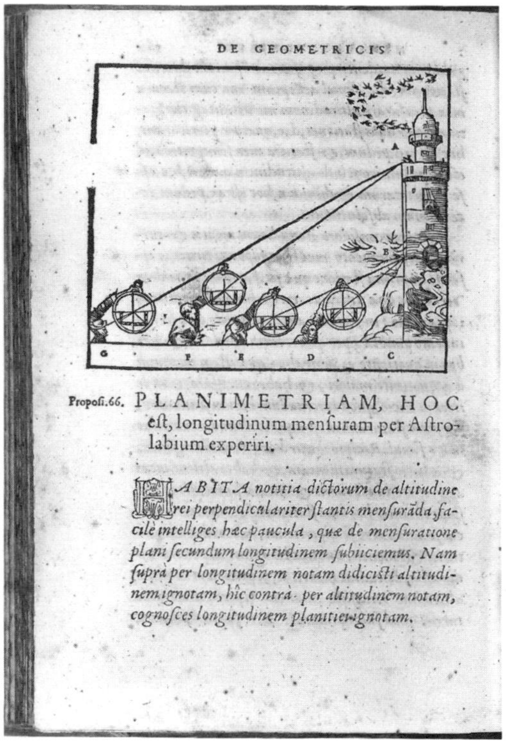

precious metals, and a hydrometer, which gave what we now call "specific weight." Like most of the printing of this early period, this represented the enthusiastic rediscovery and publication of very old texts rather than the generation of new information. Of a similar nature are three other incunabula in the Library's collections, all of which treated to some degree that oldest of the world's scientific instruments, the astrolabe. The first of these, *Tractatus astrolabii*, was printed in Ferrara in 1475. It was written by the thirteenth-century Italian astronomer and astrologer Andalo di Nigro who, besides being a world traveler, was Boccaccio's teacher in astronomy when that famed writer was in Naples. The next oldest was printed in either 1477 or 1478 in Perugia and contains a treatise on mensuration, or the geometry applied to the computation of lengths, areas, or volumes from given dimensions or angles. Written in the thirteenth century by Robertus Anglicus, it is titled *De astrolabio canones; De astrolabii compositione*. Finally, in 1498 the Venetian publisher Giorgio Valla printed a collection of translated classics entitled *Logica*. In addition to works by Aristotle, Euclid, and Galen among others, *Logica* includes two treatises on the astrolabe. The older of these by far is the work of the fifth-century Greek, Proclos, with the title *De fabrica usuque astrolabii*. Of more recent vintage is the *De astrolabo* by the fourteenth-century Byzantine historian Nicephoros Gregoras, in which he explains how to construct an astrolabe. This ancient instrument—it may date to the third century B.C.—was used by the Greeks to observe the positions and altitudes of celestial bodies, and later by medieval Europeans as a navigational instrument. Europe benefited from the Arab development of mechanical astrolabes which used gear mechanisms, and by the time of the Renaissance, the astrolabe was not only a useful measuring instrument but a symbol of scientific prowess and, at times, of high art and superior craftsmanship as well.

With the translation of Ptolemy's *Almagest* in the twelfth century, European astronomy not only revived in general but was introduced to a number of specific astronomical instruments in use during the time of that great Alexandrian astronomer. After the Venetian publication in 1515 of the *Almagest*, a wider public became aware of such instruments as the equatorial armillary, used to determine the dates of the equinoxes at Alexandria; the plinth and the meridional armillary, both of which determined the sun's altitude; the triquetrum, used to measure meridian transits of the moon and fixed stars; the dioptra, for measuring the diameter of the sun or moon; and a variety of astrolabes. The Library of Congress has this 1515 edition in its collections. As the fourteenth century closed, the medieval astrolabe had reached the peak of its popularity, proving useful for teaching astronomy and for simple astronomical calculations, and it was for just such a purpose that Geoffrey Chaucer wrote his *Tretis of the astrolabie* in 1391. Written for "litel Lowis my sone," Chaucer's work is regarded as one of the best and most understandable accounts of the astrolabe. This work was printed in London in 1542 as part of *The Works of Geoffrey Chaucer*, which the Library of Congress has in its collections. Another excellent and certainly the earliest book on the as-

The astrolabe was used in ancient times for observing the positions and altitudes of celestial bodies, and may date to third-century Greece. By the beginning of the sixteenth century, it found its way into surveying, when the problem of accurately measuring land was solved by triangulation. Simply, the area to be considered was divided into triangles, each was measured, and the total computed. Here, the height of a tower is obtained using an astrolabe equipped with a shadow square. This book was the standard work on the astrolabe throughout the sixteenth and early seventeenth centuries. *Elucidatio fabricae usuque astrolabii,* 1512. Johannes Stöffler. LC-USZ62-110336.

trolabe in German is Johannes Stöffler's *Elucidatio fabricae usuque astrolabii*, published in 1512 in Oppenheim, near Mainz. The Library of Congress has the 1570 Paris edition of this work which served to popularize the astrolabe and other measuring instruments as well. The volume's full-page woodcuts make it interesting to students of typography as well as historians of technology. Stöffler and his publisher and friend Jakob Köbel collaborated closely with the illustrator, and the results are startlingly original and beautiful woodcuts.

Although these and other such works are all significant, each was essentially backward-looking and did little to advance scientific instrumentation beyond what was already known. The following sixteenth-century works, while not uniformly innovative, did begin to reflect technology responding to the advancing demands of science. The German astronomer Peter Apian, or Petrus Apianus, was a pioneer in the design and use of both astronomical and geographical instruments. As a professor of mathematics at the University of Ingolstadt, he had written several books popularizing both mathematics and astronomy, and in 1533 he produced his *Instrument Buch*, a work devoted solely to instruments. Having designed a quadrant and an armillary sphere that were popularly used in his own time, Apianus was both a successful scholar and practitioner. The Library of Congress has this marvelously illustrated book in its first edition, published in Ingolstadt.

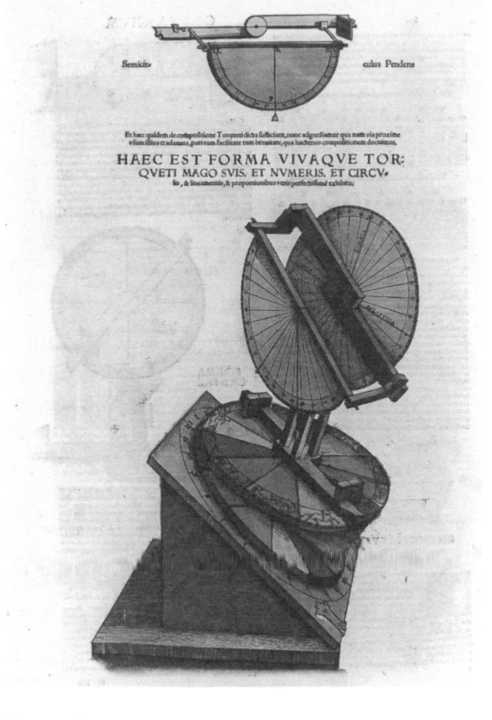

From *Astronomicum caesareum,* 1540, Petrus Apianus. See p. 104.

Pedro Nunes was a contemporary of Apianus who spent most of his career working for the Spanish court. Known as the greatest of Portuguese mathematicians, he was also a first-rate geographer, physicist, cosmologist, geometer, and algebraist as well as an accomplished poet. Nunes responded to the need of astronomers who were frustrated at not being able to measure precisely small portions of an arc. In his book, *De crepusculis liber unus* (Lisbon, 1542), Nunes describes a graduating instrument he invented to overcome this difficulty. This brilliantly conceived instrument, which later came to be known as a "nonius" in his honor, consisted of forty-four concentric auxiliary circles and was attached to an astrolabe. With it, Nunes was able to measure fractions of a degree. The nonius proved so useful it remained essentially unchanged over the centuries, although it was refined and was later called a vernier. The Library of Congress has Nunes's *De crepusculis* as it is included in his *Opera*, published posthumously in Basel in 1592.

Another sixteenth-century inventor of astronomical instruments was the northern Italian astronomer Giovanni Paolo Gallucci. Although he is hardly known today, Gallucci was active in founding in Venice one of several late-sixteenth-century scientific societies which were the precursors to the first organized scientific academies of the next century. Typical of the learned men of his age, Gallucci may have practiced some medicine and astrology. He is mentioned here not only because he was an inventor of astronomical instruments but because he produced a book which offers posterity an exhaustive survey of his century's scientific instruments used in both astronomy and applied mathematics. This comprehensive work is heavily illustrated, and there are numerous illustrations that depict these instruments being used. The Library of Congress has the rare first edition of Gallucci's work, entitled *Della fabrica et uso di diversi stromenti di astronomica et cosmografia* (Venice, 1597).

A similarly spectacular illustrated book of sixteenth-century scientific instruments in the Library's collections is *Theoria et praxis quadrantis geometrici*, published in Nuremberg in 1594. It was produced by Levinus Hulsius, an instrument maker, geographer, mathematician, and publisher who may be best known for his publication of famous voyages. The beautiful engravings of these various measuring instruments were done by Cornelius de Jode, who showed them being used in landscape settings with mountains and cities in the background.

Another designer of scientific instruments was Reiner Gemma Frisius, a Dutch contemporary of the famed Petrus Apianus. Although a mathematician and practicing physician, Gemma Frisius spent equal time learning astronomy, and at twenty-one published his version of Apianus's *Cosmographie*, "carefully corrected and with all errors set to right." Among several of his works at the Library of Congress is a collection published in Antwerp in 1581 by Jean Bellère. Titled *Cosmographie*, it also contains the relevant works of Sebastian Munster and Jean Spong, besides those of Apianus and Gemma Frisius. Among the many instruments described and illustrated in this volume is Gemma Frisius's detailed description of Jacob's staff or the cross staff, an important surveying and navigational instrument during the sixteenth century.

Early in the century a French physician—who, like Gemma Frisius, was captivated by mathematics and astronomy—was able to measure the length of one meridian degree by using an old device in an intelligent and sophisticated manner. Jean Fernel would later earn a lasting reputation for his medical work, but as a young man he used an odometer (or hodometer) to compute the circumference of the entire globe. Such a device, which measures the distance that a vehicle travels, is attributed to Ctesibius and Hero and is later mentioned by Vitrivius and even depicted by Leonardo. Although it proved a boon to accurate land surveying in the sixteenth century, Fernel saw beyond its banal use to its best application. Fernel knew that Paris and Amiens were almost exactly one degree of longitude apart, so counting the revolutions of one of his carriage wheels (which rang a bell each time), he measured his route as 111.168 kilometers. His projections to 360 degrees for the entire globe was only one-tenth of 1 percent off the true value. The young Fernel demonstrated that startling new knowledge can be gained by using existing, even archaic technology in an innovative manner. Fernel's famous global measurements are contained in his *Cosmotheria*, published in Paris in 1527. The Library has his work in first edition.

Progress in making instruments was not limited to the Continent, for about the middle of the sixteenth century, one of England's leading mathematicians combined different elements of two existing instruments and produced a new surveying instrument that would survive for two centuries. Leonard Digges's theodolite combined a horizontal circle and a vertical semicircle, thus permitting simultaneous horizontal and vertical angle

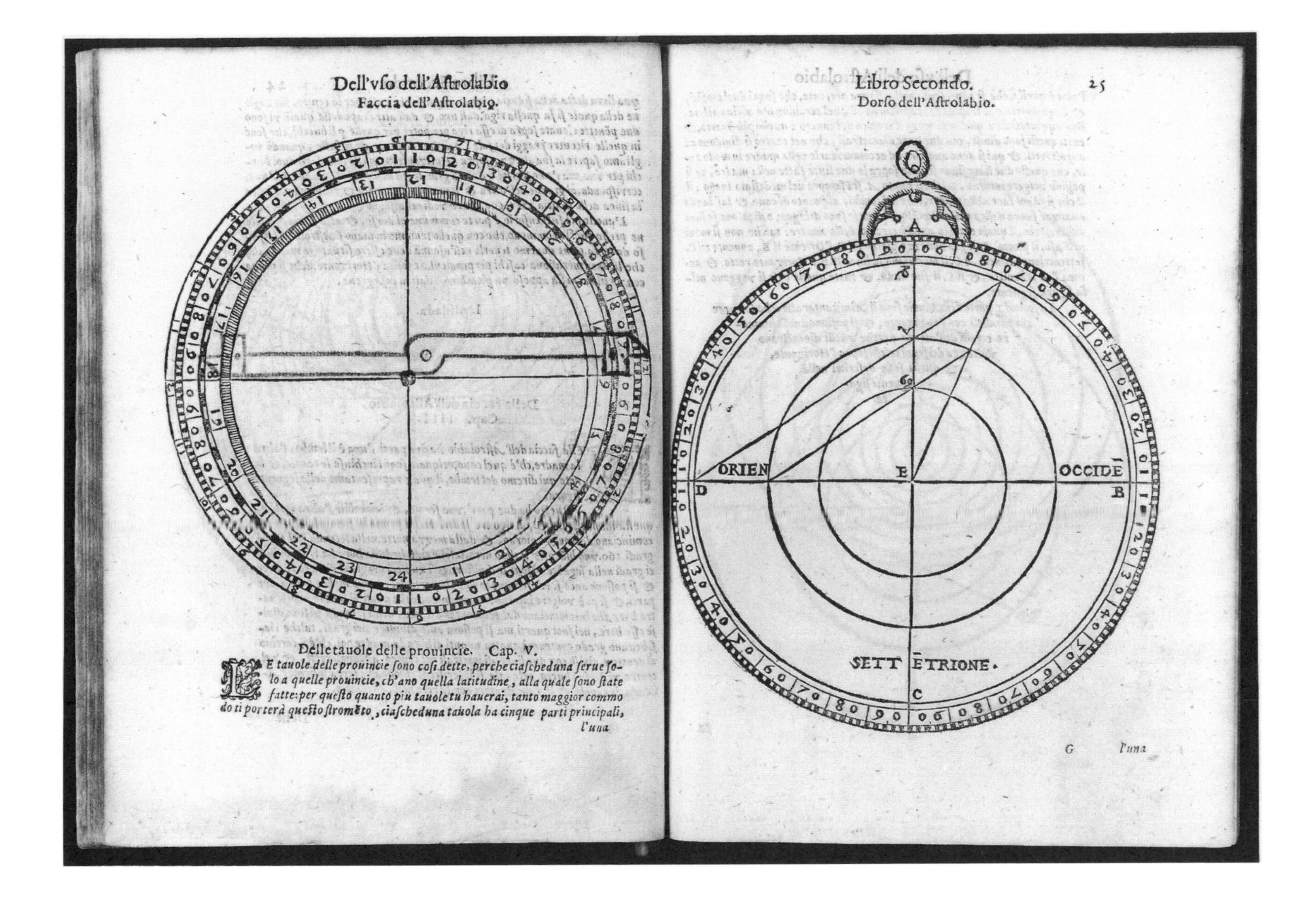
Dell'vſo dell'Aſtrolabio

Faccia dell'Aſtrolabio.

Delle tauole delle prouincie. Cap. V.

LE tauole delle prouincie ſono coſi dette, perche ciaſcheduna ſerue ſolo a quelle prouincie, ch'ano quella latitudine, alla quale ſono ſtate fatte: per queſto quanto piu tauole tu hauerai, tanto maggior commo do ti porterà queſto ſtromẽto, ciaſcheduna tauola ha cinque parti principali, l'una

Libro Secondo. 25

Dorſo dell'Aſtrolabio.

G l'una

measurements and creating the first surveying instrument that was also easy to use. Leonard Digges died in 1558, but his son Thomas completed and published his father's text which contained descriptions and illustrations of the theodolite and other measuring instruments as well. Thomas called the book *A Geometrical Practical Treatize, Named Pantometria* and published it in London in 1571. The Library of Congress has the much-enlarged second London edition of 1591. This book is not only an up-to-date surveying text as well as the first of its kind in English, but it also explicitly discusses the principles of the telescope which Digges père says he has constructed. Decades before the traditional dates for the invention of the telescope—1608 by Jan Lippershey and 1609 by Galileo—Digges states that "by glasses concave and convex of circular and parabolic forms . . . you shall discern any trifle . . . although it be distant."

Where many inventions have had more than one claimant, the inventor of the mechanical clock is completely unknown. Although the written record is sparse, there is evidence of mechanical clocks as early as the

In this survey of sixteenth-century scientific instruments, the Venetian, Giovanni Gallucci, included a section on the use of the astrolabe. The instrument on the left has engraved degree scales on its edge and a pivoted alidade attached at its center for making observations. On the right, a suspension ring is fixed by a pivot to the rim of the plate and coincides with the 90-degree mark. *Della fabrica et uso di diversi stromenti di astronomica et cosmografia,* 1597. Giovanni Paolo Gallucci. LC-USZ62-110341.

A GEOMETRICAL PRACTICAL
TREATIZE NAMED PANTOMETRIA,
diuided into three Bookes, LONGIMETRA, PLANIMETRA, and STEREOMETRIA, Containing rules manifolde for mensuration of all *Lines, Superficies* and *Solides*: with sundrie strange conclusions both by Instrument and without, and also by *Glasses* to set forth the true Description or exact Platte of an whole *Region*. First published by *Thomas Digges* Esquire, and Dedicated to the Graue, Wise, and Honourable, Sir *Nicholas Bacon* Knight, Lord Keeper of the great Seale of England. With a Mathematicall discourse of the fiue regular *Platonicall Solides*, and their *Metamorphosis* into other fiue compound rare *Geometricall Bodyes*, conteyning an hundred newe *Theoremes* at least of his owne *Inuention*, neuer mentioned before by anye other *Geometrician*.

Leonard Digges, the elder.

LATELY REVIEWED BY THE AVTHOR himselfe, and augmented with sundrie *Additions, Diffinitions, Problemes* and rare *Theoremes*, to open the passage, and prepare a way to the vnderstanding of his Treatize of *Martiall Pyrotechnie* and great *Artillerie*, hereafter to be published.

AT LONDON
Printed by *Abell Jeffes.*
ANNO. 1591.

This book was published posthumously by Leonard Digges's son, Thomas, and provided a complete introduction to all aspects of surveying and range-finding. It also describes two new surveying instruments, an early form of the theodolite and a more advanced instrument called the "topographical," which could be used to determine angles in the horizontal plane as well as elevation. It was also in this book that Thomas Digges claims priority of discovery of the telescope for his father. *A Geometrical Practical Treatize Named Pantometria,* 1591. Thomas Digges. LC-USZ62-110340.

second half of the fourteenth century in Italy and France and many believe they were in use as much as a hundred years earlier still. The mechanical clock is a device whose timekeeping is governed not by the flow of water or sand through a hole but by means of a calibrated mechanism—specifically, a verge escapement with foliot—that allows the clock's gears to operate in a slow, steady, and regular manner. Although the verge and foliot system is one of the most ingenious mechanical solutions in the history of technology—providing as it does the essential stop and go or block release at regular intervals—absolutely nothing is known about who invented it. This lack of documented or even suggestive evidence characterizes everything about the early development of the mechanical clock. In fact, most books that were published up to the end of the sixteenth century usually have more to say about sundials than about mechanical clocks. This is not as odd as it first appears since sundials were the only reliable means available to check the performance of a mechanical clock and to correct its inevitable inaccuracies. Thus, far from eliminating the need for an old technology, this new one actually stimulated the production and use of an old one. Instrument makers responded immediately to this new demand and produced all manner of small, portable sundials. Book publishers naturally followed suit, and among several of the sixteenth-century dial books in the collections of the Library of Congress are two books by the German geographer Sebastian Münster. The earliest of these is *Beschreibung der Horologien* (Basel, 1537), which describes several types of sundials. Much larger and more impressive is his *Rudimenta mathematica*, published in Basel in 1551. The first part is an introduction to geometry, while book two deals with dials and other scientific instruments. Both parts of the book contain scores of fascinating illustrations that depict all manner of specific measuring and calculating problems that are soluble by means of a knowledge of geometry and the proper instruments. Some of its illustrations have been attributed to Hans Holbein the Younger.

In stark contrast to this lavish work by Münster is an extremely thin book about the size of an open palm that was written by Johannes Dryander and published in Marburg in 1543. A physician and a colleague of Münster's, Dryander wrote the first description of the cylinder dial, a portable sundial usually a few inches high that resembles a fat pencil. Also called a shepherd's dial, it is functionally an altitude dial, which means it depends on the height of the sun above the horizon rather than the direction of the shadow cast to tell time. This type of dial was in use among the peasants in the Pyrenees until the beginning of the twentieth century. Dryander's book is titled *Cylindri usus et canones.* Another dial book in the Library, by the mathematician and cartographer Oronce Finé, is *De solaribus horologiis*, published in Paris in 1560. This modest book has a delicately illustrated and bordered title page and depicts fixed as well as portable sundials. Interestingly, in all of these sixteenth-century books on timekeeping, there is only one illustration of a mechanical clock—in the elaborate woodcut title page of Münster's *Rudimenta mathematica* which shows surveyors at work measuring a clock tower.

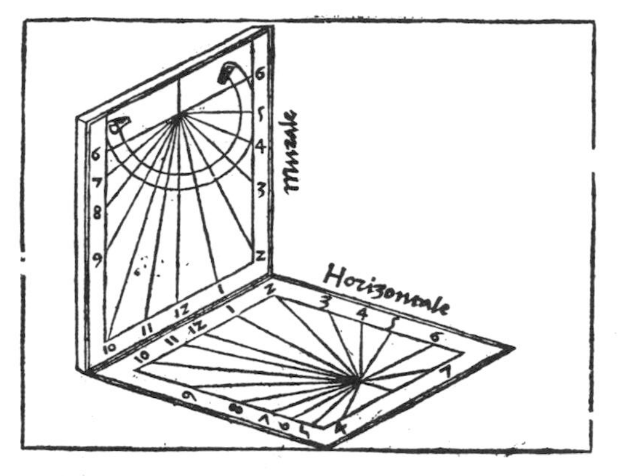

Before the early seventeenth century, it was common for travelers who wished to keep track of the time to carry portable sundials. This tablet dial is merely two small distinct sundials hinged together with a string gnomon (not shown) that stretches from a point near the top of the vertical or top dial to a point near the outer edge of the horizontal, bottom dial. The string casts a shadow across both dials and the time is read from either. When folded, the device fits in a pocket. *Rudimenta mathematica,* 1551. Sebastian Münster. LC-USZ62-110339.

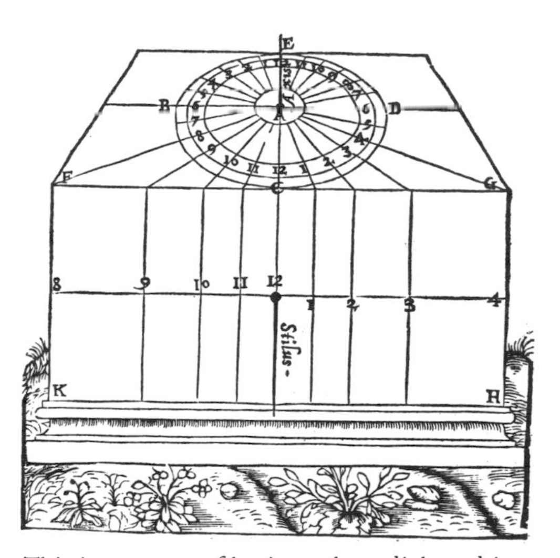

This is one type of horizontal sundial used in the sixteenth century. The horizontal is probably the most familiar type of sundial. Hour lines are inscribed on its dial plate and the gnomon at its center is a roughly triangular plate. For it to work correctly, the angle of the gnomon must equal the latitude of the place where the dial is to be used. *De solaribus horologiis,* 1560. Oronce Fine. LC-USZ62-110338.

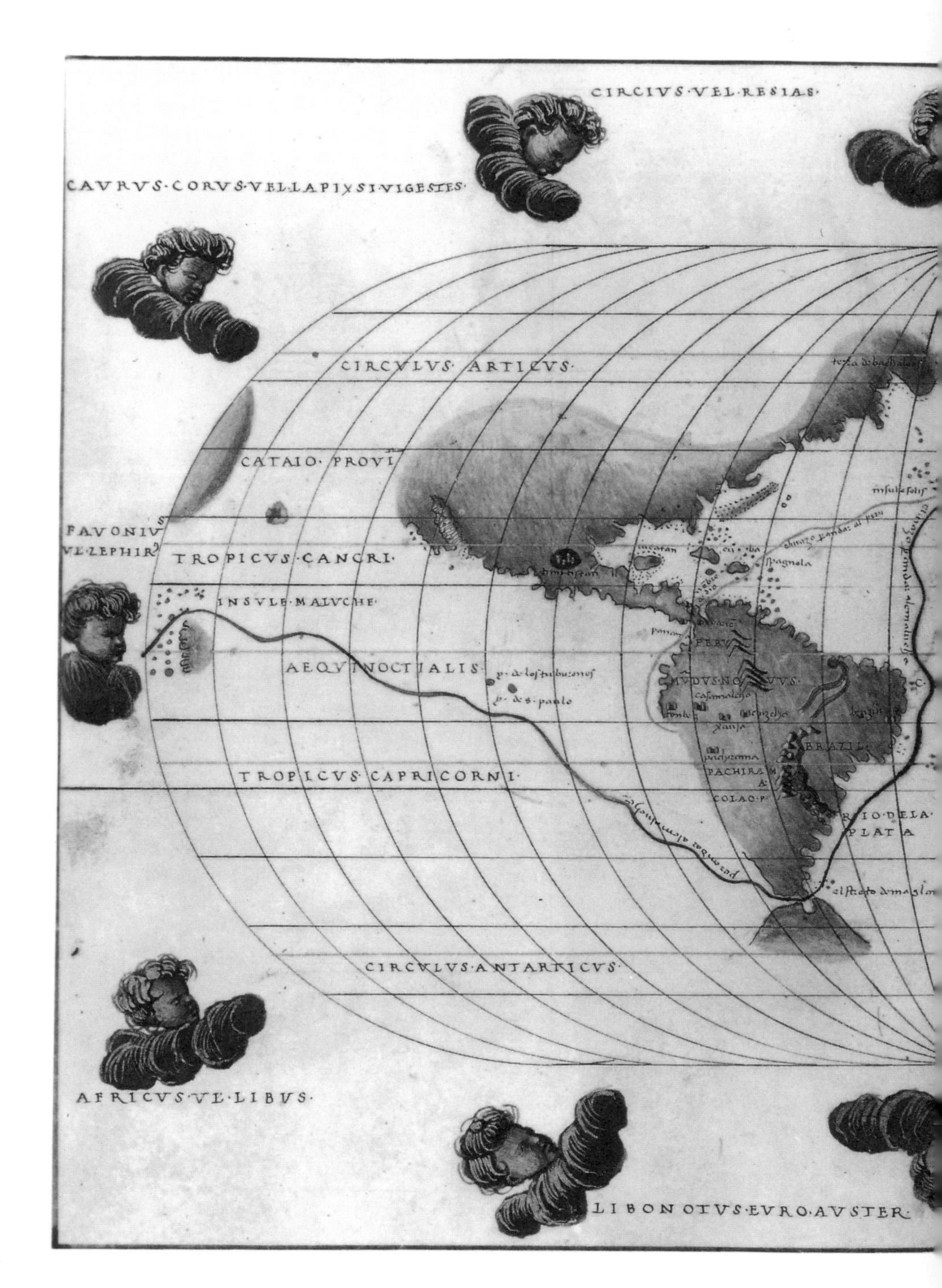

From Battista Agnese's portolan atlas of the world, ca. 1544. See pp. 106–7.

One of the most useful, essential, and powerful devices to emerge during the Renaissance, the map is not often considered to be a technological product. Usually regarded as more of an artifact if thought of at all, the map as a physical object has a technologial tale to tell that is all its own. Although maps of all kinds existed before the Renaissance for millennia—the oldest specimens are Babylonian clay tablets of around 2300 B.C.—it was during the Renaissance that the real golden age of cartography began.

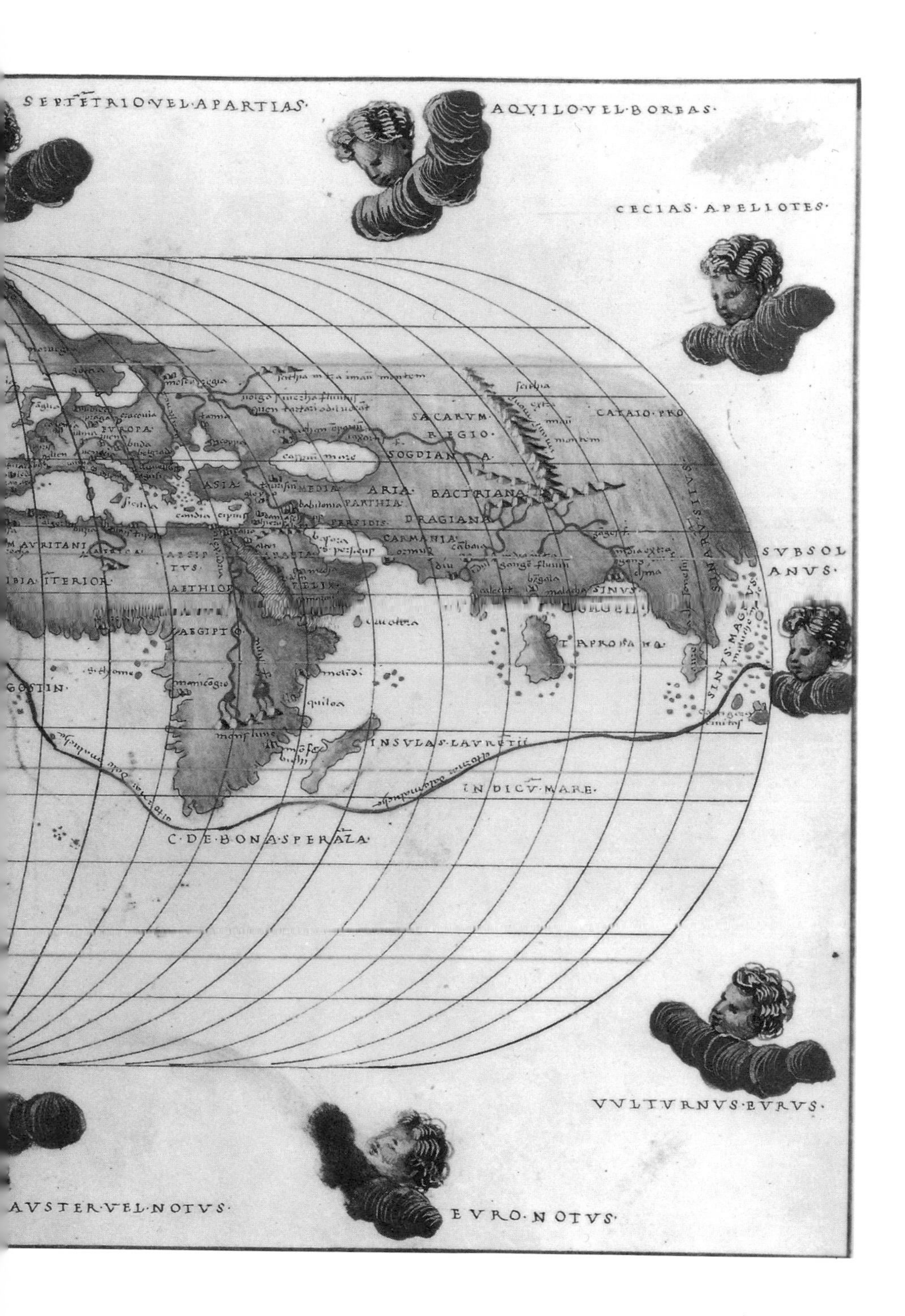

The first advances in mapmaking actually occurred during the late Middle Ages with the generation of the first accurate coastal sailing charts, called portolan charts, named for the *portolano* or graphic pilot books the Italians used. The development of the compass made these charts accurate and reliable, and they are often characterized by their striking compass roses. The charts preceded the invention of mechanical printing and were usually drawn on a single skin of parchment, generally ranging in size from thirty-six by eighteen inches (90 by 45 cm) to fifty-six by thirty inches (140 by 75 cm).

The Library of Congress has a collection of portolan charts on vellum that ranges from an anonymous fourteenth-century chart of the eastern Mediterranean and the Black Sea to several nineteenth-century Spanish charts. The most distinctive and ornamental of these is a rare manuscript atlas by Battista Agnese, done probably in 1544. Bound in contemporary brown tooled leather and stamped in gold, it has fifteen vellum leaves with ten colorfully illuminated hand-drawn maps. Among these is a world map showing an unnamed North America and a South America called "mundus novus." Agnese's charts are not typical in their beauty, color, and attention to detail, having been prepared for Venetian merchant princes and ranking officials.

With the advent of printing, maps and charts found their way into the early printed books of the fifteenth century. The first book to contain a printed map of the world—a schematic T-O map whose T divided the world into Asia, Europe, and Africa, and whose O encircled the T with water—was a medieval encyclopedia called *Etymologiae*. Written originally in the early part of the seventh century by Isidore, bishop of Seville, the book's map typifies early medieval thinking about the configuration of the physical world. The Library has the first edition, printed in Augsburg in 1472.

While the first printed map reflected long-outdated thinking, the first book to contain printed nautical charts was the very relevant and up-to-date *Isolario*, a "book of islands." Printed in Venice around 1485, this book contains forty-eight woodcuts that depict the Greek Archipelago. The Library of Congress has this Venice edition, attributed to Bartolomeo Turco (also called Bartolomeo delli Sonetti), about whom little is known.

The book that was to inspire generations and provide the greatest stimulus to cartography is Ptolemy's *Geographia*. Written by the second-century Greek Claudius Ptolemaeus, the *Geographia* offered a graphic idea of the world that was significantly more real than anything else and thus more useful to inquiring Renaissance minds. Ptolemy's work lay for centuries in the Arab world, unknown to the West, until the fall of Byzantium drove refugee scholars and their libraries westward. The *Geographia* had been translated into Latin by 1406 and thus arrived after 1453 ready to be read, understood, and copied. The first printed edition with maps was titled *Cosmographica* and was printed in 1477 in Bologna. The Library of Congress has the highly prized Rome, 1478 edition. Also titled *Cosmographia*, it surpasses the first edition in scholarship, accuracy, beauty, and the careful engraving of the maps, and outshines any other fifteenth-century edition as well. The Library's copy is the work of two German painters, Conrad Sweynheym and Arnold Buckinck, who spent years perfecting and adapting the new art of copper engraving to the printing of maps. Although this edition does not have the distinction of being the first printed Ptolemy atlas, many believe it to have been begun even before the 1477 Bologna edition was published, but that it was delayed because of the difficulties encountered with its new copper engravings.

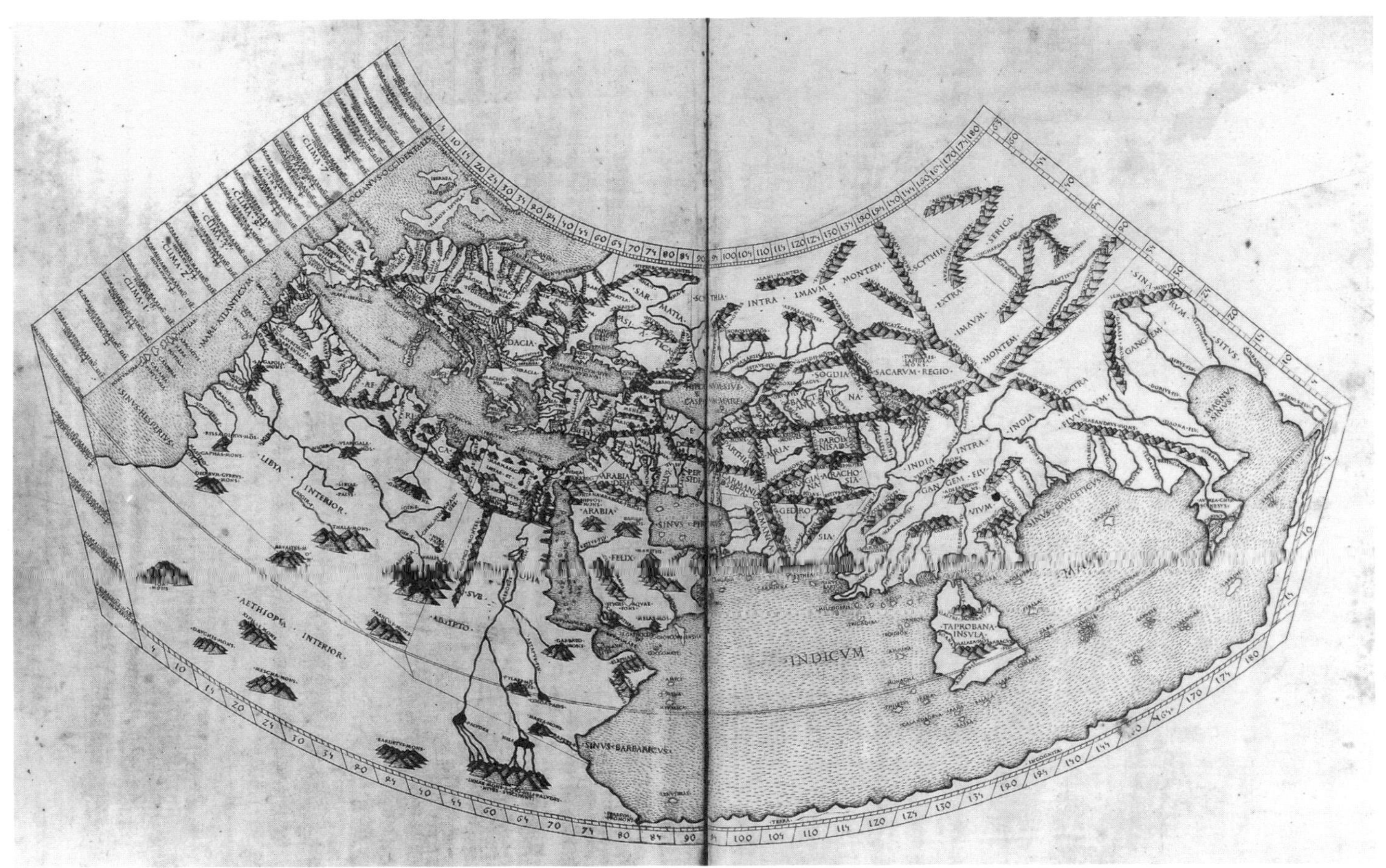

Claudius Ptolemaeus, or Ptolemy, was active in Alexandria during the period A.D. 90–168. His "Guide to Geography" reached the West via Byzantine and Arab scholars and was known to the Renaissance as the *Geographia* or *Cosmographia*. Despite his perpetuation of many mistakes of the ancients, he was correct in disputing the age-old notion that the African continent was surrounded by water, and it was he who first made technical use of the terms *parallel* and *meridian*. His adoption of a much-reduced size of the world led directly to Columbus's miscalculation and westward voyage. The volume containing this world map has twenty-seven maps printed from copperplates. *Cosmographia,* 1478. Claudius Ptolemaeus. LC-USZ62-110342.

Ptolemy's revival caused a sensation, and his emphasis on precise, systematic, and scientific methods of cartography had as much effect on Renaissance scholars as did the maps themselves (whether he drew them himself is not known). By geography, Ptolemy meant cartography, saying it "is a representation in pictures of the whole known world together with the phenomena which are contained therein."

Is is not coincidental that the Age of Discovery really begins with the revival of Ptolemy. As important as advances in shipbuilding, navigation, and even mapping were to the New World discoveries, of equal importance was the stimulus to the mind that a work like his provided. The notions of widening vistas, exotic lands, and even limitless horizons that Marco Polo's stories had engendered were only strengthened by what appeared to be hard evidence in Ptolemy's work. Polo's travels to the East between 1260 and 1295 were familiar to his contemporaries, and his story had a wide circulation in manuscript before it was first printed as *Buch des edlen Ritters und Landtfahrers Marco Polo* (Nuremberg, 1477). The Library of Congress has the 1481 Augsburg edition of this book, and another, Italian version, which was the most widely read by Mediterranean navigators.

One of these navigators, Christopher Columbus, was influenced by both Polo and Ptolemy. Columbus is known to have owned a copy of the undated Gouda Latin edition of Polo's book and he believed that the Asiatic landmass could be reached by sailing westward. Contrary to legend, Columbus was not alone in his belief that the world was round. By 1492, virtually every educated person in Western Europe regarded the earth as a sphere. As to the size of that sphere, however, there was uncertainty, and it was here that the rediscovered *Geographia* of Ptolemy played an important role. Columbus wrote to Paolo Toscanelli, a Florentine physician who had sent a map to Alfonso V, king of Portugal, that demonstrated that a westward voyage to Asia would be only 9,200 kilometers (5,700 miles). Toscanelli wrote back to Columbus, encouraging him and sending him another map. Using Toscanelli's map, which was based on Ptolemy's incorrect estimate of the earth's size, Columbus then managed, by some creative and even wishful calculations, to estimate that he would face only 4,300 kilometers (2,600 miles) of open water after he stopped at the Canary Islands. Because Ptolemy had accepted the underestimate of Poseidonius instead of the correct calculations of the earth's size made by Eratosthenes, Columbus would in turn be working from a model that was at least one-quarter too small. In fact, had the American continent not been located at roughly the distance that the Orient was supposed to be, Columbus would have faced a journey more than three times as long as he did, since the air distance between Spain and Japan is more than 15,000 kilometers (9,300 miles). Columbus died believing he had discovered the outer islands of the Indies rather than an entirely new part of the world.

As part of his bargain with Spain, Columbus had to report his expedition results directly to King Ferdinand and Queen Isabella, and these dis-

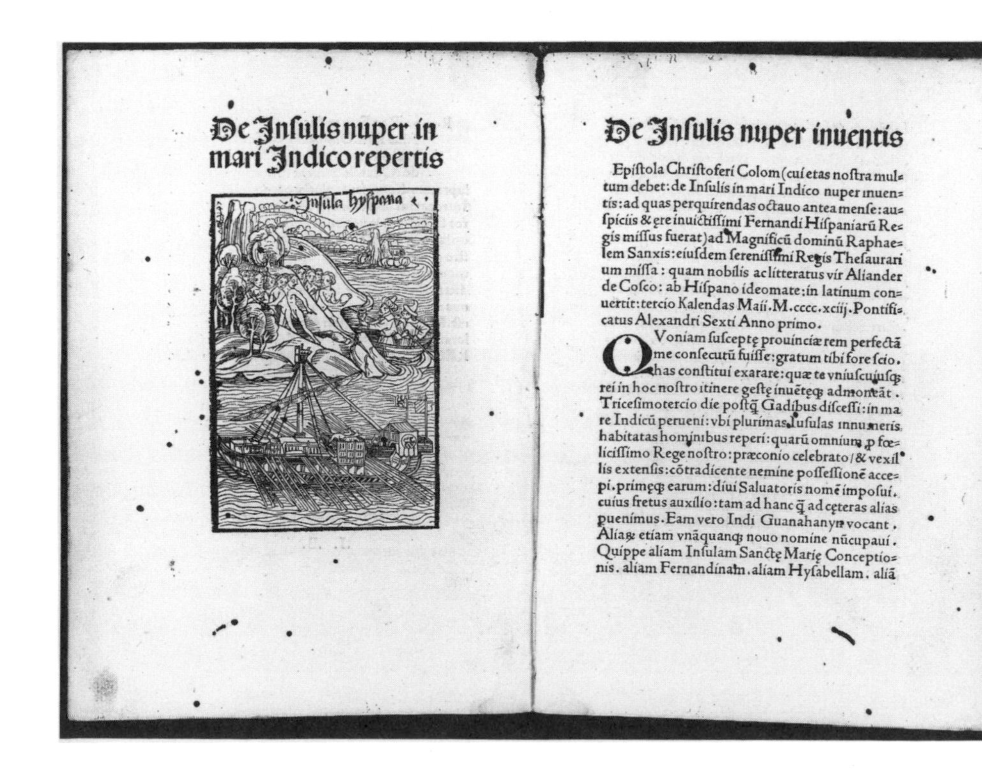
De Inſulis nuper in mari Indico repertis

De Inſulis nuper inuentis

Epiſtola Chriſtoferi Colom (cui etas noſtra multum debet: de Inſulis in mari Indico nuper inuentis: ad quas perquirendas octauo antea menſe: auſpiciis & ęre inuictiſſimi Fernandi Hiſpaniarũ Regis miſſus fuerat) ad Magnificũ dominũ Raphaelem Sanxis: eiuſdem sereniſſimi Regis Theſaurarium miſſa: quam nobilis ac litteratus vir Aliander de Coſco: ab Hiſpano ideomate: in latinum conuertit: tercio Kalendas Maii. M.cccc.xciij. Pontificatus Alexandri Sexti Anno primo.

QVoniam ſuſceptę prouinciæ rem perfectã me conſecutũ fuiſſe: gratum tibi fore ſcio. has conſtitui exarare: quæ te vniuſcuiuſqꝫ rei in hoc noſtro itinere geſtę inuẽtęqꝫ admoneãt. Tricesimotercio die poſtq̃ Gadibus diſceſſi: in mare Indicũ perueni: vbi plurimas Inſulas innumeris habitatas hominibus reperi: quarũ omnium ꝓ fœliciſſimo Rege noſtro: præconio celebrato / & vexillis extenſis: cõtradicente nemine poſſeſſionẽ accepi. primęqꝫ earum: diui Saluatoris nomẽ impoſui. cuius fretus auxilio: tam ad hanc q̃ ad cęteras alias puenimus. Eam vero Indi Guanahanyn vocant. Aliaꝝ etiam vnãquanqꝫ nouo nomine nũcupaui. Quippe aliam Inſulam Sanctę Marię Conceptionis. aliam Fernandinam. aliam Hyſabellam. aliã

Upon his return to Spain, Columbus submitted a report of his expedition to the Spanish monarchs in letter form. After the *Epistola* or letter was first published in Barcelona in 1493, ten different editions appeared in Latin. The Library has one of these as it appeared in Verardi's book. This image purportedly shows the first contact with the New World. *Historia Baetica,* 1494. Carlo Verardi. LC-USZ62-110343.

patches still survive to recount his discovery. Written originally in Spanish, the two letters that Columbus wrote were translated into Latin and printed in several editions beginning in 1493. The Library of Congress has one of these, entitled *Epistola de insulis nuper inventis*, published in Rome in 1493 by Stephen Plannck. The Library's copy of Carlo Verardi's *Historia Baetica* (Basel, 1494) also contains the Columbus letter called "De infulis nuper inventis."

Columbus was a chartmaker whose exploits would eventually lead others to the realization that Ptolemy's world maps were grossly insufficient. Once it was recognized that there was an entire landmass of unknown size between Europe and Asia, the influence of mapmakers is seen again, this time in naming that new world. When Martin Waldseemüller, a German cosmographer, was preparing a new edition of Ptolemy, he recalled the recent and well-publicized voyages of the Florentine navigator Amerigo Vespucci and created the first map in which the name "America" appeared. In his edition of *Cosmographiae introductio* (St. Die, 1507) and on the world map that accompanied it, Waldseemüller not only depicted for the first time a new, separate continent but named it America. In chapter 9 of his book he gave his reasons, saying,

> Since another fourth part has been discovered by Americus Vesputius . . . I do not see why anyone should object to its being called after Americus the discoverer, a man of natural wisdom, Land of Americus or America, since both Europe and Asia have derived their names from women.

Waldseemüller later learned of the priority of Columbus's discovery and in subsequent maps omitted the name America, but his initial christening had become so widely accepted that his efforts to change the name were fruitless. The Library of Congress has the August 29, 1507, edition of this important work, which some regard as the third edition.

The exploits of some of this era's greatest explorers, including Columbus, Vespucci, and Magellan, were collected and published in *Novus orbis regionum*. The Library of Congress has the first edition, printed in Basel in 1532, which is especially significant since it contains a map engraved by Hans Holbein the Younger.

The map as a tool made substantial advances after the mid-fifteenth-century invention of mechanical printing. The earliest of these printed maps were made from woodcuts—a dependable, traditional method. Maps were soon being engraved on copperplates, which were then inked and wiped, the ink remaining in the cut lines. When dampened paper was pressed on the copperplate and into the cut lines, the result was a much finer, and possibly more accurate, impression. This process constituted the basics of fine map reproduction until the modern invention of photolithography.

The first truly modern maps were the work of Gerhard Kremer, better known by the Latinized version of his name, Gerardus Mercator. Mercator was a Flemish geographer and maker of globes and instruments who became the foremost cartographer of the sixteenth century. His idea of

Gerardus Mercator is shown at the age of sixty-two with the tools of his trade. Having solved the problem of how to depict a spherical world on a flat surface with his revolutionary projection lines, Mercator offered navigators a mapping system in which all parallels and meridians would meet at right angles, allowing them to set a ship's course in a straight line. His son published this volume a year after his death, but it was Mercator himself who chose the name "Atlas," and the name has ever since meant a bound collection of maps. *Atlas sive cosmographicae meditationes de fabrica mundi et fabricati figura,* 1595. Gerardus Mercator. LC-USZ62-110344.

using a cylindrical projection to solve the age-old problem of how to depict a spherical surface on flat paper revolutionized map making. By means of his distortion formula, which spaced the parallels of latitude progressively wider as they moved away from the equator and toward the poles, navigators could plot bearings as straight lines. Mercator first used his projection in a 1569 world map and in 1585 began publishing a collection of maps in book form. The Library of Congress has his classic *Atlas sive Cosmographicae meditationes de fabrica mundi et fabricati figura*, a rare, three-part volume that first used the word "atlas" to designate a bound collection of maps. The first two parts were originally published separately in 1585 and 1589. The third, which in the Library's copy precedes the others, was left unfinished by Mercator, who died in 1594, and was completed, edited, and published by his son Rumold in 1595. All three were published in Duisburg. The Library's edition is the complete 1595 version, including the reissued first two parts and containing 107 maps. Mercator chose the title for his book himself, and, translated, it reads, *Atlas or the Meditations of a Cosmographer upon the Creation of the World and the Shape of That Which Was Created.* The book's introduction includes a genealogy of Atlas, the mythical Greek Titan who lost the war against Jupiter and was condemned to support the world on his shoulders, and its cover depicts that well-known scene.

In 1570, a year after Mercator introduced his new method of projection but well before he published his maps in a book, his friendly rival Abraham Ortelius compiled the first modern atlas, which he called *Theatrum orbis terrarum.* First published in Antwerp, the *Theatrum* is a collection of seventy maps which Ortelius had carefully selected from the work of others (to whom he gave credit). He did, however, reengrave each map. The Library has his work in first edition as well as subsequent editions which were revised and enlarged. Thus, although it was Ortelius who produced what is now recognized as the first modern atlas, the name for such a compilation of maps was first applied by Mercator fifteen years later.

The overall advances in technology during the Renaissance, especially the improvement of machines run by animals or falling water, had a dramatic impact on mining and metallurgy. Larger and deeper mines could be dug, more water pumped, higher furnace temperatures achieved, and better surveys made. Not surprisingly, it was in Germany that most of these advances were made and also the best and most important literature produced. The earliest printed work in the Library of Congress collections dealing in some way with mining and metallurgy was written in the thirteenth century by a southern German, Albert, Count von Bollstädt, also bishop of Regensburg, and better known as Albertus Magnus. His comprehensive *De mineralibus*, on stones, minerals, and metals, was first printed in 1476; the Library has the 1491 Pavia edition. In addition to studying in Italy and being familiar with Arabic translations, Albert traveled extensively and recorded, for example, that he visited the copper works near Paris and Cologne where he witnessed techniques that were "tested by experience." His writings are primarily a summary of metallurgical and geological knowledge up to his time.

Apart from the significant fact that this work is the first book printed with illustrations of a technical nature, it is important because it is backward-looking. Valturio's weapon systems are not of his time but are of a fairly archaic type. The drawings this book contains are closely related to the Roman Vegetius, and it is only in his fantastic weapons, such as this cannon-shooting, mobile dragon, that Valturio shows any originality. *De re militari,* 1472. Roberto Valturio. LC-USZ62-110292.

Block books were the technical forerunners to the mechanically printed, illustrated book. In them, the text as well as the images was cut on wood. Produced in large numbers, mostly in Germany and the Netherlands, they were largely religious books tedious to make but sold inexpensively. This work illustrating the Revelation of Saint John, is considered one of the great monuments of block printing. Despite its age, the coloring in the Library's copy is uncommonly brilliant. *Apocalypsis Sancti Johannis,* ca. 1470. LC-USZ62-110334.

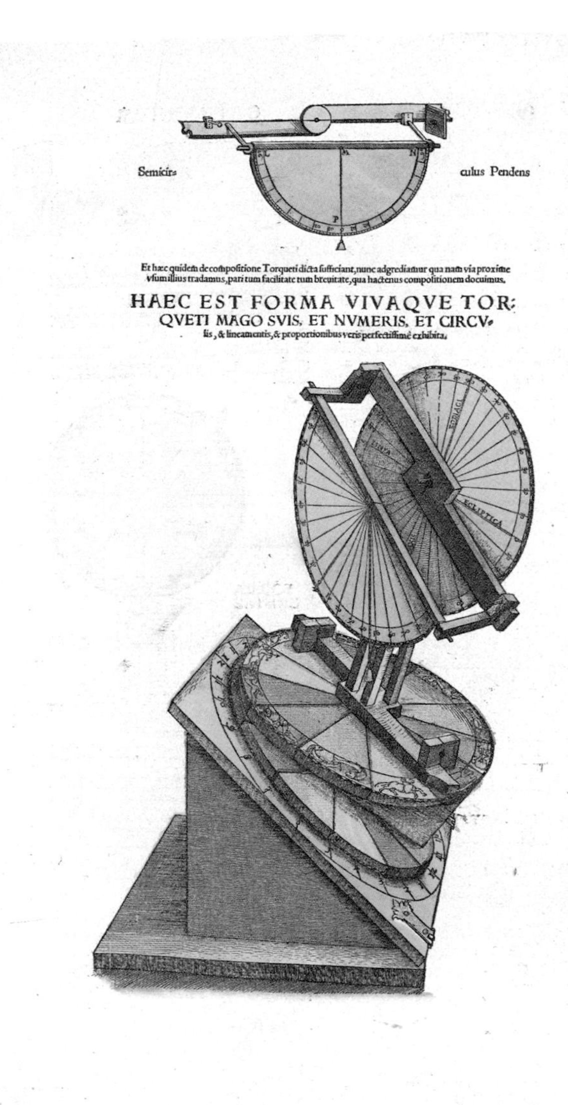

Before the invention of the telescope, astronomical observations and calculations were made with the naked eye aided by all manner of instruments and devices. Petrus Apianus, a German geographer and astronomer, preceded the great observers like Brahe, Kepler, and Hevelius by at least a generation and invented the torquetum shown here. This large device could be adjusted to measure the altitude of celestial bodies above the horizon. Apian was able to use it to resolve spherical triangles and to trace the course of comets. *Astronomicum caesareum,* 1540. Petrus Apianus. LC-USZ62-110337.

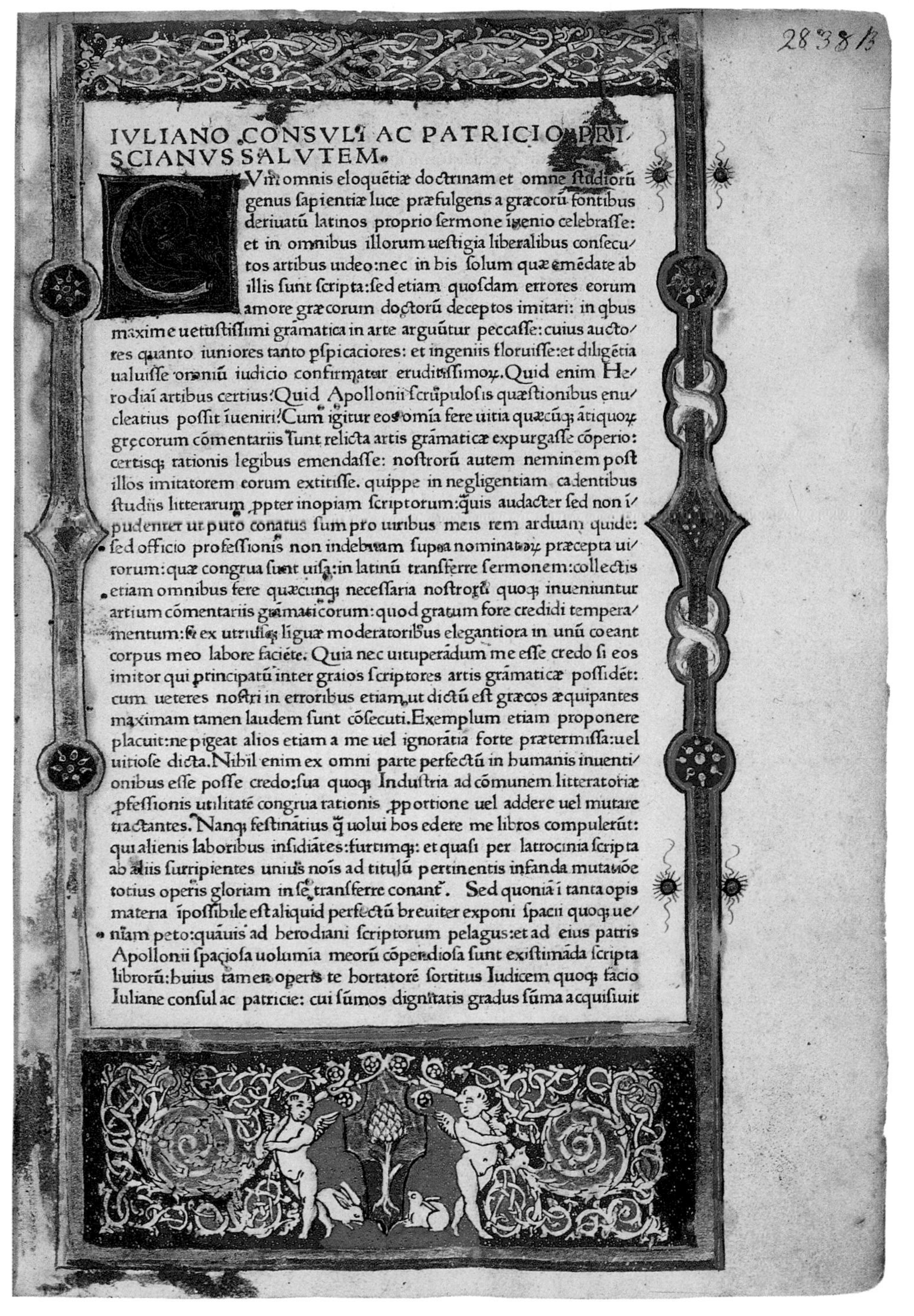
IVLIANO CONSVLI AC PATRICIO PRI
SCIANVS SALVTEM.

CVm omnis eloquētiæ doctrinam et omne studiorū
genus sapientiæ luce præfulgens a græcorū fontibus
deriuatū latinos proprio sermone ingenio celebrasse:
et in omnibus illorum uestigia liberalibus consecu/
tos artibus uideo: nec in his solum quæ emēdate ab
illis sunt scripta: sed etiam quosdam errores eorum
amore græcorum doctorū deceptos imitari: in qbus
maxime uetustissimi grāmatica in arte arguūtur peccasse: cuius aucto/
res quanto iuniores tanto pspicaciores: et ingeniis floruisse: et diligētia
ualuisse omniū iudicio confirmatur eruditissimoꝝ. Quid enim He/
rodiani artibus certius? Quid Apollonii scrupulosis quæstionibus enu/
cleatius possit iueniri? Cum igitur eos omia fere uitia quæcūq; ātiquoꝝ
grecorum cōmentariis sunt relicta artis grāmaticæ expurgasse cōperio:
certisq; rationis legibus emendasse: nostrorū autem neminem post
illos imitatorem eorum extitisse. quippe in negligentiam cadentibus
studiis litterarum ppter inopiam scriptorum: quis audacter sed non i/
pudenter ut puto conatus sum pro uiribus meis rem arduam quidē:
sed officio professionis non indebitam supra nominatoꝝ præcepta ui/
rorum: quæ congrua sunt uisa: in latinū transferre sermonem: collectis
etiam omnibus fere quæcunq; necessaria nostroꝝ quoq; inueniuntur
artium cōmentariis grāmaticorum: quod gratum fore credidi tempera/
mentum: si ex utriusq; linguæ moderatoribus elegantiora in unū coeant
corpus meo labore faciēte. Quia nec uituperādum me esse credo si eos
imitor qui principatū inter graios scriptores artis grāmaticæ possidēt:
cum ueteres nostri in erroribus etiam ut dictū est græcos æquipantes
maximam tamen laudem sunt cōsecuti. Exemplum etiam proponere
placuit: ne pigeat alios etiam a me uel ignorātia forte prætermissa: uel
uitiose dicta. Nihil enim ex omni parte perfectū in humanis inuenti/
onibus esse posse credo: sua quoq; Industria ad cōmunem litteratoriæ
pfessionis utilitatē congrua rationis pportione uel addere uel mutare
tractantes. Nanq; festinatius q̄ uolui hos edere me libros compulerūt:
qui alienis laboribus insidiātes: furtimq; et quasi per latrocinia scripta
ab aliis surripientes unius nois ad titulū pertinentis infanda mutatiōe
totius operis gloriam in se transferre conant̄. Sed quoniā i tanta opis
materia īpossibile est aliquid perfectū breuiter exponi spacii quoq; ue/
niam peto: quāuis ad herodiani scriptorum pelagus: et ad eius patris
Apollonii spaciosa uolumia meorū cōpendiosa sunt existimāda scripta
librorū: huius tamen operis te hortatorē sortitus Iudicem quoq; facio
Iuliane consul ac patricie: cui summos dignitatis gradus sūma acquisiuit

Among the works of Priscian, who lived around A.D. 500 and who is the best known of all the Latin grammarians, is a treatise on weights and measures. Contained in this first printed collection of his works, it was partly responsible for the rediscovery of information on such ancient measuring devices as the "trutina." *Carmen de ponderibus et mensuris,* 1470. Priscianus of Caesarea. LC-USZ62-110335.

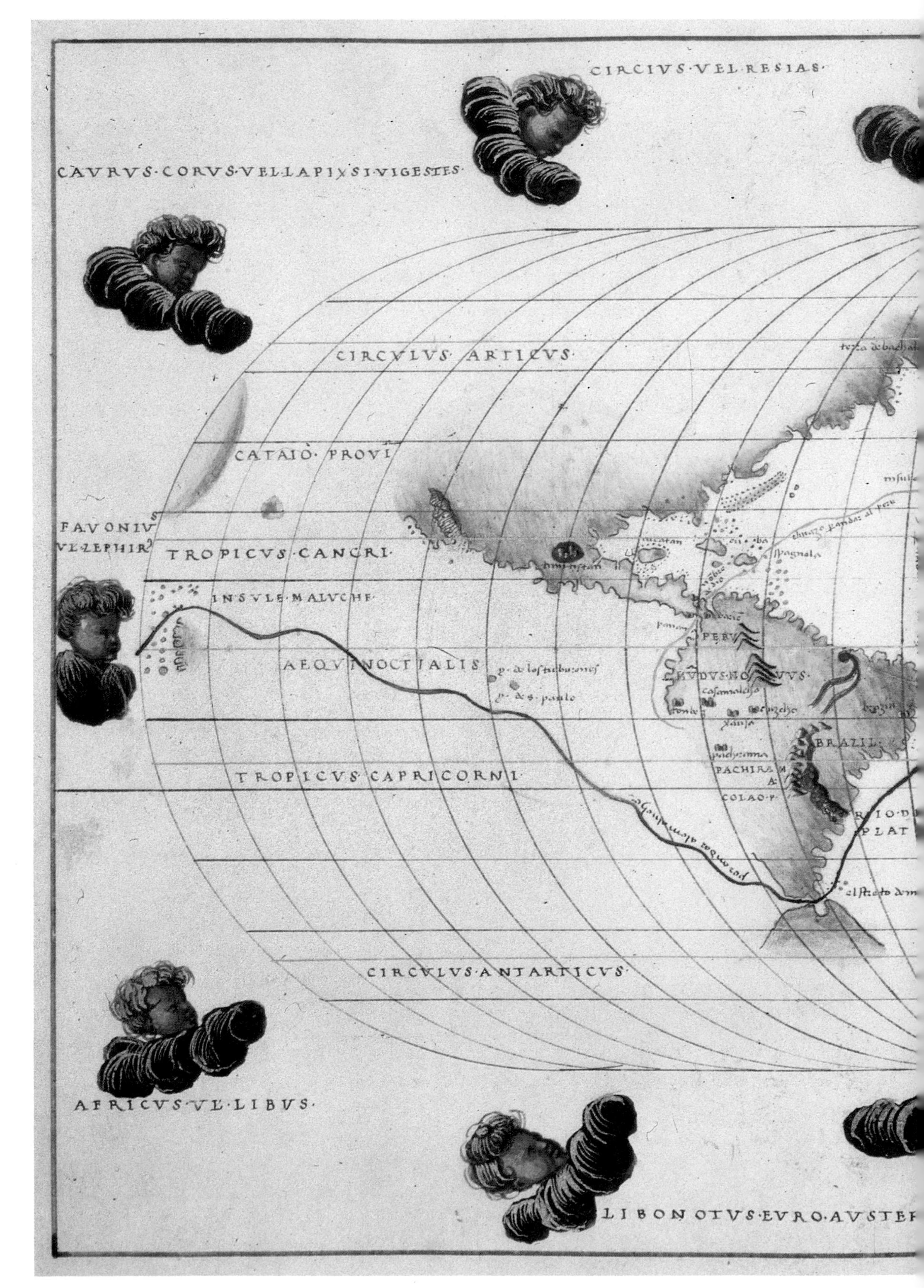

This portolan map on vellum is a map of the world showing Magellan's route. The American continent is distinctly recognizable, especially the Gulf of California. The twelve wind cherubs are all given names. All that is known about its maker, Battista Agnese, is that he was from Genoa and worked in Venice. The extravagant use of color indicates the map was made for rich Venetian merchants or high-ranking officials. Portolan atlas of the world, ca. 1544. Battista Agnese.

AQVILO·VEL·BOREAS·
CECIAS·APELIOTES·
SVBSOLANVS·
VVLTVRNVS·EVRVS·
EVRO·NOTVS·
C·DE·BONA·SPERAZA·
IN DICV·MARE·
INSVLAS·LAVRETII
SOGDIANA
BACTRIANA
DRAGIANA
CARMANIA
ASIA
EVROPA
SINARVM·SITVS·
TAPROBANA
SINVS·MAGNVS
SINVS·GAGETIS
AETHIOP
IA·SVB·
AEGIPTO·
CATAIO·PRO

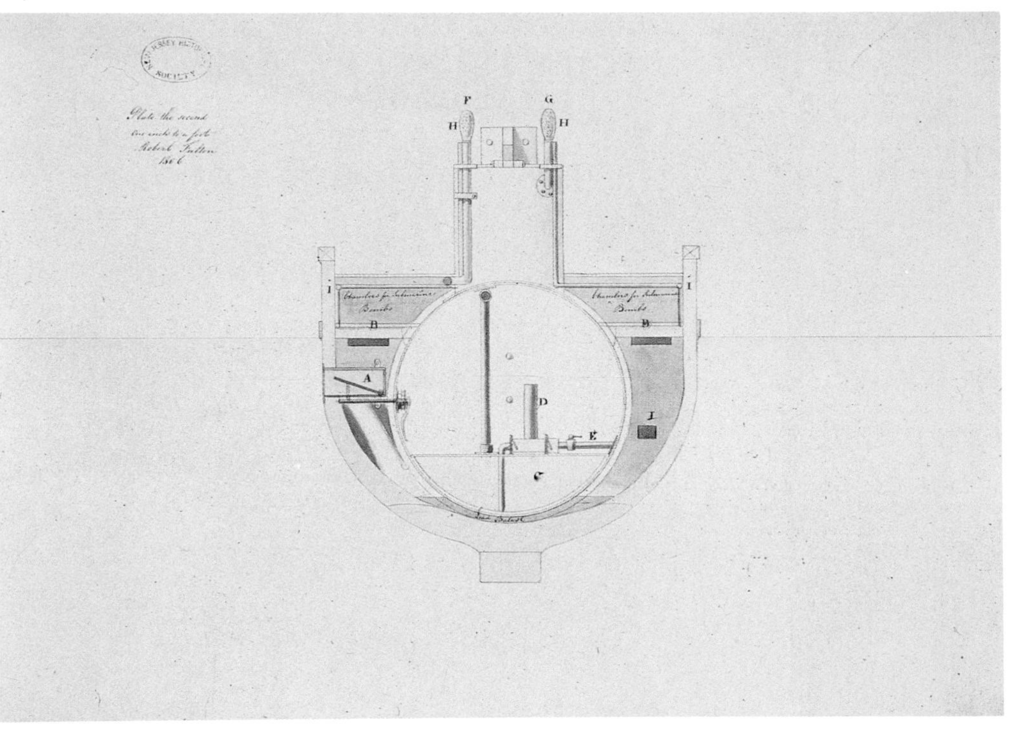

Before the successful running of his steamboat, Robert Fulton built this submarine or "plunging boat" in France. Called the *Nautilus,* the boat was built at his expense and demonstrated on the Seine in hopes of gaining French support. Fulton believed his submarine and torpedo system would aid the French in their war with Britain. But the French did not adopt his plans, nor did the British some years later, and Fulton turned his energies to a commercial steamboat. This sectional drawing was done in graphite, ink, and colored wash by Fulton himself in 1806, and shows the chambers for its submarine bombs and lead ballast. Architecture, Design, and Engineering Collections, Prints and Photographs Division, Library of Congress. Robert Fulton. LC-USZ62-110383.

It was with this locomotive that the first freight and passenger railroad began. Built by an Englishman, George Stephenson, it turned the nine-mile stretch of the Stockton and Darlington Railway into a financial success. For the first time in history, land transportation was possible at a rate faster than any horse could run. Stephenson's engine could haul ninety tons at a maximum speed of twelve miles per hour and began a transportation revolution throughout the world. *Reports on Canals, Railways, Roads, and Other Subjects,* 1826. William Strickland. LC-USZ62-110386.

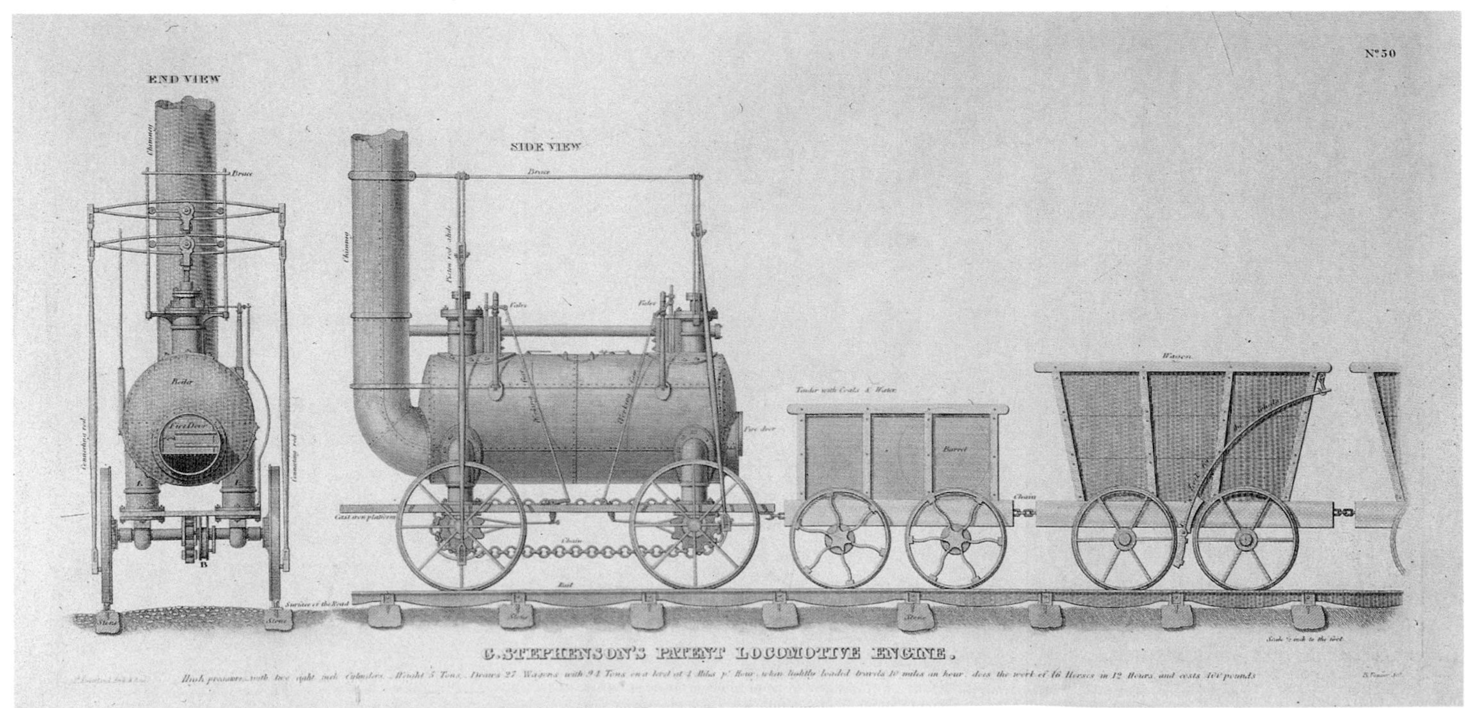

Ten days after the Montgolfier balloon inaugurated human flight, Jacques Charles, a thirty-six-year-old teacher at the Sorbonne, flew this balloon filled with hydrogen. Charles was more expert at isolating hydrogen than were the Montgolfiers, and he was accompanied on this historic flight by Nicholas Robert, who had made the rubber-impregnated silk that held Charles's hydrogen. Hydrogen has excellent lifting pioperties, and Charles's two-hour flight took him and Robert twenty-two miles from Paris. As the balloon touched down, Robert stepped out and Charles went off for another short flight of his own. Besides this watercolor, the Library of Congress has many other illustrated and manuscript materials pertinent to the early history of ballooning, among which is included the original hand-written manuscript report of this first hydrogen balloon ascent. Prints and Photographs Division, Library of Congress, LC-USZ62-8922.

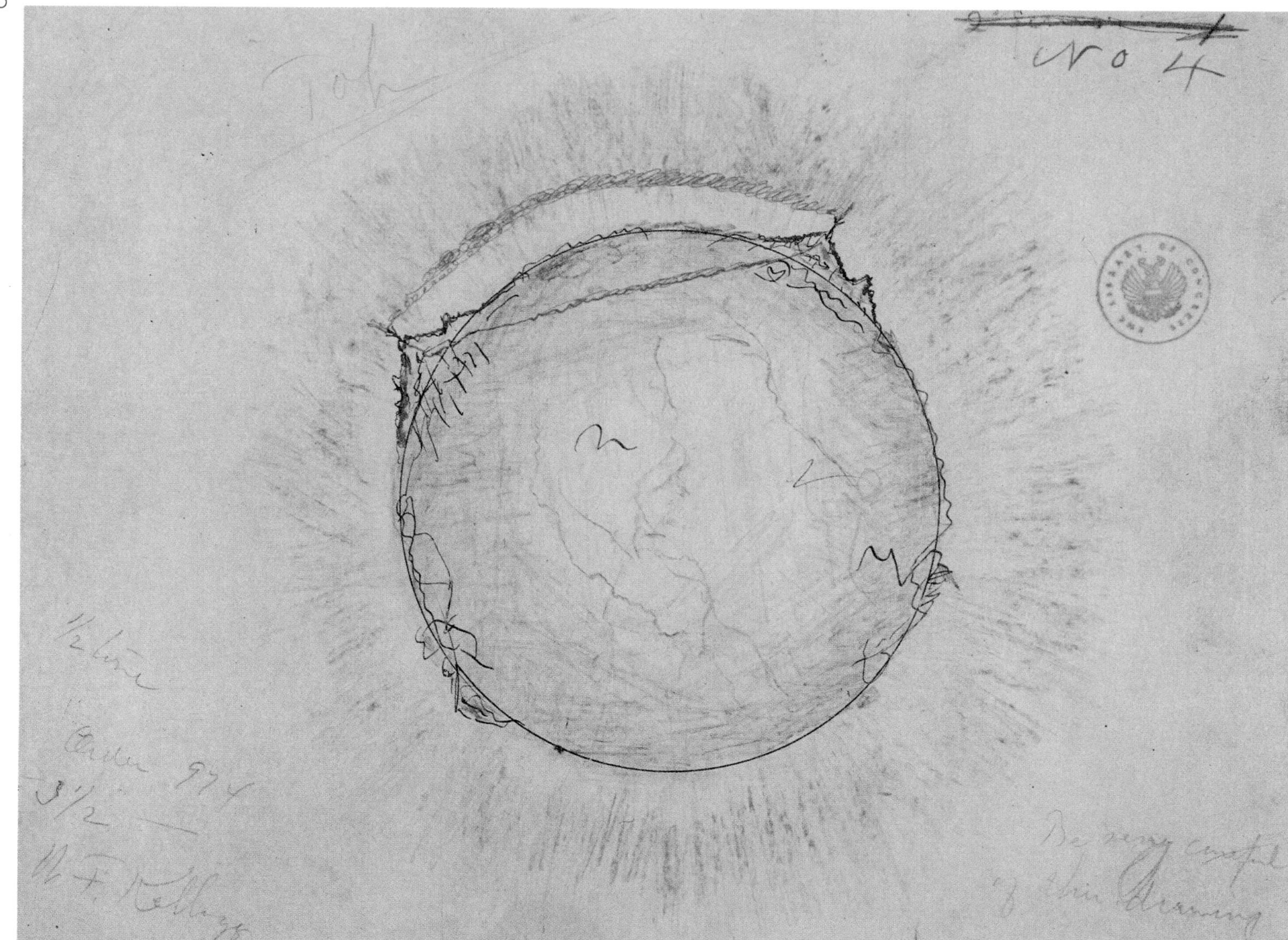

The discoveries of Hertz pushed Marconi to investigate the transmission of an electric signal without wires, and the earlier achievement of Morse and his telegraph impelled an American dentist, Mahlon Loomis, to attempt the same. In 1866, Loomis demonstrated what he called "wireless telegraphy." Kites held by copper wire instead of string were flown at about six hundred feet between two Virginia mountains, and Loomis was able to make a meter connected to one kite cause a meter on the other kite to move. Loomis knew he had transmitted something, and in 1872 he applied for and received the first wireless patent granted by the U.S. Patent Office. Loomis could not persuade the Congress to fund his research, however, and died in 1886 with his achievements virtually forgotten. The extent of his vision for radio is captured in his rough sketch showing how San Francisco might communicate via wireless telegraphy with Japan. In a speech written in 1872 and delivered to the Franklin Institute in 1881, Loomis said, "What I contemplate . . . is to send messages from continent to continent across the oceans without the use of a cable or artificial battery; or between stations on the earth however distant from each other, without intervening wires." Mahlon Loomis Papers, Manuscript Division, Library of Congress.

In America, daguerreotypes were commercially available as early as the spring of 1840. Mathew Brady opened his first "Daguerrean Miniature Gallery" on lower Broadway in 1844. One example of these early American daguerreotypes in the Library's collections shows a newly elected congressman from Illinois named Lincoln. Taken in 1846, this image is typical of its time in the straightforward, unspontaneous demeanor of the sitter. Some of this was a result of the sitters' sense that this was a solemn occasion, but the fact that he had to sit for long exposure times, head clamped and hands secured, did not make for a smiling, natural pose. Prints and Photographs Division, Library of Congress. See also p. 243.

The Lumière brothers were not only pioneers in commercial cinema but produced the first practical color photographs as well. Their Autochrome process was a real breakthrough, for it enabled color photographs to be taken with one plate using an ordinary camera. Although their system naturally became popular with amateurs, professionals like Arnold Genthe used the process, possibly favoring its effect of a pointillist color painting. The Autochrome process was expensive and required long exposures, and it was eventually replaced by the American, three-layer system marketed as Kodachrome. Here is Genthe's "Portrait of Miss S.G." Prints and Photographs Division, Library of Congress. LC-G408CT-31.

The first truly international exhibition of technology was officially called "The Great Exhibition of the Works of Industry of All Nations." Held in Hyde Park, London, in 1851 it was popularly known as "The Crystal Palace," after the central exhibition building constructed by Joseph Paxton. This iron-and-glass fantasy consisted of twenty-two acres of glass connected and supported by cast iron and wrought iron. It was essentially a prefabricated greenhouse built on an immense scale. The building was both a structural and an aesthetic success and survived the exhibition until it was destroyed by fire on December 1, 1936. The exhibition itself was wildly successful from every standard. It achieved its basic promotional purposes and generated popular enthusiasm and optimism about technological innovation. Each participating nation benefitted from the comparative look it afforded. Paxton's Crystal Palace is seen in this chromolithograph—color lithography had just been developed—view of the south side near the Princess Gate. *The Park and the Crystal Palace: A Series of Seven Picturesque Views,* 1851. Philip Brannon.

Bakelite is the registered trade name for phenol-formaldehyde resin. Although Leo Baekeland was not the first to investigate reactions between phenol and other various aldehydes, he was the first to devise the process that controlled the normally very fast reaction between phenol and formaldehyde in a way that allowed the material to be shaped and molded. His process of making "giant molecules" (those with a very high molecular weight) by joining small ones together, marked the formation of the modern polymer industry. Bakelite came to replace the older and impractical celluloid and was used for a myriad of industrial applications, especially as an electrical insulator. As this cover illustration shows, it eventually found a variety of more popular uses once its natural drabness was made bright and colorful. The German Bauhaus school of arts and crafts, whose progressive furniture design signaled its embrace of modern technology, introduced Bakelite into homes and offices. Following World War II, plastics would become synonymous with the twentieth century. *Bakelite Review,* October 1933. Courtesy of Union Carbide Corporation. LC-USZ62-110438.

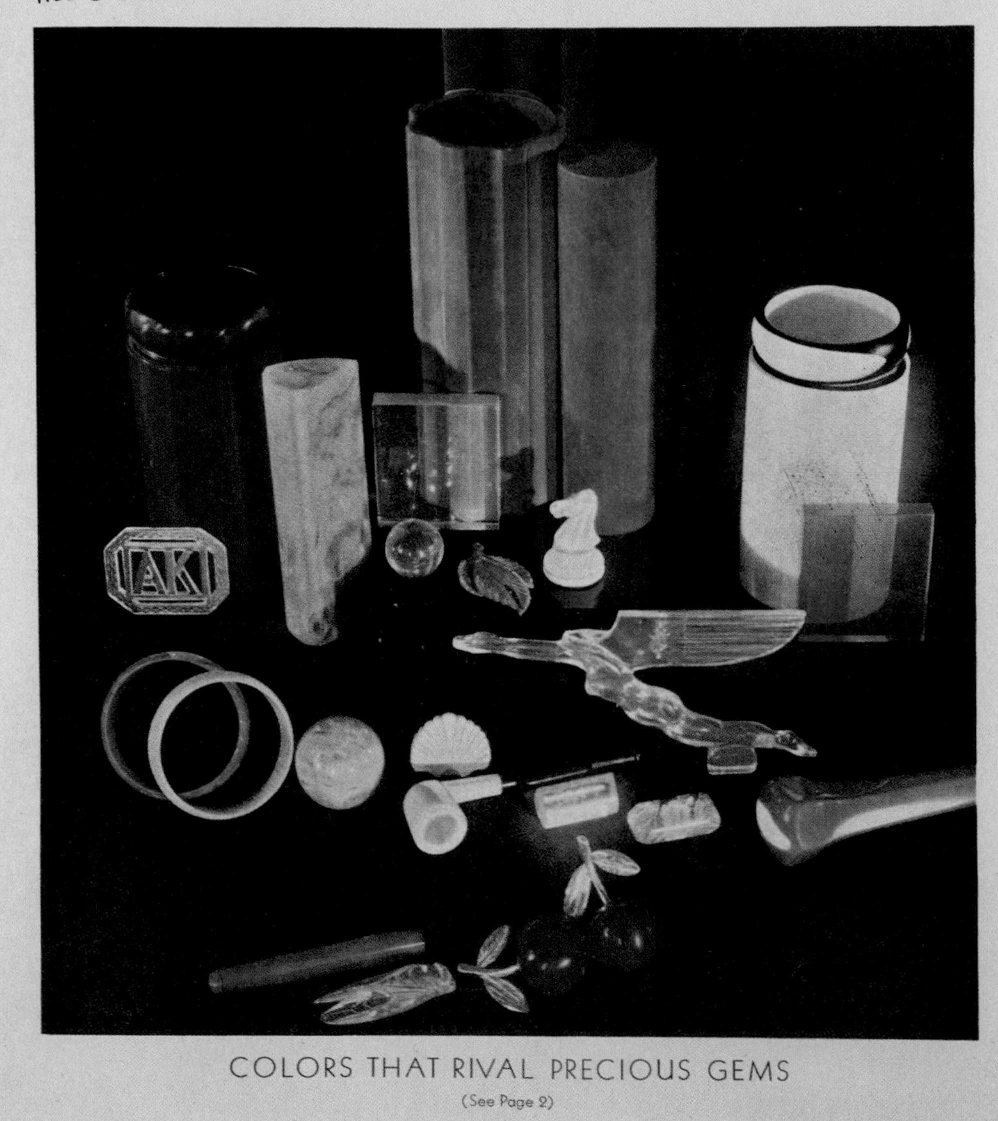

BAKELITE REVIEW

A PERIODICAL DIGEST OF BAKELITE ACHIEVEMENTS INTERESTING TO ALL PROGRESSIVE MANUFACTURERS AND MERCHANTS

OCTOBER, 1933
VOL. 5 • NO. 3

COLORS THAT RIVAL PRECIOUS GEMS

(See Page 2)

The first printed book devoted exclusively to the field of mining was *Ein nutzlich Bergbüchlein von allen Metallen*, known as *Bergbüchlein*. This "little book on ores" was written as an introduction to mining geology but contained enough factual, practical information that it also served as an excellent all-purpose mining reference book. The earliest known printed version of this work, attributed to Ulrich Ruhlein von Kalbe, is 1505. The Library of Congress has the *Bergbüchlein* as it was reprinted in a sixteenth-century book on mining law titled *Der Ursprung gemeynner Berckrecht*, published about 1520 in Strasbourg. The Library also has an English translation, with illustrations, published in 1949 by the American Institute of Mining and Metallurgical Engineers. This contemporary translation also contains a reprinting of the first printed book on assaying and metallurgy, called *Probierbüchlein*. Taken together, both books do not compare to the great works that will next be described, since they were more in the genre of affordable handbooks, but their primacy in the field as well as the practical information they contained make them both historic and notable.

The handbooks and manuals contained many metallurgical recipes and techniques, and although alchemy falls outside the subject range of this book, mention must be made of one sixteenth-century alchemical book that, as it argues against the study of alchemy, in fact contains a great deal of practical metallurgy. In his *Voarchadumia contra alchimiam* (Venice, 1530), Giovanni Agostino Pantheo discusses the metallurgy of precious metals and other materials and offers several woodcut illustrations of metallurgical furnaces and related apparatus. The book also contains a handsome woodcut bird's-eye view of Venice and its surroundings. The Library of Congress has this rare work in its first edition.

A furnace and much of the related equipment used by a Renaissance metallurgist is shown here. This rare work discusses the metallurgy of precious metals as well as many practical recipes, the use of sulphides, and the making of mirrors. Although Renaissance metallurgy saw no spectacular new discoveries or inventions, it was an important time in terms of the large-scale systematization of old ways. *Voarchadumia contra alchimiam,* 1530. Giovanni Agostino Pantheo. LC-USZ62-110345.

Ten years later, an Italian produced the first comprehensive metallurgical text. Written by Vannuccio Biringucci in his native tongue, *De la pirotechnia* was intended for the experienced metallurgist and maker of gunpowder. The book was published in Venice in 1540 and was especially clear and precise in its treatment of all aspects of smelting. The Library of Congress has this well-illustrated and often reprinted work in first edition.

The masterwork of this time was produced by a German physician, Georg Bauer, who Latinized his name to Agricola. Raised in the mining district of Saxony, he never lost his youthful curiosity about that difficult and dirty endeavor. As a young doctor in the mining town of Joachimsthal in Bohemia, he spent his free time visiting the mines and smelting works of the area. Later he would reflect on the importance of those times, saying, "Those things which we see with our eyes and understand by means of our senses are more clearly to be demonstrated than if learned by means of reasoning." A writer even in his twenties, he produced several works, four of which were published together in one volume in 1546. The last of these, *De veteribus et novis metallis*, is of interest here because it is a history of mines and metals from ancient times. The Library of Congress has *De veteribus* as it is included in Agricola's *De ortu et causis subterraneorum*. For roughly the last twenty years of his life, Agricola worked on

De re metallica, a book that was not only the first complete text on mining and metallurgy but a typographical accomplishment of the highest order as well.

Published in Basel in 1556, a year after his death, *De re metallica* contains 273 large woodcut illustrations, most of mining machines and processes, whose clarity and precision both document and instruct. Supplementing the informative text, the illustrations offer us a detailed look at the state of sixteenth-century mining, metallurgy, and its related technologies. A brief glimpse through the pictures reveals the hand tools in use, all manner of pumps, the boring of wooden pipes, construction of air shafts, and the use of treadmills, hydraulic hammers, and bucket chains, to name a few subjects. The first edition copy in the Library of Congress is in excellent condition. Given the fact that Agricola's book was the standard textbook and guide for students and experts alike for two hundred years, this is somewhat remarkable. Agricola was a wealthy man who could have lived a very different life, yet this cultured, lifelong friend of Erasmus produced one of the great works of the sixteenth century and certainly one of the first technological books of modern times. The Library of Congress also has the first English translation, done by Herbert C. Hoover and his wife, Lou Henry Hoover, in 1912. Years before he became president, Hoover had graduated as a mining engineer in Stanford University's first class, in 1895.

The last major mining or metallurgy book of this period, although not in the same class as Agricola's, was written by Lazarus Ercker, who also worked in the area of Saxony. His treatise on ores and assaying, titled *Beschreibung aller furnemisten mineralischen Ertzt unnd Bergkwercks Arten*, served as a later companion to *De re metallica* because it amplified the section on assaying. First published in Prague in 1574, this work is represented in the Library of Congress collections by the 1598 Frankfurt edition. Its illustrations show mostly mining scenes.

Like metallurgy, cooking is a kind of alchemy, and certainly a much older one. At times homely and commonplace and at other times complex and highly cultivated, cooking has a large historical body of literature. This corpus is handsomely and comprehensively represented in the Library of Congress collections primarily because of two special collections—the gastronomic libraries that once belonged to Katherine Golden Bitting and to Elizabeth Robins Pennell. Both collections have been considered in unique and erudite fashion by Leonard N. Beck in *Two "Loaf-Givers,"* a 1984 Library of Congress publication that takes the reader on a fascinating tour of the best of the collections—"the kitchen window is a good observatory from which to watch the course of history," according to Beck. Here, however, we will touch on but a few of the more interesting or significant works on cooking from the Renaissance period, including some from other collections within the Library.

The Library of Congress has in its collections what is regarded as the first printed cookbook. Bartolomeo Platina's *De honesta voluptate* is believed to have been the first, published in Rome in 1475, and the Library's copy, although dated June 13, 1475, was published in Venice and

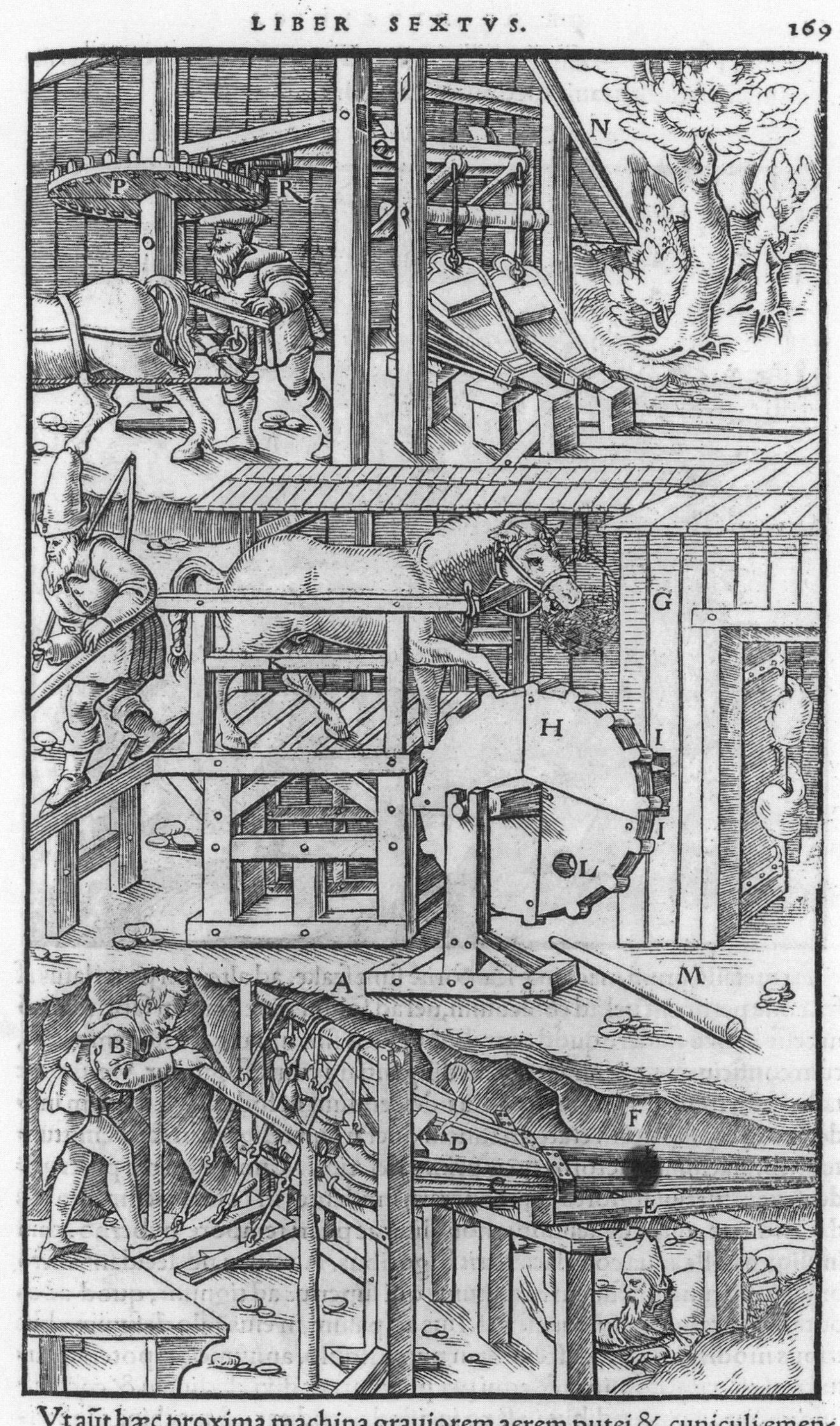
LIBER SEXTVS. 169

Vt aũt hæc proxima machina grauiorem aerem putei & cuniculi emen-
dare poteſt, ita etiam uetus ratio euentillãdi aſſiduo linteorum iactatu, quã
p explì-

Proper ventilation of mines was necessary for the health and safety of miners, and here Agricola offers three methods of renewing the stale air below ground. At bottom, a man-operated bellows draws air through the conduits E. The horse in the center turns the wheel H with his forelegs and operates a fan. At the top, the horse turns a gear wheel which operates the bellows through cams. This large work contains 273 of these informative woodcut illustrations. Its importance goes beyond its contributions to mining and metallurgy, and it can be considered one of the first technological books of modern times. *De re metallica,* 1556. Georgius Agricola. LC-USZ62-110346.

THE TRADITION OF TECHNOLOGY

is considered the second edition. This is the earliest of the five incunabular editions in the Library's collections. Platina's popular book did not originate with him, however, and he graciously acknowledges his debt to the writings of Martino of Como, "from whom I have taken nearly all of what I wrote." Maestro Martino was cook to the papal treasurer, Cardinal Ludovico Trevisan, who was also a formidable military leader. Sometime around 1450 Martino first wrote his orderly book (in reality a manuscript) called "Libro de arte coquinaria," of which only two copies are known. The Library of Congress has the earlier of these two, which is essentially individual manuscript sheets sewn together and bound between stamped calf on wooden boards. Platina himself was a humanist scholar who was twice imprisoned for his other writings and his associations. In the same year that his cookbook was published, however, he was appointed Vatican librarian by the fickle Pope Paul II.

The next earliest cookbook is an interesting assemblage of information relating to all aspects of a home or estate, including gastronomy. First published sometime before 1500, Francesco Grapaldi's *De partibus aedium* contains a glossary of terms used in building, gardening, and other domestic crafts, and takes the reader on a tour of a house, including the kitchen and dining room. The Library of Congress has the 1516 Parma edition, which is more complete than the earlier editions.

Another interesting incunabulum is the *Tractatus de vinis* of Arnold of Villanova, first published in 1478. Unlike the wide-ranging Grapaldi book, this work focuses on but one subject and is considered the earliest printed book on wine. The Library's copy is undated, but its Paris printing by Felix Baligault places it around 1500. Arnold was a thirteenth-century Spaniard to whom a large number of medical and other treatises were ascribed, and this fifteenth-century reprinting indicates that his work was regarded as a medieval classic. It is a classic in a very practical and useful sense, for *De vinis* is in all respects a medical text dealing originally with spiced or medicated wines only. Later editions, however, are embellished by the technical contributions of their commentators on how to make, preserve, and draw off wine. There is a direct connection between Arnold's prescribed "condita" and the present vermouths and other aromatic wines.

The other medical incunabulum of technical interest is the *Kleines distillierbuch* of Hieronymous Brunschwig. The Library's copy of this "small distilling book"—basically a pharmacological text containing simple remedies and recipes—is the first edition, published in Strasbourg in 1500. This volume contains a wealth of woodcuts illustrating many types of chemical apparatus, especially stills. It also contains the description of a distilling laboratory and an apothecary shop.

A collection of six individual works related to the many aspects of operating a manor was published in Paris in 1554 under the title *Praedium rusticum*. Each work in the collection was written during the 1530s by Charles Estienne, a physician, botanist, and classical scholar whose family operated a publishing house. Together, the works offered useful information primarily on the agricultural side of husbandry. Estienne later trans-

Opposite page:
A unique look behind the scenes of a Renaissance banquet is offered by Bartolomeo Scappi, Vatican cook. In several full-page illustrations, Scappi shows the superb facilities and equipment that were used in these exceptionally spacious work areas. Not surprisingly, we learn that a high degree of specialization was necessary for such large and elaborate feasts. *Opera*, ca. 1574, Bartolomeo Scappi. LC-USZ62-110347.

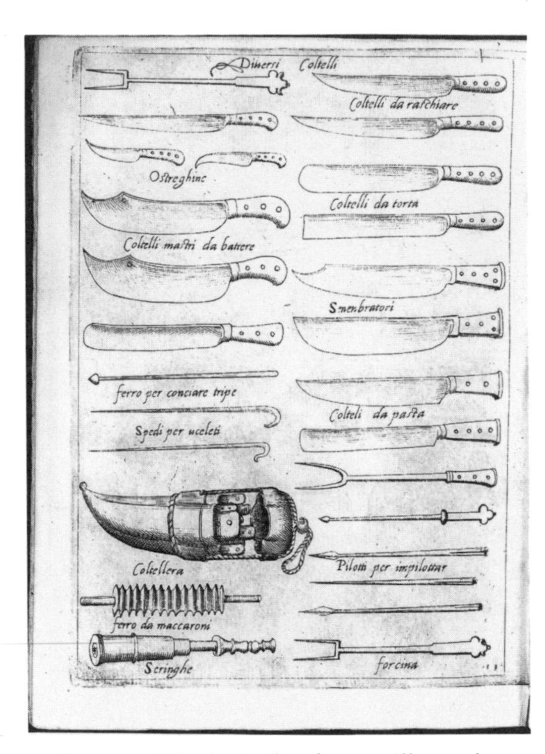

At bottom right is the first known illustration of a fork. Called a *forcina* in Italian, the implement, which evolved from the knife, originally had a single point. These early forks had two sharp, straight prongs. Although a flatter, wider version was used for serving during medieval times, forks were first used for dining in Italy during the sixteenth century. *Opera,* ca. 1574. Bartolomeo Scappi. LC-USZ62-110348.

Opposite page:
The title page of this English version of Machiavelli's work on the tactics of warfare gives some indication of the fierceness and ubiquity of armed conflict during the Renaissance. Alongside its many contributions to the art of military tactics, this book illustrates Machiavelli's blind spot with regard to the importance of artillery. In this, he reflected the general indifference to cannons of the "condottieri," or mercenary leaders who fought for the cities, republics, and princedoms of fragmented Italy. This attitude would change, and Italy would take the lead in designing new, artillery-resistant fortifications. *The Arte of Warre,* 1573. Niccolo Machiavelli. LC-USZ62-110349.

lated his work into French and enlarged the text, but he died in debtor's prison before it was published. His son-in-law eventually saw the French version to press, and its continued popularity was evidenced by the thirty-two editions known to have been published by 1702. The Library of Congress has the first edition (1554).

If a rediscovery and appreciation of classic times was essential to the Renaissance, then the greatest of the Renaissance studies of the cooking of antiquity is the *Antiquitatum convivialium* by Johann Stuck, for it considers 650 sources, each in some detail. The Library of Congress copy is the first edition, published in Zurich in 1582. Also looking back was Andrea Bacci's *De naturali vinorum historia de vinis Italiae.* Bacci was a botanist was well as doctor to Pope Sixtus V and in his book he treats all aspects of Italian wine-making in a clear scholarly manner. The Library has the only edition of this book (Rome, 1596).

The final work on cookery mentioned here contains, appropriately, several full-page illustrations of what might be called the hardware of the kitchen. Among these is the first depiction of the "forcina," the two-pronged fork used not only to aid in carving but also to bring food to the mouth. At the time of this book, the fork was being used as an eating utensil only in Italy. Written by Bartolomeo Scappi, cook to Pope Pius V, the *Opera* was first published in Venice sometime between 1570 and 1574 and contains, besides Scappi's cooking secrets, many woodcuts illustrating all kinds of kitchen furniture and utensils as well as mechanical spits and fireplaces of various shapes. Besides this first edition, the Library of Congress also has the second edition (Venice, ca. 1581), whose illustrations were made from copperplates. Both contain a double-page image showing the banquet served to the Conclave of Cardinals in 1549—an elect group that obviously ate and drank very well. This large, detailed illustration shows servants bringing the dishes and the wines to the food examiners before they are placed on the cardinals' table.

As ancient a human endeavor as the art of cookery may be, it is probably rivaled in antiquity by the technology associated with aggression and war. During the Renaissance—the age of the *condottiere* as well as the humanist scholar—the literature of warfare was vigorous if not abundant. Besides the incunabula which looked backward to classical warfare (discussed in the chapter on Greek and Roman technology), there is a body of literature which informs us primarily about the Renaissance technology of war. One of the first of these works is by the Italian statesman and political thinker, Niccolo Machiavelli. After an up-and-down career serving the political leaders of Florence, Machiavelli saw the newest ruler, Cardinal Giulio de' Medici, as the possible last chance for Italian unity and independence. He then wrote what might be called a theoretical work on warfare, which many believe laid the foundation of modern tactics. This work, *Libro della arte della guerra*, published by the famous Giunta press in Florence in 1521, eleven years before his more famous *Il Principe*, is often cited by historians as an example of an individual's inability to assess the worth and potential of a new technology. In this case, Machiavelli made much of the power of infantry and paid little notice to

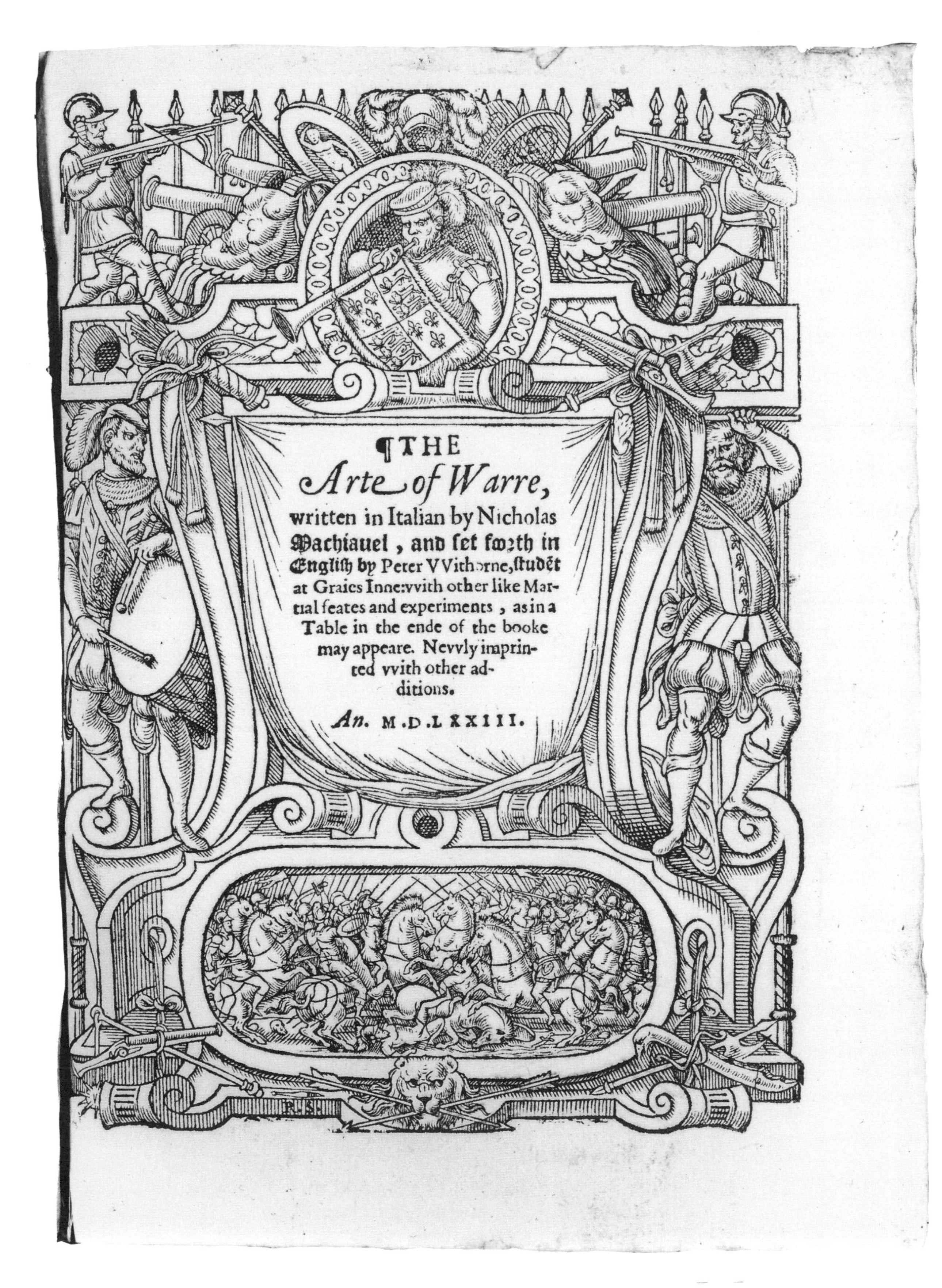

¶THE
Arte of Warre,
written in Italian by Nicholas
Machiauel, and set forth in
English by Peter VVithorne, studēt
at Graies Inne: vvith other like Mar-
tial feates and experiments, as in a
Table in the ende of the booke
may appeare. Nevvly imprinted vvith other ad-
ditions.

An. M.D.LXXIII.

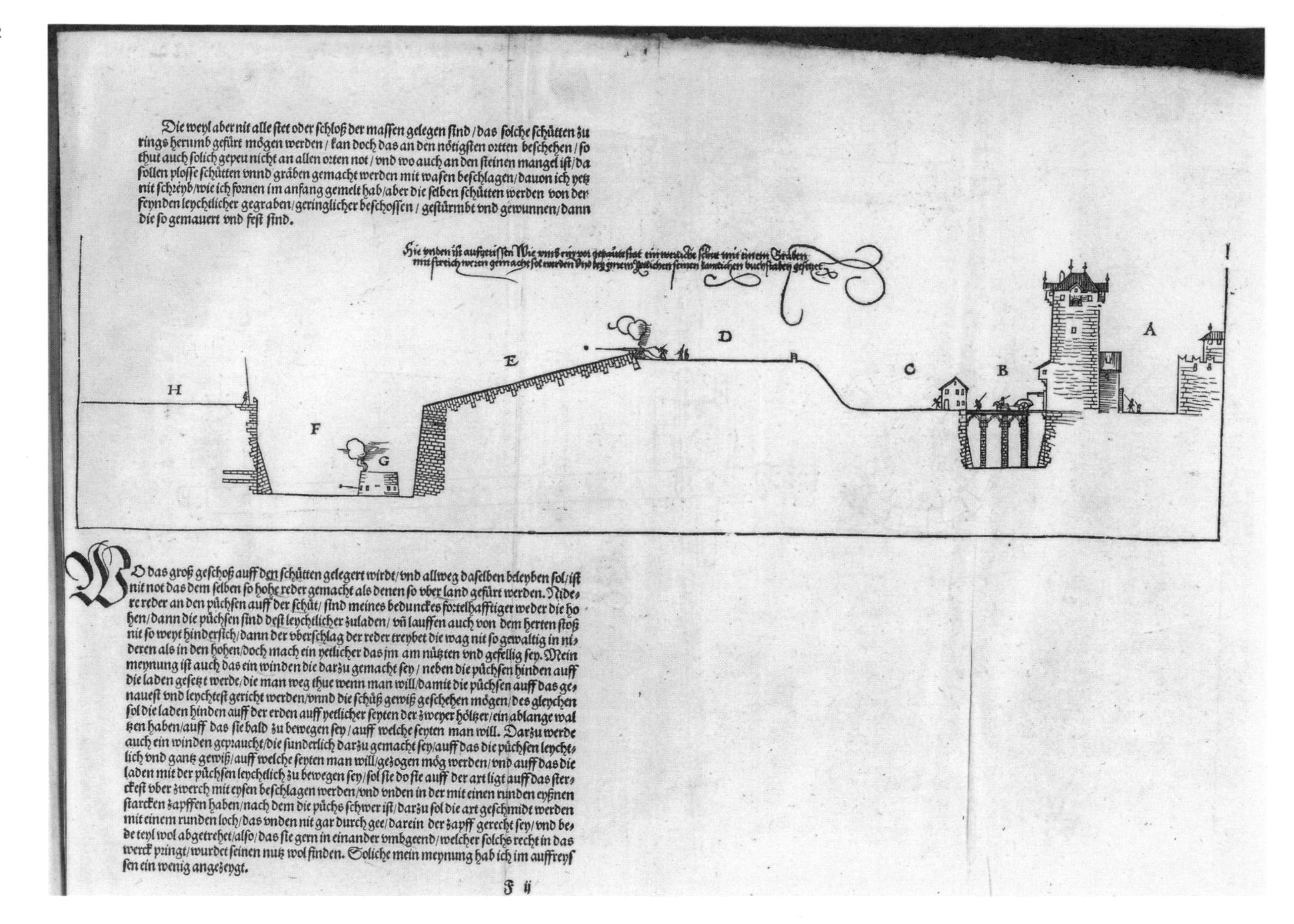

Die weyl aber nit alle stet oder schloß der massen gelegen sind / das solche schütten zu rings herumb gefürt mögen werden / kan doch das an den nötigsten ortten beschehen / so thut auch solich gepeu nicht an allen orten not / vnd wo auch an den steinen mangel ist / da sollen plosse schütten vnnd gräben gemacht werden mit wasen beschlagen / dauon ich yetz nit schreyb / wie ich fornen im anfang gemelt hab / aber die selben schütten werden von der feynden leychtlicher gegraben / geringlicher beschossen / gestürmbt vnd gewunnen / dann die so gemauert vnd fest sind.

Wo das groß geschoß auff den schütten gelegert wirdt / vnd allweg daselben beleyben sol / ist nit not das dem selben so hohe reder gemacht als denen so vber land gefürt werden. Nidere reder an den püchsen auff der schüt / sind meines bedunckes fortelhafftiger weder die hohen / dann die püchsen sind dest leychtlicher zuladen / vñ lauffen auch von dem herten stoß nit so weyt hindersich / dann der vberschlag der reder treybet die wag nit so gewaltig in nideren als in den hohen / doch mach ein yetlicher das jm am nützten vnd gefellig sey. Mein meynung ist auch das ein winden die darzu gemacht sey / neben die püchsen hinden auff die laden gesetzt werde / die man weg thue wenn man will / damit die püchsen auff das genauest vnd leychtest gericht werden / vnnd die schüß gewiß geschehen mögen / des gleychen sol die laden hinden auff der erden auff yetlicher seyten der zweyer höltzer / ein ablange waltzen haben / auff das sie bald zu bewegen sey / auff welche seyten man will. Darzu werde auch ein winden gepraucht / die sunderlich darzu gemacht sey / auff das die püchsen leychtlich vnd gantz gewiß / auff welche seyten man will / gezogen mög werden / vnd auff das die laden mit der püchsen leychtlich zu bewegen sey / sol sie do sie auff der art ligt auff das sterckest vber zwerch mit eysen beschlagen werden / vnd vnden in der mit einen runden eysnen starcken zapffen haben / nach dem die püchs schwer ist / darzu sol die art geschmidt werden mit einem runden loch / das vnden nit gar durch gee / darein der zapff gerecht sey / vnd bede teyl wol abgetrehet / also / das sie gern in einander vmbgeend / welcher solchs recht in das werck pringt / wurdet seinen nutz wol finden. Soliche mein meynung hab ich im auffreyssen ein wenig angezeygt.

F ij

This treatise on fortification was written by Albrecht Dürer, the versatile, creative German genius who is one of history's great artists. Like two other great Renaissance figures, Leonardo and Michelangelo, Dürer applied his genius to the modernization of fortifications necessitated by advances in artillery. In this side view, Dürer sketches broad, low walls, some of them sunken below ground level to minimize exposure to cannon fire. His larger drawings employed varied geometric patterns, each part of which became a separate, self-sustaining fort. Where improvements in cannon power and range had made the medieval castle vulnerable, technology now swung the balance in favor of the fort again. *Etliche underricht zu befestigung der Stett, Schloss und Flecken*, 1527. Albrecht Dürer. LC-USZ62-110350.

the role that field artillery would come to play. Despite this inability to foresee the future, Machiavelli's book was an influential work. The Library of Congress has the 1573 English version, *The Arte of Warre*, published in London and prepared in 1560 by Peter Whithorne, who added his own work, titled *Certain Waies for the Orderyng of Souldiers in Battelray*, as a second part of Machiavelli's book. Whithorne's contribution offers "new plattes for fortification of Townes" as well as other illustrations of firearms. The Machiavelli/Whithorne edition in the Library is bound with four other military tracts: Girolamo Cataneo's *Most Briefe Tables*; Vegetius's *The Foure Books . . . of Martiall Policye*; Thomas Styward's *The Pathwaie to Martiall Discipline*; and Barnabe Rich's *A Path-way to Military Practise*.

Whithorne's concern with fortification as a defense against the emerging artillery was shared by many during the sixteenth century, a time which could be called the experimental period of heavy artillery. One of the earliest non-Italian books on fortification was by the artist Albrecht Dürer. In 1527 Dürer published his *Etliche underricht zu befestigung der*

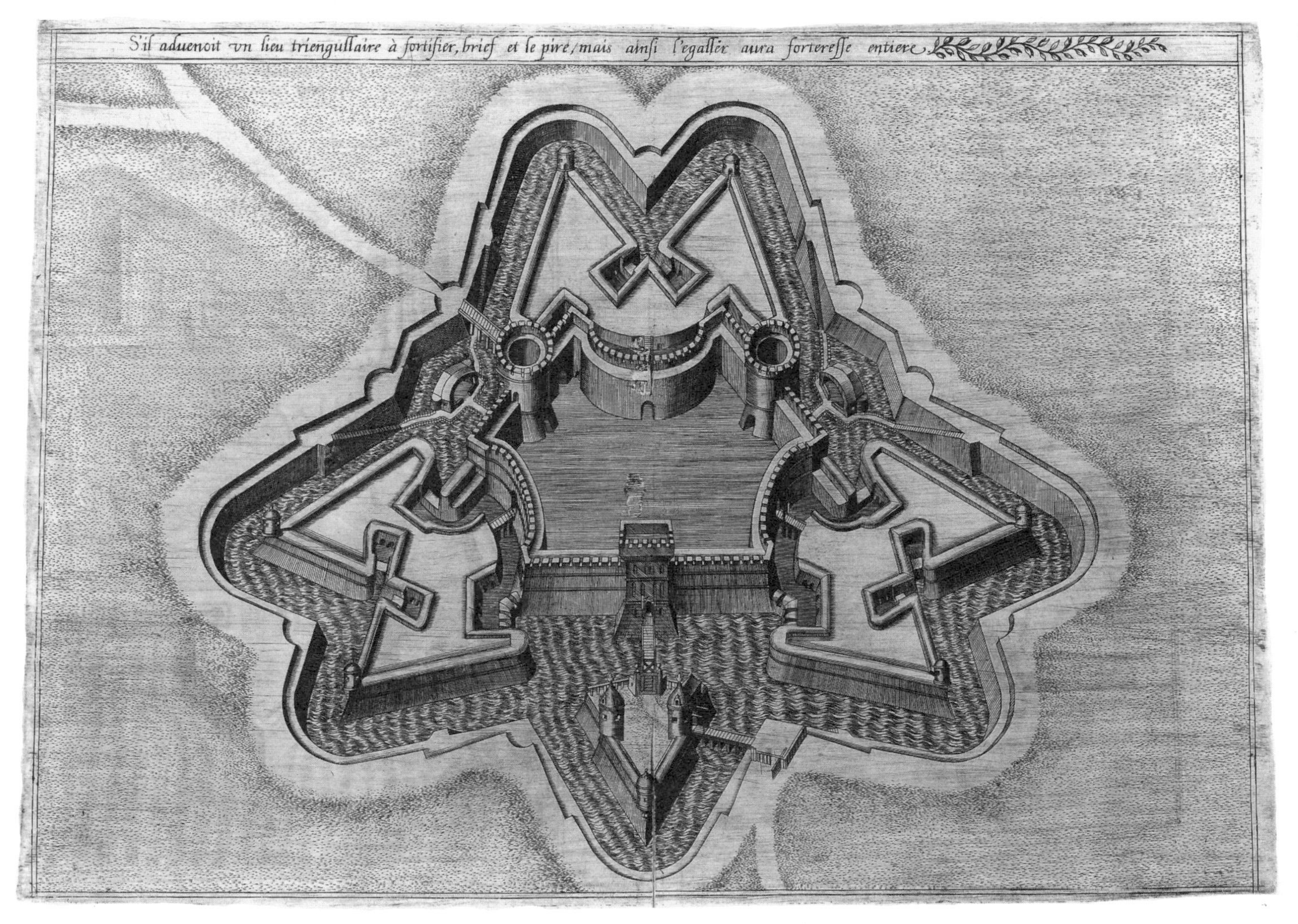

This elaborate and intricately geometrical fort is surrounded by water, adding to its defenses. By the seventeenth century, entire cities were fortified, usually with five or six bastions forming the shape of a star. One of the largest, Palmanova in Venice, was a heavily fortified polygonal town in the shape of a nine-pointed star. It still exists today. *Le Gouvernail . . . geometrie en perspective dedans l'architecture des fortifications,* 1598. Ambroise Bachot. LC-USZ62-110351.

Stett, Schloss und Flecken in Nuremberg; the Library of Congress has this in first edition. As a treatise offering instruction on the fortification of cities, castles, and towns, Dürer's plans were in reality both too massive and too costly. Despite the book's general impracticality, however, its illustrations contain many designs which were eventually implemented, such as the first practical gun casemates.

Much later, near the end of the century, two well-illustrated works on fortification were published which show significant advances. Buonaiuto Lorini was an assistant to the dean of Venetian military architects, Giulio Savorgnano. Together, they designed Palmanova—then the most powerful fortress town in all of Europe—near the eastern frontier of the Venetian republic. In 1597, Lorini put much of Savorgnano's work and plans into his *Delle fortificationi*, published in Venice. The Library of Congress has the first edition of this highly practical book. Another work of interest is Ambroise Bachot's *Le gouvernail . . . geometrie en perspective dedans l'architecture des fortifications*, published in Mehen (near Paris) in 1598. The Library of Congress has the first edition of this well-illustrated fortification book. Bachot described himself as "capitaine ingénieur du roy," but it is believed that he also was an underling of the famous military engineer Agostino Ramelli (who will later be discussed) and that he took many of Ramelli's ideas and passed them off as his own.

Fortification became a necessary concern during the sixteenth century because of regular improvements made in guns and artillery. In 1537, the first treatise on ballistics was published in Venice. Written by Nicolo Tartaglia, a mathematician with no real practical experience with cannons, *Nova scientia inventa* is of great importance to the history of firearms since it discusses various gunnery problems and experiments. The Library of Congress has the second edition, called *La nova scientia* (Venice, 1550). In 1546, Tartaglia published in Venice a larger and more ambitious work called *Quesiti et inventioni diverse*. This work, which the Library has in the 1554 Venice edition, attempted for the first time to produce a theoretical study of the motion of a projectile—in this case, the flight of cannon balls and mortar bombs. The first six parts of his book deal very specifically with such subjects as the improvement of artillery, fortification, and the line of battle. Only the seventh and eighth sections are theoretical.

At about the same time, Vannoccio Biringuccio, a Renaissance military engineer who had served the Venetian and Florentine republics and who became head of the papal foundry and director of its munitions, wrote his only book, *De la pirotechnia*. Biringuccio was as familiar with metallurgy and with making gunpowder and guns as he was with building fortifications, and his classic work gives, in particular, a full account of the manufacture of cannons. Among the eighty-three woodcuts in the book, there are several which show how to make molds for cannons as well as devices for boring cannons. The Library of Congress has the first edition of this important work, published in Venice in 1540, one year after Biringuccio's death.

The final work in this military section shows that England was not far behind the military technology of the Continent. Before he died around 1559, the mathematican Leonard Digges began writing a wide-ranging military treatise that would include, among other things, whatever arithmetic and algebra a soldier of his time might need to know. His son Thomas used his late father's notes and produced a book that was very much his own, which he called *An Arithmeticall Militare Treatise, Named Stratioticos*. Among the many military subjects he covered, Thomas attempted to examine gunnery from a mathematical perspective, much as Tartaglia had done. The Library of Congress has the first edition of this work, published in London in 1579. It does not, however, have the supplemented 1590 edition, which not only contains more on the subject of heavy artillery but also discusses the defense of England from invasion—no doubt a subject of interest just two years after the repulsion of the Spanish Armada.

It was during this final quarter of the sixteenth century that a number of books appeared that showed not only an increased technological sophistication but, perhaps even more important, a keen self awareness and an obvious appreciation of the potential of technology. Styled as a "theater of machines," a genre of technological picture books emerged which became not only very popular but extremely useful as well. These sometimes aesthetically attractive picture books of machines accomplished two things: they raised the consciousness of the generally educated reader concerning the application and potential benefits of machines, and they also spoke to the artisan who was familiar with the machines but was ignorant or unaware of the literature that offered theoretical principles upon which the machines themselves were based.

The first of these "theater" books was published around 1571 or 1572 and was the work of Jacques Besson, an Alpine Frenchman who was both a professor of mathematics and a practicing engineer—in Paris he was

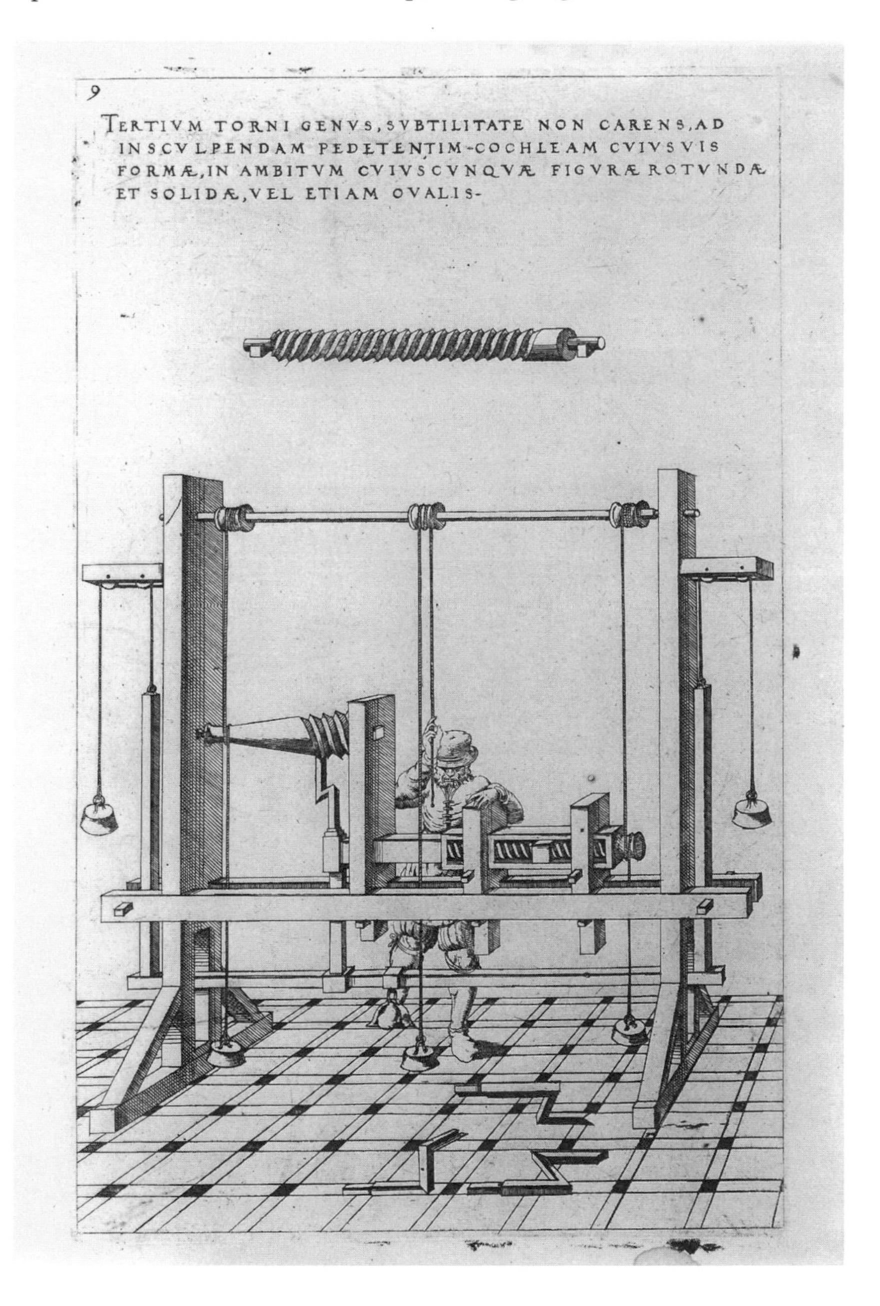

The lathe was one of the first genuine machine tools. As a turning tool which cut away unwanted material from a piece of wood (the way threads are cut), the lathe with treadle and flywheel made for greater precision cutting. Once the cutting tool could be fixed permanently and the wood turned against it, even greater precision was attainable. Here, a mounted lathe for cutting threads is shown with the lead screw at the center pulling the vertical tool that cuts the material to the worker's right. All the machine's actions are achieved by the movement of weights. *Theatrum instrumentum et machinarum,* 1578. Jacques Besson. LC-USZ62-110352.

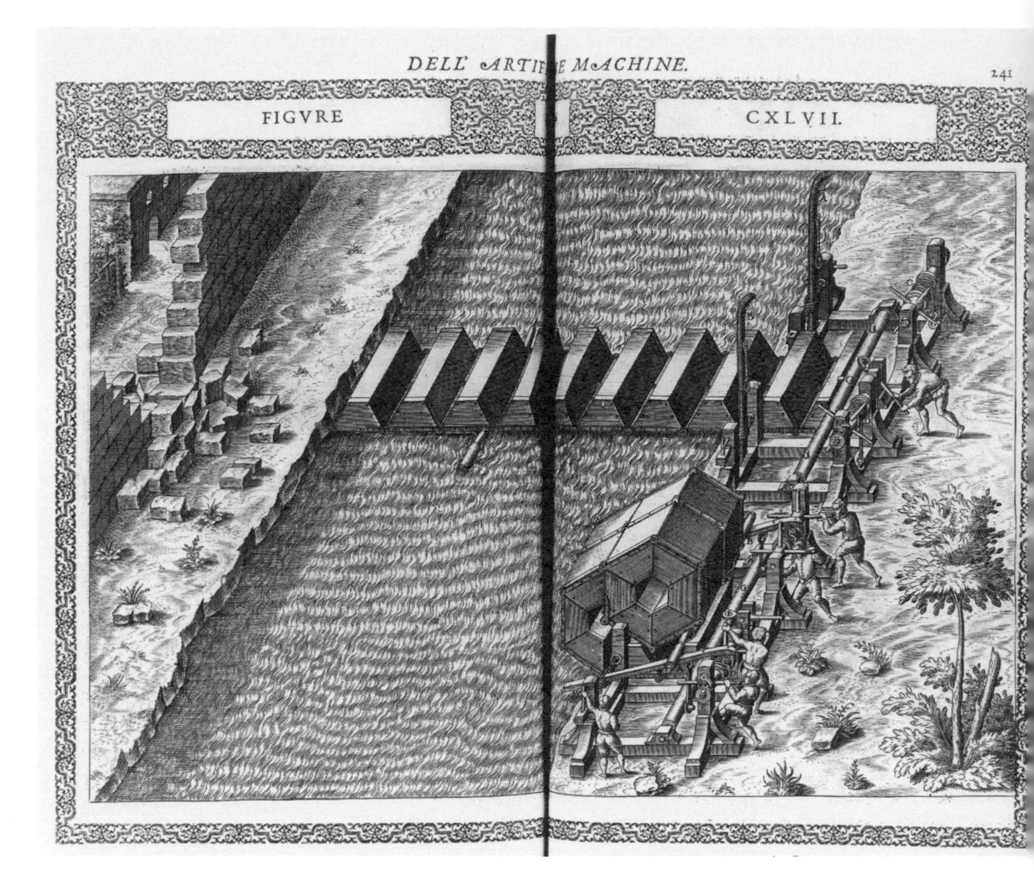

As fortifications continued to improve their defenses, engineers sought ways to overcome them. In this book of machines whose emphasis is heavily on military devices, Ramelli suggests this expanding bridge. The bottom image shows the bridge curled up and restrained by ropes. At a signal, the ropes are released and the tension springs the bridge forward, "throwing it across the moat," as seen at the top. The infantry can then hop across from section to section and enter the breach made by cannons. *Le diverse et artificiose machine,* 1588. Agostino Ramelli. LC-USZ62-110353.

"master of the king's engines." As a maker of scientific instruments and a mathematician, Besson must have been acutely aware of the growing need and demand for more accurate measuring devices. He must also have appreciated how useful the many existing illustrated books of scientific instruments were. Whatever his motivation, Besson composed the first "theater of machines" book with his *Livre des instruments mathématiques et méchaniques* (Paris, ca. 1571). This first edition being notoriously rare, the earliest edition in the Library of Congress is the 1578 Latin edition titled *Theatrum instrumentum et machinarum*, published in Lyons. Besson had died five years earlier and this new edition was put together by François Beroald, who gave it the auspicious name *Theatrum*, thus dubbing a new genre of books. The Library also has two editions in French, one published in Lyons in 1579 and one in Geneva in 1594. The sixty remarkable plates in Besson's work, engraved mostly by Jacques Androuet du Cerceau, illustrate all manner of hydraulic engines as well as various machine tools and equipment used for fighting fires and building bridges and ships.

Whereas Besson had not included in his book any machines used specifically for war, a contemporary, Agostino Ramelli, devoted nearly one-third of his much larger book to such warlike devices. Ramelli began as a professional soldier and thus always had an eye for the military application of any new invention. Little detail is known about his career, but Ramelli did call himself the "engineer of the most Christian king of France" on the frontispiece of his book. His heavily illustrated theater book, *Le diverse et artificiose machine*, has 195 handsome, full-page illustrations, each described in French and Italian. The Library of Congress has the first edition, printed in Paris in 1588. The comprehensiveness of this work, especially its treatment of any and all manner of pumps, is impressive. War machines fill about one-third of the book, reminding us again that Ramelli was primarily in the old Renaissance tradition of being a military engineer. Some have guessed at an even more intimate link with the tradition begun by Brunelleschi, since Ramelli worked for the marquis de Marignan, who may have studied under Leonardo da Vinci. Despite the fact that this is the only book Ramelli ever produced, it does not need any such pedigree to recommend it. It is the work of a professional.

The final theater book to be discussed here was published in 1607, although it was very much a book of the late sixteenth century. In fact, more than either Besson or Ramelli, the book's author, Vittorio Zonca, reflected the actual mechanical practices of his own time. His book was published posthumously in Padua by Francesco Bertelli, who mysteriously mentions that the book had "come into my hands." Unfortunately, he tells us nothing of Zonca who, it is thought, was born around 1580, making him an especially young author when he died. However, the frontispiece of his only work, *Nuovo teatro di machine et edifici*, does say that Zonca was "architect to the Magnificent Community of Padua." Compared to the works of Besson and Ramelli, Zonca's book of illustrated machines has been described as the most realistic as well as the only one that conveys any feeling of real innovation, since he often speaks of the acutal advantages realized by a particular new tool or technique. The Library of Congress has the first edition of Zonca's work, published in Padua in 1607.

Quite apart from this emerging tradition of technological theater books, there were other illustrated works whose images inform us in a less deliberate or didactic way. Typically, such works dealt with the arts and the professions and usually contained scenes depicting individuals busy at work surrounded by the telling tools of their craft. Two books in the collections of the Library of Congress, published exactly one hundred years apart, are good examples. The first is a sort of encyclopedia of human life called *Speculum vitae humanae*. The Library of Congress has the earliest of all its many editions—published in Rome in 1468, without illustrations—as well as later, illustrated editions. It was written by a learned Spanish ecclesiastic, Roderigo Sanchéz de Arévalo, known by the Latin name Rodericus Zamorensis. He was bishop of Zamora as well as later governor of Castel Santangelo in Rome and treasurer of the Holy See. Among the many woodcuts in the heavy, iron-clasped German edition (1475–78) are those showing a blacksmith at work, an undershot windmill, an abacus and geometrical compass and square in use, and details of the cloth trade. Given its early date, the book's late medieval emphasis is understandable.

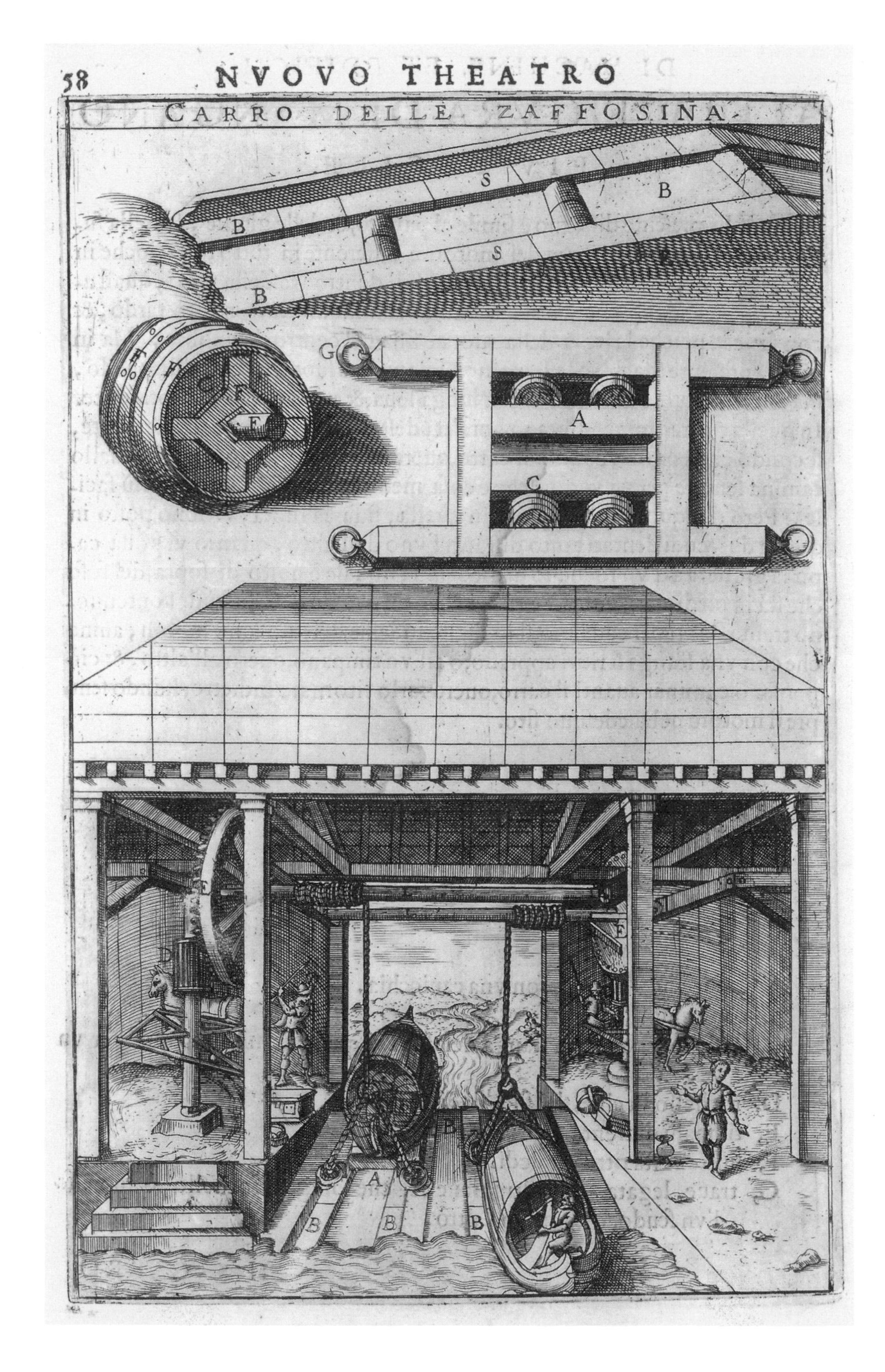
58 NVOVO THEATRO
CARRO DELLE ZAFFOSINA
B
S
S
B
B
G
A
C
A
B
B
B

A century later, a book that typified the character of the high Renaissance was published in Frankfurt with the title *Panoplia omnium illiberalium mechanicarum aut sedentariarum artium*. Written by Hartmann Schopper—about whom much less is known than about its illustrator, Jost Amman—this well-illustrated little book has been described as the first book of trades. Like the Zamorensis work a hundred years before, this book's full-page woodcuts depict the essence of a trade or craft by showing a typical practitioner busily employing his skills and naturally surrounded by the paraphernalia of his occupation. The Library's copy was published in 1568, in Latin; there is also a German version, published the same year by Hans Sachs, titled *Eygentliche Beschreibung aller Stände auff Erden*, which the Library does not have. While neither this book nor the Zamorensis compilation has the illustrative precision of the theater books, they do serve more than an evocative purpose, since each offers us real insight and hard specifics into the lesser technologies and techniques of the everday trades of this important period.

Another genre of books popular during the late Renaissance was the compilation of inventions. Although the earliest of these was probably Pliny's *Historia naturalis*, the first book devoted solely to a list of inventions was compiled by the Renaissance humanist and Vatican librarian Giovanni Tortelli. Unlike Pliny's book, which only lists the inventions and their supposed inventors, Tortelli's work provides background, commentary, and historical references in a way that gives a true feeling of actual continuity and progression. Called *Orthographia*, Tortelli's book was first published in Venice in 1471 by Nicolaus Jenson and then in Rome during the same year. The Library of Congress has both of these editions, with the huge, Venice edition the more spectacular of the two. Although virtually unillustrated, it is a beautiful production containing a large, ornamental initial and a trilateral border with a miniature in gold and colors on the first page. Tortelli organized the book alphabetically by the name of the invention (it begins with "Abacus"), and he explores each name essentially from a philological basis. Interestingly, this literary and historical inquiry into the origins of the names of "new" (post-classical) inventions renders a greater service to the history of technology than do many a text that set out to do just that.

Tortelli's book, however, has been overshadowed by another, similar work published nearly three decades later. Often mistakenly considered the first of such lists of inventions, *De inventoribus rerum* was written by an Italian cleric, Polidoro Vergilio, who served two English kings (Henry VII and Henry VIII) and thus anglicized his name to Polydore Vergil. *De inventoribus rerum* was first published in Venice in 1499. The earliest Latin version in the collections of the Library of Congress is the 1512 Strasbourg edition; the earliest English edition in the Library was published in London in 1546. Vergil by no means restricted himself to things mechanical or technical in his book, for he only considers them in his last chapter. What he does include there and the order in which he does it together make a good case for the argument that he owed much to Tortelli's earlier work. Another work that may have served in some way as a

Calcearius. Der Schuwmacher. Cordomer

QVi ventosa regunt hominum vestigia paßim,
Calceolos docta consuo sutor acu.
Glorior humano quicquid de semine cretum
Esse vides, artis semper egere meæ?

Me fine per glaciem quis rusticus ire niualem
Audeat, intactas aut violare niues.
Arte laboratis peronibus vtitur omnis
Qui sapit, immensas & parat ire vias.
Nam merito glacies illi secat aspera plantas,
Quisquis opem duro respuit ore meam

Pilea-

Even the humble trade of shoemaking had its own tools and technology, and during the Renaissance shoemakers' guilds flourished. Once the medieval style of an exaggerated pointed toe had passed by the end of the fifteenth century, shoes became more rounded, and the broad-toed shoes seen hanging in this picture became the norm. *Panoplia omnium illiberalium mechanicarum avt sedentarium artium,* 1568. Hartmann Schopper. LC-USZ62-110355.

Opposite page:
Slipways, or inclined surfaces over which boats were hauled from one body of water to another, were not unusual during the Middle Ages. The suggestion put forth in this illustration, however, marks a significant step forward in transportation. The main innovation here is the boat-carrying carriage or sled whose wheels or rollers move through tracks. In principle, it is a type of railway. Small boats could move in both directions, says Zonca, in this manner: a boat is pulled up by horse-winch until its keel rests in a groove in the sled (on rollers), which is then pulled across tracks to the other side. *Nuovo teatro di machine et edifici,* 1607. Vittorio Zonca. LC-USZ62-110354.

model for Vergil was the mid-sixteenth-century publication of a two-hundred-year-old manuscript that was written by a friend of Petrarch, Guglielmo da Pastrengo. Although primarily bibliographical and in some ways a type of encyclopedia, it does have a section on men who accomplished or created famous firsts. The Library of Congress has a copy of Pastrengo's work, titled *De originibus rerum* (Venice, 1547).

Up to 1600, two general works stand out as being not only fascinating storehouses of information—some of it pertinent to the history of technology—but also especially characteristic of the Reniassance and its interest in all things natural and man-made. The earliest of these, Gregor Reisch's *Margarita philosophica*, has simple woodcuts that evoke times more medieval than Renaissance. Despite this, Reisch's attempt to cram everything from grammar and rhetoric to music, geometry, and astronomy into his encyclopedia, along with details on mining, geology, anatomy, and of course philosophy, evinces a Renaissance fervor for gaining knowledge about anything and everything. The Library of Congress has several editions of Reisch's encyclopedia, with the first edition (Fribourg, 1503) one of the best. In addition to several full-page woodcuts which often contain telling details, the book's chapters on geometry, astronomy, and music offer the best images from a technological point of view, especially of devices and instruments.

The second general work, Giovanni Battista della Porta's *Magiae naturalis*, typifies all the headiness and the enthusiastic spirit of inquiry of the High Renaissance. Where Gregor Reisch was unquestionably a man of the fifteenth century, Porta not only lived much later (he died in 1615) but was also by nature a most forward-looking individual. Having been raised by a worldly and wise uncle, Porta had composed essays in both Latin and Italian by the age of ten and made the grand tour of Italy, France, and Spain with his brother. *Magiae naturalis* was first printed in Naples in 1558 and was his first published work. It contained only four books and discussed as many half truths and marvels as it did demonstrable facts. The Library of Congress has the 1561 Antwerp printing of this rare, early version. Thirty years after its initial printing, Porta issued a much expanded version—twenty books—of his first effort, and the work's inclusion of new, decidedly scientific material shows his professional growth. The Library also has this fuller version, which was published in Naples in 1589. Of particular interest here are its sections on metals and metallurgy, cookery, hunting, magnets, and a clear description of the camera obscura. Porta's book may not have been entirely his, and is possibly the product of what is now recognized as a precursor to our contemporary scientific societies. In Porta's Naples home a group of *otiosi* (men of leisure) met and formed the Accademia Secretorum Naturae. The condition of membership was that each man must have contributed a new fact or discovery in the field of natural science. Although this club did not exist very long (in fact, Porta was called to Rome by Pope Paul V to explain away charges that the group was dabbling in the occult sciences), it was eventually succeeded by the Accademia dei Lincei, among whose members were both Porta and Galileo. Porta's expanded

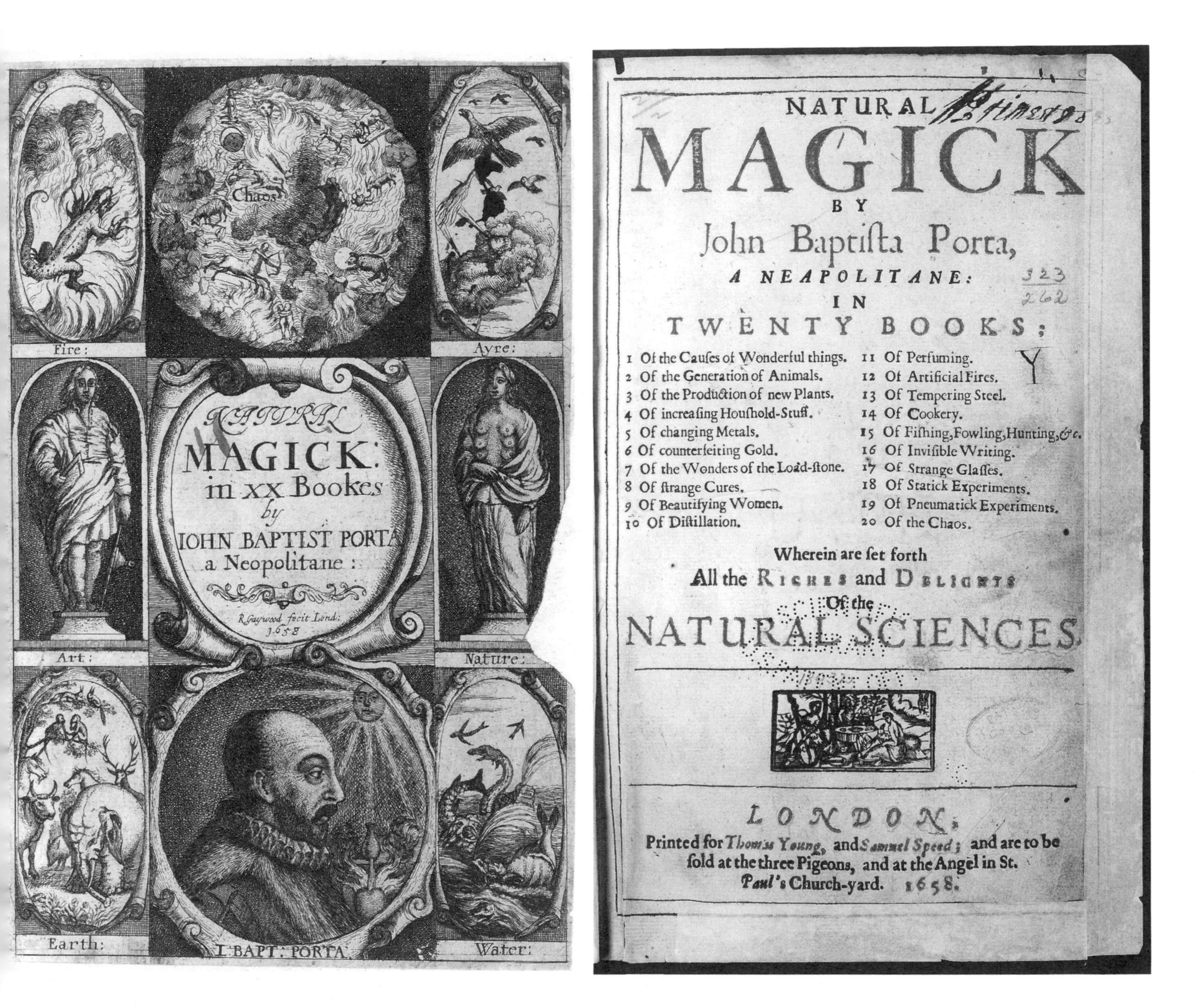

NATURAL MAGICK: in XX Bookes by IOHN BAPTIST PORTA a Neopolitane:

R Gaywood fecit Lond: 1658

Fire: Ayre: Art: Nature: Earth: Water: Chaos

I. BAPT: PORTA

NATURAL

MAGICK

BY

John Baptiſta Porta,

A NEAPOLITANE:

IN

TWENTY BOOKS;

1 Of the Cauſes of Wonderful things.
2 Of the Generation of Animals.
3 Of the Production of new Plants.
4 Of increaſing Houſhold-Stuff.
5 Of changing Metals.
6 Of counterfeiting Gold.
7 Of the Wonders of the Load-ſtone.
8 Of ſtrange Cures.
9 Of Beautifying Women.
10 Of Diſtillation.
11 Of Perfuming.
12 Of Artificial Fires.
13 Of Tempering Steel.
14 Of Cookery.
15 Of Fiſhing, Fowling, Hunting, &c.
16 Of Inviſible Writing.
17 Of Strange Glaſſes.
18 Of Statick Experiments.
19 Of Pneumatick Experiments.
20 Of the Chaos.

Wherein are ſet forth
All the RICHES and DELIGHTS
Of the
NATURAL SCIENCES.

LONDON,
Printed for *Thomas Young*, and *Samuel Speed*; and are to be ſold at the three Pigeons, and at the Angel in St. *Paul's* Church-yard. 1658.

The title page of this English translation of a work done one hundred years earlier captures explicitly the transformation that was taking place in both technology and science. Its author, the confident Neapolitan Giovanni Battista della Porta, was a prototype of what we now call a Renaissance man. The twenty separate subjects listed in the book's contents give some idea of the variety of his interests. What is significant about the translated title page, however, is its explicit definition of "natural magick" as now being "natural science." What was deemed magical during the Renaissance, no matter how natural or understandable, was now being termed scientific. *Natural Magick,* 1658. Giovanni Battista della Porta. LC-USZ62-110356.

The order of Pope Sixtus V to relocate this 368-ton obelisk from its ancient position in the Circus Nero to the Piazza of Saint Peter's can be considered the height of Renaissance arrogance. But it can also exemplify the absolute belief in things human and the firm will that all things imaginable are indeed possible. In actuality, the feat proved eminently doable. The Vatican architect, Fontana, first erected a tower around it, lowered the obelisk onto a sled with wheels, pulled this 260 yards across an elevated railway, and then raised the obelisk once again in its temporary tower to stand at its new location. It took one year. *Della trasportatione dell'obelisco Vaticano,* 1590. Domenico Fontana. LC-USZ62-110358 through LC-USZ62-110362.

version of his book in 1589 was immensely popular and went through many later editions in Latin, Italian, French, and German. The first English edition, titled *Natural Magick* (London, 1658), is in the Library of Congress. A perusal of the titles of its twenty books indicates the limitless vision of Porta and his group, although some might call it indiscriminate: Of Artificial Fires, Of Pneumatick Experiments, Of the Wonders of the Loadstone, and Of Beautifying Women. Following the listing of these and other subjects on the title page is a statement which aptly sums up the attitude of these Renaissance inquirers into both science and technology: "Wherein are set forth All the Riches and Delights of the Natural Sciences." This feeling and belief that by learning about the natural world one would experience "riches and delights" is perhaps one of the more estimable attitudes of the Renaissance. It is perhaps not exaggerating to say that virtually all our great scientists and inventors to this point in time have, at least at the beginning of their careers, experienced this intangible and probably ineffable reward.

The final book in this chapter on Renaissance technology could have been included in the earlier discussion of architecture or engineering but was reserved to close this chapter for many reasons. First, its subject is a great technological achievement and as such, it is an expression of Renaissance pride, ambition, and ingenuity. Not only does the book offer us an excellent picture of the technology and methods of its time with a wealth of detail but it further informs us of the prevailing culture and

customs of the late sixteenth century. The technological undertaking it documents was the technical event of its century and came to symbolize the growing ability of man to manipulate his landscape. Whether as an engineering feat, a symbol of Renaissance energy, or a sign of things to come, the physical relocation of the stone obelisk, ninety-two feet high, from the Circus of Nero to where it stands today in the center of the Piazza of St. Peter in Rome is perhaps the consummate act of this entire exciting period of technological history.

The architect who masterminded the overall relocation plan and laid out even the smallest details was Domenico Fontana. To commemorate this event and to preserve the record, Fontana wrote *Della trasportatione dell' obelisco Vaticano*, a monumental work published in Rome in 1590. The Library of Congress has Fontana's work in first edition. Among its thirty-eight plates are images that not only convey the engineering details and the level of technological complexity but also capture the essence of a stirring event. These magnificent engravings, by Bonifacio Sabenico, present almost a tableau vivant of all the stages of operation. The pope who gave his authority to Fontana was the same man who would change Rome into a beautiful baroque city. In his five years as pope, Sixtus V rebuilt the Lateran, Quirinal, and Vatican palaces, completed the great dome of St. Peter's Cathedral, and laid out the streets to converge on the great basilicas, transforming Rome from a medieval city to a thriving capital.

The technological challenge that Sixtus V laid before Fontana was this: remove a ninety-two-foot stone obelisk weighing 340 tons that had stood in place over fifteen hundred years and reestablish it, undamaged, in a new location 260 yards away. Fontana responded with a carefully thought-out plan to encase the entire monument in a wooden frame, build a triangular tower around it to raise it up, and then lower it horizontally onto a rolling platform. On this it would be drawn to the newly prepared site, the tower rebuilt, the stone obelisk raised vertically, and set into place. The technology available to Fontana was little more than what Archimedes might have used—a complex combination of pulleys and windlasses operated by nine hundred men and seventy-four horses. By the time the sheathed obelisk was ready to be lifted, it weighed about 368 tons. To transport it on its rolling carriage, a raised earthen causeway retained at the sides by timbers was built connecting the two sites. The entire project took one year, from September 21, 1585, when the receiving pit began to be dug. The tower was completed on April 28, 1586, and the raising began two days later. By May 7, the lowering onto the rolling platform had begun. The transport and reconstruction of the tower took all summer, and raising started on September 10, 1586. On September 28, 1586, the secured, standing obelisk was ritually exorcised of its pagan past, purified, consecrated, and topped with a cross.

This reference to the monument's past allows us to put into perspective just how far (or not so far) technology had advanced since ancient times. This great stone ornament was cut during the reign of Noncoreo, king of Egypt, during the tenth century B.C. Nearly a millenium later, it was transported from the city of Heliopolis (an ancient religious site called "pillar city," now a suburb of Cairo) to Rome in A.D. 41 by orders of the Emperor Caligula. This entailed a land journey to the nearby Nile River, a trip by barge up the river to the Mediterranean, and a sea voyage of half the length of the Mediterranean to reach Rome, where the land transport began again. In this manner, at least twelve such obelisks were transported to Rome by various Roman emperors. In this light, Fontana's dramatic accomplishment, which held all Rome spellbound, should not be diminished but should rather be regarded as an act of reconnecting with the technical past, reestablishing the continuum of technological history, and thus reaffirming that indeed, a real renaissance was taking place in the West.

Renaissance technology, like everything else about this romantic period of revival, is simply bursting with vigor, excitement, and creativity. Following upon the plodding gradualism of medieval times and the successive shocks of famine, pestilence, and war, the Renaissance technological tradition came to embody the new and closely embraced idea of humanism. Celebrating all that was new, practical, and worldly, this revivifying spirit gave technology a solid, human-centered rationale as well as a knowable, achievable goal. Renaissance engineers consequently strove to work and improve on the present, discarding the weighty and lugubrious medieval attitudes that considered the hereafter as much as the present. Disdaining the abstract and the impractical, they can be said to have laid the founding stones for the beginnings of the modern technological tradition which might be described, in contrast to medieval empirical groping, as guided or informed groping.

Besides its foundation of humanism, the technological tradition of the Renaissance is characterized by change—constant and pervasive alteration and improvement made for its own sake. This, along with a fledgling respect for the need for incrementalism and piecemeal experimentation, set the stage for the scientific revolution to follow.

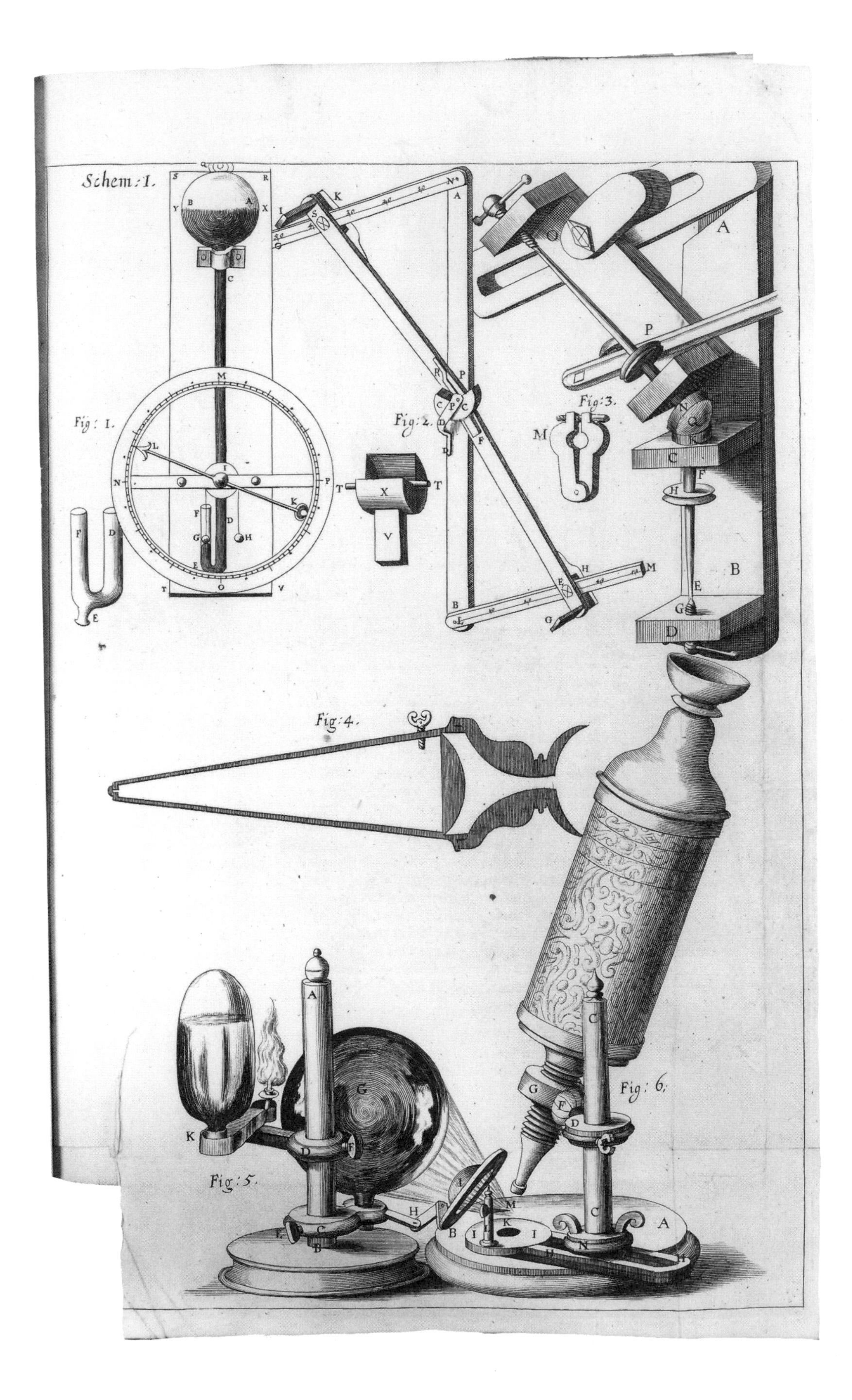
Schem: I.
Fig: 1.
Fig: 2.
Fig: 3.
Fig: 4.
Fig: 5.
Fig: 6.

4. *Technology and the Scientific Revolution*

Technology participated in the scientific revolution most importantly by coming to embrace its essential idea of a knowable universe. The new science told technology that the world was not unpredictable and chaotic but functioned according to knowable laws and contained calculable forces. In another sense, science moved somewhat closer to technology by seeking the latter's more limited and soluble "how" of a problem and departed from its exclusive search for the more complicated and elusive "why." It was thus during this time that science began its move toward a more patient, positivc, and ultimately productive incrementalism, and began a real methodological and conceptual revolution. Increasingly but still gradually, science and technology would come to inform each other.

If there were one technological achievement during this period that could be considered the most significant, it would have to be the development of the steam engine. Knowledge of the motive power of steam goes as far back as Hero of Alexandria (ca. A.D. 50), whose scientific novelties operated by using steam power (see chapter 1). It was not until the twelfth century that another reference to "heated water" is found, and still another three centuries would pass until Leonardo da Vinci described what might be called a steam gun. During the mid-sixteenth century, the polymath Girolamo Cardano wrote of steam power and how condensing steam can produce a vacuum. The Library of Congress has his *De rerum varietate* (Basel, 1557) which contains this description. The theater machine books of his contemporaries Jacques Besson and Agostino Ramelli (see chapter 3) also contain references to steam power. The Library does not have *Pneumaticum* (Naples, 1601) in which the ever-curious Giovanni Baptista Della Porta describes a machine for raising a column of water using steam pressure, but it does have the 1605 Paris edition of *Les Elemens de l'artillerie* in which David Rivault Flurance describes how heated water enclosed in a bombshell would burst it, no matter how thick the shell walls. Ten years later, the French engineer and physicist Solomon de Caus published his *Les Raisons des forces mouvantes* (Frankfurt, 1615), which also described raising water by steam. All of these accomplishments were somewhat preliminary to what was about to happen.

By the middle of the seventeenth century, many of the best minds were dealing with the problems surrounding the fascinating properties of air, and by 1654 Otto von Guericke had developed the vacuum pump. Shortly thereafter, Robert Boyle, Robert Hooke, and Christiaan Huygens were all experimenting with different aspects of air and air pressure, and while these three each went off in a different experimental direction, it

Opposite page:
The beginnings of a real connection between technology and science occurred as the assumptions of the revolution going on in science—that the world is understandable and its physical laws are knowable—informed technology and offered it new, specialized, and more focused goals and methods. In turn, technological advances were taken up by science, which put these new tools to good use. The invention of the microscope perhaps best exemplifies how science increasingly would come to benefit from technological advances as an entire new, miniature world began to open up. Here in this first treatise devoted entirely to microscopy is Robert Hooke's improved compound microscope. It has a hemispherical lens objective and a plano-convex lens eyepiece. *Micrographia; or, Some physiological descriptions of minute bodies made by magnifying glasses,* 1665. Robert Hooke. LC-USZ62-110443.

The early use of the power of steam in the seventeenth century evokes the playful plans of the Greek, Hero of Alexandria. This design for a garden statue that "shall send forth a Sound when the Sun shineth upon it" indicates a direct connection with Hero and shows how fascinated by automata were the men of the scientific revolution. Caus was an engineer and architect from Normandy who worked for the British royal family. Some of his designs found expression in the garden at Wadham College. *Les Raisons des forces mouvantes,* 1615. Solomon de Caus. LC-USZ62-110444.

was an assistant of Huygens's, Denis Papin, who offered the next major advance toward a real steam engine.

After working with Huygens in Paris, Papin went to London and established himself with Robert Boyle. He also worked for Robert Hooke, who was the Royal Society's "curator of experiments." It was during these five years in London that Papin made his most significant contribution to the development of steam power. He devised a working "steam digester," or pressure cooker, that had a safety valve should the internal pressure get too high. Papin used his invention not only to show how it could soften the hardest of bones but also to cook some excellent meals. Papin documented his invention in a 1681 London publication, *A New Digester or Engine for Softening Bones*. The Library of Congress has the first edition of this work. As a French Huguenot, Papin found the France of Louis XIV a difficult place to live and he never returned after 1680, living in Italy and later in Germany. It was during his years as a professor of mathematics at the University of Marburg that he demonstrated how a piston could be moved upward by low pressure steam and returned downward by atmospheric pressure. He published his findings in a 1690 article titled "Nova methodus ad vires motrices validissimas levi pretio comparandas." The Library has this article in a facsimile of Leibniz's journal, *Acta Eruditorum*. Papin spent his last years in England, but his friends Huygens, Boyle, and Hooke were all dead, and he himself died penniless and unknown. Although Papin had conceived the basic design for a steam engine (steam driving a piston in a cylinder), he never built an engine himself.

In the application of this emerging technical understanding to a practical need, one major technological problem presented itself to everyone concerned, in England at least. Since the British countryside had been deforested to the point where the nation relied increasingly on its plentiful deposits of coal, the efficient extraction of that resource was of utmost importance. The problem that plagued the mining industry, however, was that water would constantly seep into the mines and collect, requiring its removal before any coal could be mined. To this very real problem an industrious military engineer, Thomas Savery, brought his skills. On July 25, 1698, Savery's successful patent for a steam-actuated pump initiated the modern usage of steam power. During the following year, he demonstrated a working model of his steam engine to the Royal Society, which published his article and made him a member. The Library of Congress has Savery's article, "An Account of Mr. Tho. Savery's Engine for Raising Water by the Help of Fire," as it was first published in the Society's *Philosophical Transactions* in June 1699.

Savery called his machine the "miner's friend," but in fact it was inefficient, expensive, and very dangerous. It operated using steam at high pressure with no safety valve; it had to be placed very close to the level of the water to be pumped; and the high temperatures it reached caused its solder to melt and joints to split. Despite the fact that it was actually little used to pump water, Savery's device was a landmark achievement in the conversion of energy. As the first real, useful energy conversion to go

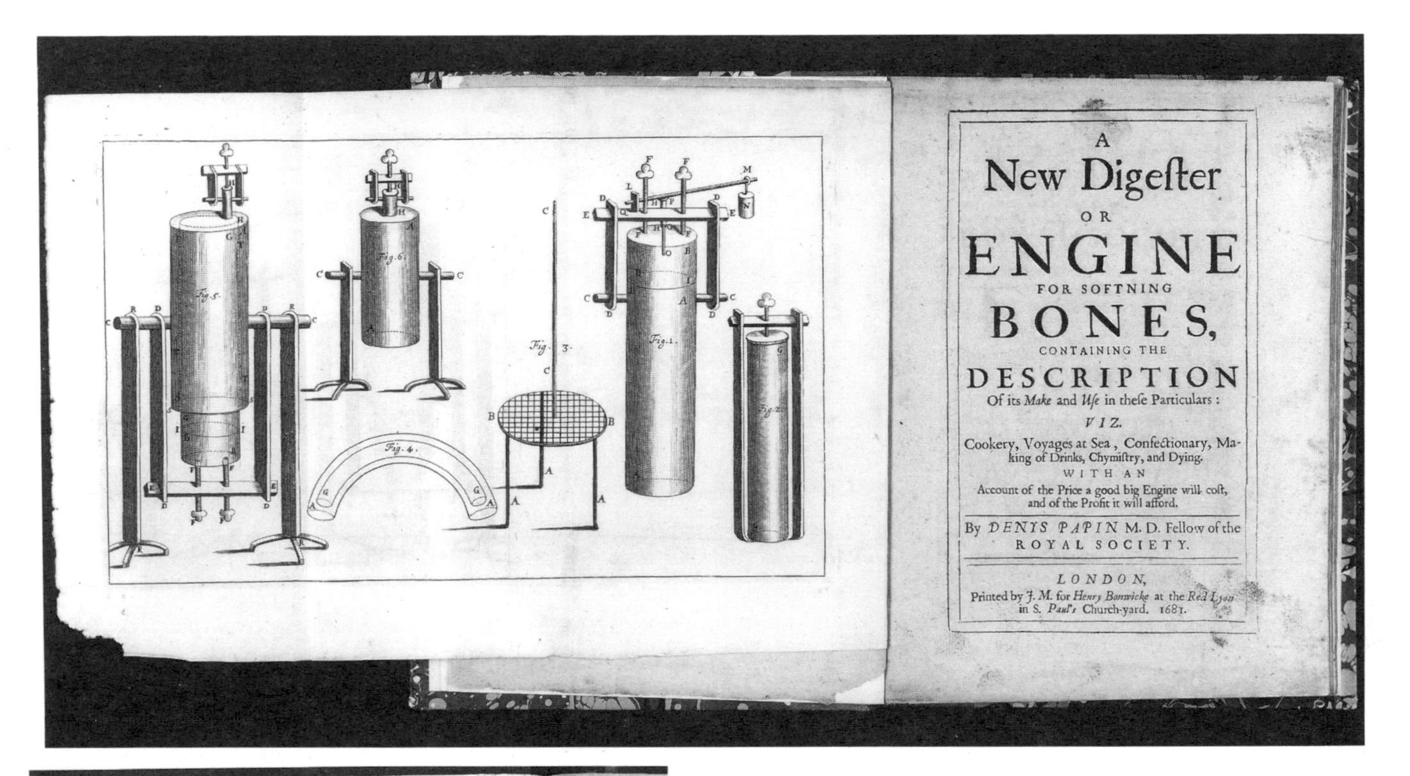
A
New Digeſter
OR
ENGINE
FOR SOFTNING
BONES,
CONTAINING THE
DESCRIPTION
Of its *Make* and *Uſe* in theſe Particulars:
VIZ.
Cookery, Voyages at Sea, Confectionary, Making of Drinks, Chymiſtry, and Dying.
WITH AN
Account of the Price a good big Engine will coſt, and of the Profit it will afford.
By *DENYS PAPIN* M. D. Fellow of the ROYAL SOCIETY.
LONDON,
Printed by *J. M.* for *Henry Bonwicke* at the *Red Lyon* in S. *Paul's* Church-yard. 1681.

The first suggestion that steam could drive a piston in a cylinder was made in a book proclaiming the invention of a pressure cooker (steam digester). Papin had worked with Christiaan Huygens and Robert Boyle, and although he made a basic design for a steam engine, he never built one of his own. He did, however, prepare meals with his invention for both the Royal Society and King Charles II. *A New Digester or Engine for Softening Bones,* 1681. Denis Papin. LC-USZ62-110445.

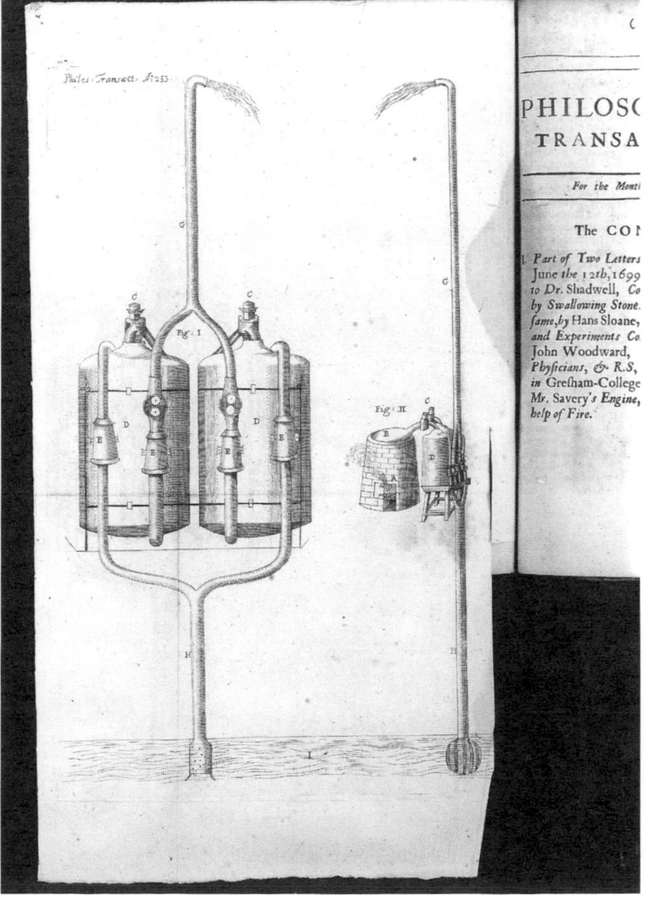

The modern use of steam power began with Thomas Savery's steam pump, which he called the "miner's friend." This drawing shows how his condensing type engine was supposed to draw water out of flooded mines. Two receivers were alternately filled with steam which, when condensed, created a vacuum and thus "drew" water up and out the connected pipes. Savery's high pressure pump was dangerous and inefficient, but it was the first practical conversion of energy to go beyond the traditional water wheel or windmill. "An Account of Mr. Tho. Savery's Engine for Raising Water by the Help of Fire," *Philosophical Transactions,* 1699. Thomas Savery. LC-USZ62-110446.

beyond that of the traditional water wheel and windmill, power production by steam offered the essential precondition for the technological revolution that was to come. This new ability to tap sources of inanimate energy and transform them into mechanical energy would cause a profound and irrevocable break with the past.

This coming break with the past was given great impetus by the ingenuity and doggedness of a little-educated English ironmonger named Thomas Newcomen and his friend, a plumber named John Calley. Together they were to transform Savery's difficult engine into one that was embraced at home and abroad. Writing in 1824 on the history of steam power, the historian Robert Stuart Meikleham said that "science . . . never had anything to do in the matter. Indeed there is no machine or mechanism in which the little that theorists have done is more useless. It arose, was improved and perfected by working mechanics—and by them only." While this may overstate the case somewhat, Newcomen and Calley may be the best examples of how technology can progress with little or no help from science. Indifferent to theory, they experimented for years on Savery's crude pump and eventually produced a totally different engine that was an ingenious assemblage of existing elements and techniques. Their low steam pressure, valved engine was cheap and safe, though inefficient, and, perhaps most important, it had moving parts that could transfer their motion in order to work other machinery. As with many craftsmen or artisans, neither man preserved a record of his experimental work or published a word.

If the harnessing of steam power that began during this period signalled the real beginning of man's ability to reshape his world, the invention of the telescope gave him a tool to learn more about that world and its relation to other worlds. Like the story of steam, the invention of the telescope had little to do with science. As the ancient Greeks and medieval Arabs were well acquainted with the motive power of steam, so they also were well aware of the magnifying property of glass and polished stones. By 1600, the art of polishing glass was flourishing in the Netherlands, where lens-making for spectacles was an established industry. It was about this time that the discovery was made that the alignment of two of these lenses could make distant objects seem closer.

Who did this first has always been a point of contention, with three Dutch spectacle-makers staking the best claims. Besides James Metius and Zacharias Jansen, Hans Lippershey is acknowledged by most because of the priority of his patent application made on October 2, 1608. Ironically, the States-General, the governing body of the Netherlands, attempted to keep Lippershey's invention a secret since it was then at war with Philip II of Spain. Lippershey was given a money award but no patent and was required not to divulge anything about his invention. As might be expected, news of such a device spread so fast that telescopes with ordinary spectacle lenses were on sale in Paris in the spring of 1609. In August of that same year, Galileo Galilei presented to the doge of Venice a nine-power telescope that could identify ships at sea two hours before they could be seen by the naked eye of trained observers. By the

In the frontispiece to Galileo's collected works published in 1655 (thirteen years after his death), the great man is seen on his knees, humbly offering his telescope. The seated, crowned woman in the center may be Urania, one of the nine Muses of the Greeks and the patron of astronomy. Galileo's left hand points to the heavens, where some of his discoveries—Jupiter's moons, Saturn's "ears" or rings, the irregular shape of our Moon, and the existence of sunspots—are seen. The lenses used by Galileo for his tubelike telescope had a necessarily weak convex lens for the objective or front lens, combined with a strong concave lens. Only this unlikely combination placed at opposite ends would magnify a distant object. Such lenses were not plentiful during Galileo's time, since it was the exact opposite combination—strong convex and weak concave lenses—that were used in correcting vision problems. *Opere di Galileo Galilei,* 1655. Galileo Galilei. LC-USZ62-110447.

next year, Galileo had adapted the device to science and scanned the moon's craters and discovered the moons of Jupiter. Although he had never seen a telescope, Galileo was able to make one and then improve upon it simply because he was given news of its existence. Because of his insight, opportunism, and effective scientific use, what had been called the "Dutch trunk perspective" or "cylinder" became known after 1611 as the Galilean telescope.

Less than a year after he first heard a rumor about the Dutch invention, Galileo published, in March 1610, a small book in which he documented his astronomical discoveries. So brief as to be more of an announcement than a full disclosure of his findings, the book was called *Sidereus nuncius,* which translates into the poetical *Starry Messenger*. In this twenty-four-page book, which he astutely dedicated to Cosimo II de' Medici, the fourth Grand Duke of Tuscany, Galileo barely discussed the invention that enabled his discoveries. He offered only one small proportional diagram which illustrates how a telescope multiplies the size of objects at a distance, and barely two pages to a physical description of what he called a "perspicillum," or spyglass. After briefly telling of convex and concave lenses aligned in a lead tube, he says impatiently, "Let it suffice for the present, however, to have touched on this so lightly and to have, so to speak, tasted it only with our lips, for on another occasion we shall publish a complete theory of this instrument." From there he rushes enthusiastically toward a description of what he obviously considered to be his real achievement—his astronomical discoveries. As a scientist, Galileo knew even at this early stage that the implications of what he was seeing in the heavens would usher in a new age and profoundly disturb the comfortable notions of the old one. *Sidereus nuncius* is a rare piece and is not in the collections of the Library of Congress. It is, however, found in the Library's copy of *Opere di Galileo Galilei*, published in 1655 in Bologna.

Galileo sent his *Starry Messenger* to his northern contemporary, Johann Kepler, who in 1610 held the badly paying but impressive-sounding position of imperial mathematician of the Holy Roman Empire. Kepler had been working on vision and optics ever since he collaborated with Tycho Brahe, and when he received Galileo's book (and more importantly, obtained a telescope for himself so he could see Jupiter's moons firsthand), he promptly applied his optical theories to explain how a telescope works. In his major work on optics, *Dioptrice*, published in Augsburg in 1611, he not only offered a correct theory of vision (which refuted the traditional notion that light went from the eye to the object seen) but also gave several practical suggestions on how existing telescopes might be improved. One of his ideas was a telescope with two convex lenses which offered a larger field of view but inverted the image. The Library of Congress has Kepler's *Dioptrice* in first edition.

While Kepler's suggestion proved impractical for everyday use, the inversion problem mattered little to astronomers, and his improvements were incorporated in the practice of others. One of the first to construct this new kind of telescope was a Jesuit astronomer, Christoph Scheiner, who also later added a third lens which righted the image again. Scheiner was an inventive, active cleric who taught Hebrew as well as mathematics and also practiced astronomy. He observed sunspots at about the same time as Galileo and was involved in a controversy with the great man regarding the philosophical implications of an imperfect universe. The Library of Congress has Scheiner's oddly titled *Rosa ursina*, published in Bracciano (west of Rome) between 1626 and 1630, which gives an overall account of his astronomical work.

A generation after Galileo first sketched the moon and made the telescope a workable and eventually indispensable instrument to astronomy,

Opposite page:
The German Jesuit astronomer Christoph Scheiner offered the reader of his book an elaborate, instructive two-part frontispiece. Among the early telescope observers, Scheiner had probably the best equipment and the most time to use it. The Ingolstadt observatory where he did most of his observing is pictured in the top circle. To the left of the two towers is a man using a telescope, while another uses a helioscope to the right. Two others are projecting the sun's disc in order not to gaze at it directly. At the bottom, Scheiner shows himself seated at a table studying the sun indirectly from its reflection in a camera obscura. Scheiner perfected this method and was the first to see sunspots. His assistant aligns a telescope so the rays of the sun can pass through and cast a picture onto a paper screen. Scheiner could not reconcile the spots he saw with his religious belief concerning a perfect sun and the purity of its sunlight and was the subject of a furious attack by Galileo. *Rosa ursina,* 1626–30. Christoph Scheiner.

150.
C. Plinius. Sec: Lib. 2. c. 108.
IMPROBVM AVSVM
VERVM ITA SVBTILI ARGVMENTATIONE COMPREHENSVS
VT PVDEAT NON CREDERE
Immissione naturali.
Transmissione Refractoria simplici.
Teliescopio
Helioscopio
Reflexione.
Immissione Refractoria composita.
Maculæ et Faculæ ex uariis obseruandj modis, stabiliuntur.

the son of a wealthy Danzig brewer would use telescopes of much greater magnifying power to produce a magnificent map of the moon's surface. Johann Hevelius had the good fortune to be able to pursue his astronomical interests to the fullest, and by the middle of the seventeenth century, he possessed probably the finest and grandest observatory in Europe. Industrious as well as wealthy, Hevelius was a skilled lens grinder who devised a new style of lathe that enabled him to grind very large lenses. When his moon atlas, *Selenographia*, was published in 1647, none of his telescopes exceeded 12 feet and about fifty magnification. By the time he published his *Machinae coelestis*, which described many of his instruments and telescopes, Hevelius had built telescopes of 60 and 70 feet in length, with one reaching an amazing 150 feet. The erection of this giant telescope with its supporting 90-foot tower atop his observatory must have been an impressive sight in Danzig. Hevelius built a platform that spanned the roofs of three houses he owned and mounted there his giant telescope with its accompanying ropes and pulleys, as well as his other equally outsized instruments like a 6-foot sextant. In the buildings below the platform, he placed his workrooms, library, and private printing press. While Hevelius did achieve greater magnification with his increasingly longer telescopes, they were in fact rather impractical instruments that performed well only under the best of conditions. They were naturally beset by major problems of alignment as well as by the outdoor elements of humidity and wind. When Hevelius was sixty-eight years old, his world collapsed when fire destroyed much of his observatory, books, and equipment. Among the published works of Hevelius in the collections of the Library of Congress are the *Selenographia* (Danzig, 1647) and *Machinae coelestis* (Danzig, 1673–79), which contain the most information about his contributions to the telescope.

Christiaan Huygens was a much younger contemporary of Hevelius and a Dutchman of real genius who combined deep theoretical powers with the touch and manual skills of a craftsman. At twenty-six years old, he discovered a better method of grinding lenses while helping his brother work on improving a telescope. Using these new lenses in his own telescopes, Huygens made one astronomical discovery after another, among them the surface markings on Mars, the satellite of Saturn (which he named Titan), and the fact that Saturn appeared to have a ring around it. Huygens's early training was in mathematics and it was in this vein that he sought to use the telescope to gain more universal knowledge. Attempting to translate his telescopic findings into something that was quantitatively useful led him to seek ways to measure both space and time, which he accomplished by inventing the micrometer and the pendulum clock. In the hands of a genius like Huygens, the telescope was a powerful inspirational device as well as a useful tool, as it had been to Galileo fifty years earlier. In the very large context of all of his varied scientific and technological writings, Huygens actually devoted little space to the development of the telescope. His *Systema Saturnium* contains an account of his telescopes, but the Library of Congress does not have this small book in its collections. However, the Library does have his

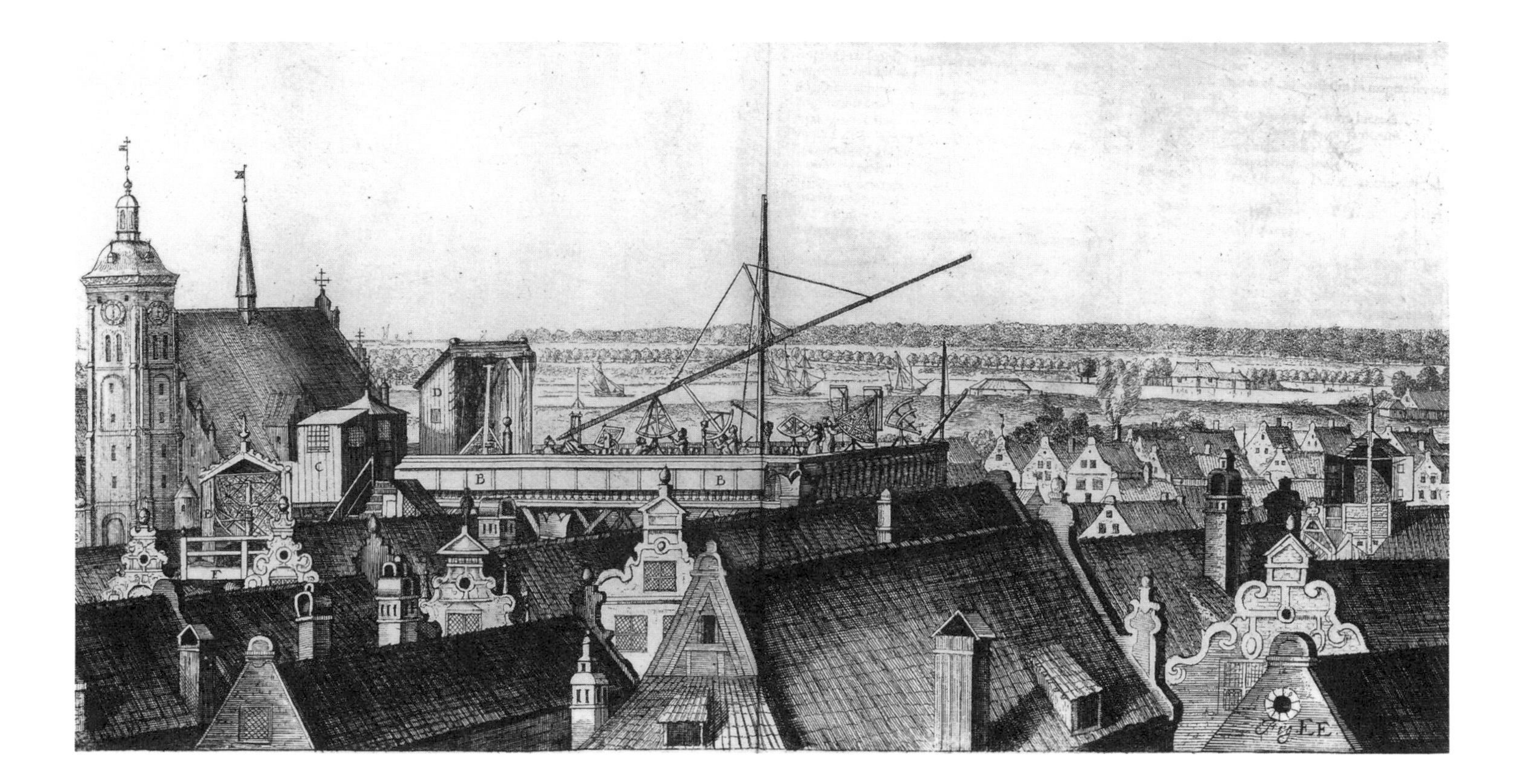

Dioptrica as it is contained in the 1703 Leiden edition of his *Opuscula posthuma*. *Dioptrica* discusses refraction and other optical principles as well as the construction of telescopes, and Huygens called it "Tractatus de refractione et telescopiis" when he first wrote it in 1653. It was never published separately. The Library also has his "Astroscopia compendiaria" as it appeared in *Philosophical Transactions of the Royal Society* in 1684, in which he briefly describes his astronomical instruments, among them his famous aerial telescope.

Unlike Huygens, many seventeenth-century astronomers depended on others to make their telescopes, and two of the most skilled craftsmen of this period were the Italian rivals Giuseppe Campani of Rome and Eustachio Divini of Bologna. Although neither published very much, both were famous, prolific, and influential men of their times, having designed and made telescopes for and collaborated with some of the best astronomers of their age. To this point, most advances in the telescope were elaborations on the basic technology of the very first one. But the final major advance of this period—the invention of the reflecting telescope—was to come about as the direct result of scientific inquiry and exertion and not just more tinkering.

Isaac Newton, in the course of his highly original experiments on the nature of light and color, determined that the chromatic aberration of the refracting telescopes then in use was inescapable, no matter how large the telescopes became. Once he realized, via his prism experiments, that sunlight (or white light) is composed of a combination of colors, he came

At mid-seventeenth century, the finest astronomical observatory in Europe was owned by the German astronomer Johann Hevelius. Although he used telescopes, Hevelius was the last of the great astronomers to insist on naked-eye measurement of the stars. Here is his rooftop observatory called "Stellaeburgum" in Danzig. Stretching across the roofs of his three houses, a platform holds a number of sextants and quadrants as well as an extremely long telescope. The buildings below, which housed the substantial library, workrooms, and a printing press, were all destroyed by fire in 1679. *Machinae coelestis,* 1673–79. Johann Hevelius. LC-USZ62-110448.

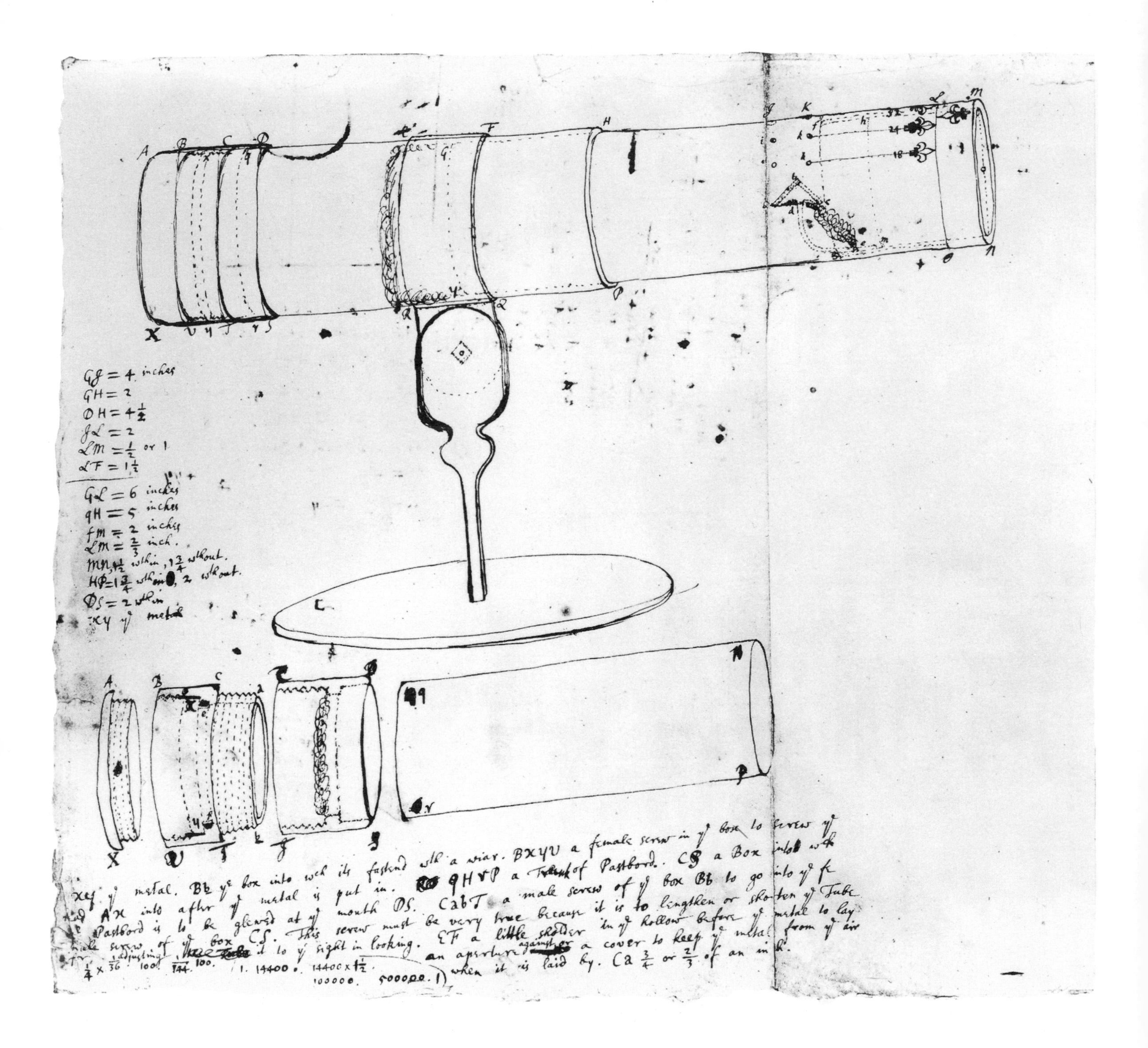

to believe that the blurred colored rims—the chromatic aberration—that distorted the objects seen with a refracting telescope were inevitable as long as light passed through a lens. Although this is not true and is a rare example of Newton actually being wrong about something, it led him nonetheless to search for an alternate method of magnification. It was then he hit upon the idea of light being concentrated by a mirror (reflection) rather than by a lens (refraction). The idea itself was not a new one, having been put forth by the Scottish mathematician James Gregory in 1663. But although Gregory wrote of his idea, he was unable to make a reflecting telescope himself or find someone who could.

Newton, however, could and did. In 1668, his first reflecting telescope magnified thirty to forty times despite its small size—a mere six inches long and one inch wide. With it, he observed the satellites of Jupiter and the phases of Venus. Its concave mirror was made of an alloy called bell metal—six parts copper and two parts tin—to which Newton added one part arsenic. This provided a bright, white surface similar to silver, although the heavy copper content meant it required frequent polishing. Although Newton built his first reflecting telescope in 1668, it was not until 1671 that the Royal Society learned of his work and requested information about it. So in the fall of 1671, Newton built a larger telescope, nine inches long and two inches in diameter, which he presented to the Royal Society, where it remains today, and wrote his famous letter on light. Published by the Royal Society in its *Philosophical Transactions* in 1672, "A Letter . . . Containing his New Theory about Light and Colors" contains an account of his optical discovery that led him to the reflecting telescope. The Library of Congress has that journal as well as a first edition of Newton's *Opticks* (London, 1704), in which he provided more details about his invention of some thirty-two years before.

The final advance in telescopes made during this period was John Hadley's version of the reflecting telescope, which had paraboloidal mirrors made of speculum metal. Hadley's much larger and improved instrument sat on a stand, which made focusing and direction much easier, and although only 6 feet long, it attained close to the 200 magnification of Huygens's 123-foot refractor. From then on, the reflecting telescope became the useful, practical instrument it is today. Hadley described his Newtonian reflector in "An Account of a Catadioptrick Telescope." The Library of Congress has Hadley's article as it first appeared in *Philosophical Transactions* in 1723.

The history of the microscope is closely tied to that of the telescope, for in a simplified way a microscope is but a telescope turned around. As Galileo quickly discovered and as any inquisitive young mind knows today, looking through the objective and out the eye lens of a telescope makes small, close objects appear larger. As a telescope makes distant objects appear larger and clearer, so a microscope makes minute objects more easily seen.

Microscopes can be simple or compound, according to whether they have one lens or more. The simplest microscope, with one convex lens, is

Opposite page:
Isaac Newton's solution to the chronic blurring of refracting telescopes of his time was typically direct and unique. His answer was to use a concave mirror, rather than a lens, to collect and focus light. His knowledge of light informed him that while a simple lens focuses different colors at different points, a mirror focuses all the colors of white light at the same point (since mirrors do not split light into its component colors). In this rough, unskilled sketch, Newton shows his reflecting telescope in its essentials. The concave mirror is located on the left side of the barrel, and a second tiny mirror is set to its right at a 45-degree angle and reflects the gathered light at right angles into an eyepiece above (K). The entire telescope was less than twelve inches long. *The Correspondence of Isaac Newton,* vol. 1, 1661–75. Isaac Newton. LC-USZ62-110449.

what we could call a magnifying glass and has a history that goes back at least to the ancient Greeks. The compound microscope with its use of more than one lens was, however, discovered approximately at the same time as the telescope, since it involves the same principle of aligning lenses. And as with the very early history of the telescope, in which references to it can be found before its first documented appearance (such as telescope claims made for Digges in 1571 and for Porta in 1589), there are at least two references to pre-1600 use of a compound microscope. In one of these, Zacharias Jansen, a Middleburg spectacle-maker and rival of Lippershey's as inventor of the telescope, is credited with developing the microscope around 1590. The second claim was made in a book first published a century later, in 1691, in which Filippo Buonanni stated that a Frankfurt entomologist named George Hufnagel had used such a microscope in his 1592 work on insects. The Library of Congress has Buonanni's book, *Micrographia curiosa*, in an edition published in Rome in 1703. A similar belated claim is made by Francesco Fontana. Writing in a chapter entitled "De microscopio" in his *Novae coelestium terrestrium rerum observationes* (Naples, 1646), Fontana says he made a compound microscope in 1618. The chapter contains no drawings of his handiwork. The Library has Fontana's work in first edition.

Although he published none of his microscopical research, Galileo undoubtedly was the first to use a real, compound microscope regularly for scientific purposes. It is known that throughout his career he remained interested in microscopes and at one point studied the legs and wings of insects as well as their compound eyes. But given his mathematical background and obvious passion for the energy and forces of mechanics, his preference for the telescope as an astronomical tool over the microscope as an instrument of natural history makes perfect sense. Galileo was a member of the Accademia dei Lincei, founded by Duke Frederico Cesi and also called an "academy of curious men." It is no surprise then that Galileo would bring his new instrument to his academy friends (in 1624 he gave three microscopes away), who systematically took to studying living things with microscopes. It was because of the academy that the first printed work to contain microscopic illustrations is found in a book of satirical poems called *Persio, tradatto in verso sciolto e dichiarato* by Galileo's colleague, Francesco Stelluti. In 1625 Stelluti had produced a single printed sheet showing all anatomical aspects of a honeybee in great, microscopic detail. In 1630, Stelluti used the same illustration to dedicate his book of poems to Cardinal Barberini, whose family crest showed three bees. The Library of Congress has the first edition (Rome, 1630) of this very important work.

While development of the microscope was slower than that of the telescope and lagged behind it in popularity, by the middle of the seventeenth century it was ready to become the scientific instrument in vogue. The great René Descartes, in the "Dioptrique" published as part of his famous *Discours de la methode*, had discussed not only a simple microscope but a hypothetical one that used a mirror and worked from a stand taller than a man. The Library has the first edition of Descartes's *Discours*

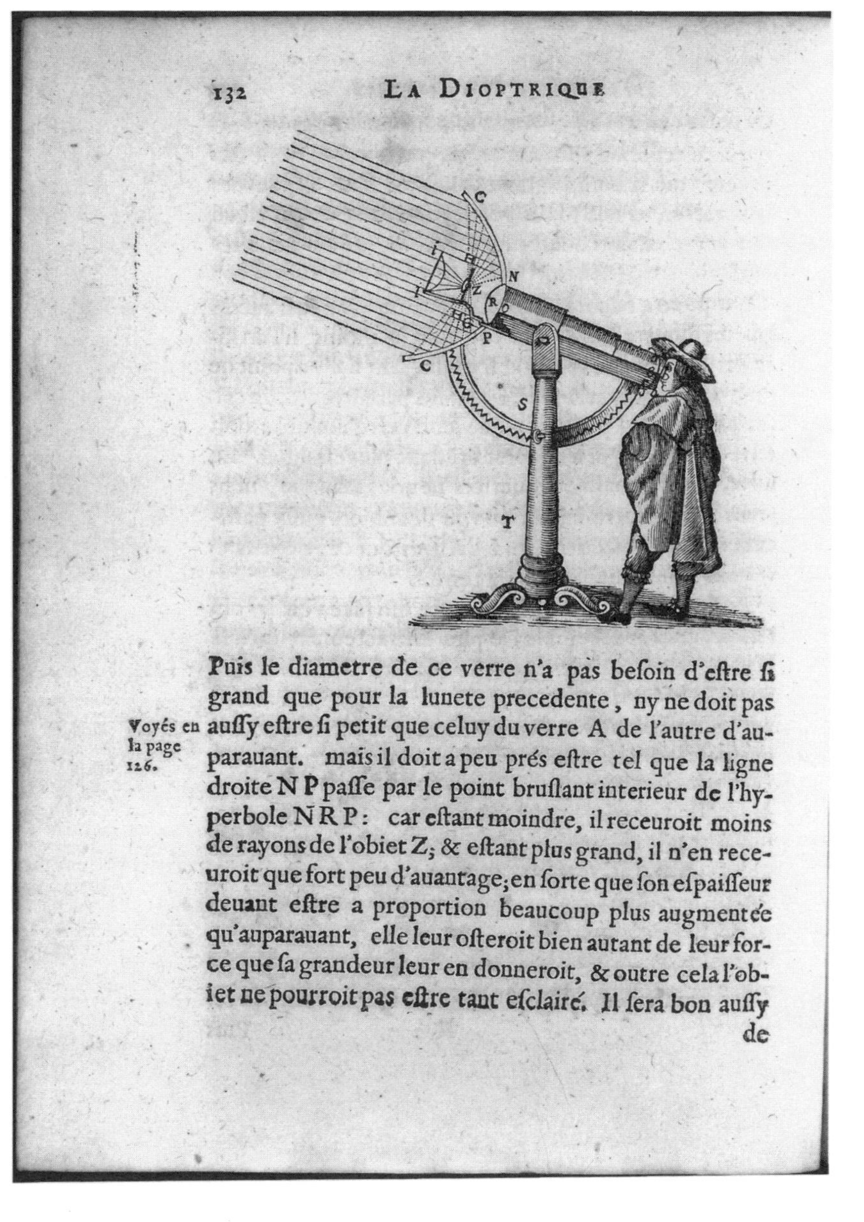
132 LA DIOPTRIQUE

Puis le diametre de ce verre n'a pas beſoin d'eſtre ſi grand que pour la lunete precedente, ny ne doit pas auſſy eſtre ſi petit que celuy du verre A de l'autre d'auparauant. mais il doit a peu prés eſtre tel que la ligne droite N P paſſe par le point bruſlant interieur de l'hyperbole N R P: car eſtant moindre, il receuroit moins de rayons de l'obiet Z; & eſtant plus grand, il n'en receuroit que fort peu d'auantage; en ſorte que ſon eſpaiſſeur deuant eſtre a proportion beaucoup plus augmentée qu'auparauant, elle leur oſteroit bien autant de leur force que ſa grandeur leur en donneroit, & outre cela l'obiet ne pourroit pas eſtre tant eſclairé. Il ſera bon auſſy de

Voyés en la page 126.

Often mistaken for a telescope, this is considered one of the earliest drawings of a compound microscope. This confusing image in Descartes's "Dioptrique" has intrigued many over the years, since any microscope whose illuminating source was a reflector this large would incinerate the object in focus. Most now feel that Descartes's original design was for a small instrument, and that an engraver later inserted the observer, thus giving the picture an unreal scale. *Discours de la méthode,* 1637. René Descartes. LC-USZ62-110450.

(Leiden, 1637). It also has the 1671 Amsterdam edition of Athanasius Kircher's *Ars magna lucis et umbrae*, first published in 1646. Kircher was a well-connected Jesuit whose work with microscopes served to stimulate more research and also contained some fertile speculation concerning the roles of tiny living creatures in the process of disease and decay.

But the greatest popularizer by far was the publication of Robert Hooke's *Micrographia*, published in London in 1665. As the earliest treatise devoted entirely to microscopical observations, it probably would have been a popular work anyway, but its success has little to do with mere priority. The book sold out rapidly and exerted its influence beyond England to the Continent, despite being written in English. Altogether, the book excels at many levels—scientific, technological, artistic—and

Schem. XXXII
C
D
D
A
B
C
A
I
I
I

contains qualities rare in a landmark book of science. It is approachable, even easy to read, and the author's enthusiasm is obvious and unrestrained. While many scientists may greatly enjoy their work, few reveal in their writings that they are actually having fun. Robert Hooke knew no such restraint. Using his own compound microscope, Hooke turned his new instrument upon objects of the everyday world—a grain of sand, snail's teeth, the point of a needle, a flea, a louse, and a house fly, the hair of a hog, mold, and the beard of a wild oat—and discovered worlds never before seen by anyone. Hooke described and sketched all these minute, detailed marvels with the clinical eye of a scientist, but always his sense of wonder came through. The text of Observation 53, which describes a common flea, begins, "The strength and beauty of this small creature, had it no other relation at all to man, would deserve description." Two pages earlier, he casually introduces his observations of a crab-like insect with the words, "Reading one day in September, I chanced to observe a very small creature creep over the book I was reading, very slowly; and having a Microscope by me . . . ."

Hooke was described by John Aubrey as "the greatest mechanick this day in the world," and was without doubt a brilliant and talented craftsman as well as a skilled scientist. He built his improved, compound microscope using the technology of his time as he knew it, and gave a detailed description of his instrument in the preface of his book. The very first illustration he shows us is that of his microscope—an elegant instrument. Besides the real, substantive scientific and technological contributions to microscopy contained in Hooke's famous work, the contagious zeal, enthusiasm, and sheer enjoyment it conveys to readers make it a singular achievement. The Library of Congress has his *Micrographia* in first edition.

Despite Hooke's dominance of the field of microscopy, the first documented, scientific discovery made using a microscope was that by Marcello Malpighi, a professor at the University of Bologna. In 1661, Malpighi wrote two letters to his colleague in Pisa, Giovanni Borelli, about his discoveries of capillary blood flow in the lungs of a frog, the first evidence that capillaries connect arteries with the veins. These letters were published in Bologna the same year under the title *De pulmonibus*. They are also contained in Malpighi's *Opera omnia* (Leiden, 1687), which the Library of Congress has in its collections.

Where Robert Hooke was too impatient, curious, and multi-talented to stay very long with one concern, moving on to physics, biology, and astronomy, a little-schooled, isolated Dutch draper, Anton van Leeuwenhoek, devoted his long life to one thing—microscopy—and in doing so, reaped the harvest of most of the major microscopical discoveries of the seventeenth century. It is commonly believed that Leeuwenhoek's remarkable and plentiful microscopical insights were achieved using inventive techniques and a simple microscope whose tiny lenses he had patiently and with great delicacy ground by hand. Using his simple microscope, he pioneered the observation of protozoa and bacteria.

From the first, Leeuwenhoek's startling achievements have been re-

Opposite page:
The cantankerous polymath Robert Hooke used his compound microscope to examine and describe "minute bodies" of plants and animals. Here he shows an ant that was a particular problem to sketch, since his normal methods of immobilizing but not killing the insect would not work. "If its feet were fetter'd in Wax or Glew, it would so twist and wind its body, that I could not any wayes get a good view of it." He solved the problem by drenching it in brandy which, he said, "knock'd him down dead drunk." *Micrographia; or, Some physiological descriptions of minute bodies made by magnifying glasses,* 1665. Robert Hooke. LC-USZ62-110451.

Liber Sextus. 55

citiſſima [illegible] hac igitur ampulla nihil ſit præter aerem, in parte tamen inferiori A C. ſit aqua colore aliquo ſatis perſpicuo tincta, quæ ex ſuppoſito vaſe D. ſurſum aſcendat, ac deſcendat. Iam ſi prædictæ ampullæ capiti B. manum aut digitum admoueris, illico videbis aquam A C. deſcendere: cauſa eſt calor manus, quo aer incluſus citiſſimè rareſcit, fitque maior, ac proinde dilatatur per fiſtulam E C. ſicque aquam deorſum pellit. quod ſi ampullæ aliquod frigidum admoueris, aer incluſus ſtatim condenſabitur, ac ſeſe contrahet, ac propterea ne vacuum exiſtat, aqua A C. aeris receſſum ſupplens, ſurſum aſcendet. auxilio huius inſtrumenti, quod ego Thermoſcopium libenter appellarem, multa ad aeris naturam ſpectantia, indagari poſſunt. audiui Doctorem quendam Medicum Patauij degentem, qui Santorius cognominatur huius eſſe inuentorem.

4 Ventorum agitatio nihil Aſtronomicum ſapit, ideo Philoſophicis diſquiſitionibus reliquanda eſt.

De Aeris figura. *Cap. III.*

IAm dictum ſuperius eſt in quo ſphęra, & orbis diſcrepent; quod nunc in memoriam reuocandum eſt. Aio igitur aerem orbis figurę pręditum eſſe: cum enim circa terrenum globum effuſus ſit, eumque vndique ambiat, neceſſario concauam ſuperficiem habebit, eamque ſphęricam iuxta ſphæricitatem terreni globi, circa quem configuratur. pariter eum habere ſuperficiem ſupremam, & conuexam, quæ ſphærica ſit, inde colligi poteſt, quia aer cum ſit fluidus, ac leuis & proinde ſurſum ad omnes partes æqualiter aſcendat, neceſſario in ſphæricam figuram deſinit, quia a centro Mundi æqualiter vndique aſcendens recedit: ſicuti enim è contrario aqua quia fluida eſt, & grauis deſcendendo ſphæricitatem aſcendendo conſequi.

De Aeris illuminatione. *Cap. IIII.*

AEr purus, ideſt, abſque vlla exhalatione, aut vapore eſt omnino diaphanus, & tranſparens, quare lumen Solis nullo modo ſiſtit, ſed illud pręterire permittit: vnde ſequitur eum nullo modo, quamuis toto lumine profundatur, fieri conſpicuum, ſeu videri poſſe.

Aer vero impurus qui terrę proximior eſt, ob terreſtres halitus, qui ei perpetuo admiſcentur, impurior ac craſſior euadit, vnde aptè Atmoſphæra, ideſt, halituum ſphæra nominatur; hæc inquam Atmoſphæra, cum imperfectè tranſpareat, abundetque prædictis halitibus, quæ opacitatem aliquam illi inferunt, fit vt lumen Solis per ipſam diffuſum, partim tranſmittat, partim detineat ac reflectat, eoque illuſtretur, ac proinde reddatur conſpicuus, lucemque Solis, quæ diem efficit, vniuerſæ terræ communicet. imo crepuſculum, quod diei [illegible] eſt, nihil aliud eſt quã huiuſmodi halitis in extremo, orientalique horizonte illuminati, vt ſequenti cap. patebit. quæ ex opticorũ doctrina deſumpta ſunt.

This is the first published illustration of a thermoscope, or a thermometer without a scale. It is described as containing colored water by the writer, who also says that with it, "many things may be found out about the nature of the air. I have heard that the inventor of this is a certain Doctor of Medicine called Santorius, who lives in Padua." *Sphaera mundi,* 1653. Giuseppe Biancani. LC-USZ62-110452.

garded with awe, since no one has been able to duplicate his results using a simple microscope such as he devised. In 1973, however, a British scientist, Brian Ford, argued that Leeuwenhoek had used a second, hand-held lens to achieve much greater magnification. Thus, argued Ford, the inventive and secretive Dutch pioneer used what could be described in principle as a compound microscope, achieved by holding his prefocused simple microscope, with specimen attached, in one hand and a second lens close to his eye with his other hand. Ford makes a circumstantial case at best, but his intriguing argument is certainly worth considering and does no harm to Leeuwenhoek's reputation.

Although he lived to be ninety and gained international fame, Leeuwenhoek never wrote a book or scientific paper, instead documenting his research via informal letters written in Dutch to the Royal Society. This correspondence began in 1673 and continued almost to the end of his life. The Library of Congress has these letters in three forms: as they were translated and appeared individually in the Royal Society's *Philosophical Transactions* over a long period of time; as they were first collected and published in Dutch in four volumes, titled *Ontledingen en ondekkingen . . . brieven* (of which the Library's set is variously composed of first through third editions, published in Leiden and Delft between 1696 and 1718); and as they appeared in a Latin collection of selected letters titled *Arcana naturae detecta* (Delft, 1695).

There naturally were others who helped establish the microscope as a valid tool of scientific investigation, among them the Dutch naturalist Jan Swammerdam and the English botanist Nehemiah Grew. And as the names of Campani and Divini dominated the actual construction of telescopes, so the names of John Marshall and Edward Culpepper are synonymous with the art of making compound microscopes in this early stage of their development.

The ancient Greeks were familiar with the optical principles behind both the telescope and the microscope, and they were also aware of the key principle at work in what became the thermometer—that air expands as it gets warmer. And as with the modern beginnings of those optical instruments, the origins of the thermometer focus again on Galileo and his coterie of curious men. Most historians attribute the invention of the thermometer to Galileo, although they can do so only by the testimony of his friends and colleagues who date his work on it to about 1592. What he may have built would actually have been a thermoscope—a device that reacts to and indicates temperature change but does not have any scale on it to calibrate or measure that change. Claims for priority have also been made for Robert Fludd, a mystical philosopher and Welsh physician, as well as for the Dutch inventor Cornelius Drebbel.

While Galileo may have invented the thermoscope, it was his contemporary, Sanctorius, who was the first to make use of the thermoscope as a scientific instrument by providing it with a scale. In 1612, Sanctorius, a professor of medicine at Padua whose full name was Santorio Santorio, wrote in his *Commentaria in artem medicinalem Galenii* about "a marvelous way in which I am accustomed to measure, with a certain glass instrument, the cold and hot temperature of the air of all regions and places, and of all parts of the body." What he was describing is the first clinical thermometer. The Library of Congress has only one of his major works, a 1676 English translation of his *De statica medicina*.

The thermometer remained mostly an object of curiosity until the 1640s, and the first published illustration of one in 1620 was actually of a thermoscope. It appeared in Giuseppe Biancani's *Sphaera mundi*, first published in Bologna in 1620. The Library of Congress has the 1653 Modena edition. The first time the word "thermometer" appeared in print was in 1626 in a book titled *Recréation mathématique*, published in Pont-à-Mousson. This book by the Jesuit Jean Leurechon (who wrote under the pseudonym H. Van Etten) also contains an illustration contrasting the Italian and the Dutch versions, with only the latter having a scale. The Library has this work in first edition.

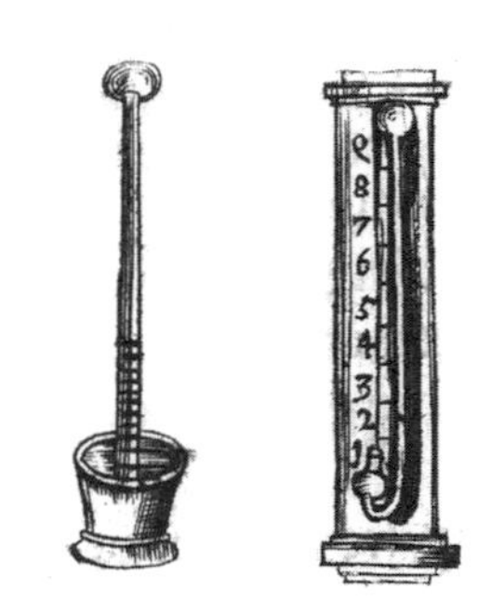

The first use of the word *thermometer* was made in a 1626 book written by the French mathematician Jean Leurechon. In this 1653 English translation, he shows the Italian type of thermoscope (on the left), and contrasts it to the Dutch thermometer (right), which is a two-bulb version with a scale. *Mathematical recreations*, 1653. Jean Leurechon. LC-USZ62-110453.

Until about the 1640s, it was an air thermoscope or air thermometer that was used, consisting essentially of a glass tube open at the bottom and ending in a closed bulb at the top. The lower, open end sat in a vessel of colored water. Changes in bulb temperature produced expansion or contraction of the air in it, which in turn changed the level of the liquid in the tube. In 1632, however, an obscure French physician, Jean Rey, reversed the air and water arrangement, filling the bulb with water and the stem with air, and used the liquid as an index of the temperature. Rey communicated this invention in a letter to Marin Mersenne, the famed "postbox of Europe," but his invention of the liquid thermometer went little noticed until 1777, when the letter was appended to a reissue of his book called *Essays* (1630). The Library of Congress has Rey's work only as it was reissued in an 1896 Paris edition (*Essais de Jean Rey*) and an 1895 Edinburgh/London edition (*Essays of Jean Rey*).

The next major improvement was made, most believe, by the Grand Duke Ferdinand II of Tuscany, who was the first to seal the open end of a calibrated tube, thus making the instrument impervious to barometric pressure. He may have done this as early as 1641, but certainly well before 1657, when he and his brother, Prince Leopold de' Medici, founded the small group of Florentine physical scientists that came to be known as the Accademia del Cimento. Leopold had been a pupil of Galileo's, and this group of Florentine experimenters was very much his own court group. Membership was small and select, and all were dedicated to systematic, experimental investigation in the physical sciences. They avoided debate, rhetoric, and theory and had a penchant for new instruments which the Medicis readily provided. After a little less than ten years' existence, though, the academy was dissolved and a partial record of the group's experimental work was published in Florence in 1666. Titled *Saggi di naturali esperienze fatte nell' Accademia de cimento*, this elegant folio contains many plates which precisely depict the instruments used in their experiments. The first instrument shown is a thermometer, and the first experiment described is "to measure the degrees of heat and cold in the air." The engraving shows, no doubt, Ferdinand's sealed, calibrated thermometer, and the text not only details the experiment, but tells how to make such a device. The Library of Congress has this work in its original edition. The academy produced at least four different types of thermometers, and its workmen became highly skilled at reproducing them, to the point that Florentine thermometers soon spread across Europe. Its members were also among the first to try mercury, although they preferred to use wine because it expanded more in response to temperature increases.

After 1665 or so, many of the more famous names of European science were using liquid-in-glass thermometers in one way or another. It is possible to document such work in the writings of Boyle, Guericke, Hooke, Huygens, Halley, and Newton. It was apparent to all that what was needed next was an absolute thermometric standard, and virtually all offered their own ideas about how and upon what such a standard should be based. None of these suggestions took hold or met with success until Gabriel Daniel Fahrenheit, a skilled instrument maker, perfected what would become the modern thermometer. First, Fahrenheit invented a way to clean mercury so it would not adhere to the walls of the tube. Mercury was known as an ideal indicator because its boiling point was well above that of water and its freezing point also below that of water. Next, he added the vital "fixed point" by calling zero the point at which water mixed with salt would freeze. Toward the other end of the scale were two normals—96 degrees for body temperature and 32 degrees for freezing water. Later adjustments brought normal body temperature to 98.6 degrees. Fahrenheit reported his inventions to the Royal Society in 1724, and they were published in five brief papers between then and 1726. The Library of Congress has the most significant of these, "Experimenta circa Gradum Caloris," which first appeared in the society's *Philosophical Transactions* in 1724.

Although the thermometer became a very important instrument to science, it did not have that rare, extra dimension of effecting what might be called social consequences. Some devices or technologies not only influence the way we live but have actually altered our scientific concepts and have shaped the way we think about the world. The telescope was one of these, eventually contributing to the demise of a dominant, earth-centered cosmology. The microscope would also, in another way, help overthrow the long-held doctrine of spontaneous generation and validate the germ theory of disease.

One of the major debating topics of this exciting first half of the seventeenth century centered on the validity of a revered but seemingly unprovable truism, Aristotle's dictum that "Nature abhors a vacuum." It was partly in their attempts to disprove it, by investigating air pressure and actually trying to create a vacuum, that experimenters gave us the air pump and the barometer. As with the telescope, microscope, and thermometer, it appears that the prodigious Tuscan, Galileo, was again significantly involved with the origins of at least one more major scientific tool, the barometer. Interestingly, the first barometer and the first man-made vacuum were created by Evangelista Torricelli during his experiments to find if air had weight, and it is not surprising that it was the aged and blind Galileo who pointed him in the direction of the elusive vacuum.

Opposite page:
The remarkable Florentine institution called the Accademia del Cimento (Academy of Experiments) flourished for ten years and laid the foundations of experimental physics. Founded by pupils of Galileo and funded by two Medici brothers, Grand Duke Ferdinand II and Leopold, it conducted experiments on temperature measurement, and the vacuum, and magnetism. Here are some examples of the different sealed or Florentine thermometers it used. Number 4 is a thermometer with a helical scale, and a number 6 is a hygrometer. Much of the success of the academicians should be shared with the duke's glassblower, Mariani, an apparently consummate workman of extraordinary skill. *Saggi di naturali esperienze fatte nell'Accademia del cimento,* 1666. LC-USZ62-110454.

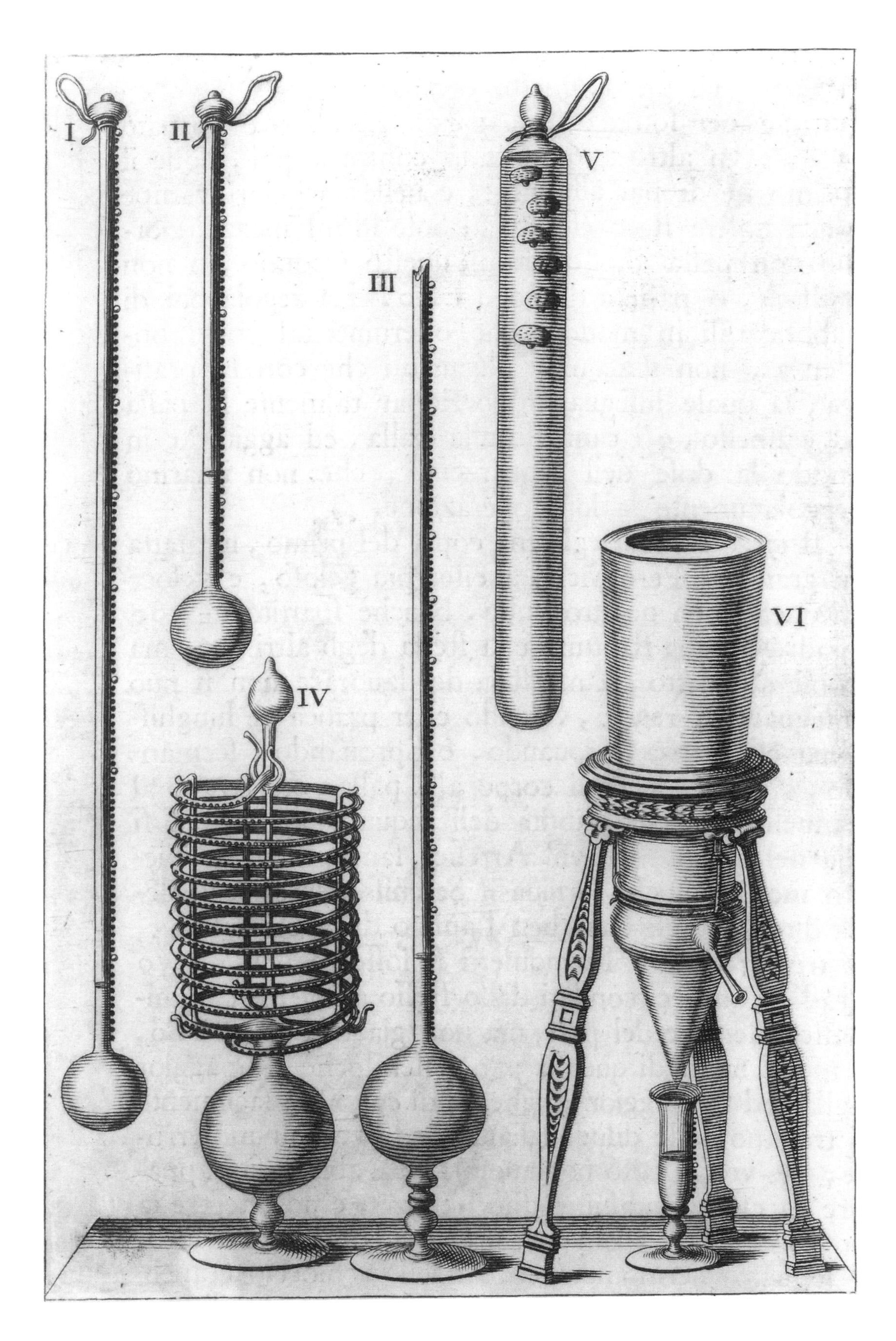
I
II
III
IV
V
VI

The young Italian professor of mathematics, Gasparo Berti, apparently did everything correctly in this large-scale air pressure experiment, but he could not prove the existence of a vacuum. Here in Berti's experiment, a long lead tube is attached to his house. It is closed at the bottom with a valve that is submerged in water. The globe at the top contains a bell and clapper. When the bottom valve was opened and the water dropped to about thirty feet, no sound should have been heard from the bell in the globe. Nevertheless, the bell was heard. Although Berti concluded that a vacuum had not been created, later experimenters felt that the sound had been transmitted by the metal attachment of the bell. *Technica curiosa,* 1664. Gaspar Schott. LC-USZ62-110455.

Galileo believed what every mining engineer of his time did—that the phenomenon of suction that made suction pumps work was due to nature's abhorrence of a vacuum. But, like miners, he was also aware that despite nature's need to fill a space lest a vacuum be created, water could not be raised by suction pumps more than thirty to thirty-three feet. It was this apparent contradiction that the seventy-seven-year-old Galileo suggested the young Torricelli investigate just before the old man died in 1642. But sometime between 1640 and 1643, another young Italian, Gasparo Berti, already had conducted a full-scale experiment, trying to create a vacuum by attaching to the facade of his house a sealed lead tube over thirty feet high, filling it with water and then opening the bottom tap into a barrel. Not all of the water flowed out, and Berti contended, though could not demonstrate, that the sealed space in the tube above the remaining water was a vacuum. Berti died the next year and left no writings of his own, but his Roman friends documented his work and one, Athanasius Kircher, illustrated the experiment in his book, *Musurgia universalis* (Rome, 1650). The Library of Congress has this work in first edition as well as another, Gaspar Schott's *Technica curiosa* (Nuremberg, 1664), which also illustrates Berti's experiment.

It was left to Torricelli, who, with his friend Vincenzio Viviani, shared the same intellectual milieu and colleagues as Berti, to repeat Berti's work in a controlled, manageable manner. Torricelli used mercury instead of water, since that metal is nearly fourteen times as dense as water, and a glass tube only one meter high. Torricelli inverted the filled tube in a bowl and, predictably, the mercury flowed out only to a certain level and stopped. Here he demonstrated his genius. He knew that the vacated space in the tube was a vacuum and was able to prove it, but he was more interested in the properties of the air itself, postulating that it was a physical effect—the weight of the air pressing down on the mercury in the bowl—that created this phenomenon. Air was a force then, external to the tube and having nothing really to do with the vacuum within it. In June 1644, Torricelli wrote his friend Michelangelo Ricci the famous, almost poetical words, "we live submerged at the bottom of an ocean of air." Finally, Torricelli also was astute enough to notice that the height of the mercury in the tube would vary slightly from day to day, a fact he attributed to the variations in atmospheric pressure. Torricelli had invented the first barometer.

As with Berti, who died when he was forty-three years old, Torricelli succumbed at thirty-nine to typhoid fever. During his lifetime, he saw only his *Opera geometrica* published, and a copy of its first edition (Florence, 1644) is in the Library's collections. Nearly seventy years after his death, the more complete *Lezione accademiche* was edited by Tommaso Bonaventuri and published in Florence in 1715. The Library also has this work, which includes Torricelli's writings on his famous mercury experiment.

Among these writings was his logical speculation that if the atmosphere had weight and it were all atop us (his analogy of being at the bottom of an ocean of air), the higher up we went, the less air there would be on us. This decrease in the weight of the atmosphere should then be detectable by his barometer. In a letter to his friend Ricci, Torricelli wrote of his experiment and his hypotheses, the logical extension of which is that if you go up high enough above the atmosphere, a vacuum would be found. Ricci sent a copy of this letter to a Minorite priest in Paris, Marin Mersenne, who performed the valuable service of regularly circulating the most current scientific information among his scores of scientific contacts throughout Europe. A mathematician himself, Mersenne knew the value of first-rate minds stimulating and cross-fertilizing one another, and took it upon himself to communicate, encourage, and even guide his absent coterie, all to the advancement of science. It was appropriate then that when he received Ricci's letter, he communicated its contents, via Pierre

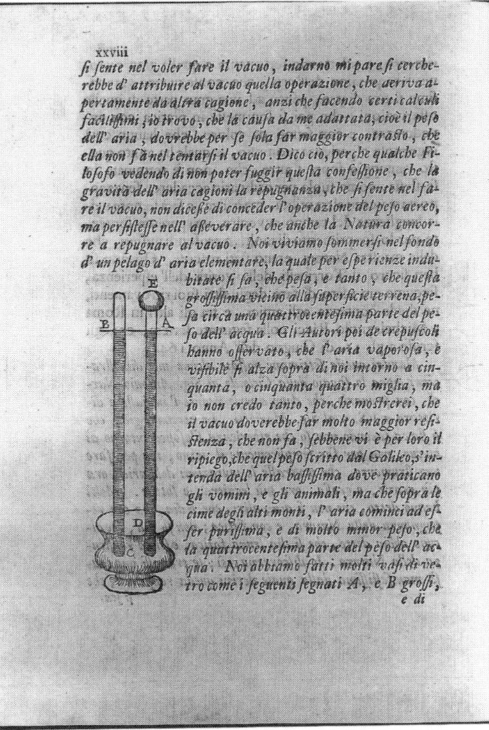
xxviii

ſi ſente nel voler fare il vacuo, indarno mi pare ſi cercherebbe d' attribuire al vacuo quella operazione, che deriva apertamente da altra cagione, anzi che facendo certi calculi faciliſſimi, io trovo, che la cauſa da me adattata, cioè il peſo dell' aria, dovrebbe per ſe ſola far maggior contraſto, che ella non fà nel tentarſi il vacuo. Dico ciò, perche qualche Filoſofo vedendo di non poter fuggir queſta confeſſione, che la gravità dell' aria cagioni la repugnanza, che ſi ſente nel fare il vacuo, non diceſſe di conceder l' operazione del peſo aereo, ma perſiſteſſe nell' aſſeverare, che anche la Natura concorre a repugnare al vacuo. Noi viviamo ſommerſi nel fondo d' un pelago d' aria elementare, la quale per eſperienze indubitate ſi ſa, che peſa, e tanto, che queſta groſſiſſima vicino alla ſuperficie terrena, peſa circa una quattrocenteſima parte del peſo dell' acqua. Gli Autori poi de crepuſcoli hanno oſſervato, che l' aria vaporoſa, e viſibile ſi alza ſopra di noi intorno a cinquanta, o cinquanta quattro miglia, ma io non credo tanto, perche moſtrerei, che il vacuo doverebbe far molto maggior reſiſtenza, che non fa, ſebbene vi è per loro il ripiego, che quel peſo ſcritto dal Galileo, s' intenda dell' aria baſſiſſima dove praticano gli uomini, e gli animali, ma che ſopra le cime degli alti monti, l' aria cominci ad eſſer puriſſima, e di molto minor peſo, che la quattrocenteſima parte del peſo dell' acqua. Noi abbiamo fatti molti vaſi di vetro come i ſeguenti ſegnati A, e B groſſi,

e di

As an old man, Galileo had recognized the genius of the young Torricelli and suggested that he investigate the vacuum problem. In his famous mercury experiment, however, Torricelli went beyond the creation of a vacuum and showed that he was truly an inspired experimenter. Seeing that the mercury flowed out of this tube only to a certain level and then stopped, he took the vacuum in the evacuated space for granted and postulated further that it was the weight of the air outside the tube, pressing down on the mercury in the bowl, that was responsible. The invisible, intangible air we breathe is a force, he said, and has weight that is measurable. Torricelli's measuring device became the first barometer. Four years after his experiment, he was dead at age thirty-nine of typhoid fever. His eloquent words remain, however: "We live submerged at the bottom of an ocean of air." *Lezione accademiche,* 1715. Evangelista Torricelli. LC-USZ62-110456.

Petit, to the man he knew would use it best—the eccentric, chronically ill, but brilliant young mathematician, Blaise Pascal.

Pascal initially doubted that a vacuum had been created, but after re-creating Torricelli's experiment using mercury and then water, he was convinced. Thus he was persuaded by the arguments of both Mersenne and René Descartes that the only way to decide with finality was to test Torricelli's hypothesis by taking a calibrated thermometer up a mountain. Pascal then arranged for his robust brother-in-law, Florin Perier, to conduct just that experiment. On September 19, 1648, Perier and his friends took one mercury-filled tube up the Puy de Dôme and left another at the mountain's base to be monitored. After climbing up about one mile and setting up the barometer, Perier noticed the mercury had fallen three inches. He repeated this reading five more times under various conditions, always with the same results. Upon descending, he learned that the control barometer left behind had not changed at all. Pascal had proven Torricelli correct and had demonstrated the existence of both air pressure and the vacuum. Like Torricelli, Pascal would also die at thirty-nine. His posthumously published work, *Traitez de l'equilibre des liqueurs*, published in 1663 in Paris, contains his writings on both the weight of air and the vacuum. The Library of Congress has this thin volume in first edition, which also contains Pascal's 1648 detailed, twenty-page account of the experiment titled *Recit de la grande expérience de l'équilibre des liqueurs*. Even at this early point in time, fertile minds were already speculating on other uses for the barometer, such as determining elevation or as something called a "weather glass."

Experiments on air pressure also led to the invention of the next major scientific instrument of this time—the air pump or vacuum pump. This important device was constructed as early as 1647 but probably no later than 1654 by Otto von Guericke, who was then mayor of Magdeburg. Trained in law and mathematics and experienced in engineering, Guericke was a far-sighted tinkerer whose ideas ranged from the mechanical to the metaphysical. By the time he learned of Torricelli's experiments, he had already been working on the problem of the vacuum which, he felt, might be what the planets move through. Guericke was as much an engineer as he was a confirmed Copernican, and it was as an engineer that he approached his theoretical dilemma in a mechanical manner. Using the available technology to suit his own needs, Guericke developed the first air pump, for which he had a very specific purpose. His early mechanical, hand-operated pumps were styled somewhat like the fire extinguishers of his time and created their suction by using a cylinder and piston to which he added two flap valves. Beginning with beer casks from which he pumped water, he progressed eventually to hollow copper spheres from which his brass pumps were able to evacuate virtually all of the air.

Guericke was a thorough experimentalist with a keen eye for detail, and he quickly took advantage of the fact that he had created a vacuum in a space much larger than Torricelli's glass tube. He was able to determine that within the evacuated container, mice would die, candles were

extinguished, and bells would ring silently. Having created a vacuum at will and determined some of its properties with his new pump, he also learned that air has weight, since an evacuated sphere weighed less than one filled with air. Demonstrating what he had learned brought out his politician's public flair, and in 1657 he conceived one of science and technology's more dramatic experiments. Having used his pump to create a vacuum in a large copper sphere (actually two hemispheres that fit snugly together), he attached a team of eight horses to opposite sides of the sphere and had them try to pull the globe apart. Their inability to do so, and the ease with which the spheres came apart once a stopcock was opened and air introduced, was a powerful and graphic example of the power of a vacuum.

The German emperor Ferdinand III witnessed this drama and was so impressed he ordered a Jesuit professor of physics and mathematics at Wurzburg, Gaspar Schott, to document Guericke's work. So the earliest

This nineteenth-century drawing recreates one of the most famous experiments in the history of physics. Planned and organized by Blaise Pascal, the experiment completed Torricelli's work by demonstrating that indeed, the higher one went, the less air there was pushing down on one. As the men carried the large barometer up the Puy de Dôme in the Auvergne, the mercury progressively fell. It rose again as they descended. Today we call this barometric pressure. *Les Merveilles de la science,* 1867–69. Louis Figuier. LC-USZ62-110457.

Here the well-dressed experimenter gracefully points to a glass globe on the left from which the air has been evacuated. At its bottom, it is connected by a stopcock to a bottle full of air. This experiment demonstrates the force of air pressure. Once the stopcock is opened allowing the air from the bottle to rush into the empty globe, the bottle is usually shattered by the force the outside air exerts on it. The round shape of the globe resists this outside pressure, even when evacuated, but the flat four-sided bottle cannot when it suddenly loses its own air. As scientific experiments became more exacting and sophisticated, they came to depend increasingly on technology to provide them with precise technical equipment. *Experimenta nova (ut vocantur) magdeburgica de vacuo spatio,* 1672. Otto von Guericke. LC-USZ62-110458.

published account of Guericke's pump and his experiments was contained in Schott's *Mechanica hydraulico-pneumatica*, published in Frankfurt in 1657. The Library of Congress has this work in first edition as well as Schott's second work, *Technica curiosa*, published in Nuremberg in 1664, which contains additional information on Guericke's work. Guericke did not publish until fifteen years after his famous experiment, and the Library of Congress has his *Experimenta nova (ut vocantur) magdeburgica de vacuo spatio* (Amsterdam, 1672). Guericke's experimental work on pumps stimulated Huygens, Papin, and Boyle and eventually led to the development of steam power, already discussed. It also gave science another tool with which to study the physical properties of gases.

It is known that Robert Boyle, the former child prodigy now grown to thirty in 1657, was directly influenced by Schott's account of Guericke's experiments. Boyle was a wealthy aristocrat and one of the founders of the Royal Society. Upon reading of Guericke's pump, he directed his laboratory assistants (among whom he numbered Robert Hooke and Denis Papin) to design a more practical pump to be used for experiments. He specifically wanted one with a glass container as well as one into which large objects could be easily inserted. The inventive Hooke complied, and he and Boyle produced a series of improved pumping systems. Boyle then carefully conducted a range of landmark experiments on the physical nature of air that proved conclusively the correctness of Torricelli's and Guericke's theories. Boyle first wrote of his pump and his findings in 1660 in his *New Experiments Physico-Mechanical, Touching the Spring of Air, and Its Effects*. Although the Library of Congress does not have this first, English edition, it does have the 1680 Latin version, *Nova experimente physico-mechanica*, published in Geneva. It was through this edition that Boyle's work became widely known. The Library also has his 1669 book, *A Continuation of New Experiments, Physico-Mechanical*, published in Oxford, which further documents many of his later air pump experiments. Following his 1660 publication, both Guericke and Christiaan Huygens produced improved air pumps incorporating Boyle's advances.

The final major technical invention of this era to be discussed at length here is the pendulum clock. Once again, the story begins with Galileo and involves another very familiar name of the period, Christiaan Huygens.

While still a teenager in Pisa, Galileo had discovered the principle of isochronism—that a pendulum swings in constant time, irrespective of the width of its swing. Nearly sixty years would pass before the blind and dying Galileo would suggest to his son, Vincenzio, that such a concept could be used to power a clock, by having a pendulum escapement replace the traditional verge and foliot. Vincenzio died in 1649, seven years after his father, and never completed the plan.

In 1656, Huygens began to develop the idea of using a pendulum as a regulator in a clock. It is believed he pursued this notion unaware of Galileo's priority. The similar needs of two very different communities—astronomers and navigators—may have motivated his research, for both badly needed an accurate time-keeping device. Now that astronomers could track the planets and their satellites with telescopes, they realized the old, spring-driven clocks were too imprecise to measure the planets' regular movements. Dutch sailors who were opening up the East felt the same, for now that they were sailing east or west in appreciable distances, they needed to keep accurate time to be able to calculate and adjust their longitude tables correctly.

In the winter of 1656–57, Huygens inaugurated the era of accurate timekeeping with his invention of the pendulum clock, which he patented in 1657. Many would have been satisfied with such an accomplishment, but Huygens's theoretical abilities surpassed even his mechanical skills. Thus, after inventing the "grandfather clock," he proceeded to try to understand it. Further theoretical study on curves and oscillation led him to conclude that for absolute accuracy, the pendulum's arc had to be slightly off the path of a perfect circle—an effect he was able to achieve through attachments at the pendulum's fulcrum. Huygens described his perfected clock in his *Horologium oscillatorium*, published in Paris in 1673. The Library of Congress has this first complete treatise on the modern clock in first edition. Besides being a major technological milestone, this work contains sufficient original theoretical insights and theorems on curves and the dynamics of circular motion to rank on a par with Newton's *Principia* of fourteen years later.

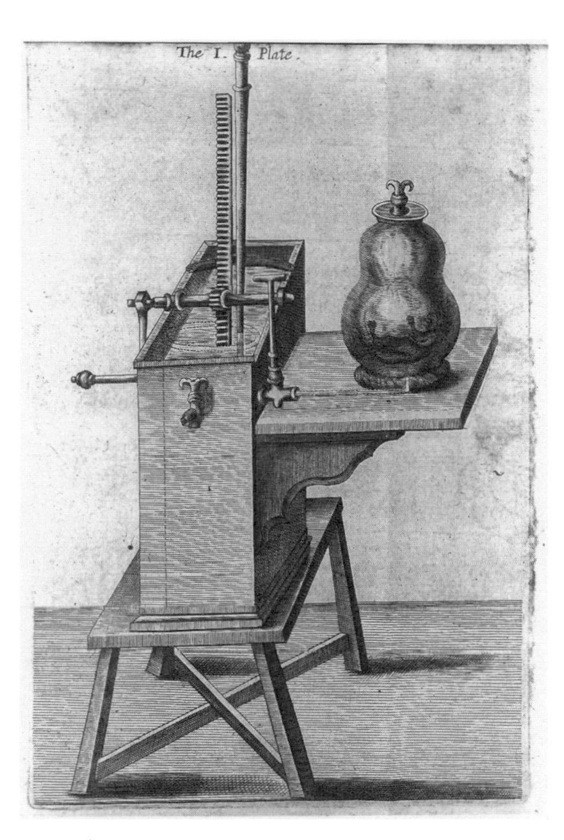

In 1659, Robert Boyle and Robert Hooke collaborated in building an improved air pump, shown here connected to a glass receptacle for experiments. The pump's cylinder is placed vertically and is operated from below. A fitted winch and rack make the pump's piston easy to move. An oiled brass plug has been fitted into a small hole at the top of the glass as a valve, and a cock above it regulates the entry of air. The forethought and technical precision required to construct this device show an increasing degree of technological understanding and competence. This allowed Boyle to demonstrate that a small animal will die in a vacuum (right). *Continuation of New Experiments, Physico-Mechanical*, 1669. Robert Boyle. LC-USZ62-110459.

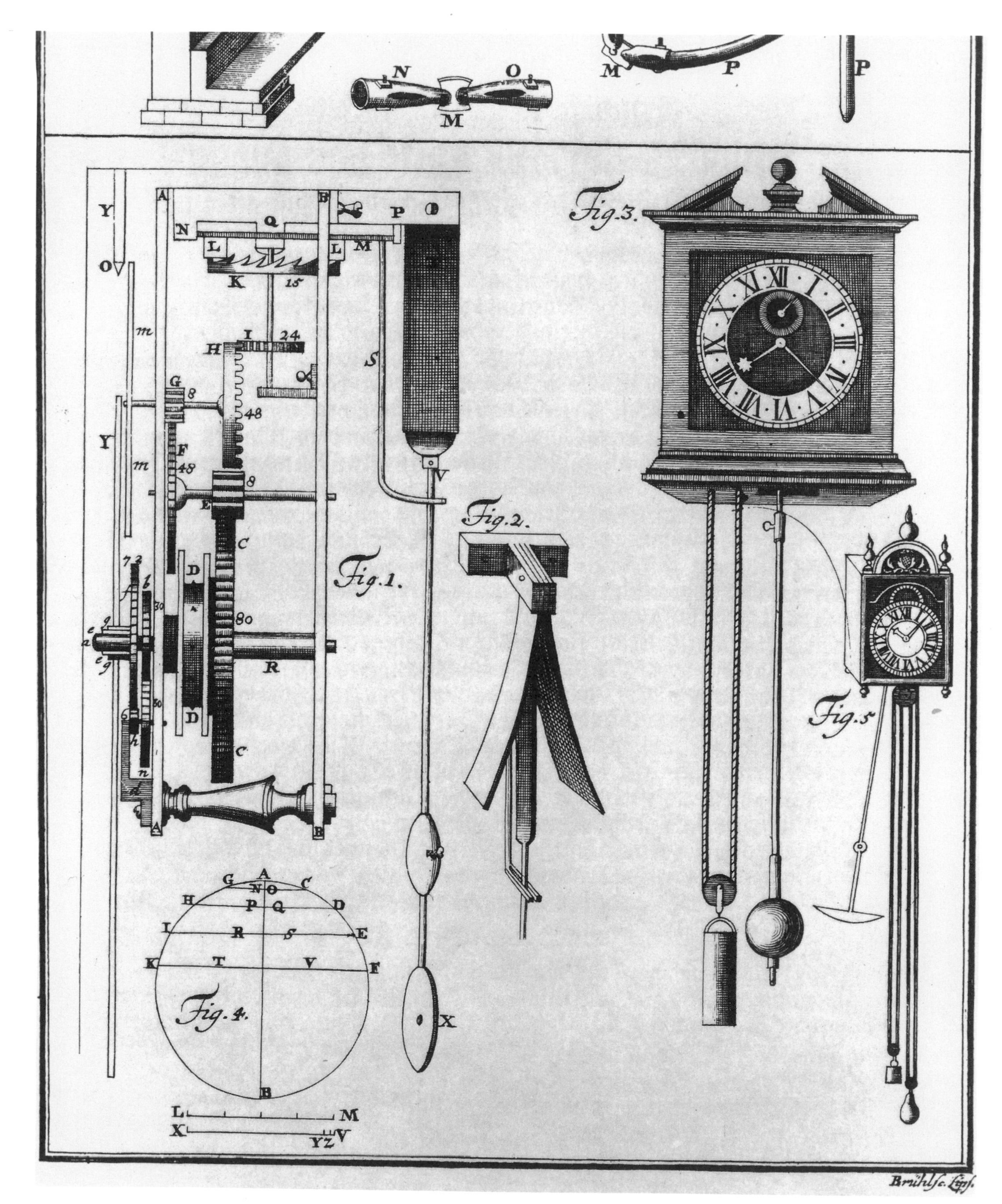
Fig.1.
Fig.2.
Fig.3.
Fig.4.
Fig.5
Brühl sc. Lips.

An inventory of available devices, instruments, and machines at the beginning of the seventeenth century would be a very short list when compared to what was produced by century's end. Steam pumps, air pumps, telescopes, microscopes, barometers, thermometers, and pendulum clocks are only the major technological accomplishments of the age. To this list might be added, in the area of navigational aids alone, the marine clock, sounding instruments, seawater samplers, magnetic dipping needles, wind gauges, and hydrometers. In other areas, the first (frictional) electrical generator, the hygrometer, and the beginnings of the first calculating machines are worthy of note.

None of these devices simply appeared, but were, rather, the result of many individuals' ingenuity, insight, hard work, and perseverance. Each new device or instrument had a creative human mind behind it and a concrete, human purpose as an objective. If there were one individual who personified the technical accomplishment and inventive passion of this time, however, it would be an Englishman, Robert Hooke. Hooke's perception of the natural world perfectly embodied the prevailing Cartesian notion of nature as a great machine. A contemporary described Hooke's mechanics as "his first and last Mistress." From the founding of the Royal Society in 1662, Hooke was its curator of experiments for the next fifteen years, and it was during this demanding time that he produced his most creative work. Besides his major and very popular book on microscopy, Hooke worked with and contributed something to every major device and instrument of his time. In addition to the *Micrographia*, there are several works in the Library of Congress collections that demonstrate the true inventiveness of this "Newton of mechanics." Five titles are bound together with the Library's copy of Hooke's *An Attempt to Prove the Motion of the Earth* (London, 1674), and each shows his mechanical range and accomplishment. These are: *Animadversions on the First Part of the Machina coelestis* (London, 1674); *A Description of Helioscopes and Some Other Instruments Made by Robert Hooke* (London, 1676); *Lampas, or Descriptions of Some Mechanical Improvements* (London, 1677); *Lectures and Collections: Cometa, Microscopium* (London, 1678); and *Lectures De potentia restitutiva; or, Of Spring, Explaining the Power of Springing Bodies* (London, 1678).

In the field of architecture and its related disciplines, two treatises bracket this chapter's time period, although each stretches the arbitrary 125-year boundary somewhat. Both, however, are part of the Vitruvian or classical tradition. The earlier book was first published in 1570 and is the work of the last of the great humanist architects, Andrea Palladio. At the other end of this time frame is the work of another Italian, Piranesi, whose book was first published in 1748.

Andrea Palladio was a builder like his predecessors, the Roman Vitruvius and the Renaissance architect Alberti. And like them, he also produced an architectural treatise of lasting influence. In his case especially, his writings account for his fame more than do the many palaces and villas that he built. He began his career apprenticed to a sculptor and then went to work for a stonemason in Vicenza at sixteen. There he was noticed by the humanist poet and scholar Count Gian Giorgio Trissino,

Opposite page:
A major advance in the precise measurement of time was made when Christiaan Huygens built the first pendulum clock (left). Its construction followed only after Huygens understood the theory of the pendulum (already known by Galileo). Essentially, he learned that the time required for a complete swing (period) depends only on the length of the pendulum and is almost independent of the extent of its arc. His invention revolutionized the clock industry, which soon made clocks with short, weight-driven pendulums that hung on walls (right). The more precise measurement of time would also prove a boon to science. *Theatrum machinarum molarium,* 1725. Jacob Leupold. LC-USZ62-110460.

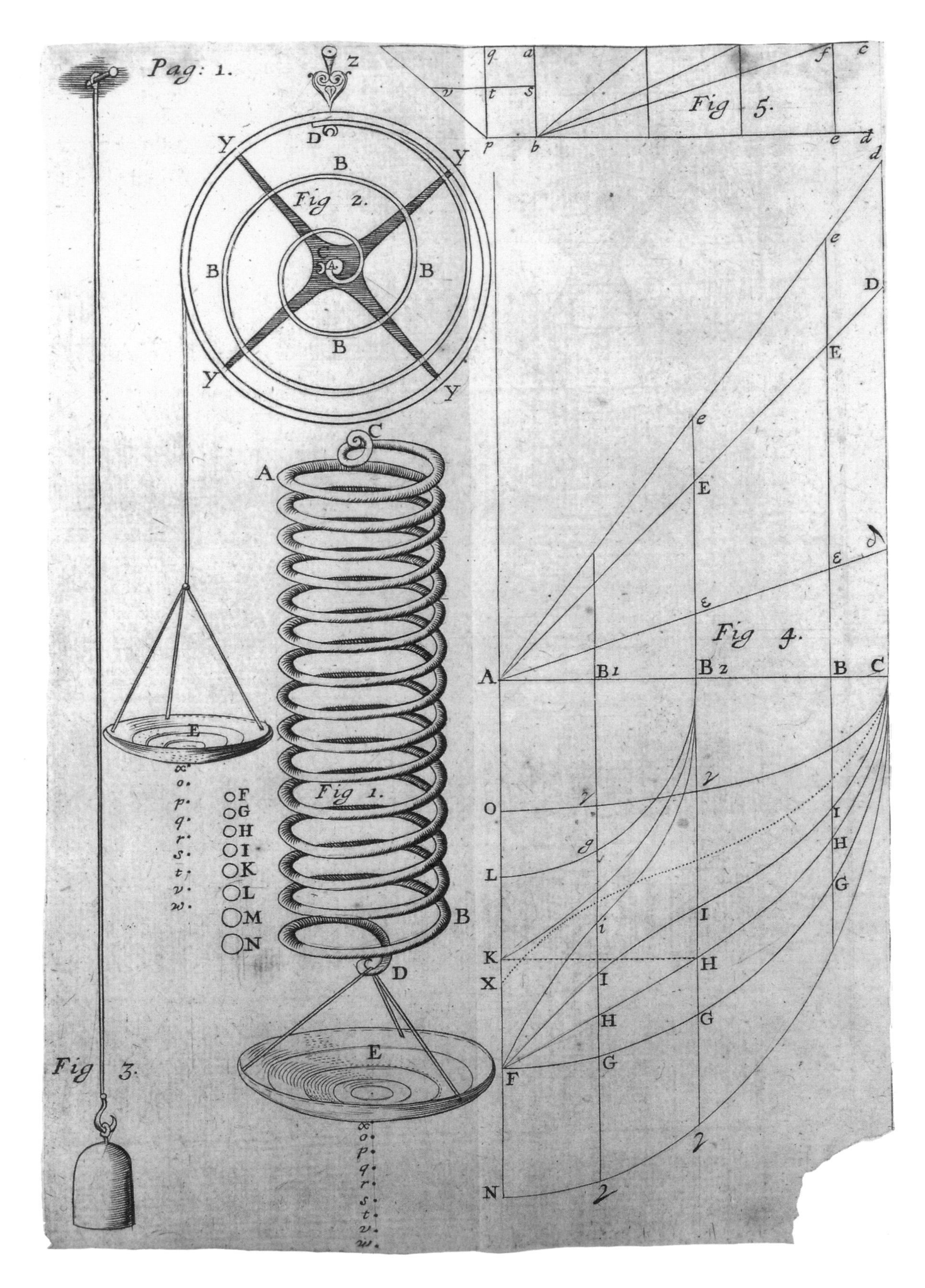
Pag: 1.
Fig 2.
Fig 1.
Fig 3.
Fig 4.
Fig 5.

who was building his version of a Vitruvian villa, and who took the talented Andrea under his wing. Trissino not only educated and influenced the youth but gave him a new name as well. He changed the boy's christened name of Andrea di Pietro della Gondola to "Palladio," after a character who symbolized classical architecture in one of his epic poems. Following twenty years of intensive building, Palladio published his lasting work, *I quattro libri dell'architettura* (Venice, 1570). Quite possibly the most influential architectural book ever printed, this summary of his work and studies of classical architecture would have a profound influence far beyond his native land for more than two centuries. The Library of Congress has this work, which might best be described as a manual of classicizing design. Two later editions (1581 and 1601) are also in the Library's collections.

Palladio has been described as having achieved the mission of the Renaissance for architecture by rediscovering, reviving, and reinterpreting the essence of antiquity. His appealingly clear and simple designs are soundly based on the classical imperatives of symmetry, order, and mathematical proportionality, all presented in a monumental manner. Through his book, he was particularly influential in England, as seen in the work of Inigo Jones, the founder of the English classical tradition of architecture. Jones not only designed the Queen's House at Greenwich in the classical style, but copiously annotated his own copy of *Architettura*. These notes were included in the first English translation of the work, made by Giacomo Leoni. The Library of Congress has the third edition of this work, titled *The Architecture of A. Palladio* (London, 1742).

It was also in England, in the 1680s, after the devastating plague of 1665 and the great fire of the next year, that a revolutionary book was published whose subject was not the grand designs of classic architecture but the mundane and practical concerns of the humble tradition of building. Written by Joseph Moxon, a printer, cartographer, globe maker, and mathematician, *Mechanick Exercises or the Doctrine of Handy-Works* was far ahead of its time in many ways. Its most unconventional aspect was Moxon's deliberate intent for the book—to break away from the old monopoly of the "mystery" of the crafts and to lay bare the details of carpentry, turning, smithing, and the like for all who could read English. Although the genre of technological picture books had already begun to come into its own, Moxon's book was quite different from these pictorial surveys. First, it was written in plain, everyday language and was intended to be used as a handbook. This meant its text and illustrations had to be simple, direct, detailed and effective. Second, it was intended for masters to use in instructing their apprentices, covering in a matter of weeks what had normally required a much longer period. Third, it was issued initially as a serial publication, coming out monthly in segments and costing fairly little (six pence). Obviously, Moxon was making sure that his open exploration and demonstration of the crafts in an illustrated book would not only reach a large audience but would also be available and affordable to the people for whom it was intended.

Since the invention of mechanical printing, few printers or publishers

Opposite page:
As one of the most technically creative and able "mechanics" of all time, Robert Hooke was so described by Samuel Pepys, who said, "Mr Hooke, who is the most . . . promises the least, of any man in the world." This drawing is from his work on springs in which he measured stress in springs and posited what became "Hooke's Law." This law enabled technicians to calculate the strength of springing mechanisms, from a bow to a watchspring. Hooke is also credited with such technical innovations as the anemometer, reflecting quadrant, wheel barometer, gear cutter for watch wheels, and universal joint, among many others. *An Attempt to Prove the Motion of the Earth,* 1674. Robert Hooke. LC-USZ62-110461.

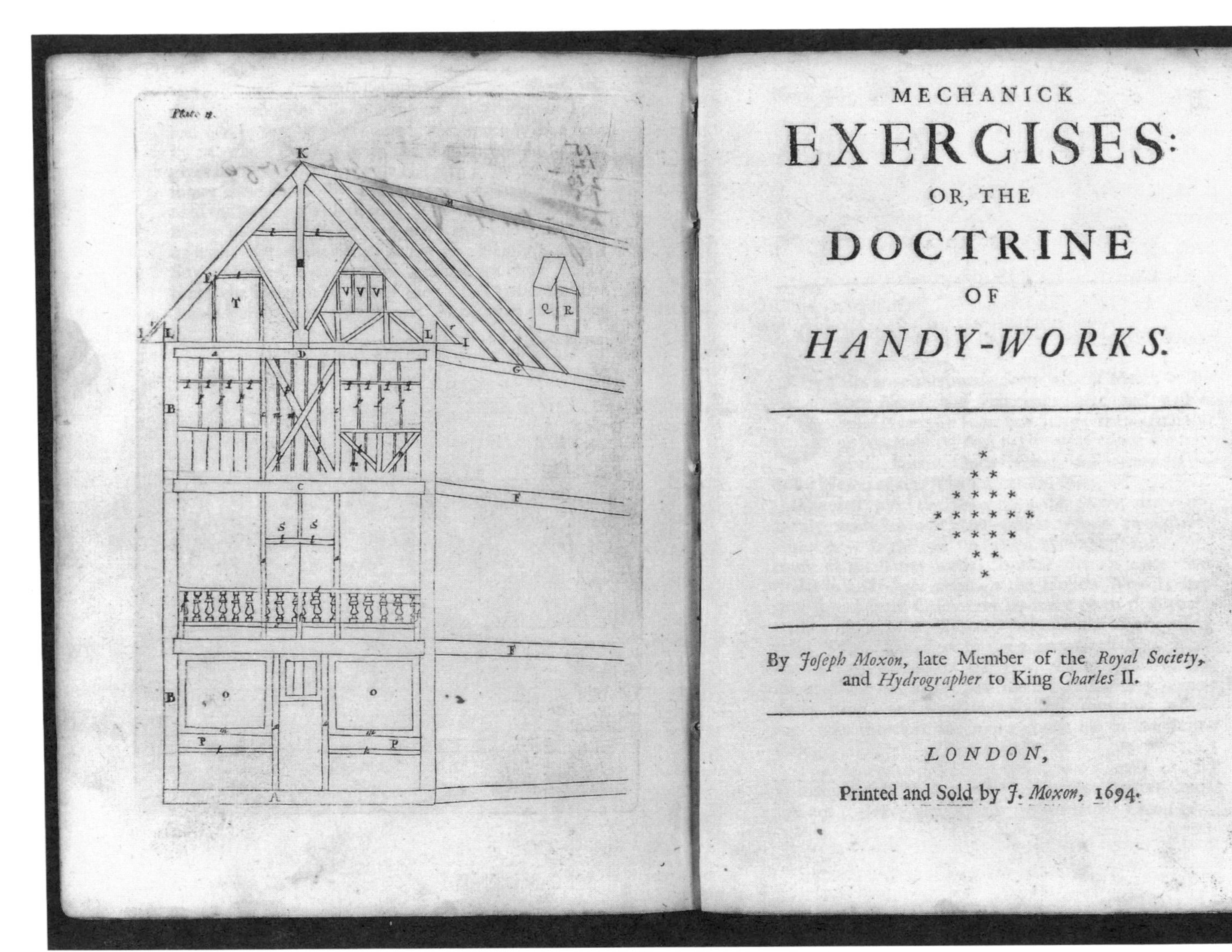

MECHANICK
EXERCISES:
OR, THE
DOCTRINE
OF
HANDY-WORKS.

By *Joseph Moxon*, late Member of the *Royal Society*, and *Hydrographer* to King *Charles* II.

LONDON,
Printed and Sold by *J. Moxon*, 1694.

There is nothing grand about the simply planned framed building shown in Moxon's frontispiece. Since his book dealt with the humbler, more ordinary aspects of everyday construction, the drawing is appropriate. Moxon's plain and direct prose matched his mundane subject as he detailed all aspects of working with wood—describing and showing the tools and techniques of the carpenter, the turner, and the joiner, among others. In its time, the book was useful and affordable; three centuries later, it is both a valuable record for the history of technology and a revealing document of ordinary life. *Mechanick Exercises,* 1694. Joseph Moxon. LC-USZ62-110462.

had taken advantage of the new medium to try to communicate with a new, large class of readers. Although André Félibien's 1676 work, *Des Principes de l'architecture, de la sculpture, de la peinture, et des autres arts qui en dependent*, might seem to be the immediate precursor and inspiration of Moxon's effort, it differed substantially in purpose and content. Where Moxon said he wrote for "Any Man, tho' of an Ordinary Capacity," Félibien's book explained the principles of the beaux arts to gentlemen "who do not wish to apply them in depth." Thus where Moxon's book was a useful handbook, Félibien's was more for the dilettante. The Library of Congress has Félibien's illustrated work as it was first printed in Paris in 1676.

The Library's copy of Moxon's *Mechanick Exercises* is considered to be the first edition. Since the book began as a series of monthly issues, its first compilation was issued in London in 1677, consisting of fourteen numbers. A second volume of this first edition containing additional numbers was published in 1683. The Library of Congress has this two-volume work as well as the second London edition, published in 1694, which has only the first fourteen numbers. The Library also has the enlarged third edition, with twenty-four numbers, published in London in 1703.

It is now believed that Moxon died in 1691 and that the J. Moxon who added later numbers, specifically on bricklaying, was his son, James. This is significant, for it gives us insight into why the house plan rendered by the elder Moxon gives details of a seventeenth-century house, while the plan offered in the son's bricklaying section is definitely that of a London home (or one in Philadelphia or New York) of the eighteenth century. Altogether, Moxon's innovative work not only offers posterity the most telling details on the architect's supporting crafts as they existed in the late seventeenth century, but also gives insight, in text and pictures, into the culture of that period.

Ranging somewhere between the high art of a Palladio treatise on architecture and the more mundane techniques of Moxon's handbook were a number of illustrated books that became very important to the history of technology. The theater machine books of Besson, Ramelli, and Zonca discussed in the previous chapter were the vanguard of such books, most of which were heavy on engineering and mechanics. Dealing with devices and techniques that pertained primarily to architecture, the "Queen of the Arts," which was regarded as encompassing all other practical arts, these aesthetically attractive books of mechanical inventions document in word and picture the technology of this interesting and important time.

The style and composition of these rooflines and the classical design of the window treatments contrast sharply with Moxon's straightforward illustrations. Where Moxon's book was intended as a handbook for instruction and actual use, this work was aimed at the monied gentleman who wished only to be informed of the principles and methods of the art of building. Félibien was secretary of the Royal Academy of Architecture, and the plates for his book served as models for those used in the *Description des arts et metiers*. They later were used as prototypes for the plates in the *Encyclopédie*. *Des principes de l'architecture*, 1676. André Felibien. LC-USZ62-110463.

One of the earliest of these seventeenth-century books is *Theatrum machinarum*, by Heinrich Zeising. This small, oddly shaped book (it is nearly square) contains twenty plates, nearly all of which are redrawn versions of the images that appeared earlier in the books of Besson, Ramelli, and Zonca. Many of the illustrations pertain to military engineering or to ways of using water power. Published in Altenberg in Meissen, the Library's copy of Zeising's book is bound with *Musica mathematica* of Abraham Bertolus, published in 1614.

Printed about the same time was the machine book of Faustus Verantius, known also as Fausto Veranzio. Verantius came to his interest in the technology of machines fairly late in life and was an amateur. A Croatian who spent most of his life as a diplomat and administrator, Verantius is believed to have been influenced by Giovanni Ambrogio Mazenta, the first biographer of Leonardo da Vinci, when he moved to Rome in 1606. Mazenta, the keeper of Leonardo's manuscripts, may have shown them to his friend Verantius. Although his book is undated, most agree that Verantius published his *Machinae novae* in 1616 in Venice, a year before his death. The Library of Congress has this now-rare work in first edition. It contains forty-nine plates of Verantius's machines and "inventions." It is thought that he drew them himself, since they are not very polished or artistic. Many illustrate various machines powered by water and wind that are simply variations on a very old theme, but his drawings of agricultural devices and implements may indeed have been original. Giving

It is thought that Verantius may have seen the drawings of Leonardo, for among his derivative illustrations of machines and inventions is this image of what he calls a "flying man." This man's parachute is actually similar to Leonardo's design and conveys the Renaissance enthusiasm that all things are possible. *Machinae novae,* 1616. Faustus Verantius. LC-USZ62-110464.

credence to the influence of Leonardo, via Mazenta, is his design for a parachute. *Machinae novae* is a very large book (36 cm tall) and some of the plates cover two pages. It differs from many of the other theater machine books in that its text is limited solely to summaries of or captions for the images. As a linguist, Verantius was able to offer his readers the same explanatory text in five different languages (Latin, Italian, Spanish, French, and German).

In 1618, Octavius de Strada published a similar collection of machines that, while dealing with mechanical engineering in general, is heavy on mills and water works. Although Octavius says the drawings are the work of Jacobus de Strada, an antique dealer and art merchant, it is not clear whether it is Octavius the son or the grandson of Jacobus who made the claim. The Library of Congress has the French version of this book, *La première partie des desseins artificiaulx de toutes sortes des moulins, à vent, à l'eau, à cheval & à la main*, published in Frankurt am Main in 1617.

Another book in the Library's collections is similar to that of Verantius and Strada in both the amateur status of its author and the crudity of its sometimes fanciful designs. Written in Italian, Giovanni Branca's *Le Ma-*

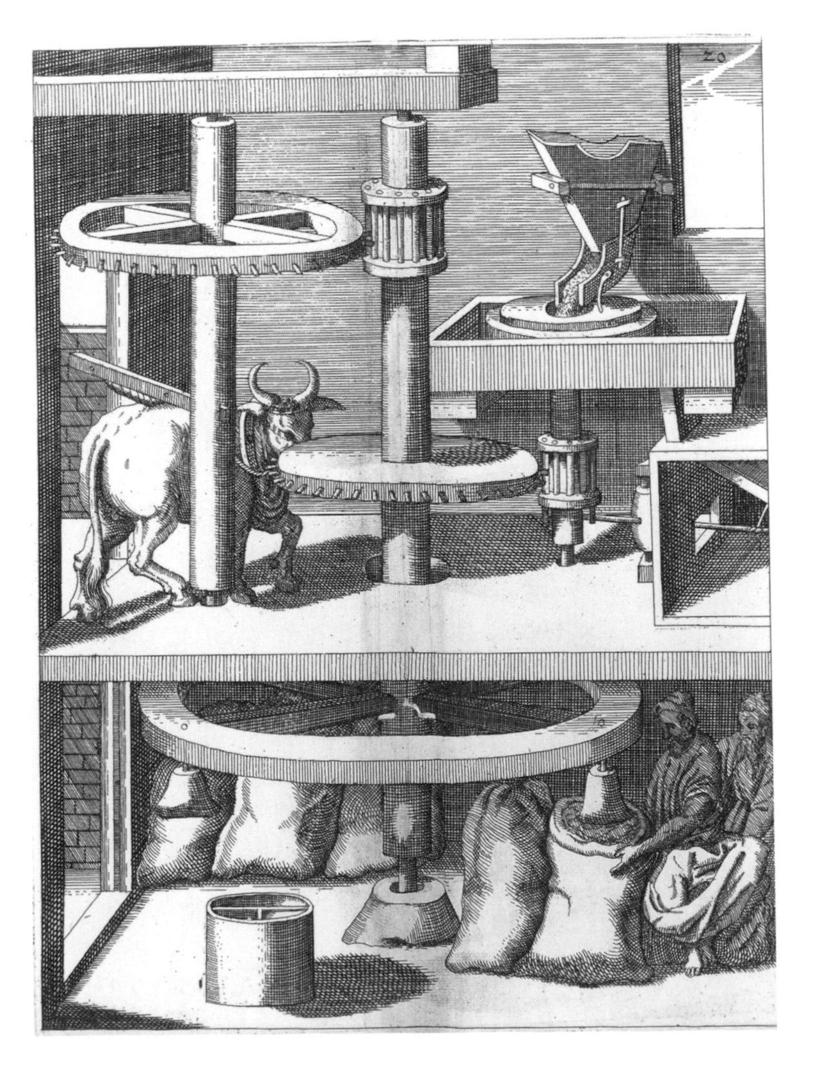

More in keeping with the subject and tone of traditional "theater of machine" books is Strada's early seventeenth-century effort. Although many of its illustrations document the mills and water works of his time (like this ox-powered mill), none is original with Strada, who, in fact, was an art and antiques merchant. He is now considered the first to make a successful living collecting paintings, sculpture, and antique coins from all over Europe for his wealthy patrons. *La première partie des desseins artificiaulx,* 1617. Octavius de Strada. LC-USZ62-110465.

chine was published in 1629 in Rome. Of its seventy-seven full-page plates, some are particularly noteworthy, but in general the book's draftsmanship and the author's imagination are well below the high level achieved by the early trio of Besson, Ramelli, and Zonca. One of Branca's more famous illustrations, however, shows for the first time how steam is applied to do useful work. Like Verantius, Branca restricted his text to fairly short paragraphs describing each illustration.

It is generally thought that following the best of the machine books—by Besson, Ramelli, and Zonca—produced during the last quarter of the sixteenth century, there was a general decline in quality. However, the publication in 1661 of Georg Andreas Boeckler's *Theatrum machinarum novuum* again showed how lively and informative a machine book could be when done well by a real professional. Boeckler was a German engineer and architect who left to posterity one of the great decorative books of the

The illustrations for this book by Boeckler, who was a German engineer and architect, indicate his professionalism. Among the 154 full-page engravings of the working machines and engineering practices of his time is this image of a large hand pump atop a reservoir being used to put out a fire in a three-story building. This method was actually very inefficient, since most pumps of Boeckler's time gave only a feeble stream of water, and the apparatus had to be positioned dangerously close to the fire. In this case, the water stream is actually very short, and the fire is reached by a very long—and probably impractically heavy—tube-like extender made of bronze, brass, and iron segments. *Theatrum machinarum novuum,* 1662. Georg Andreas Boeckler. LC-USZ62-110466.

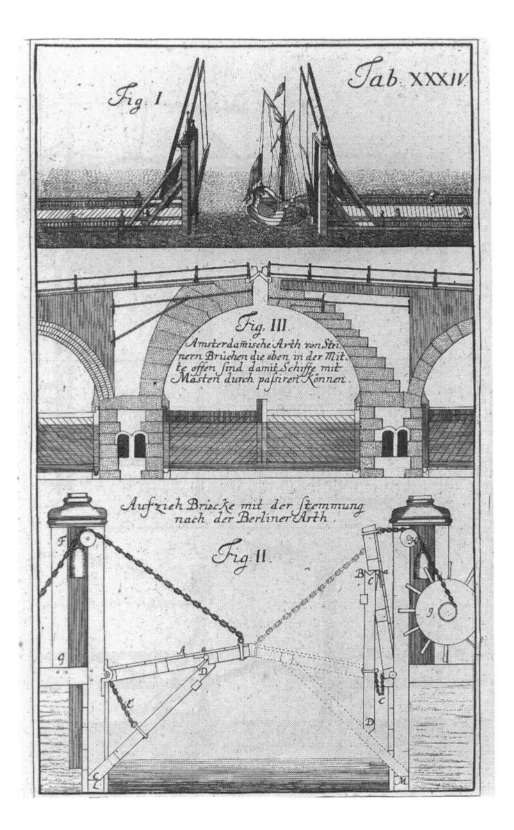

In his encyclopedic summary of mechanical engineering, Leupold writes with authority of the entire presteam era, and his series does include a treatment of steam power up to his time. His massive nine-volume work describes every conceivable technology of the time. He writes with special authority in this treatise on bridges, which is the most comprehensive of the early German bridge books. *Theatrum pontificale, oder Schau-Platz der Brucken und Brucken-Baues,* 1725. Jacob Leupold. LC-USZ62-110467.

baroque era. Despite its Latin title, the 1661 first edition was written in German. The Library of Congress has this important work both in its first edition (Nuremberg, 1661) and its Latin version (Cologne, 1662). Both contain 154 full-page etchings of great detail and interest. This tour de force shows many actual, working machines and engineering practices of the time. The majority of the plates depict machines for raising water and grinding grain, powered variously by wind, water, or men and animals. The last plate in the book shows a team of fire fighters spraying water on a blaze using a hand pumper on a sort of sled.

Following Boeckler, the last great similar effort of this time period was produced by another German of a later generation. Between 1724 and 1739, the massive nine-volume *Theatrum machinarum* by Jacob Leupold was published in Leipzig. Each of the nine volumes carries its own title, but the series was bound into ten volumes. Complete sets are very rare, and the Library of Congress has all nine titles in the series, although three are of a slightly later edition.

Leupold was a mechanical engineer and mathematician, and he established a workshop in Leipzig to manufacture mathematical and mechanical instruments. His *Theatrum machinarum* is much more than a picture book and should be described as the first systematic treatment of mechanical engineering. Leupold writes with authority on many of his subjects, which range from dikes and hydraulic measuring instruments to water wheels, turbines, lifting gear, weights, balances, bridge construction, calculating machines, and mills. Together, the nine volumes contain about 525 finely detailed and often beautiful plates. It is said that James Watt learned German in order to read Leupold's work.

The final book in the area of architecture and engineering falls somewhat beyond the cutoff date of this chapter, but it is intellectually very much a part of it. By the first part of the eighteenth century, some architects were already rebelling at the decorative excesses of the late baroque and rococo style and were countering it with neoclassicism. Foregoing the Renaissance interpretation of antiquity, neoclassicism sought to confront and be inspired by the ancient world itself, without an intermediary, and the city of Rome became the goal and pinnacle of regular artistic and architectural pilgrimages. Giambattista Piranesi was a Venetian architect and printmaker who was particularly attracted to the Roman aspect of the classical style, and he championed a version of neoclassicism which downplayed the contributions of the Greeks to the advantage of the Romans' native predecessors, the Etruscans. The nationalistic excesses of Piranesi aside, he left behind a unique body of etchings that captured the grandeur of Roman architecture in both their drama and their unparalleled accuracy. His four-volume *Le Antichita Romane*, published in folio size, played an important and influential role in the course of neoclassicism. First published in 1748, Piranesi's book depicts Rome not in its glory but in a kind of decaying grandeur. His Roman ruins are huge, desolate, dark and compelling, ravaged and even menacing. The Library of Congress has the 1756 edition of Piranesi's original and inspired work, published in Rome. His engravings of the remains of ancient Rome offer the rare combination of exactitude and inspiration, in that they both document and interpret, with neither aspect detracting from the other.

Returning to the more practical but no less inspired technological accomplishments of this period, it is here that the direct ancestors of the modern computer are found. Although the history of mechanical aids to calculating actually begins with the abacus—an ancient device probably of Babylonian origin that was commonly and universally used in medieval Europe—the beginnings of modern computing are found, not surprisingly, in the time frame of the scientific revolution. Interestingly, it is a mathematical discovery and not a mechanical invention that marks the beginnings of modern computing. When, in 1594, a Scottish lord named John Napier turned his energies from inventing destructive devices to be used in the war with Spain to more peaceful pursuits, he invented an entirely new method of calculation. Once he discovered that all numbers can be written in exponential form (the number 4 can be written as 2^2, and 16 as 2^4), Napier realized that multiplication and division could be no more complicated than addition and subtraction.

Following this discovery, he spent twenty years studying and gradually elaborating his computing system in which roots, products, and quotients could be simply and quickly determined from his tables, which showed powers of a fixed number used as a base. In 1614, at the age of sixty-four, he published his *Mirifici logarithmorum canonis descriptio* in Edinburgh to the great joy and surprise of the scientific community of the time. Napier's logarithms were recognized almost immediately and embraced, especially by astronomers, as a quick, simple, and accurate way of avoiding the onerous task of routine calculations. The book had a stimulating impact on Europe. The Library of Congress has a first edition copy of this work which contains Napier's logarithmic tables and the rules for their use.

Three years later, in 1617, the year of his death, Napier published in Edinburgh his *Rabdologiae, seu numerationes per virgulas*. It was in this book, which the Library has in first edition, that he described his invention of a calculating device, the virgula or rod, that became commonly known as "Napier's bones." This ingenious device was the forerunner of the slide rule. It was made of ten movable rods or oblong pieces of wood, horn, or ivory and looked like a chessboard with numbers in its squares. An improved version mounted the rods on a rotating cylinder arranged in a box. Essentially a mechanical multiplication table, the device quickly became well known, mostly because of its inclusion in Gaspar Schott's wide correspondence. As a Jesuit and a pupil of Athanasius Kircher, Schott functioned for two decades as the center for an informal international network of Jesuits. It was to him they communicated the most current scientific idea or invention. Over a period of about ten years, Schott eventually published this accumulated collection of the technical wonders of his day. It took over ten volumes to contain it all. It was through Schott's 1657 book, *Mechanica hydraulico-pneumatica*, that most of Europe learned of Guericke's vacuum pump, and so it similarly learned more about Napier's device through Schott's *Organum mathematicum*. The

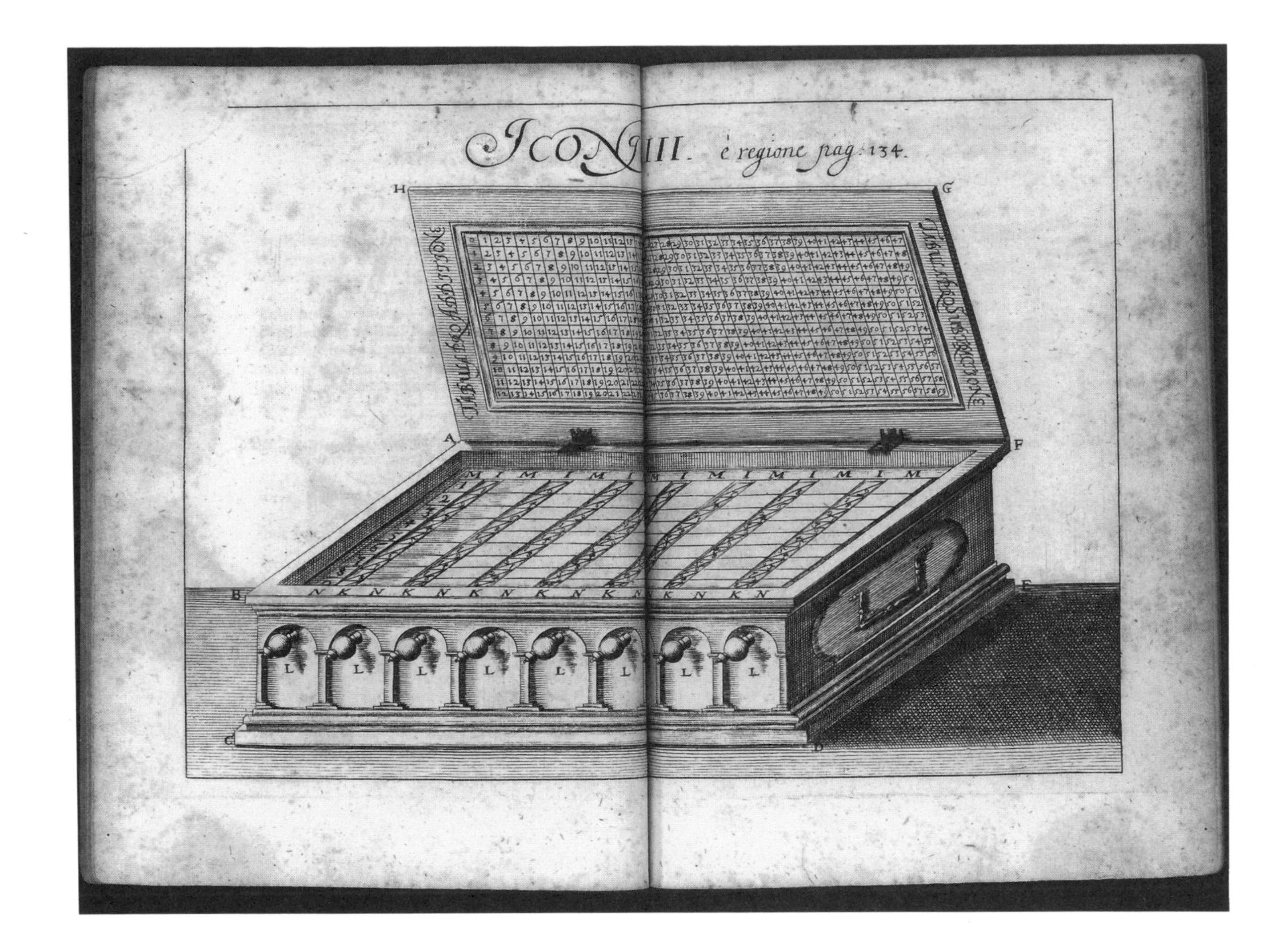

John Napier is best known for his invention of logarithms, but his "numbering rods" or "multiplying rulers" were a fundamental advance in the history of mechanical aids to calculation. Popularly called "Napier's Bones" because often they were made of ivory, horn, or bone, these strips or rods formed a matrix-like grid of numbers when laid together. Calculations were done by rotating the rods. This drawing shows an attempt at improving Napier's device that was made by the Jesuit mathematician Gaspar Schott, who took the simple collection of rods and mounted them in a box so they would stay properly aligned and still be easily turned. *Organum mathematicum,* 1668. Gaspar Schott. LC-USZ62-110468.

Library has a copy of *Organum mathematicum* as it was first published in Wurzburg in 1668. It contains an illustration of Schott's version of Napier's "bones"—a container shaped like a cigar box with a lid and handles and ten knob-like protrusions in the front.

In 1642, the teenaged genius Blaise Pascal knew of Napier's device and began work on what would become the first mechanical digital calculator to do addition and subtraction. Not only was he able to solve the theoretical problem of how to mechanize computation, but he designed a machine capable of doing it easily and simply. In 1645, he produced the definitive model of what he proudly called "la Pascaline." The documentation that detailed the machine's purpose, operating principles, and capabilities took the form first of a letter and then of an eighteen-page pamphlet that had a rather ephemeral existence. It was, however, reproduced in many editions of Pascal's collected works, *Oeuvres de Blaise Pascal*; the oldest edition in the collections of the Library of Congress was published in 1819 in Paris. The Library also has an original description and detailed illustration of Pascal's machine as it first appeared in two French sources published nearly a century after Pascal lived. The first is in *Ma-*

The precocious Pascal designed the first of his fifty different calculating machines when he was only nineteen years old, and all were elaborations of the ideas embodied in that 1642 machine. The mechanism within "la Pascaline," as it came to be called, was a complex of toothed wheels that performed the operations of addition and carrying. It was designed specifically for use by accountants, who would rotate its small wheels with a stylus until they came to a stop, much like a dial telephone. Although it could not subtract—its gears would not go backwards—Pascal ingeniously designed it to use complements, and to subtract by addition. A complement is the difference between ten and an integer from zero to nine. To subtract four from seven, six (the complement of four) is added to seven, yielding thirteen. The one is dropped and the answer is three. *Oeuvres de Blaise Pascal,* 1819. Blaise Pascal. LC-USZ62-110469.

chines et inventions approuvées par l'Académie royale des sciences (Paris, 1735–77). The other is in the famous *Encyclopédie; ou, Dictionnaire raisonné des sciences, des arts, et des métiers*, published by Diderot in Paris between 1751 and 1765. In the *Encyclopédie*, Pascal's "machine arithmetique" appears in volume 1 under the alphabetical listing "Arithmetique (machine)" and it is described at length. Also, in the fifth volume of the accompanying plates, a detail of the machine's inner moving parts is found under the listing, "Algebre et arithmétique. Machine Arithmétique de Pascal."

In 1652, two years before Pascal turned away from science completely and embraced an extreme, ascetic religious fervor, he sent one of his calculating machines to Queen Christina of Sweden. Having become queen elect at the age of six when her father, King Gustavus Adolphus, died in 1632, the young queen distinguished her reign by her lavish patronage of the arts and sciences. She was the patron of composers like Scarlatti, architects like Bernini, scholars like Grotius, and natural philosophers or scientists like Descartes. Therefore, it was not unusual that an ambitious Pascal would send his newest invention to such a distinguished and singular court. It was at this court that an Englishman, Samuel Morland,

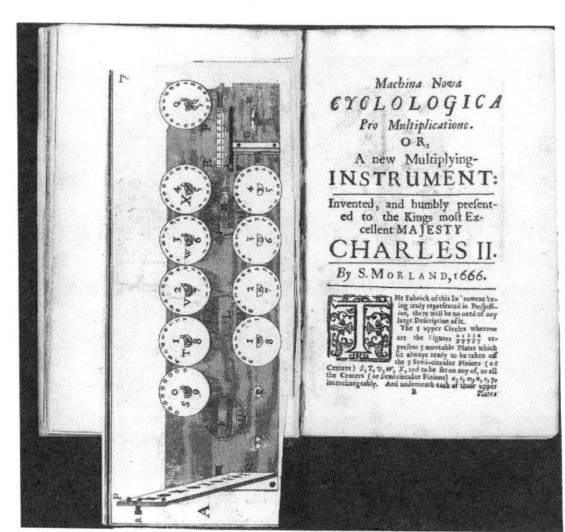

The small calculating machine shown here in Samuel Morland's 1673 book was a circular or disk version of Napier's Bones. It was a multiplying machine that came equipped with thirty separate disks that the operator placed on pins and rotated. It was a much smaller device than Pascal's and much less complex, but it did require attentive operation. *The Description and Use of Two Arithmetick Instruments,* 1673. Samuel Morland. LC-USZ62-110470.

Opposite page:
The intricacy and complexity of the inner workings of the Leibniz calculator are clear in this detailed illustration. His machine went further than any of his time and could perform all four fundamental arithmetic operations, with multiplication accomplished easily and directly. This is attributed to its major innovation—the use of a stepped drum, or pins of a different length. His principle of the stepped-drum gear remained the only workable device for calculating operations until 1875. *Theatrum arithmetico-geometricum,* 1725. Jacob Leupold. LC-USZ62-110471.

presumably saw Pascal's machine and began work on his own version. Morland was a member of the English ambassador's party and, although a supporter of Oliver Cromwell during the English Civil War, it is believed that he was in fact a spy for the exiled Charles II. With the restoration, Morland was appointed "Master of Mechanicks" by the king.

Sometime during the 1660s, Morland devised two calculating machines. One was a simple, five-digit adding machine, and the other, a multiplying machine, was a mechanical version of "Napier's bones." It was not until 1673 that Morland published a small illustrated book, *The Description and Use of Two Arithmetick Instruments*, that described both machines and detailed the types of operations they could perform. The Library of Congress has the first edition of this work, published in London in 1673. Unlike Pascal's mechanically complex instrument, Morland's machines were essentially hand calculators with gear wheels operated by a stylus. Further, they were small enough (3 by 4 inches) to be carried in a pocket. Despite their simplicity and size, the machines did not sell in great numbers. Morland's friend, Samuel Pepys, later remarked candidly in his diary that Morland's machine was "very pretty but not very useful."

During the same year as Morland's publication (1673), René Grillet published in Paris his *Curiositez mathématiques de l'invention du Sr. Grillet.* In this book, which the Library has in first edition, Grillet describes his calculating machine which, although still not very sophisticated, did have an eight-digit capacity in its adding mechanism. One of the illustrations in his book shows a rectangular box with a set of twenty-four concentric circles which the operator would manipulate to perform calculations. Grillet, about whom little is known other than that he was at one time the clockmaker to Louis XIV, published a short article in 1678 in the *Journal des Scavans* that described his machine. The Library has the issue of the pioneering French journal which includes Grillet's article, entitled "Nouvelle machine d'arithmetique."

While credit for the first calculating machine must go to Blaise Pascal, it was another prodigy, Gottfried Wilhelm Leibniz, who built what was to become the precursor of the modern calculator. Most importantly, Leibniz devised the essential stepped-drum mechanism that became the key part of all calculating machines for nearly two centuries. A universal genius whose thought dominated the field whether it was mathematics or metaphysics, Leibniz began work on his machine sometime in 1671 at the age of twenty-five and completed his final version two years later. Whereas Pascal's machine would only add and subtract, Leibniz invented a device that could perform all four fundamental operations of arithmetic. Essentially, it would multiply and divide by the mechanical repetition of adding and subtracting. Although he presented his invention to the Royal Society during his first visit to London from January to March 1673, it was not until 1685 that he attempted to document his machine. That year, he wrote a manuscript in Latin, "Machina arithmetica," that was never published. It remained in the Royal Library of Hannover until it first appeared in 1897 in volume 26 of *Zeitschrift für Vermessungswesen*, both in its original Latin and in German. In 1929 the historian of

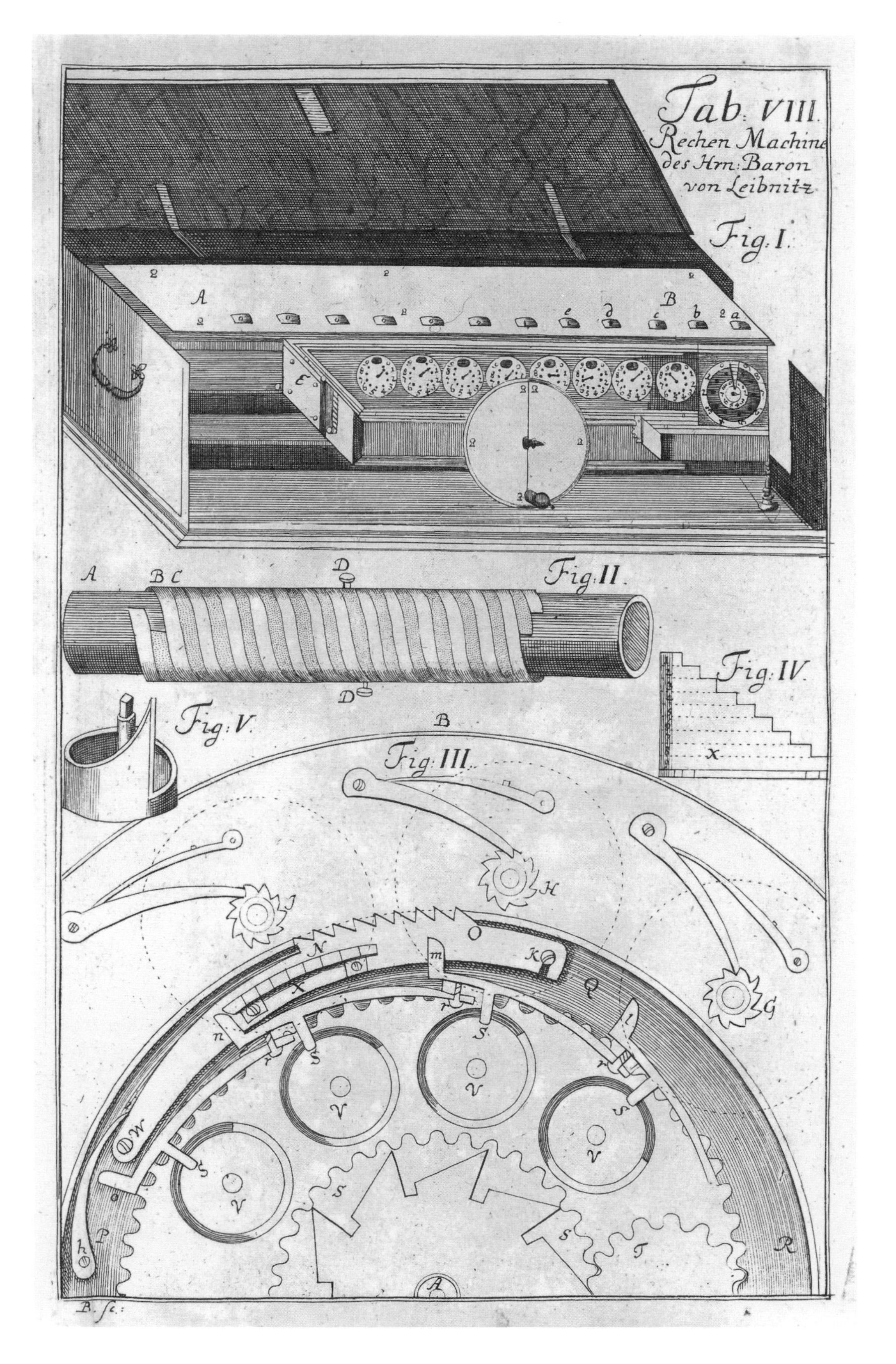
Tab: VIII.
Rechen Machine
des Hrn: Baron
von Leibnitz
Fig: I.
Fig: II.
Fig: III.
Fig: IV.
Fig: V.

mathematics, David Eugene Smith, included an English translation in his *Source Book of Mathematics*. The Library of Congress has both Smith's book and the 1897 Latin and German versions in its collections. Although Leibniz's machine is pictured in the 1897 article, it was also illustrated in a much older and more interesting source. In the multivolume *Theatrum machinarum* produced by Jacob Leupold from 1724 to 1739, volume 8 is titled *Arithmetico-Geometricum* and contains plates showing the machine invented by Leibniz (in chapter 10, tabula 8, figure 3, and in chapter 11, tabula 9, figure 3). Elsewhere, Leupold shows not only his own invention of a calculating machine but also "Napier's bones" and Grillet's machine as well. The Library has Leupold's volume 8 in first edition.

As this chapter spans a time of scientific awakening and major technological accomplishment—a proud period in the history of mankind—so it also contains a dark period when Europe seemed to be tearing itself apart. The Thirty Years' War, which raged for actually fifty years (1610–60), was a religious and political conflict that reached new extremes of intolerance, bloodshed, and inhumanity. From the standpoint of weaponry and military practice, the growing use of firearms, the standardization of weapons and technique, and the steadily increasing size of the armies made this war a precursor of mechanized, unrestricted, mass warfare. It was in this conflict that muskets became the standard infantry weapon and killing from a distance became an easily achievable and almost mechanical act. In a religious conflict exacerbated by extreme political divisions, no quarter was given. Prisoners were slaughtered, civilians routinely killed, and whole cities leveled by siege. No doubt, much of this would have occurred with or without the improvements in firepower made during this time, but technology certainly contributed to the efficiency and lethality of the bloodletting.

Among the many volumes of this period in the Library of Congress collections that deal with the technology of warfare, one of the earliest and most influential was Jacob de Gheyn's work, *The Exercise of Arms*. Gheyn was the artist commissioned to illustrate the step-by-step technique for loading and firing a musket that had been devised by Maurice of Nassau, prince of Orange and son of William the Silent. Maurice had reformed and reorganized the Dutch army, turning it into a professional, highly lethal machine. His detailed rationalization of the forty-three steps required to ready a musket provided drillmasters a way of teaching a standardized procedure that could be repeated rapidly. The reforms and efficient techniques of Maurice were quickly adopted by others, as the republishing and even pirating of his book show. Published originally in Dutch in 1607 with the title *Wapenhandelinghe van Roers, Musquetten ende Spiessen*, it appeared in several languages within a year. The Library of Congress has the book in English as *The Exercise of Arms*, published in The Hague in 1608, and in French as *Maniement d'armes, d'arquebuses*, published in Amsterdam in 1607. In 1614, it was pirated entirely by Johann Jacob Wallhausen, whose *Kriegkunst zu Fuss* even used the same copperplates as Gheyn's work. The Library does not have this German

Opposite page:
By 1550, every unit of every army in Europe had adopted some form of hand weapon that discharged shots by gunpowder. Exercise manuals were then needed to instruct soldiers in proper technique, and one of the best and most influential of its time is illustrated here. The technique of firing a musket is shown by a Dutch soldier, whose weapon is a typical firearm used in the Netherlands around 1600, with a characteristically deep thumb-hole. The musket-rest resembles a walking stick fitted with an iron tip on one end and a solid iron fork on the other. The soldier also carries a leather bandolier holding twelve leather-covered, wooden powder charges. A length of glowing match or punk is connected to a mechanism called the "serpentine." This device ignites the charge when the trigger is pressed. *Maniements d'armes,* 1607. Jacob de Gheyn. LC-USZ62-110472.

version but it does have a Russian translation published in Moscow in 1647 with the romanized title *Uchenie i khitrost' ratnago stroeniia pekhotnykh liudei*.

One of the more complete treatments of the artillery techniques and military engineering of this time period is Jean Appier Hanzelet's *La Pyrotechnie*. Published originally in 1620 with François Thybourel as coauthor and titled *Receuil de plusiers machines militaires et feux artficiels pour la guerre*, Hanzelet's *La Pyrotechnie* was published in 1630 as a new edition

MARTE
ET ARTE
LA
PYROTECHNIE
Appier
DE
HANZELET LORRAIN
ou sont representez les plus
rares & plus appreuuez
secrets des machines &
des feux artificiels
Propres pour assieger battre
surprendre & deffendre
toutes places.
AU
PONT A
MOUSSON PAR
I. & Gaspard
Bernard.
1630
auec
permission

which contained many additions and more illustrations. Appier took the name Hanzelet to distinguish himself from his father, Jean Appier, who was an engineer for Charles III of Lorraine. Hanzelet was "Master of Artificial Fire" to the duke of Lorraine. His book is both a summary of all the relevant knowledge, techniques, and regulations pertinent to artillery and an inventive foray into all manner of speculative machines and bombs. The Library of Congress has the 1630 edition which was published in Pont-a-Mousson. It has 133 engravings, some of which are missing.

In 1639, the Polish historian Szymon Starowolski wrote what became a virtual encyclopedia of military science for the Thirty Years' War. Titled *Institutorum rei militaris* and published in Cracow, this work generally reflects the views of the great Lithuanian military leader Jan Karol Chodkiewicz. Its wide scope treats the nature of war, the proper qualities of leadership, discipline, battles and attacks, naval warfare, strategy, and the defense of cities. As secretary to Chodkiewicz, Starowolski no doubt witnessed at least some of Chodkiewicz's regular military miracles overcoming great odds, whether against Turks or Swedes. The Library of Congress has this work in first edition.

Although the Thirty Years' War involved most of northern and central Europe, France was engaged in that conflict for some time as well as in other wars long afterwards. Indeed, Louis XIV, who was born during the war, confessed to having "loved war too much," and it is not exaggerating to say that war was the normal state for France during his reign. It was during his war with the Spanish Netherlands that a well-placed architect, François Blondel, presented him in 1675 with a manuscript he had written called "L'art de jetter les bombes." As was typical of a monarch who sought to control everything from military tactics to menus, Louis XIV determined that Blondel's work was too important to be published and revealed to the enemy, and he did not allow it to be published until peace was concluded. Blondel, an architect in the classical, Vitruvian style, eventually saw his book published in 1683. The Library of Congress has a first edition copy of Blondel's work, published in Paris.

One of the best examples of the role of naval warfare at the end of this decade was written by Paul Hoste, a Jesuit mathematician and chaplain in the French navy. Published in Lyons in 1697, his *L'art des armées navales, ou Traité des evolutions navales* was essentially two books in one, discussing in remarkable detail and splendid illustrations both war at sea and the construction of ships. Hoste detailed and depicted naval maneuvers used in attack, defense, ambush, and even retreat and much of this formed the basis for later work in this field. The second part of the book was the first to apply mechanical and hydrodynamic principles to the construction of vessels. The Library of Congress has this important work in its later English version, *Naval Evolutions: or, A System of Sea Discipline*, published in London in 1762.

One final noteworthy book in this section on military concerns included in the Library of Congress collections is the first book written by an American author on what might be called military science. Written by Nicholas Boone and published in Boston in 1701, *Military Discipline;*

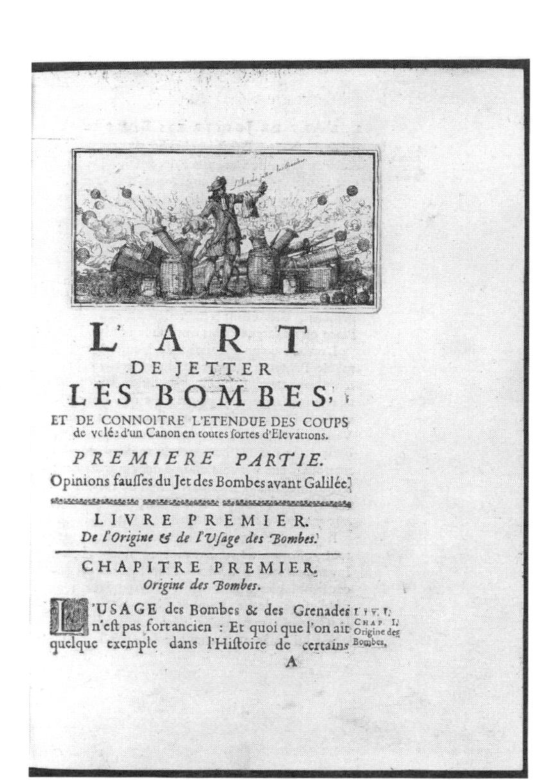

L'ART
DE JETTER
LES BOMBES,
ET DE CONNOITRE L'ETENDUE DES COUPS
de volée d'un Canon en toutes sortes d'Elevations.

PREMIERE PARTIE.

Opinions fausses du Jet des Bombes avant Galilée.

LIVRE PREMIER.

De l'Origine & de l'Usage des Bombes.

CHAPITRE PREMIER.

Origine des Bombes.

L'USAGE des Bombes & des Grenades n'est pas fort ancien : Et quoi que l'on ait quelque exemple dans l'Histoire de certains

LIV. I. CHAP. I. Origine des Bombes.

A

When Louis XIV was presented with the manuscript for this book, he would not allow it to be published for fear it would aid the enemy. Although its accurate firing tables were made known only to his troops, most French gunners stubbornly continued using the old, faulty tables. The talented Blondel was in the infantry at seventeen (where he was appalled at the incompetent French bombardiers), succeeded Gassendi as lecturer in mathematics at the College Royal, and eventually became a naval general. *L'Art de jetter les bombes*, 1683. François Blondel. LC-USZ62-110474.

Opposite page:
There is little doubt about the subject matter of this treatise, as its title page seems to be exploding with cannon fire. The liveliness of the initial page continues throughout the book, as its author fairly revels in discussing and illustrating all manner of explosive devices, real and imagined. *La Pyrotechnie*, 1630. Jean Appier Hanzelet. LC-USZ62-110473.

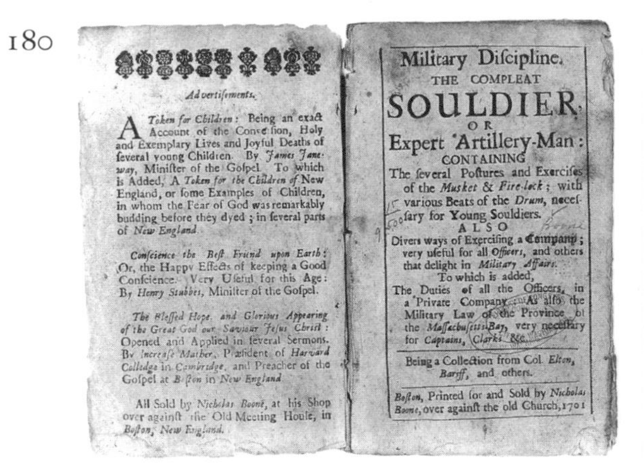

Advertisements.

A *Token for Children*: Being an exact Account of the Conversion, Holy and Exemplary Lives and Joyful Deaths of several young Children. By *James Janeway*, Minister of the Gospel. To which is Added, A *Token for the Children of New England*, or some Examples of Children, in whom the Fear of God was remarkably budding before they dyed; in several parts of *New England*.

Conscience the Best Friend upon Earth: Or, the Happy Effects of keeping a Good Conscience. Very Useful for this Age: By *Henry Stubbes*, Minister of the Gospel.

The Blessed Hope, and Glorious Appearing of the Great God our Saviour Jesus Christ: Opened and Applied in several Sermons. By *Increase Mather*, President of *Harvard Colledge* in *Cambridge*, and Preacher of the Gospel at *Boston* in *New England*.

All Sold by *Nicholas Boone*, at his Shop over against the Old Meeting House, in *Boston*, New England.

Military Discipline.
THE COMPLEAT
SOULDIER,
OR
Expert Artillery-Man:
CONTAINING
The several Postures and Exercises of the *Musket* & *Fire-lock*; with various Beats of the *Drum*, necessary for Young Souldiers.
ALSO
Divers ways of Exercising a Company; very useful for all *Officers*, and others that delight in *Military Affairs*.
To which is added,
The Duties of all the Officers, in a Private Company: As also the Military Law of the Province of the *Massachusetts-Bay*, very necessary for *Captains*, *Clarks* &c.

Being a Collection from Col. *Elton*, *Barriff*, and others.

Boston, Printed for and Sold by *Nicholas Boone*, over against the old Church, 1701

The author of this early American book on military science was at one time a bombardier who saw action in Nova Scotia. He was a constable and a bookseller as well, and therefore printed his own books, which sold from his shop in Boston located "over against the old Church." *Military Discipline, the Compleat Soldier, or Expert Artillery-Man,* 1701. Nicholas Boone. LC-USZ62-110476.

The illustrated naval maneuvers shown here described how to attack, defend, ambush, and retreat at sea and formed much of the basis of many later works. The second part of this work deals with naval construction. One of the first to apply mechanical and hydrodynamical principles to that end, Hoste was a Jesuit mathematician and a naval chaplain before writing this work. *Naval Evolutions: or, A System of Sea Discipline,* 1762. Paul Hoste. LC-USZ62-110475.

the Compleat Souldier, or Expert Artillery-Man shows that military matters were of grave concern on this side of Europe as well.

As much of this period was dominated by the European bloodbath known as the Thirty Years' War and the resulting political anarchy and population collapse in northern Europe, the words of its definitive historian best serve. According to C. V. Wedgwood, "The war solved no problem. Its effects, both immediate and indirect, were either negative or disastrous. Morally subversive, economically destructive, socially degrading, confused in its causes, devious in its course, futile in its result, it is the outstanding example in European history of meaningless conflict."

It is arguable that technology owes more to this period—the scientific revolution—than to any before or since. This is not so simply because the process of technological development began to involve the use of scientific principles, but more importantly because of its embrace of the prevailing wisdom of the time. That wisdom, as espoused and demonstrated by Galileo, said that mankind lived in an understandable world in which things happened because of reasons which were discernable and knowable. The new science called for a focused, specialized method of experimentation, arguing that the world can be understood incrementally, one law at a time.

Gradually, the technological tradition of this age came to embrace this new, incremental philosophy, and its practitioners began to investigate how to achieve practical solutions to discrete, individual problems. While Galileo turned his telescope on the moon, the skilled mechanic Robert Hooke used his improved microscope to investigate the intricate details of the common flea. The tradition of technology was entering its more mature, premodern phase.

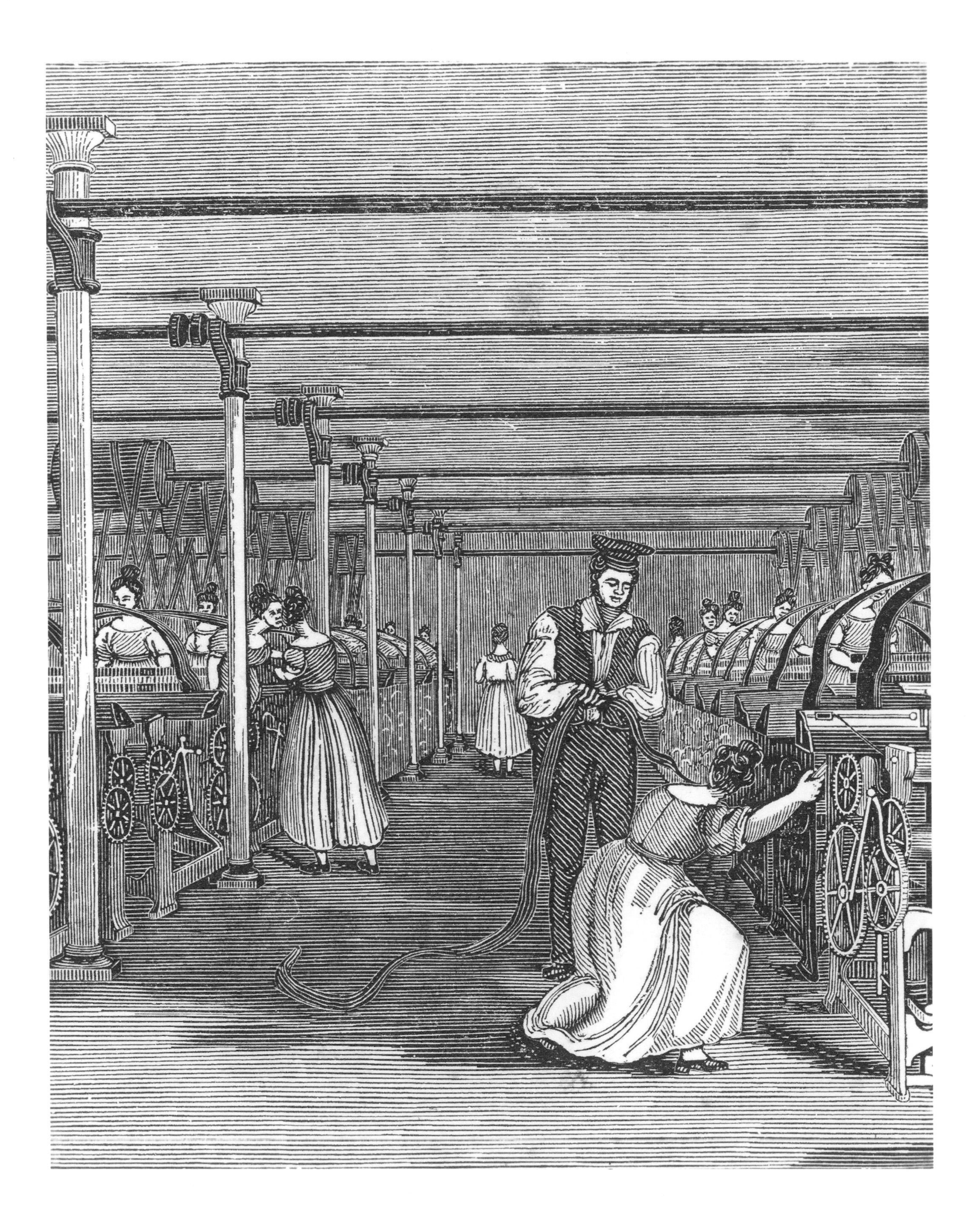

5. *Technology in the Industrial Revolution*

If the Industrial Revolution can be described as the process in which an agrarian or craft economy was changed into one dominated mainly by machine manufacture, then the one thing essential to that change would be an efficient, dependable, and portable power source. The first successful device that fits this description was James Watt's steam engine.

James Watt was the son of a Scottish shipwright and maker of nautical instruments. When he was a youngster, Watt's health kept him from school, so his father made his workshop available to his talented and curious son, setting the boy on the path to invention. By twenty-one, he had become "mathematical instrument maker" to the University of Glasgow. This position gave him access to a workshop but, more importantly, also exposed him to some of the leading scientific minds and ideas of the time.

It was while he was working on a model of the Newcomen steam engine that was used in the university lectures on natural philosophy that Watt became interested in steam engines. After some study and experimentation, he hit upon the reason for the engine's well-known inefficiency—a major loss of energy at every cycle occurred since the same chamber had to be reheated just after it had been cooled down. His solution was simple in principle. He would keep the working chamber permanently hot by introducing a separate cold chamber or "condenser" into which the steam could be led. Watt arrived at this conclusion in 1765, but did not produce a working, full-scale engine with a separate condenser until 1768. He patented his invention in 1769.

After entering a successful beginning venture to develop his new engine commercially, first with John Roebuck and later with Matthew Boulton, Watt worked hard on further improving his invention. As his business flourished, Watt produced one remarkable improvement after another. In 1781, he developed the sun-and-planet gear which translated the engine's reciprocating or back and forth movements into the rotary movement of a wheel, thus allowing the engine to power all manner of devices. In 1782, he patented the double-acting engine in which the piston both pushed and pulled. In 1788 he invented a "centrifugal governor" that automatically controlled the engine's output of steam. And in 1790 he produced an "indicator" or gauge for determining pressure in the cylinder during the cycle.

By that year, Watt's steadily improving engines had completely replaced the older Newcomen engines and become integral to the emerging factory system of manufacture. For where the old engines were little more than pumps, the new engines that Watt produced were real, modern en-

Opposite page:
In Samuel Slater's memoirs, published a year after his death, the image "Power Loom Weaving" was used to illustrate the chapter titled "Advancement of Machinery." As the founder of the U.S. cotton textile industry, Slater (1768–1835) was the one who began the mechanization of work in America. Having emigrated to the United States from England in 1789, Slater duplicated Arkwright's spinning machine from memory and established the first successful cotton mill in the United States. Although this image seems quaint and almost gentile, it contains the essence of a revolution that would transform the young nation's agrarian and handicraft economy into one dominated by machine manufacture and heavy industry. Slater did employ women as shown here, but most of his workers were children. *Memoir of Samuel Slater,* 1836. George S. White. LC-USZ62-110375.

gines. They did not need to be built near rapidly moving water because they burned coal for power and so they could be placed anywhere and would produce useful energy on the spot. The factory system of manufacture, built in the cities and powered by steam, thus came into being very rapidly. The Industrial Revolution had begun in England.

Watt was no doubt a real genius, but he was by no means simply an inspired mechanic who fiddled while ignorant of any scientific principles. His longtime friend, John Robison, first met Watt when the young genius was only twenty-two, and Robison later wrote, "I saw a workman and expected no more: I found a philosopher." Indeed, throughout his life, Watt was anything but a narrow specialist. He was a cultured man who was passionate about music and poetry and had mastered French, Italian, and German so as to stay current with scientific development.

Although he wrote a great deal, he actually published little. Watt's only publication concerning his inventions came about as extensive corrections and additions to an *Encyclopaedia Britannica* article written in 1797 by his late friend, John Robison, whose writings were being edited for separate publication by David Brewster. Robison's work was published in Edinburgh in four volumes (plus one of plates) in 1822 as *A System of Mechanical Philosophy*, and volume 2 contains Watt's long detailed accounts of his many inventions. The Library of Congress has the first edition of this work.

Volume 2 had also been first published separately in a very small quantity in 1818 for presentation to Watt and his friends. This rare separate, which the Library of Congress does not have, was titled *The Articles Steam and Steam-Engines . . .* , and was published in Edinburgh.

The next major development of the steam engine came about as a result of returning to Savery's seventeenth-century use of high pressure steam and doing away with Watt's condenser. The prudent James Watt had stayed away from the explosive dangers of high-pressure steam engines, but the young and incautious Richard Trevithick saw the economies that high pressure promised. Trevithick grew up amidst the tin mines of Cornwall, the son of a mine manager, and emerged from the village school barely literate. His engineering talents were noticed early, however, and at nineteen he became an engineer to several local mines.

Watt's patent monopoly on low-pressure steam drove Trevithick to search for ways to circumvent it, and he believed that he could build boilers capable of safely withstanding high pressures. He also knew that such high-pressure engines would provide the same power as Watt's but would be considerably smaller, lighter, and possibly even portable. In 1797, Trevithick built several working models of high-pressure engines that were so successful he soon built a full-scale high-pressure engine for hoisting ore that was used at Coalbrookdale. Within two years, he had built thirty engines, each small enough to fit in an ordinary farm wagon to be transported to the mines.

Trevithick was an intuitive engineer whose vision for steam went beyond mere pumping and hoisting. From the beginning, his working models were of both stationary and locomotive engines, and in 1801 he

Opposite page:
The Watt steam engine is depicted in an 1822 illustration. This engine works by transforming the vertical action of the piston rod (P) into rotary motion through the flywheel (V). During the twenty-five years of the Boulton and Watt monopoly (1775–1800), their firm built nearly five hundred steam engines, which the company would assemble wherever the customer wished. *A System of Mechanical Philosophy,* 1822. John Robison. LC-USZ62-110376.

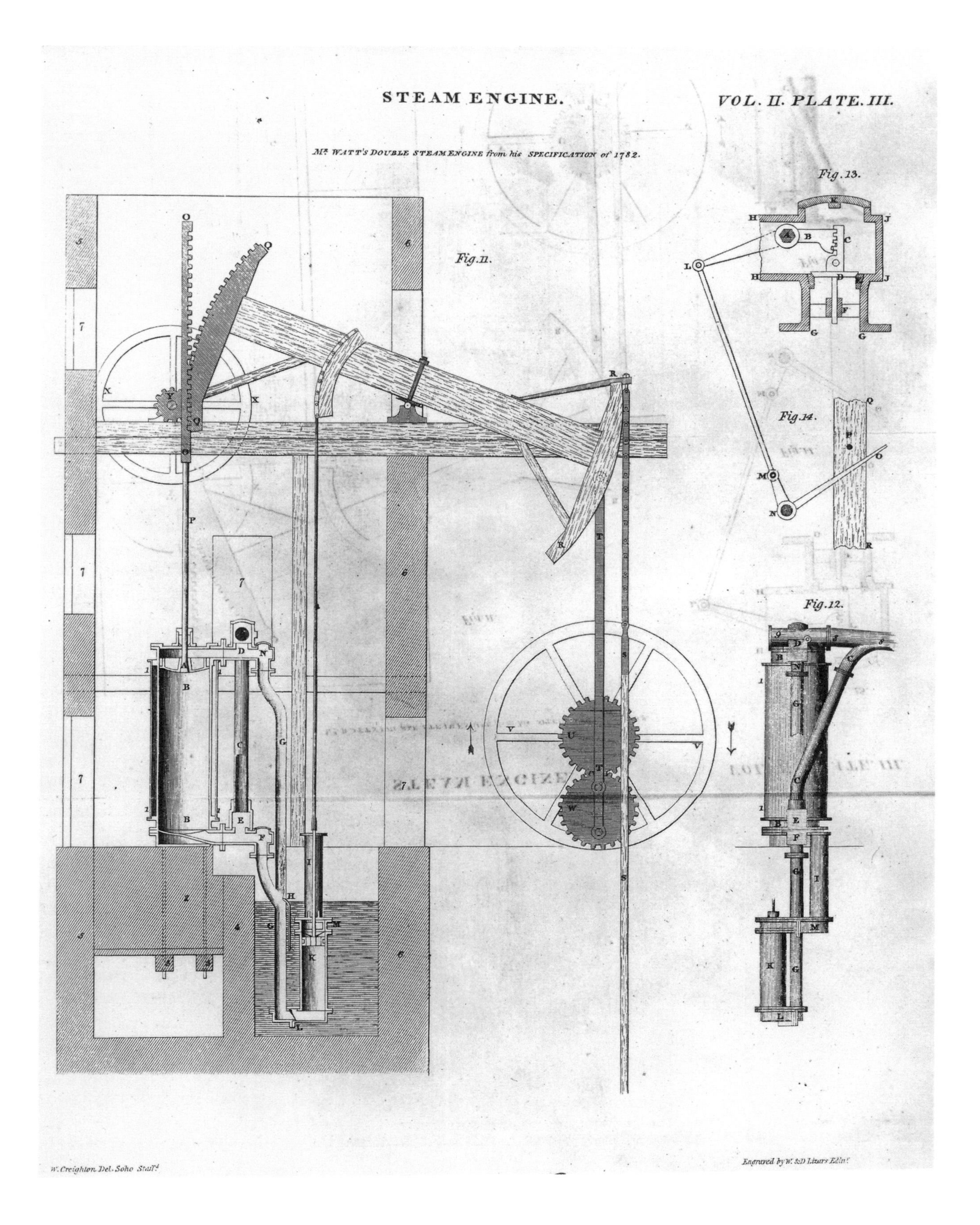
STEAM ENGINE.
VOL. II. PLATE. III.
Mr. WATT'S DOUBLE STEAM ENGINE from his SPECIFICATION of 1782.
Fig. 11.
Fig. 12.
Fig. 13.
Fig. 14.
W. Creighton. Del. Soho Staffd.
Engraved by W. & D. Lizars Edinr.

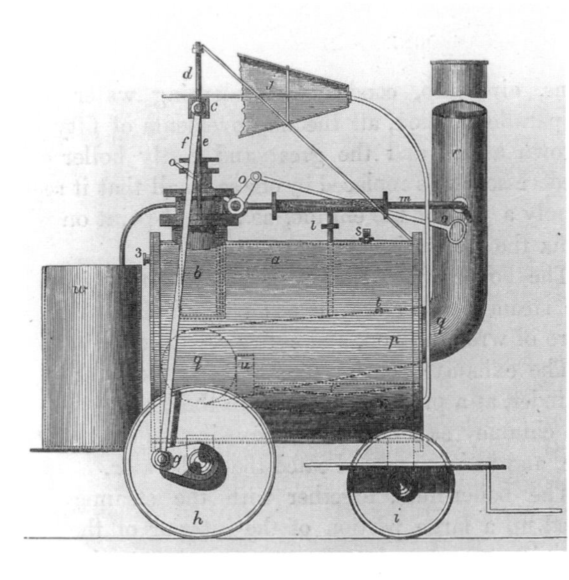

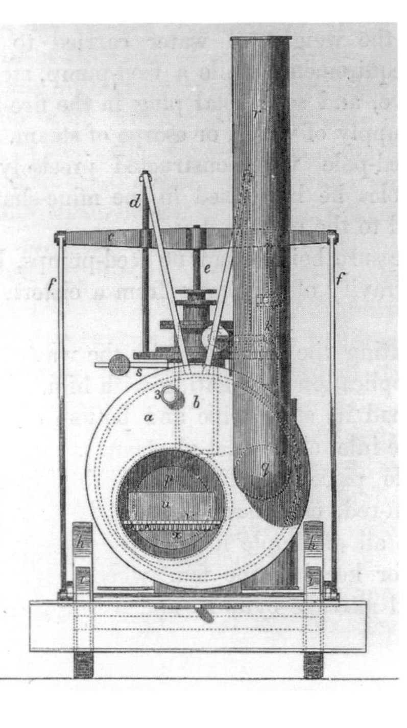

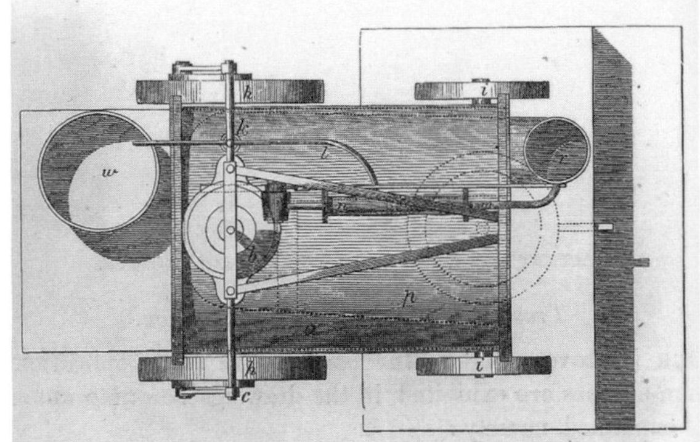

Three views of Richard Trevithick's (1771–1833) first passenger-carrying locomotive. When the Cornish engineer built and patented a high-pressure stationary steam engine that was small enough, he placed it on four wheels and took several friends and enthusiasts for a ride on Christmas Eve 1801. Trevithick's biggest problem was the lack of a good road surface, for his engines kept overturning. *Life of Richard Trevithick,* 1872. Francis Trevithick. LC-USZ62-110377.

built his first steam road carriage which he drove with some friends up a hill in Cornwall on Christmas Eve. In March 1802, he took out a patent on his high-pressure engines for stationary and locomotive use, and in the following year drove another road carriage through the streets of London at ten miles per hour. On February 21, 1804, he won a 500-guinea wager when his locomotive hauled a load of ten tons and seventy men along a ten-mile tramway that had been built in South Wales. This successful ten-mile journey marks the first steam locomotive to run on rails.

Where Trevithick's daring and impulsive imagination served him well in harnessing high-pressure steam, it proved a liability in trying to translate his inventions into a marketable product. James Watt was fortunate to have allied himself with a savvy businessman like Matthew Boulton, but Trevithick's partner, Andrew Vivian, apparently was no Boulton. Although Trevithick's fertile mind kept producing all manner of applications for steam, he was never able to stay with one specific idea long enough to see it through to the end. He eventually spent eleven years in South America, mostly in Peru, where his engines were used in the silver mines. He died back in England at sixty-two, never having made his fortune.

Aside from his letters and patents, Trevithick left no published writings documenting his life's work. However, he was considered important by the writers of his time and they treated him seriously. The earliest of these was Abraham Rees, whose *The Cyclopaedia*, a forty-one-volume encyclopedia (plus six volumes of plates), gave Trevithick's work a full two pages in the revised version of its first American edition published in Philadelphia from 1810 to 1824. In 1824, Robert Meikleham, writing under the name Robert Stuart, published in London *A Descriptive History of the Steam Engine* in two volumes. Meikleham included a detailed engraving of the essentials of Trevithick's "simple and powerful machine," and expounded the virtues of his "bold deviation from the beaten track."

The work which contains the most drawings and details, both technical and personal, is the two-volume *Life of Richard Trevithick, with an Account of His Inventions*, written by his son Francis and published in London in 1872. The Library of Congress has this work in first edition as well as Meikleham's 1824 history of steam. It also has Rees's complete *Cyclopaedia* in its 1810–24 edition.

At virtually the same time that the Englishman Trevithick was working on his invention, high-pressure steam was being investigated by Oliver Evans on the other side of the Atlantic. Oliver Evans was one of the first real American technological geniuses, having invented a textile machine to cut and mount wire teeth on leather for an improved carding device, as well as a fully automated production line for a flour mill by the time he was thirty years old. Born in Delaware in 1755, he joined the Pennsylvania militia company along with his brothers during the Revolutionary War, but saw no action fighting the British. The mechanically inclined Evans had been apprenticed to a wheelwright when he was sixteen, and not long after that he became intrigued by the possibilities of high-pressure steam. He envisioned harnessing high-pressure steam both to drive stationary engines for industry and to run engines for land and water transportation. He would do both.

By 1801 he built a working steam engine whose boiler was made of cedar wood. Instead of building his planned steam wagon, he used his first high-pressure engine to grind gypsum into plaster of Paris, used for fertilizer and cement. This engine served as a prototype for his larger versions which he attempted to market for grain mills, sawmills, and even steamboats. During his life, however, Evans would meet with little success in trying to get other Americans to share his future plans for steam.

In 1804, Evans obtained a new patent for a high-pressure steam engine and boiler and built his version of a working steamboat, the *Orukter Amphibolos* or amphibious digger. This was a flat-bottomed boat with wheels. The engine drove its ground traction wheels on land and the paddle wheel on the water. It also powered a bucket chain, since Evans had built the *Orukter* to dredge in the Schuylkill River. Before clearing the Philadelphia waterfront, Evans drove his steam vehicle down Market Street to Center Square, making it the first powered road vehicle to operate in the United States.

Despite relatively little acceptance or financial success, Evans nonetheless did well enough to begin his Mars Works factory where, during the next ten years, he built scores of steam engines for all manner of manufacturing uses. He died in 1819, the same year his Mars Works burned to the ground.

The Mechanic.

JULY, 1834.

[For the Mechanic.]

STEAM-CARRIAGES.

[Evans' Steam-Engines. See page 196.]

HAVING traced the steam-engine from its first invention to the successful application of its power for the purpose of navigation, by Fulton, it now remains for us to continue the subject, by a brief history of its application to land-carriages on common roads, and its superior advantages on railways. Two individuals, in particular, are claimants for the honor of this invention, both of whom also claim the invention of steam-boats; but as all claims to the invention

17

This unlikely looking vehicle resembling a paddleboat on wheels is exactly that. It was built in 1804 by the American technological genius Oliver Evans (1755–1819), who had obtained a contract from the Philadelphia Board of Health for a steam engine to be used in dredging and cleaning the docks. Evans decided to construct an amphibious vehicle that he could drive on the streets to the water. His invention became the first wheeled vehicle to move under its own power in the United States. It was also the first vehicle to use high-pressure steam with success. *Boston Mechanic and Journal of the Useful Arts and Sciences,* July 1834. LC-USZ62-110378.

Throughout his career, Evans fought numerous legal battles to obtain patent protection from Congress and the courts, often hearing from the courts that a patent was an infringement of public rights. Much of this is documented in the collections of the Library of Congress, which has several copies of Evans's petitions to Congress as well as proposed bills for relief and other committee reports. The Library also has first editions of Evans's published works. His earliest book, *The Young Mill-Wright and Miller's Guide*, was published in Philadelphia in 1795. Its usefulness is evidenced by its fifteen editions, appearing as late as 1860. Evans also

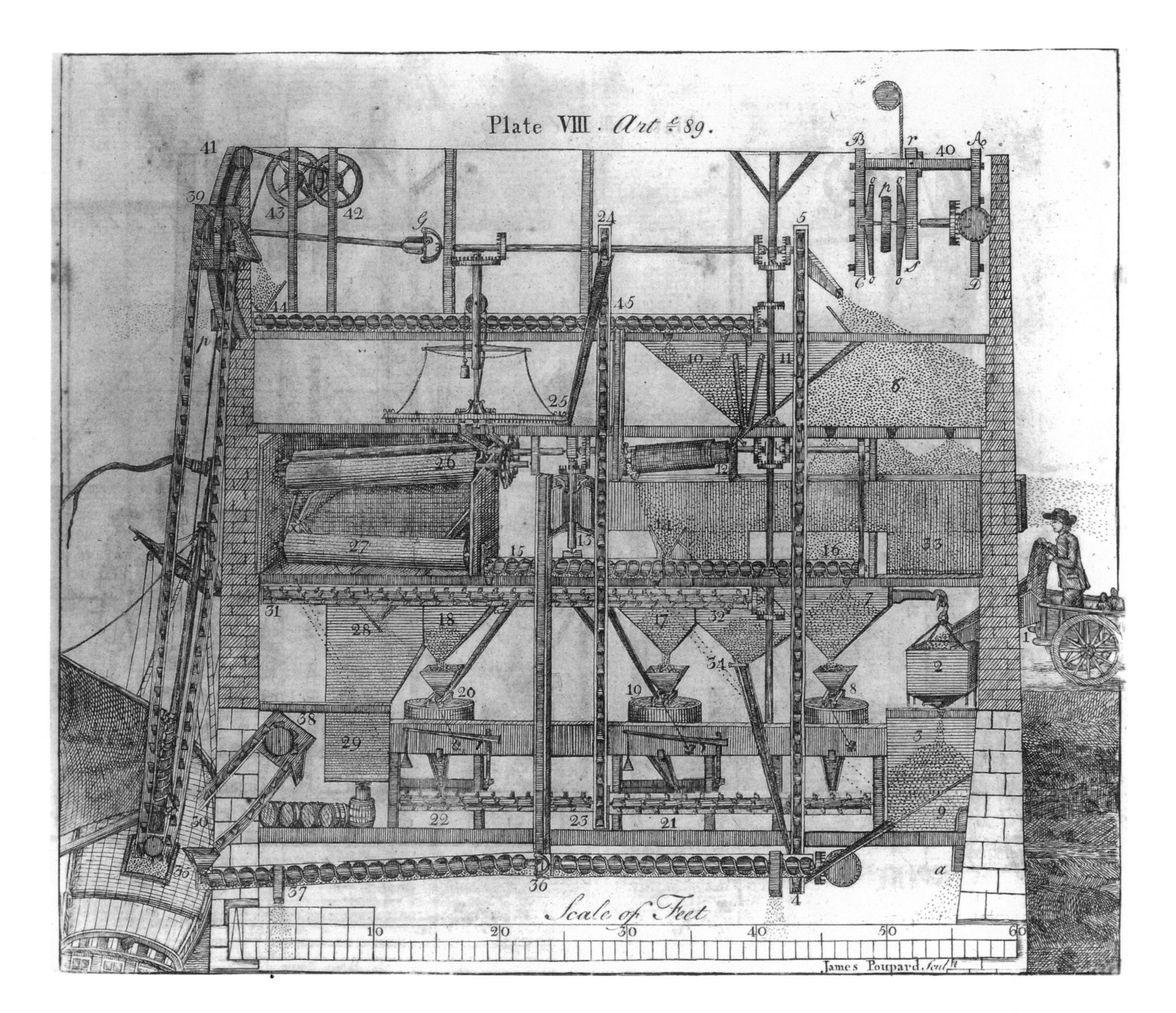

wrote a companion book on steam whose pointed title, *The Abortion of the Young Steam Engineer's Guide*, reflects the bitterness he felt after Congress had failed to renew one of his flour mill patents. The Library of Congress has the first edition of this work (Philadelphia, 1805). Besides Evans's account of steam, the book also contains his ideas for artificial refrigeration as well as a forecast of the need for governmental support of technology. Although Evans should occupy the first place in a long line of American technical geniuses, posterity seems to have ignored him, just as his contemporaries did.

It was also in the newly formed United States that the first practical steamboat was created. Since the steam engine was invented in Europe

In this diagram, Oliver Evans details his automated mill, which needed only one person to operate it. Evans devised a system of bucket chains and Archimedean screws powered by water that performed its five essential operations. The man on the far right needed only to place his wagonload of grain near the elevator and it would be unloaded, raked, aerated, ground, separated, and reloaded. Everyone marveled at Evans's automated factory, but no one would invest in it. *The Young Mill-Wright's & Miller's Guide,* 1795. Oliver Evans. LC-USZ62-110379.

This 1737 illustration of a steam tugboat is important mainly as evidence of how early in the century inventors were trying to propel a boat with steam. Although there is little evidence that Jonathan Hulls ever built and tested this craft, he did obtain a patent for it in England. Hulls's tugboat has a stern paddle wheel that is driven by a Newcomen atmospheric engine. He suggested that a continuous rotation of the wheel could be obtained by the use of ratchet wheels driven by ropes from the piston. This is taken from a nineteenth-century reprint. *A Description and Draught of a New-Invented Machine for Carrying Vessels or Ships Out of, or Into Any Harbour,* 1855. Jonathan Hulls. LC-USZ62-110380.

and its application to boats and ships was discussed or considered by many Europeans as a potential application, it seems odd at first that America should claim the honor. But Europe, specifically England and France, approached steam propulsion on water from a different perspective than did America. Both older nations had efficient, well-run systems of inland waterways and consequently thought of steam primarily in terms of ocean-going naval or merchant vessels. This meant large vessels and long voyages. America, on the other hand, would seek to build smaller boats to navigate its great rivers of the west. These trips would be short, quick, and cheap.

The early European literature of the steamship is not well documented in the collections of the Library of Congress, although the Library does have a copy of one of the first books to describe and illustrate a proposed steamboat. In 1736 Jonathan Hulls, a Gloucestershire clockmaker, took out a patent for a stern-wheel tugboat powered by a Newcomen atmospheric engine. There is no evidence that the boat was ever built, but Hulls did publish in 1737 *A Description and Draught of a New-Invented Machine for Carrying Vessels or Ships Out of, or into Any Harbour*. The Library has this forty-eight-page, illustrated work as it was reprinted in London in 1855.

Following later unsuccessful attempts at actual construction and operation, such as American William Henry's 1763 Watt-type engine paddleboat and the unsatisfactory steamers built by the French marquis de Jouffroy d'Abbans in 1774 and 1776, it fell to the most unlikely of candidates—John Fitch—to create the first practical steamboat. Fitch was a silversmith and surveyor, the details of whose life were sad and melo-

dramatic. Insecure but highly principled, high-strung, energetic, and ambitious, Fitch was often his own worst enemy. In 1785, Fitch became possessed by the notion of steam propelling both boats and carriages—despite the fact he was ignorant of the very existence of steam engines. Nonetheless, by 1787, the *John Fitch* was navigating the Delaware River at Philadelphia. This was a fifty-foot stern-wheeler powered by a steam engine built by his friend Henry Voight, a local clockmaker. Later, in 1790, Fitch would actually operate for one brief summer a regular steamboat service between Pennsylvania and New Jersey. During its brief summer run, his steamboat traveled between two thousand to three thousand miles and achieved speeds as high as seven or eight miles per hour. The service never turned a profit, however, and Fitch would never again have the chance to make it work.

Fitch's historic 1787 steamboat was one of several improved versions which followed his first attempt made on the Delaware in July 1786. During that month Fitch hit upon the idea of attaching paddles to arms

Almost twenty years before Robert Fulton would become known as the inventor of the steamboat, another American, John Fitch, built and actually operated a steam ferryboat. Here is one of his early designs suggesting that steam be used to duplicate the action of human paddlers. Fitch encountered nothing but bitter disappointment in both his personal and professional life, and his possible suicide may have been related to alcoholism. *Columbian Magazine,* December 1786. LC-USZ62-110381.

moved by a crank, which was eventually powered by steam. And it is just such a system that is pictured in the December 1786 issue of the new journal *Columbian Magazine*. Fitch had written a short "Description of a New Invented Steam-Boat" for that magazine in the form of a letter, and his paddle system is shown clearly on a "prefixed plate" opposite its first page. In the preface to the first issue, September 1786, the editor states that he intends "to furnish novelty, entertainment, and instruction to his readers," and his publication of Fitch's claims demonstrates that he was trying to do exactly as he said. The Library of Congress has the *Columbian Magazine* from its first appearance in 1786.

The Library also has the papers of John Fitch, a collection of seven hundred items that include his correspondence for the years 1784–94, his personal diaries, financial accounts, surveying notes, printed matter (such as handbills, pamphlets, and broadsides), and, most important, many sketches and drawings related to his steamboat. Most of these drawings by Fitch are related to engines, although there is one full drawing of his paddleboat that is more complete than the one in the *Columbian Magazine*, in that it appears to include the engine which powered the crank. The Library also has the papers of Dr. William Thornton, Fitch's steadfast patron, and this collection contains documents dealing with Fitch's later years.

Finally, the Library of Congress also has the rare first edition of one of Fitch's few publications, *The Original Steam-Boat Supported, or a Reply to Mr. James Rumsey's Pamphlet*. This was published in Philadelphia in 1788 and is a somewhat convoluted rebuttal by Fitch to Rumsey's claim of priority. The Library has Rumsey's side of the argument, *A Short Treatise on the Application of Steam* (Philadelphia, 1788). The Library also has William Thornton's 1814 publication, *Short Account of the Origin of Steam Boats*, printed in Washington.

Few people know the name Fitch or any of the other predecessors of the man most popularly associated with the first steamboat, Robert Fulton. While Fulton obviously cannot claim to be the sole inventor of the steamboat, nonetheless it was he who made it a real, going concern. Until Fulton, no one was able to produce an efficient, reliable steamboat that would actually turn a profit. If Fulton had a genius, it was for building, logically and systematically, on the work of others and for not only understanding the principles involved in this complicated process but also learning from his mistakes. Where Fitch's approach was hit-and-miss because he did not fully understand everything he was doing, Fulton's method was systematic, based on his full comprehension. For Fulton, a failed experiment was another step in the learning process. For Fitch, it was potential disaster.

Like Fitch, who was a craftsman before he took up steam propulsion, Fulton began his career in another field, originally as a portrait and landscape painter. Having discovered in Europe that his art was no more than average, Fulton turned to what was then the rage in Europe—canals—and offered some new designs for inland water transportation. From here his interests moved to submarines, but his underwater craft, *Nautilus*, did

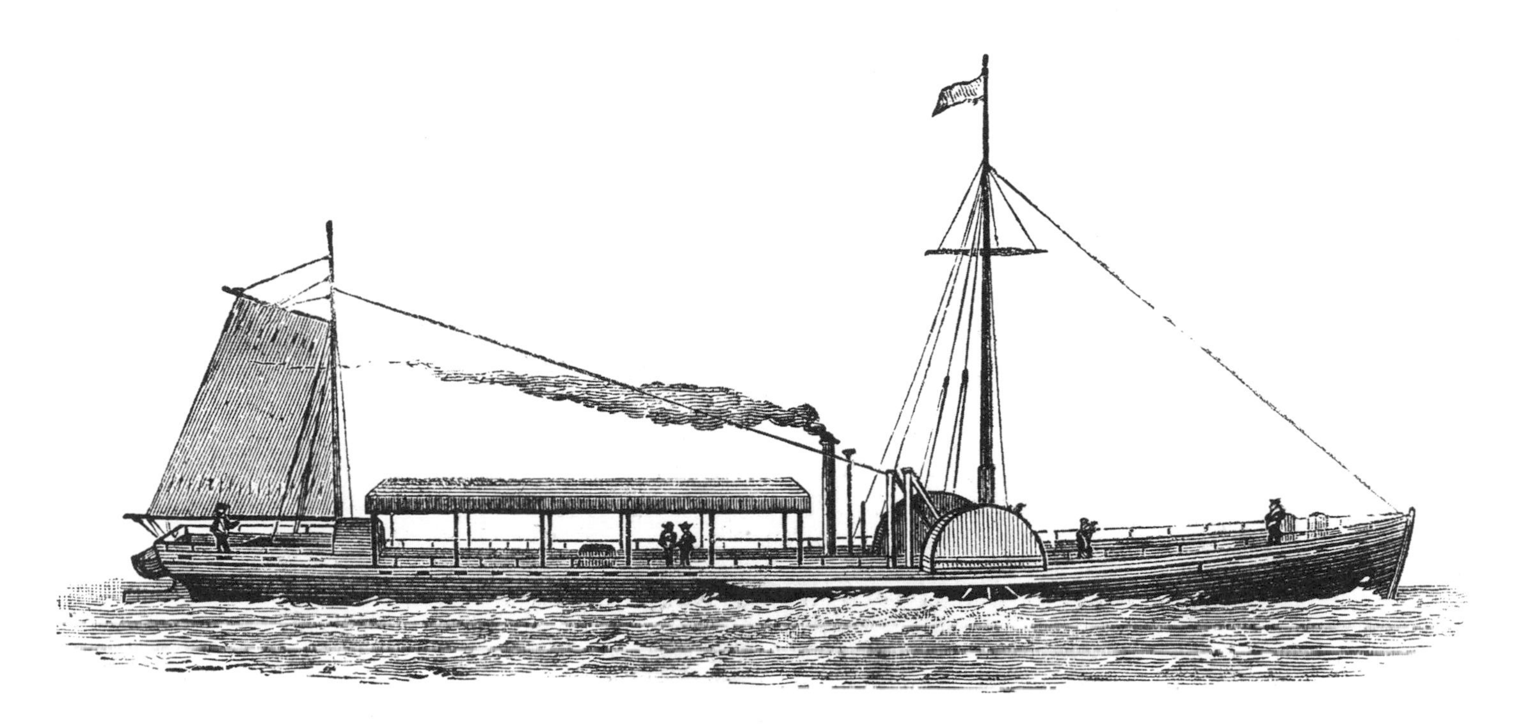

Of the many depictions of Robert Fulton's *Clermont,* this is one of the more accurate. His steamboat was 150 feet long, displaced 100 tons, and moved at about 5 miles per hour. The paddle system is seen at the front third of the boat. It was powered by a rectangular boiler 20 feet long and 8 feet wide. The significance of this vessel lies not primarily in its technology but in its commercial viability. *The Steam Engine and Its Inventors,* 1881. Robert L. Galloway. LC-USZ62-110382.

not perform well and the French rejected his requests for financial support. In 1801, however, he met Robert R. Livingstone, American minister to France, who had obtained a twenty-year monopoly for steamboat navigation within the state of New York. The two men agreed to share the expense of building a steamboat. By early August 1807, Fulton's 150-foot vessel, called simply *Steamboat*, was ready for trials. The 150-mile run from New York to Albany was made in thirty-two hours, compared to four days required by sailing sloops. Commercial trips began in September, improvements followed, and by 1808 his famous *Clermont* was afloat. Fulton eventually married Livingstone's niece and became both rich and famous despite having spent most of his fortune defending his patents in court. A year before he died, he produced the world's first steam warship, *Demologos*, for the United States. The war with the British ended before it could see action.

Although diligent, opportunistic, and systematic when it came to steamboats, Fulton's real passion was for submarines. It is safe to say that he viewed steamboats as a practical and potentially lucrative way of paying for his submarine research, which indeed they proved to be. Ironically, the documentation Fulton generated for his submarine work far exceeds anything that exists for his steamboats. The Library of Congress has among its small collection of Fulton letters a seventy-one-page manuscript signed by Fulton and dated August 10, 1806. This original manuscript, entitled "On Submarine Navigation and Attack," contains introductory text and detailed descriptions of Fulton's own pen-and-wash drawings of his submarine plans and prototypes. The Library also has the sixteen numbered drawings signed by Fulton that accompany the text. Altogether, these unique drawings touch upon almost every aspect of Fulton's work in the fields of submarine and surface naval warfare. Also in the Library's collections are the first edition of a rare monograph by

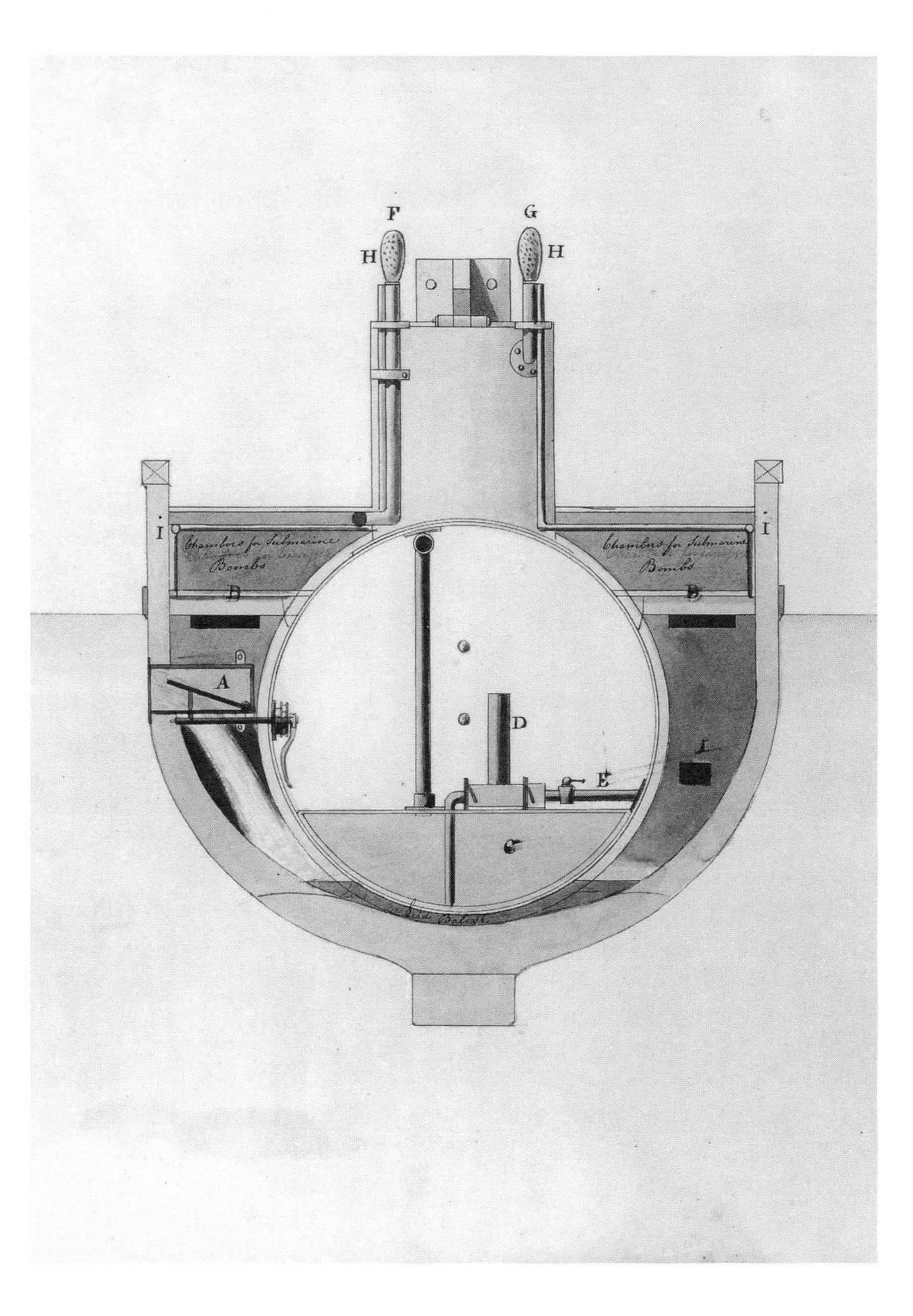

Robert Fulton's sectional drawing of a submarine, 1806 (detail). Prints and Photographs Division, Library of Congress. See p. 108.

Fulton, *Torpedo War and Submarine Explosions*, published in New York in 1810, and his earlier *Treatise on the Improvement of Canal Navigation* (London, 1796).

As Fulton had a steamboat predecessor in Fitch, so did he have one for submarines. In 1771, as a twenty-nine-year-old freshman at Yale University, David Bushnell experimented with underwater explosives. By 1775, he had constructed a six-foot-high submersible one-man vessel he called the *American Turtle* that exhibited many of the principles of the modern submarine. Bushnell wanted his invention to contribute to the war effort against the British, and in 1776 his craft conducted what must have been

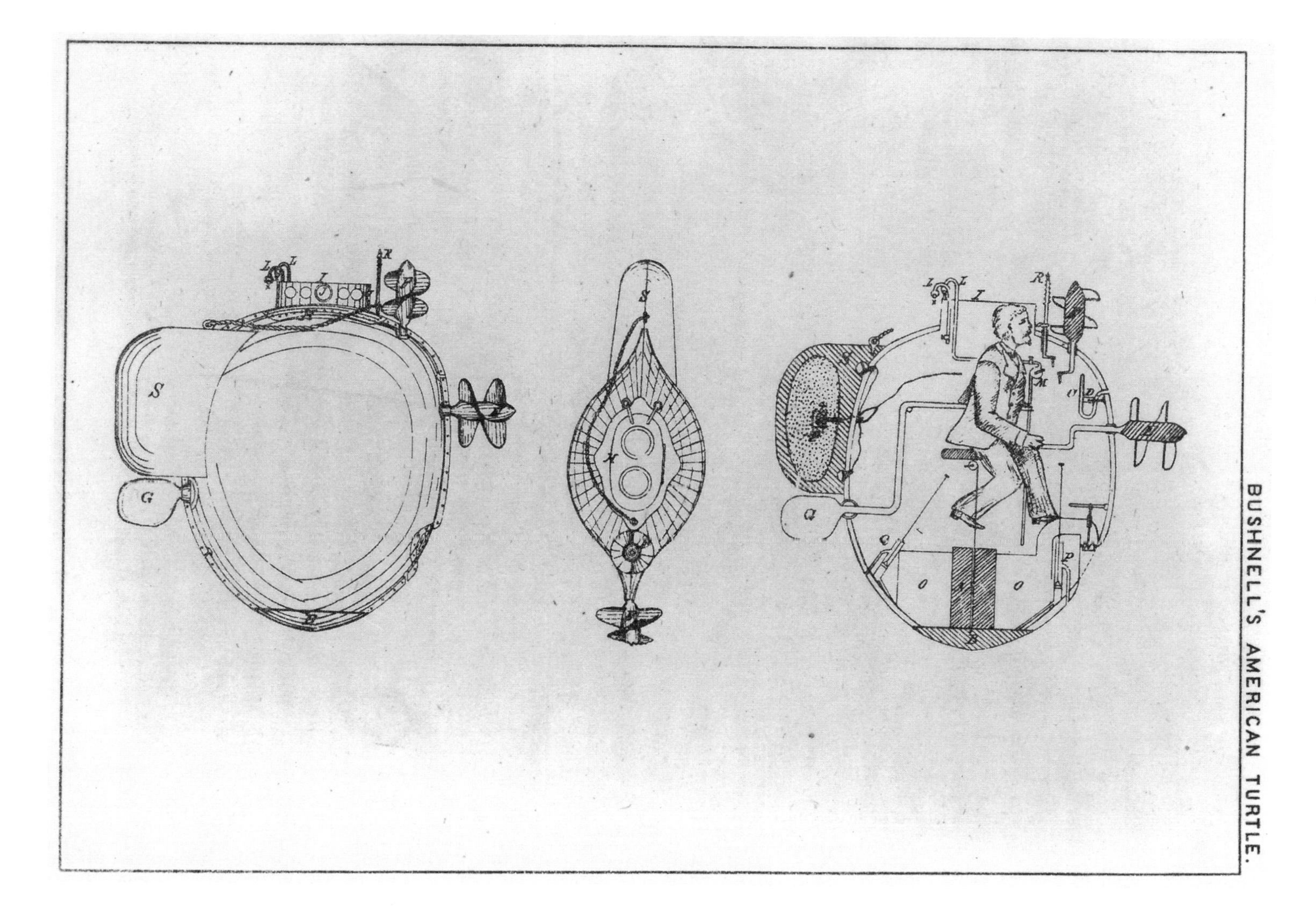

David Bushnell built his ingenious one-man submarine, called the *Turtle,* to help America win its Revolutionary War with Britain. He intended for its operator to approach a British man-of-war stealthily underwater and to screw a time bomb into its hull. Despite the operator's inability to attach the bomb, Bushnell's submarine was able to do those things essential to any submarine. It could dive, remain submerged, withstand increased pressure, proceed on a compass course at modest speed, deliver secretly an explosive, and retreat submerged and undetected. In his 1785 letter to Thomas Jefferson, George Washington called the failed craft "an effort of genius." "The Beginning of Modern Submarine Warfare," *Professional Papers of the Engineer School of Application of the United States Army,* vol. 1, no. 3 (1881). Henry L. Abbot. LC-USZ62-110384.

the first military submarine mission. Bushnell's *Turtle* saw real action when it actually got beneath a British ship in New York harbor. The operator attempted to attach a bomb to the ship's keel but could not attach the clockwork device. Following this failure and a later accidental sinking, Bushnell put his invention to rest when peace came.

Nonetheless, in October 1787, he wrote to Thomas Jefferson in France, describing his apparatus and experiments. This letter was eventually published twelve years later in the *Transactions of the American Philosophical Society* for the year 1799. The Library of Congress has Bushnell's ten-page essay, titled "General Principles and Construction of a Sub-marine Vessel," as it appeared in the *Transactions*, roughly twenty-five years after Bushnell first built his craft.

Most historians now believe that Fulton had at least a general knowledge of Bushnell's work well before the *Transactions* article appeared, and that he obtained this information from his friend Joel Barlow, who was at Yale with Bushnell. Fulton's 1801 submarine *Nautilus* thus literally owed a great deal to someone else, as did his steamboat of six years later.

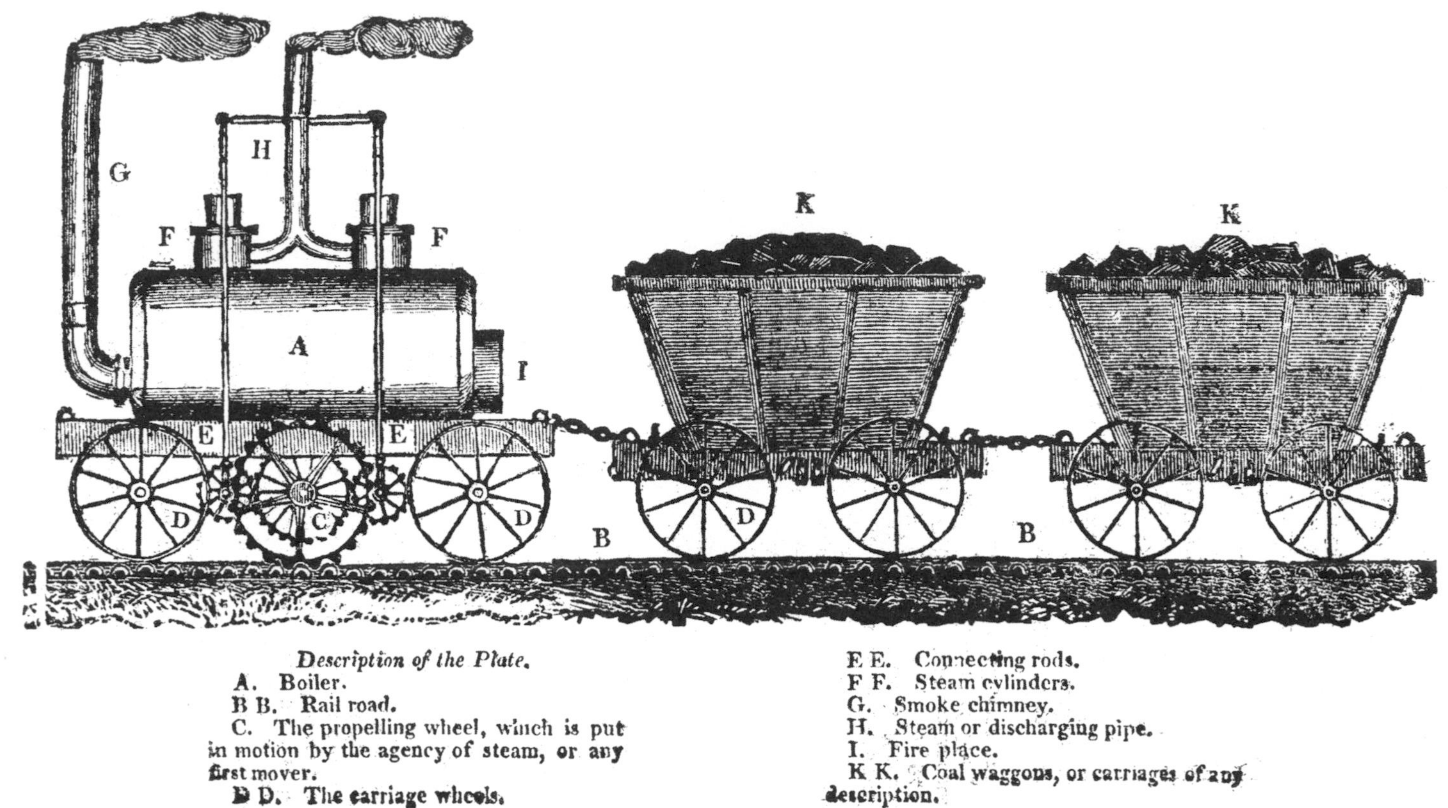

Many of the early engineers who worked on turning steam power into some sort of rail system believed that smooth metal driving wheels would never be able to get a grip on smooth metal rails. One of these was the Englishman John Blenkinsop, who built the first commercially successful locomotive. His locomotive was driven along by a toothed wheel which engaged in a rack on the rail. Here, Blenkinsop's "propelling wheel" (C) is obviously toothed. *Monthly Magazine,* June 1, 1814. LC-USZ62-110385.

The railroad locomotive followed hard upon the steamboat. Between Richard Trevithick's successful 1804 steam locomotive on rails and George Stephenson's achievement of the first practical passenger railway in 1825, there were some efforts by others at sorting out and testing all the new technologies that a successful railway involved. Of the two most notable individuals who worked on these problems during this time—John Blenkinsop in 1812 and William Hedley in 1813—only Blenkinsop is represented, and in a minor way, in the collections of the Library of Congress. In the June 1, 1814, edition of the London journal the *Monthly Magazine*, there was a letter to the editor from Blenkinsop dated March 26, 1814, which was essentially an advertisement for his "Steam Carriage." The letter describing the locomotive was accompanied by a detailed sketch, published with the letter, of a locomotive pulling two coal wagons. The Library has a complete run of the *Monthly Magazine.* As might be expected from a busy engineer more intent on doing than on documenting, nothing else that Blenkinsop published, if he ever did, can be found. It also appears that Hedley published nothing concerning his vehicle called "Puffing Billy." Both inventors employed their locomotives to haul coal from the mines to the wharves, and both improved on Trevithick's work to some degree.

It fell to George Stephenson—a man who only learned to read in his late teens and who obtained the fundamentals of an education by doing his son's schoolwork along with him at home—to create the first real,

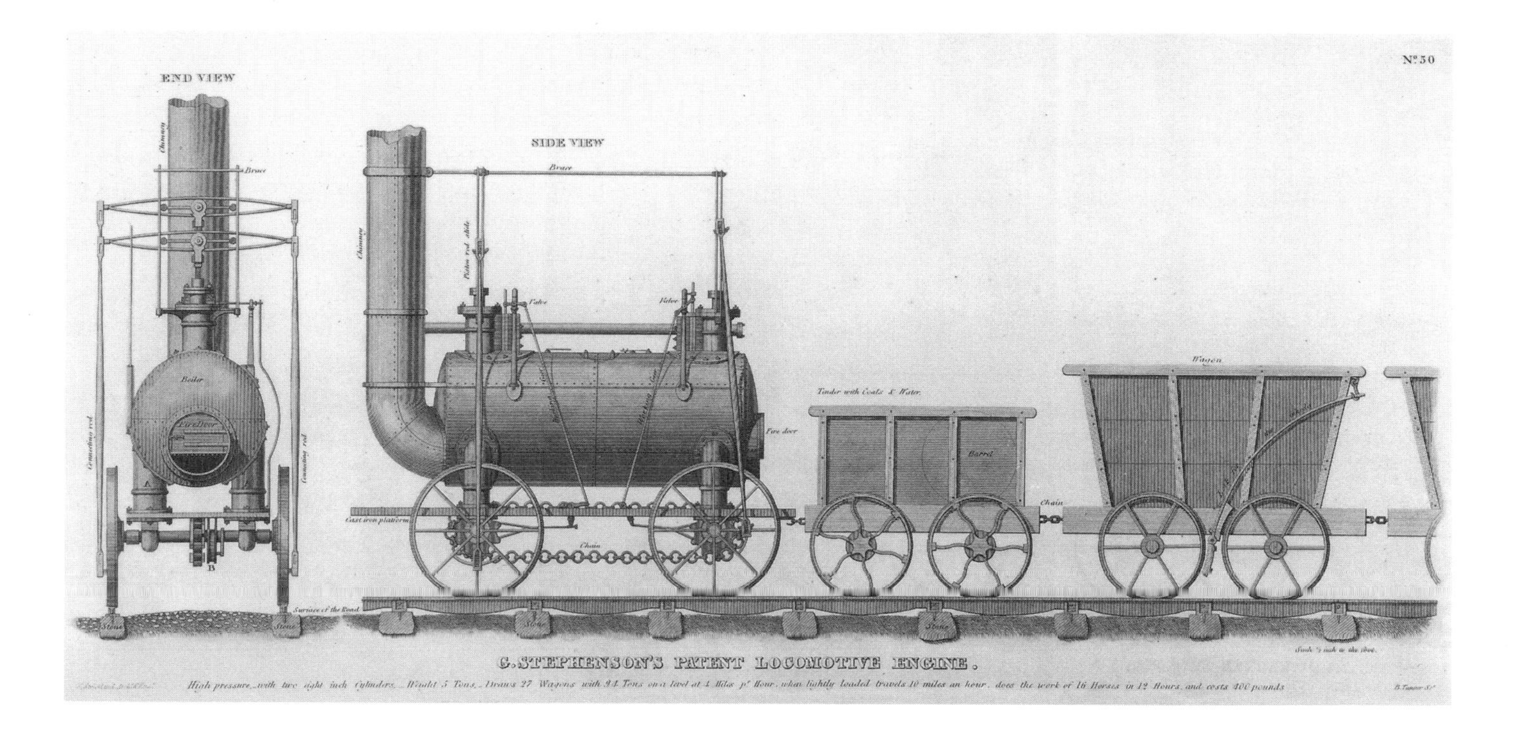

From *Reports on Canals, Railways, Roads, and Other Subjects,* 1826. William Strickland. See p. 108.

practical railway system. Stephenson had been working on what might be called a steam boiler on wheels as early as 1813, and over the next decade he continued to rebuild and refine his designs. When he heard of a planned horse-drawn railroad line to be built from Stockton to Darlington to transport coal, he saw his opportunity and persuaded the owner to use steam. Thus, railroad transportation was born on September 27, 1825, when his locomotive made the run at fifteen miles per hour while carrying 450 people. Stephenson and his son, Robert, continued to be leaders in the field for many years, and such notable features as flanged wheel design and standard rail gauge are attributable to them. As for existing documentation, neither man left any writings save letters and company records. Nonetheless, there are two books, both published in 1825, that together constitute the first comprehensive works on railway engineering. The first of these is Thomas Tredgold's *A Practical Treatise on Rail-roads and Carriages*, published in London. The other, larger work is Nicholas Wood's *A Practical Treatise on Rail-Roads*, also published in London. Tredgold's work is especially informative, since it contains 125 engravings of steam engines and early locomotives, including Stephenson's. The Library of Congress has both the Tredgold and Wood works in first edition.

During the beginnings of the Industrial Revolution, steam did not define and dominate the textile industry the way it would other industries. That is not to say that the industry was not mechanized, for it certainly was, but rather that its power source was still generally such a traditional or conventional one as moving water.

Reduced to its essentials, textile manufacture is spinning yarn or thread and weaving it into fabric or cloth. During the first half of the

Opposite page:
An inventor's obsession with creating a pick-proof lock spawned the British machine tool industry. In 1784, the ingenious cabinet-maker Joseph Bramah exhibited his lock (at the top) in his shop window, advertising a reward to anyone who could pick it open. The offer stood for sixty-seven years. More importantly, Bramah needed well-designed and precisely engineered machine tools to build more of his complex locks, and it was his alliance with the young engineering genius Henry Maudslay that would lead eventually to the manufacture of identical, interchangeable metal parts necessary for mass production. *Cyclopaedia,* vol. 3 (plates), 1802–20. Abraham Rees. LC-USZ62-110388.

eighteenth century, spinning was essentially a cottage industry that differed in technique and output very little from Renaissance times. There was a constant bottleneck in the overall production process since it took eight to ten spinners working full time to keep up with one weaver. Such a situation provided the natural and necessary incentives for invention, and by 1790, spinning technology had so improved output that the situation had entirely reversed—weavers were now in short supply.

What brought this about was a series of technical inventions and improvements in England that would transform a backward industry into one of the most modern. The first of these was John Kay's "flying shuttle," patented in 1733, which sped up the weaving process and accelerated the search for similar improvements on the spinning side. These then came from James Hargreaves's "spinning jenny" in 1764 (a multiple-spindle spinning wheel); Richard Arkwright's "spinning water frame" in 1769 (a water-powered, continuous spinning machine); and Samuel Crompton's "spinning mule" of 1779 (an improved Arkwright machine that increased yarn quantity and improved its quality). All of these improvements in production technique were made before steam ever got to the factory. Two subsequent developments in America and France, Eli Whitney's cotton gin in 1793 (which cleaned and separated cotton from its pod) and Joseph Marie Jacquard's new silk loom in 1801 (which ingeniously combined the perforated card system of Falcon with the needles-and-hooks system of Vaucanson), are examples of technical develop-

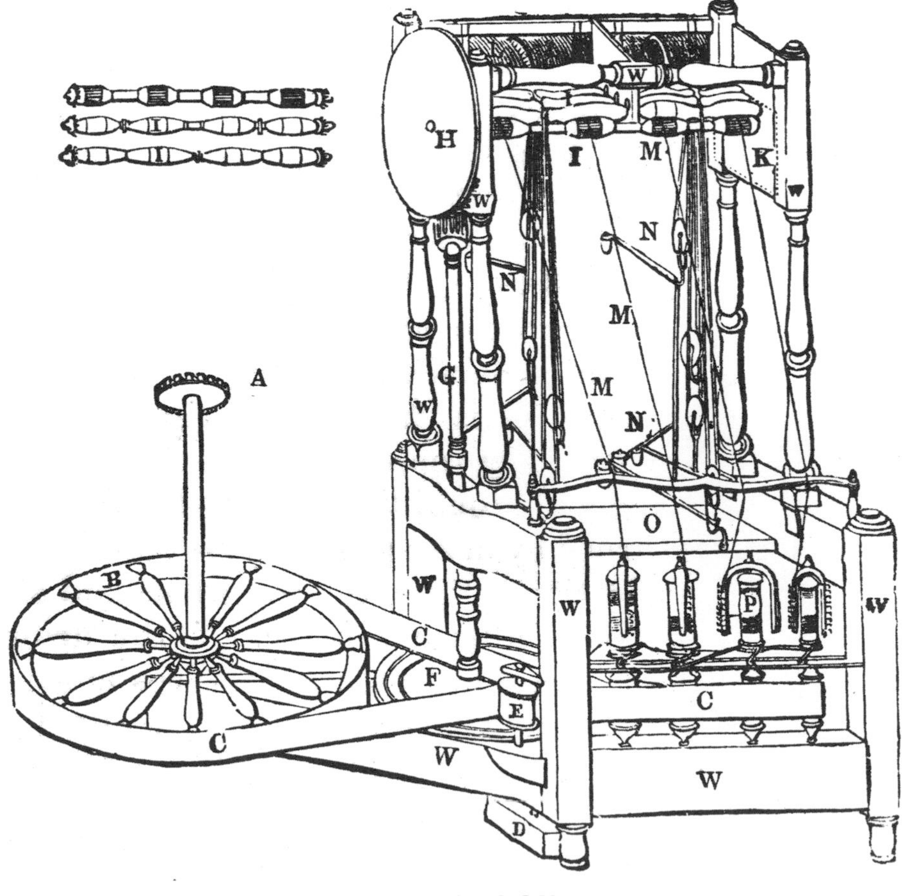

Spinning thread and weaving it into cloth was an integral part of nearly everyone's daily life before the Industrial Revolution. Although Kay's flying shuttle and Hargreaves's spinning jenny significantly speeded up the process, each was small enough to still be used in the home. The invention by Richard Arkwright of a large spinning frame that required power, however, moved the spinning process from the home to the factory, and began its transformation from a cottage industry into a factory system. Arkwright's machine was first driven by horsepower and later by waterwheels. *Cotton Spinning: Its Development, Principles, and Practice,* 1909. Richard Marsden. LC-USZ62-110387.

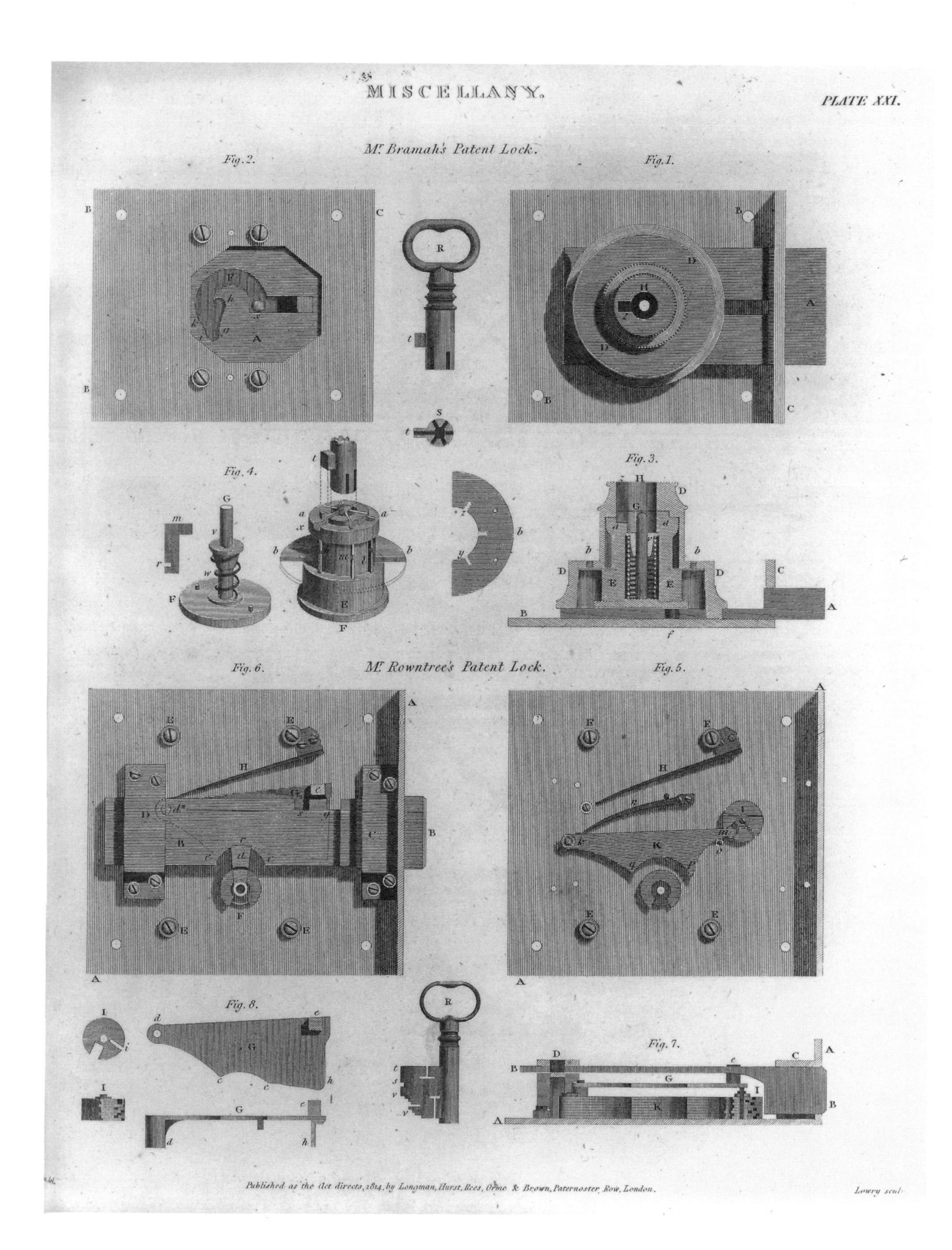
MISCELLANY.
PLATE XXI.
Mr. Bramah's Patent Lock.
Fig. 2.
Fig. 1.
Fig. 4.
Fig. 3.
Fig. 6.
Mr. Rowntree's Patent Lock.
Fig. 5.
Fig. 8.
Fig. 7.
Published as the Act directs, 1814, by Longman, Hurst, Rees, Orme & Brown, Paternoster Row, London.
Lowry sculp.

ments in textiles that had major effects beyond their particular industry. "King Cotton" would dominate the American and British politics of its time, and Jacquard's punch-card device would become important to the programming of early computers in the twentieth century.

Virtually every individual associated with one of these technical advances in textiles apparently spent most of his time and energy on the device itself, with little time to document his invention in a lasting way. And those who did write about their invention and how it came to be, seemed always to be composing in a legal context, attempting to protect their patent from piracy. It is no surprise, therefore, that in searching for documentation contemporary with the person or invention, a source like Abraham Rees's *The Cyclopaedia; or, Universal Dictionary of Arts, Sciences, and Literature* (London, 1819) is uniquely valuable. This work is the most detailed English source for information on the technology of this period, and its plates are superb. Although its publication date is 1819, this English edition began to be issued in 1802 and was completed in 1820. Altogether, it has thirty-nine volumes with six additional volumes of plates. Lacking both a table of contents and an index, the book contains articles arranged alphabetically. This can be a problem at times since, for example, a topic such as Kay's flying shuttle is located within an article on cotton. The Library of Congress has Rees's complete work in first edition.

It could be argued that the inventions of Arkwright, Whitney, and many others created a demand for accurate machine tools. Conversely, it could be said that the existence of certain tools made the inventions possible. There has probably always been this cycle of cause and effect between the tool and the invention throughout the history of technology, but it was never more apparent than during the Industrial Revolution. Rather than attempt to discern which came first in a particular mechanical process, it sometimes is better to simply acknowledge the phenomenon described as "the propagating power of tools," and grant that machine tools and the machines themselves have often shared a sort of symbiotic relationship. One example of such an intertwined story is Wilkinson's boring machine, which is known to have been essential to Watt in building his steam engine, which in turn brought about a new source of power, which in turn led to an increase in the number of machine tools. In the final accounting, however, what was perhaps the most important accomplishment during the early period of the Industrial Revolution was the new-found ability to work with metals in a measured and economical manner. Discarding wood as the prime substance for tangible inventions, the craftsmen were developing tools to do to metal what had only been done to wood. Thus boring machines, lathes, shapers, and planers were now metal instruments to be used on metal itself, to make more and better machine tools and superior machines themselves.

As with the nearly anonymous textile inventions and inventors, so the creators of machine tools—men who in their day were credited with the highest respect—are nearly anonymous today. Besides failing to publish any accounts of their own, they are cursed by the fate of the extremely

specialized. Thus, few today recognize the acknowledged "father of the English machine tool industry," Henry Maudslay, who invented the metal lathe (1800) and was the first to realize the critical importance in a machine shop of accurate plane surfaces for guiding tools. Other notable tools and toolmakers are John Wilkinson's already mentioned boring mill (1774) with its unequaled precision in boring cylinders for Watt's engines; Richard Roberts's metal planer (1817); and such useful and essential larger tools as Joseph Bramah's hydraulic press (1795) and Marc I. Brunel's block-making machine (1808).

The work of these and other skilled inventors and makers of new tools is not so much documented as revealed in such cumulative works as the 1819 *Cyclopaedia* by Rees. A similar compilation in the collections of the Library of Congress, although published later, is Charles Tomlinson's two-volume *Cyclopaedia of Useful Arts, Mechanical and Chemical*, published in London and New York in 1854. Organized alphabetically, Tomlinson's work is an industrial arts dictionary, beginning with "abbatoir" and ending with "zirconium." Like Rees's volumes, it does not have an index and the work of an individual can only be located within a specific area or

The first large-scale plant using machine tools for mass production was a block-making factory. Once a machine was designed by Marc I. Brunel and constructed by Henry Maudsley that could make wooden blocks for block and tackle, the British Admiralty set up such a factory. The plant consisted of forty-five machines of twenty-two different types that could make 100,000 blocks a year. With machines like this shaping machine, 10 unskilled men could do the work of 110 skilled blockmakers. The iron frame around the moving parts was made to protect the operator who moved the levers (G) and (l). *Cyclopaedia of Useful Arts, Mechanical and Chemical,* vol. 1, 1854. Charles Tomlinson. LC-USZ62-110389.

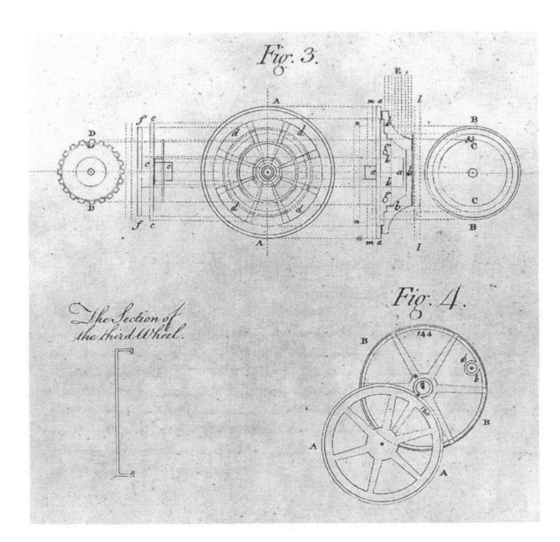

The invention of a successful marine chronometer earned its inventor, John Harrison, the phenomenally large reward of 20,000 pounds in 1773. This large money prize had been offered ever since 1713 by the Board of Longitude of the British Parliament, which viewed the invention of a clock that could keep accurate time at sea (and thus determine longitude by comparing local time with that of the Greenwich meridian), as critically important to British maritime interests. Through diligence, great skill, and unsurpassed precision, Harrison built four successively improved chronometers whose pendulum action could not only compensate for the pitch and roll of a ship but could adapt itself and adjust to extremes of temperature and humidity. Harrison devised a bimetallic method in which he alternated strips of brass and steel. Since each metal expanded and contracted at different temperatures, he could arrange them in such a way that their size changes would balance one another. His fourth chronometer was the size of a pocket watch and never gained or lost more than two minutes on even the longest sea voyage. Here are his detailed drawings of part of its inner workings. *The Principles of Mr. Harrison's Time-Keeper,* 1767. John Harrison. LC-USZ62-110390.

subject. Thus Henry Maudslay's inventions are discussed under the subject "blocks." There, however, one discovers not only everything about how blocks (as in pulleys or block and tackle) are made, but also illustrations of the tools used. In this section alone, Tomlinson shows a boring machine, mortising machine, shaping machine, scoring engine, and corner saw. Another much more focused work that discusses the machine tools and makers of its time is Robertson Buchanan's *Practical Essays on Mill Work and Other Machinery*, first published in London in 1808. This work is represented in the Library's collections by the two-volume 1823 second edition published in London. The Library also has the larger third edition, published between 1841 and 1850.

The eighteenth century was an enthusiastically practical time, so it should be no surprise that even the tools or instruments of science should have a decidedly practical bent. One major concern of astronomy during this period was the very real problem of determining longitude at sea. Discovering one's latitude or north-south location relative to the equator had not been a great problem since the fifteenth century. It became even easier and more reliable with the invention of James Hadley's quadrant, whose mirrors allowed the mariner to measure the angle between two observed objects.

Determination of longitude, which is measured in an east-west direction from an arbitrary datum point, proved a much more intractable problem. To navigators, the solution seemed simple. If one only had a clock that could keep accurate time at sea, the navigator could set it for Greenwich time (as Flamsteed had earlier proposed), easily figure the time difference, and then calculate longitude by the sun's meridian or high point. The British government saw the need for such a sea-going clock as so paramount that in 1714 it offered a prize of up to twenty thousand pounds for a solution.

It was for this prize and the joy of doing what he loved that John Harrison, a provincial, self-trained mechanic with a preternatural mechanical sense, produced a series of five clocks, each smaller and better than the preceding one. The problem was not only the obvious one of a ship's sway upsetting the pendulum, but the temperature and humidity extremes also affected its materials and, therefore, performance and accuracy. Harrison designed a remarkable clock, mounted in an ingenious way that eliminated sway as a factor. He then went further, however, and invented a pendulum composed of several metals which expanded and contracted differently in response to the weather. In effect, these reactions negated each other, and the pendulum's beat was unaffected.

Harrison's chronometer is well documented in the collections of the Library of Congress. First, the Library has a copy of the 1714 act as it appeared in the 1702–14 law and statute compilation, *Anno regni Annae Reginae Magnae Brittanniae*, published in London in 1714. This act has been described as one of the earliest examples of a government deliberately attempting to advance science and technology. The Library of Congress also has the first edition of *The Principles of Mr. Harrison's Time-Keeper* (London, 1767). This volume not only describes his solution to the

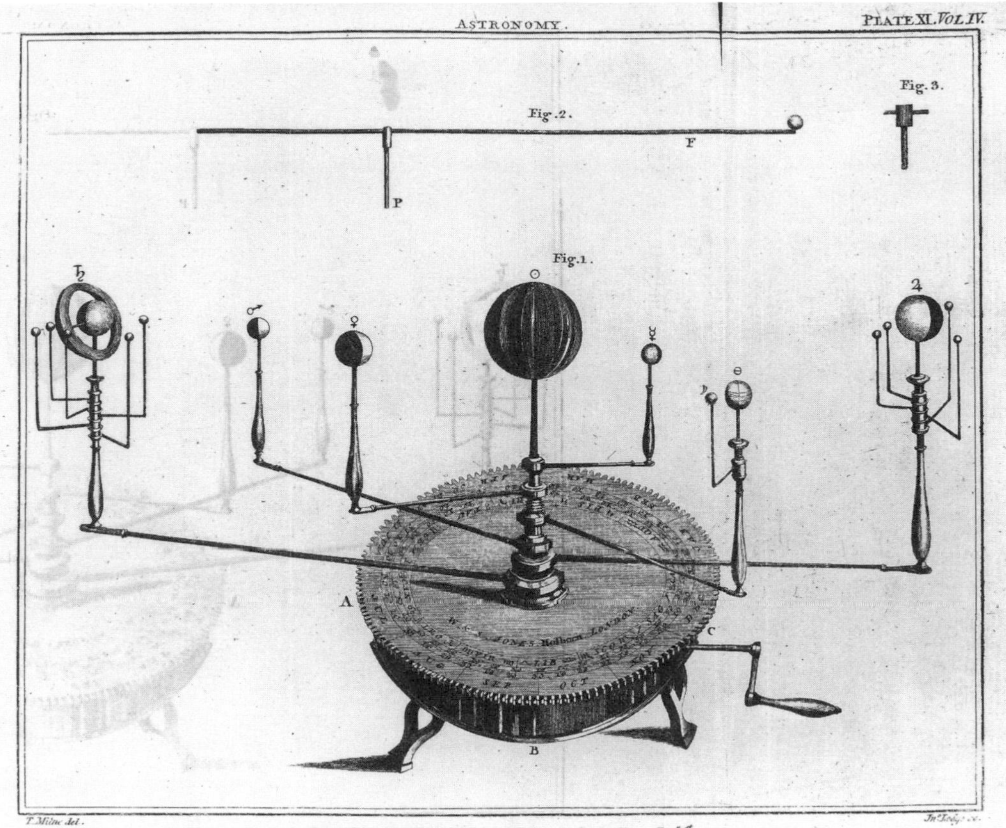

An orrery or planetarium is a mechanical model of the solar system used to demonstrate the motions of the planets about the sun. First made by two British clockmakers, George Graham and Thomas Tompion, the model got its name after being presented to Charles Boyle, the fourth earl of Cork and Orrery. Also called a "tellurium," an eighteenth-century orrery like this one made of brass and steel by George Adams was usually an expensive device, although more modest ones were made at a cheaper price. *Astronomical and Geographical Essay,* 1812. George Adams. LC-USZ62-110391.

problem of longitude, but has ten plates taken from his original drawings of the clock. Beginning in 1728, when Harrison produced his first clock, the British Parliament repeatedly changed its standards for awarding the prize to Harrison, and it was only in 1773, through the direct intervention of King George III, that Harrison was paid in full.

Another astronomical instrument that is characteristic of the eighteenth century is the orrery. Designed specifically to represent the Copernican or heliocentric universe, the orrery was usually a brass and steel apparatus whose clockwork gears showed the relative positions and motions of the planets by moving balls about. According to its specialized design, it was also called a lunarium, tellarium, or planetarium. Once set properly according to the planetary positions found in the *Ephemeris* or British *Nautical Almanac*, the orrery was an excellent three-dimensional teaching device. Orreries sold in such great numbers, however, that they must have been prized simply as elegant possessions. Two books in the collections of the Library of Congress are fairly representative of the interest in orreries. The earliest is John Harris's *Astronomical Dialogues between a Gentleman and a Lady*, published in London in 1719. The other was edited by William Jones and is the sixth edition of George Adams's *Astronomical and Geographical Essay*, published in London in 1812. Both works describe the orrery and similar instruments and tell how to use them, while also offering images of the instruments themselves.

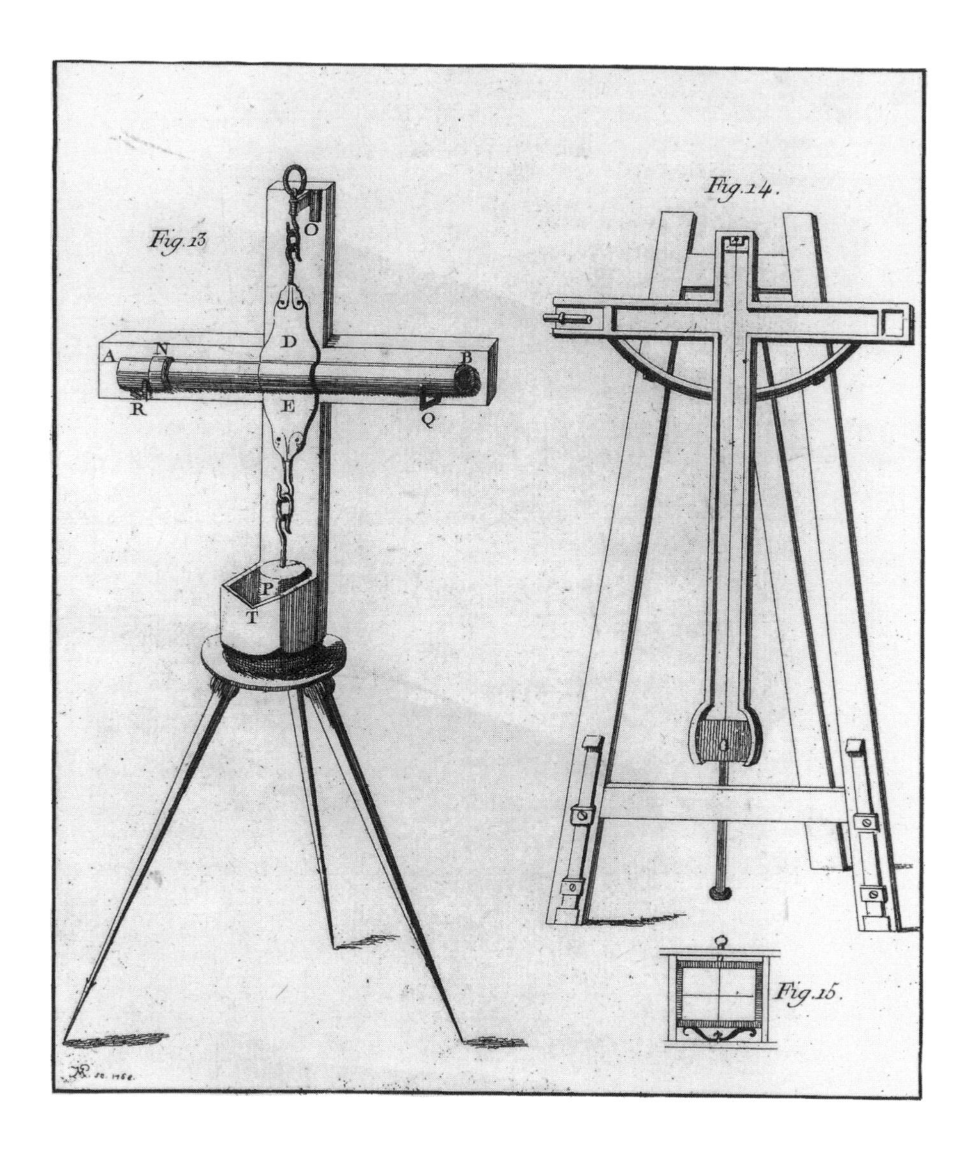

From Roman times until the seventeenth century, surveying tools remained essentially the same: they were rulers held in a horizontal position by a plumb line or some weight, and the sighting was done through small holes. By about 1700, however, telescopic sights and bubble or spirit levels were added, and surveying accuracy greatly improved. The figure on the right shows Jean Picard's level in the shape of a cross. A plumb bob passes through the vertical tube and makes it possible to regulate the horizontal position of the eyepiece. The small, square figure 15 below shows Picard's cross-hair invention for the eyepiece. The figure on the left is Christiaan Huygens's level, which uses a suspended telescope. *Traité sur la théorie et la practique du nivellement,* 1768. Augustin, comte Lespinasse. LC-USZ62-110392.

More down-to-earth, the art and practice of topographical surveying was relatively ignored and unimproved until land itself became a source of revenue. When fields and estates were enclosed and registered for purposes of taxation, surveying had to become more accurate. Around the middle of the eighteenth century, a surveyor's tools might include a bubble level, theodolite, clinometer, and surveyor's compass. The first of these was difficult to manufacture, requiring glass-blowing skill, and was slow to reach the surveyor. Leveling, or finding a true level on undulating terrain, had been done the classical way with plummet and graduated square or quadrant, until the bubble or spirit level was devised in the 1660s by the French savant Melchisedech Thevenot. Improved by Hooke and Huygens, it was still not a common instrument for decades. Jean Picard's book, *Traité du nivellement*, first published in Paris in 1684, exhaustively discusses this new tool, but it was not until nearly forty years later that the bubble level became an everyday device for surveyors. In the first four editions of *The Compleat Surveyor* by William Leybourne, there is no mention of it, but it does appear in the fifth edition in 1722. The Library of Congress has this fifth edition, published in London. Although it does not have the first edition of Picard's work, the Library does have a later German translation, *Abhandlung vom Wasserwägen*, published in Berlin in 1749.

By about this time, the theodolite had been improved by the addition of a sighting telescope, making it a very sophisticated surveying tool. This standard instrument, usually mounted on a circular base, was often combined with a level and compass, and gave the mid-eighteenth-century surveyor a handsomely impressive and highly useful instrument. This and other tools of the trade are clearly depicted and described in *The Practical Surveyor* by John Hammond, which the Library of Congress has in an edition published in London in 1725.

Probably the one invention that was key to improving these and nearly all the small precision instruments that would follow was Jesse Ramsden's "dividing engine." Before his machine, the scale markings on any tool's

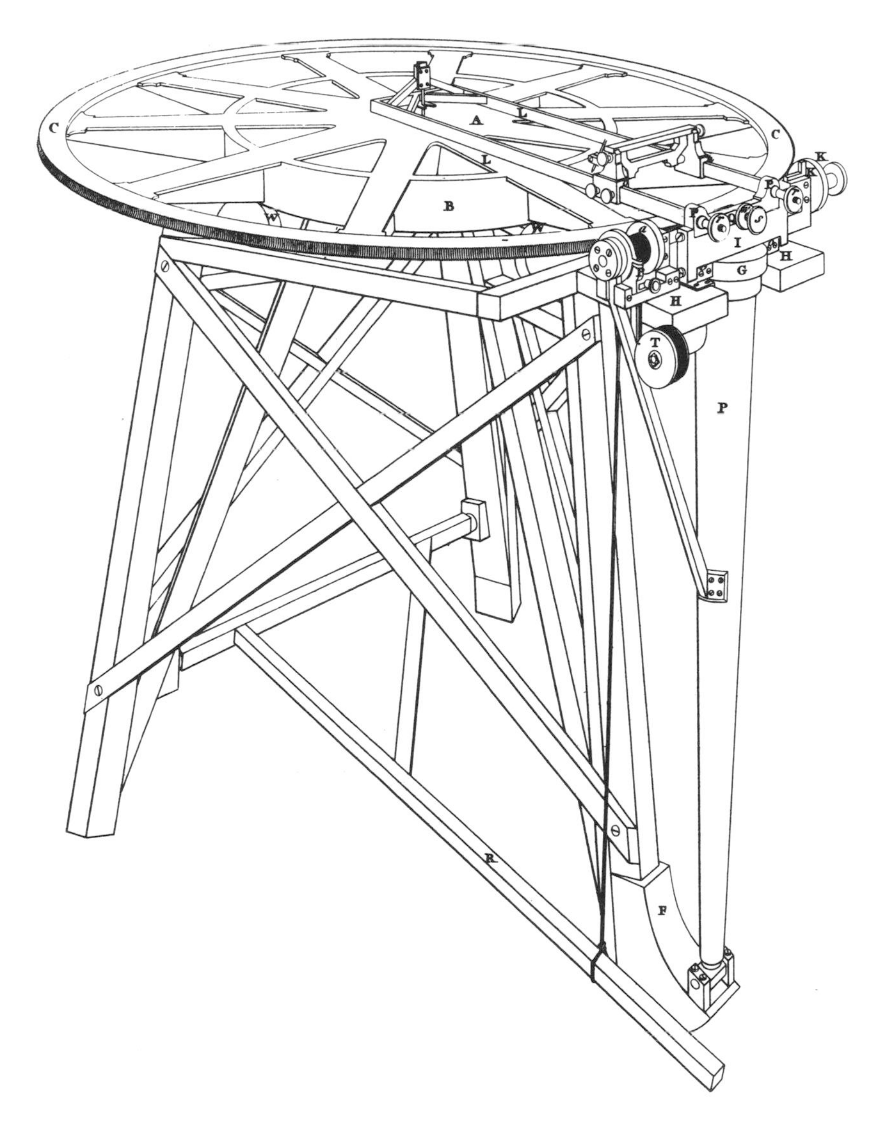

The career of Jesse Ramsden reflected the eighteenth-century change in the role of the precision instrument-maker from handicraftsman to inventor of accurate machinery for the production of interchangeable, standardized machine parts. In 1763, Ramsden produced his circular "dividing engine" shown here, which could calibrate scales on other circular measuring instruments. Ramsden discovered that the accuracy of any dividing engine depends on the uniformity of the threads in their adjustment screws. His invention of the screw-cutting lathe, which could cut uniform threads in metal screws, produced a revolutionary leap in accuracy. *Cyclopaedia,* vol. 2 (plates), 1802–20. Abraham Rees. LC-USZ62-110393.

The first cast-iron bridge spanned the Severn at Coalbrookdale, Shropshire, England. From the sixteenth century, cast iron was used in Europe for cannon shot, grave slabs, and fireplace backs, but never for anything structural. With Abraham Darby's discovery in 1709 of a method of using coke instead of scarce charcoal for smelting iron, large-scale production of high-quality cast iron began. Completed in 1779, the bridge's roadway is supported by an arch of five iron ribs, each of which was cast in two halves. Darby's successful advertisement for iron freed bridge builders from their exclusive dependence on timber, stone, and brick and opened a range of new building possibilities. Ironbridge carried traffic as late as the 1950s, and became a national monument following its restoration. Prints and Photographs Division, Library of Congress, lot 4451.

metal ring had to be marked by hand. Not only was this generally inaccurate, but it made very small divisions—down to the newly necessary one-sixth degree—impossible. While any engineer knew what was needed to obtain such a level of detail—simply 2,160 marks or one for every tenth minute of arc on a complete 360 degrees—this was a fearsome level of precision to try to achieve, using mid-eighteenth-century tools. After fourteen years of effort, Ramsden produced in 1773 just such a machine for dividing. It consisted of a large bronze, toothed wheel that was laid horizontally and operated by manual gears. The plate to be divided or scored was placed atop this spoked wheel and received one precise bite from each of the wheel's 2,160 teeth. Interestingly, there is little record of this invention that would transform the construction of many instruments and allow modern small, precision instruments to come into being. Since Ramsden had been competing for a money prize offered by the British Board of Longitude and received several hundred pounds in 1774, he was unable to take out a patent and was obligated to publish the details of his machine. This he did by pamphlet in 1777, producing a rare, fourteen-page document which the Library of Congress does not have.

In the field of optics, advances and refinements were made during this time that were built upon the inventions of the previous century. Thus, the Newtonian reflecting telescope would come to be replaced by the achromatic telescope whose compound lenses produced a far clearer image that was also free of unwanted colors. Microscopes similarly improved and became specialized, such as the late-eighteenth-century botanical microscope. Two books in the Library's collections document many of these optical developments. For those emerging from the early part of this era, Robert Smith's two-volume work, *A Complete System of Opticks* (Cambridge, 1738), serves well as a comprehensive account of optical instruments and improvements. Documenting a century's progress after Smith's account is *An Introduction to Practical Astronomy* by William Pearson, published in London between 1824 and 1829. This two-volume work goes beyond astronomical instruments and surveys a range of optical instruments, including microscopes and micrometers.

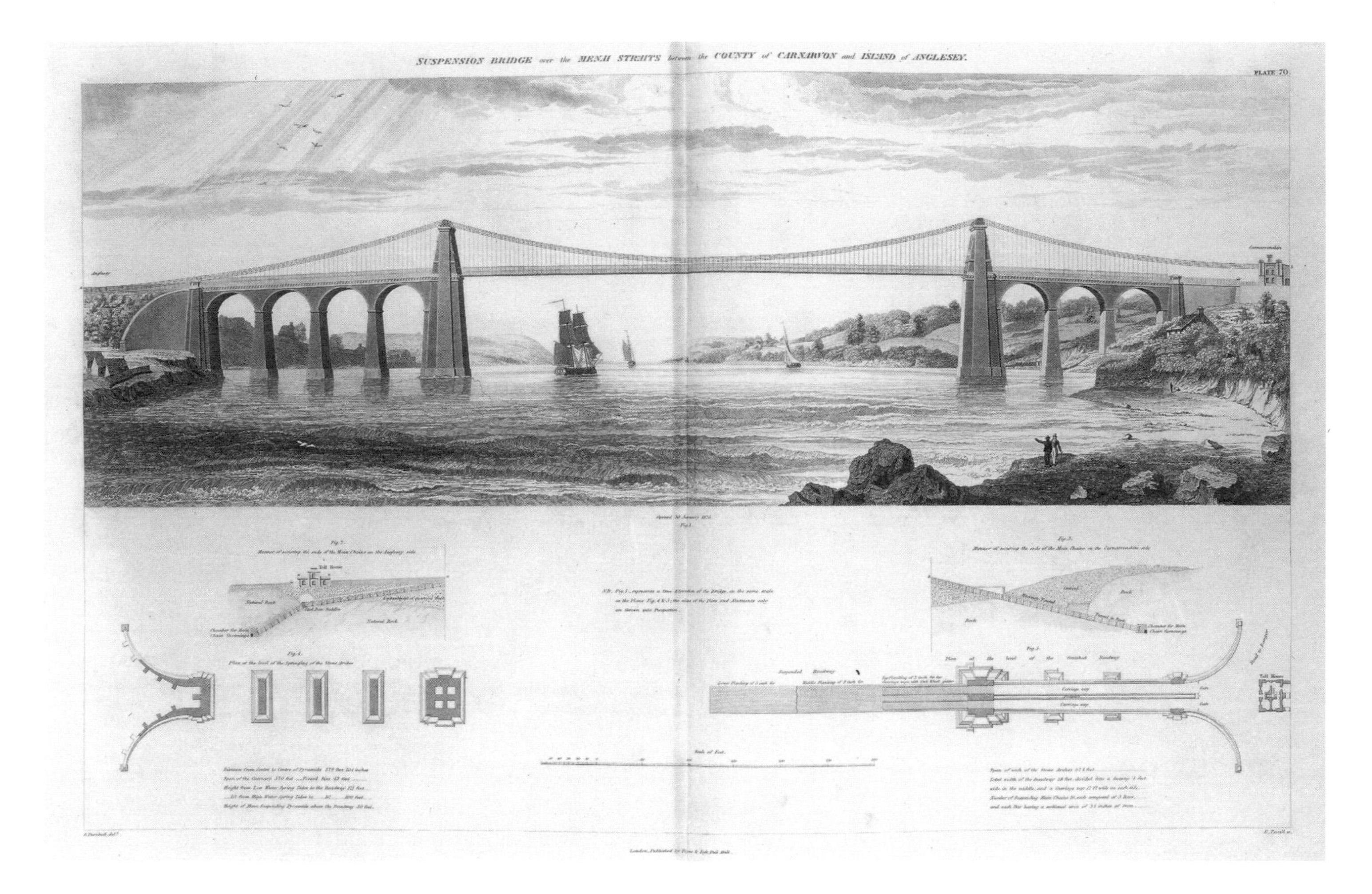

If the Industrial Revolution was founded on iron, coal, and mechanical engineering, as has been said, then it can also be said that a modern bridge embodies all three of these essential components. The first complete and successful application of cast iron as a principal structural material is seen in the bridge called Ironbridge that crosses the Severn River near Coalbrookdale, England. Attributed to John Wilkinson but actually designed and built between 1776 and 1779 by Abraham Darby III, it spans 100 feet and 6 inches and still stands today.

Where documentation is scarce concerning this first iron bridge, the work of a Scottish bridge pioneer who was inspired by the Coalbrookdale accomplishment is well represented in the Library of Congress. Thomas Telford built aqueducts, bridges, canals, harbor works, and roads throughout Wales. In 1819, he began his masterpiece, the Menai Bridge, spanning the Menai Strait in Wales. This was a suspension bridge—then a novelty—which would span 580 feet, with a deck hung by Telford's masterly wrought iron links. His great achievement in iron was also built high enough to allow free passage of ships underneath. The Library of Congress has two works that document Telford's Menai accomplishment.

The British engineer Thomas Telford (1757–1834) was instrumental in creating the transportation network that would come to serve Britain so well during the Industrial Revolution. As a builder of roads, bridges, canals, and harbors, Telford excelled at adapting known techniques in creative ways. Here is his largest bridge, the cast-iron suspension bridge over the Menai Straits off the coast of northern Wales. Built between 1820 and 1826, it spanned 580 feet, and its oak roadway hung by sixteen wrought iron chains from two stone towers. The bridge linked central England and Ireland and was used for 120 years, until it was rebuilt in 1940. *Life of Thomas Telford, Civil Engineer,* 1838. John Rickman.

The earliest, *An Historical and Descriptive Account of the Suspension Bridge Constructed Over the Menai Strait*, has seventeen plates besides text. Written by William Alexander Provis, it was published in London in 1828 (six years before Telford died) and has details on Telford's Conway suspension bridge as well. The other is a two-volume work edited by John Rickman and published in London in 1838, entitled *Life of Thomas Telford, Civil Engineer*. The complete title claims that it was "Written by Himself," that is, by Telford. Of more importance, however, is a very large folio atlas of eighty-three copper plates that accompany the work and show in excellent architectural detail exactly how wide-ranging his engineering talents were.

Bridge building in France at the beginning of the eighteenth century was responding to the same requirements for improved means of communication and transport that were pushing developments in England. As both trade and manufacture were expanding, so was the need for better highways and for bridges where only ferries had been. In France in 1716, the Corps des Ponts et Chaussées was formed to advance such civil engineering projects. In 1747 the Corps formed a school, the École des Ponts et Chausses, whose first director, Jean-Rodolphe Perronet, would become one of the greatest bridge builders ever, and certainly the greatest designer of stone arches. Although he worked in stone and not iron, Perronet's work often inspired others, like Telford, who copied his elliptical arches. Perronet's two great bridges are the bridge across the Seine below Paris at Neuilly (1768–74), and the Pont de la Concorde in Paris, which was completed in 1791. Perronet's bridges were startling in comparison to those of his predecessors, and looked almost unsafe with their flatter arches, shallower arch rings, and thinner piers. His Neuilly masterpiece, as well as several other construction projects he undertook during his long career, are documented in the collections of the Library of Congress by a large volume titled *Description des projets et de la construction des ponts de Neuilli, de Mantes, d'Orléans, de Louis XVI*, published in Paris in 1788. Accompanying this volume is an extremely large folio of seventy-three plates which show in the greatest detail the intricacies and beauty of Neuilly and other bridges Perronet built in France.

The final major transportation advance of this time period occurred in 1815 with the appointment of Scotsman John Loudon McAdam as surveyor general of British roads. McAdam had long been experimenting with different materials and techniques, trying to improve the terrible condition of Britain's highway system. At the time, carriage roads were paved with round stones and gravel—a combination that made for deepening holes as the weight of the coach wheels literally forced the stones out. McAdam offered instead a road made of crushed rock. Properly built with allowances for drainage built in, his method produced a smooth, artificial surface that only became more solid under the pressure of carriage wheels. McAdam wrote of his experiments and new road system in his book, *Remarks on the Present System of Road Making*, first published in 1816 in London. The Library of Congress has his popular book in its seventh edition, published in London in 1823.

At about the same time that these important but very much down-to-earth advances in transportation were being made, an ancient dream was being realized. During a short span of months in 1783, aerostation, or lighter-than-air navigation, was born and then rapidly advanced. In 1782, the notion of a man traveling through the air and arriving safely elsewhere, out of sight, seemed the stuff of fiction, but by the end of 1783, man had made his first successful journey into the atmosphere. Balloon flight was rapidly becoming a popular, romantic, and ubiquitous adventure.

The traditional story of the invention of the hot-air balloon correctly attributes the Montgolfier brothers as its discoverers. They first put to practical use the long-known phenomenon of the lifting properties of hot air above a fire. During 1782, the brothers Jacques-Étienne and Joseph-Michel experimented with several small models, and on June 5, 1783, they exhibited a large linen bag, thirty-five feet in diameter, in the center of their town of Annonay. They inflated it by holding it over a fire of damp straw and wool and, upon release, it floated upward fifteen hundred feet and covered a mile and a half in ten minutes. Although the brothers had no idea why hot air made their balloon rise, by late summer they were in Paris doing the same exhibition and then, in September, before the king at Versailles.

On November 21, 1783, a Montgolfier balloon lifted two passengers, a noted French scientist, Jean François Pilâtre de Rozier, and the marquis d'Arlandes, on a free flight over Paris. This historic first manned flight was followed immediately by the flight of Jacques-Alexandre César Charles, whose buoyant, hydrogen-filled, valved balloons were to become

Opposite page:
This section of the Pont de Neuilly under construction across the Seine exemplifies the design breakthroughs made by the French civil engineer Jean-Rudolphe Perronet (1708–1794). His stone bridges have flatter arches, shallower arch rings, and thinner piers than anything that went before. Once Perronet discovered that the sideways thrust of an elliptical (flat) arch is passed along to the abutments at the ends of a bridge, he realized that the intermediate piers would have only to support the deadweight of the arches and their superstructure. This enabled him to design elegant, convenient, and safe bridges whose piers obstructed much less of the waterway. Perronet had a long and distinguished career, starting from his appointment at thirty-nine years of age as director of the world's first engineering school (Ecole des Ponts et Chaussées) to his construction of the Pont de la Concorde, which he began to build when he was eighty years old. *Description des projets et de la construction des ponts de Neuilli, de Mantes, d'Orléans, de Louis XVI,* 1788. Jean-Rudolphe Perronet.

Above left:
Hydrogen balloon flown by Jacques Charles, 1783. Prints and Photographs Division, Library of Congress. See p. 109.

Above right:
On September 19, 1783, the Montgolfier brothers sent up this balloon from Versailles in the presence of the royal family. It carried a duck, a rooster, and a lamb in the small wicker basket seen dangling from the balloon. This was their successful precursor to the first manned balloon flight that would take place on November 21, 1783. From the beginning, the Montgolfier brothers sought "to enclose a cloud in a bag," and after failing with steam and later with impure hydrogen, they settled on ordinary smoke or hot air from a fire. Prints and Photographs Division, Library of Congress, LC-USZ62-42858.

the preferred flying vehicles of most aeronauts. Charles also deserves credit for the method of suspending a wicker basket or gondola from the net around the balloon.

The earliest record of the Montgolfiers' historic November flight was recorded and published in the same year as the first aerial voyage. Written by the eminent French geologist, Barthélemy Faujas de Saint-Fond, who was both the Montgolfiers' financial backer and chronicler, this is a two-volume work titled *Description des Expériences de la Machine Aérostatique de MM. de Montgolfier*, published in Paris, 1783–84. The first edition copy of this work in the collections of the Library of Congress carries the ex libris of Antoine Lavoisier.

The strength of the Library's overall collection in aeronautics is especially demonstrated by how well this seminal time is documented. In addition to the Faujas de Saint-Fond effort, the Library has first accounts by Pilâtre de Rozier—who was both the first to fly and, later, the first to die in a balloon—and by hydrogen balloon pioneer Charles. Pilâtre de Rozier's *Premiere expérience de la Montgolfière construite par ordre du roi* was published in Paris in 1784, and Charles's *Représentation du globe aérostatique que s'est élevé de dessus l'un des bassins du jardin royal des Thuilleries* was first printed in Paris in 1783. The Library has both in first edition.

From the same time period, the Library also has first edition accounts of the balloon's spread to England. Vincent Lunardi's *Account of the First Aerial Voyage in England* was published in London in 1784 and tells how the young secretary of the Neapolitan ambassador became the first man to fly in England. Thomas Baldwin used Lunardi's balloon in September 1785 and later related his aerial observations during that flight in *Airopaidia* (Chester, 1786). And the Library also has the first edition of the account of the first American to ascend in a balloon—John Jefferies's

SIC ITUR AD ASTRA

45th Ascention and the first made in America January 9th 1793. at Philadelphia 39° 56' N. Latitude by Mr. J. P. Blanchard.

45e ascension et la premiere faite en Amerique Le 9 Janvier 1793 a Philadelphie 39 56' Latitude N. par. Mr. J. P. Blanchard.

JOURNAL

OF MY

FORTY-FIFTH

ASCENSION,

BEING THE FIRST PERFORMED IN

AMERICA,

ON THE NINTH OF JANUARY, 1793.

Æthereum tranabit iter, quo numine BLANCHARD?
Impavidus, ſortem non timet Icariam.

PHILADELPHIA:
PRINTED BY CHARLES CIST, NO. 104. NORTH
SECOND-STREET, M,DCC,XCIII.

Narrative of the Two Aerial Voyages. Printed in London in 1786, the book describes Jefferies's and Jean Pierre Blanchard's first aerial crossing of the English Channel (Dover to Calais) on January 7, 1785. The same Blanchard, a Frenchman, also made the first balloon ascension in America. On January 9, 1793, Blanchard traveled fifteen miles from Philadelphia into New Jersey, carrying with him a passport from President Washington which asked American citizens to help the balloonist. The Library of Congress has Blanchard's report of this voyage, *Journal of My Forty-Fifth Ascension*, as it was first printed in Philadelphia in 1793.

Beyond these first early years, the history of ballooning in Europe is documented exhaustively by the Library's unique Tissandier Collection. Purchased by the Library in 1930, this collection contains eighteen hundred books, six thousand manuscript items, and seven hundred watercolors, prints, and photographs all relating to lighter-than-air travel from the late eighteenth century to the beginning of the twentieth century.

The first balloon flight in America was not made by an American but rather by the daring Frenchman Jean-Pierre Blanchard. By 1793 when he made this first American flight, Blanchard had established himself as "the greatest of the early aeronauts." This reputation was justified, for Blanchard had made forty-four flights in several European countries before his U.S. trip, with one of them being the first flight across the English Channel. On his balloon trip from Philadelphia, the forty-year-old Blanchard took along only a small black dog and traveled fifteen miles across the Delaware River, landing in the woods east of Woodbury, New Jersey. Among those who saw him off were President George Washington. Blanchard carried a let-
(over)

(cont.)
ter from Washington which guaranteed his safety, since Blanchard spoke no English. Blanchard died from injuries received in a balloon crash in Europe in 1809, and his wife, Marie, died ten years later in a fall from a burning balloon. *Journal of My Forty-Fifth Ascension, Being the First Performed in America,* 1793. Jean-Pierre Blanchard. LC-USZ62-110394.

Assembled by the Tissandier brothers, Albert (1836–1906) and Gaston (1843–1899), who were themselves balloonists, dirigible builders, and aeronautical historians, this collection not only documents their own work and correspondence but contains what was described in 1930 as "approximately all the worth-while aeronautic literature published in France up to 1900."

It was during this time that progress began on heavier-than-air flying machines with the work of George Cayley. In fact, it could be said that aviation reached its high point for the entire nineteenth century in his work. Although still generally unknown, the prodigious Cayley was the real founder of the science of aerodynamics and was the first to build a successful glider capable of carrying a man. He was the first to do away with the counterproductive notion that man could fly by using movable or flapping wings. From there, he developed what is essentially the basic configuration of the modern airplane—fixed wings, a fuselage, a tail unit with elevators, and a rudder for flight control. He studied all the essential problems of flight using models, and predicted the development of the internal combustion engine and its application to powered flight.

Cayley did not spend his entire life contemplating the problems of flight. A Yorkshire baronet, he was a multi-talented man with diverse interests and activities who basically spent two parts of his long life on aeronautics. During his late twenties and thirties, he published a triple paper titled "On Aerial Navigation" that appeared in three installments in the *Journal of Natural Philosophy, Chemistry, and the Arts* (November 1809, February and March 1810). This three-part paper laid the foundations of modern aerodynamics. It further discussed parachutes, stability and control, and streamlining. This periodical is often referred to as *Nicholson's Journal*, since it was edited by William Nicholson. The Library of Congress has the beginning years of this short-lived journal (which eventually merged with *Philosophical Magazine*), but it does not have the years 1809 and 1810. The article was, however, reprinted by American James Means in the first issue of his *Aeronautical Annual*, published in Boston in 1895. Although much time had passed since Cayley's article was first published, Means's reprinting is the only one up to that time to have used the original article in *Nicholson's Journal* as the source. All other reproductions reflect alterations, omissions, and even additions that crept in afterwards. Cayley also wrote three papers on airships that were published in 1816 and 1817 in *Philosophical Magazine*. These years are not in the Library's collections.

It was not until nearly twenty years later that Cayley again published anything on aviation. Beginning in 1837 and ending in 1852, when he was seventy-nine years old, Cayley published four papers on airships and gliders in *Mechanics' Magazine*. The Library has all of these as they first appeared.

Cayley's aeronautical writings, no matter how imaginative or prescient, had little real effect during or immediately after his lifetime. It could also be argued that in terms of everyday life and business during this time, even the sudden enthusiasm for balloon flight created by the Montgolfiers and their successors had no real effect on how people lived their lives or on the quality of those lives. What did affect them, however, were inventions and processes that in modern times are barely given a second thought. Many can be described as nearly anonymous, since the person behind the technology is seldom popularly known and usually published little to which we can refer today.

One of these major shapers of everyday life was the inventor of the Argand burner or the oil lamp. In the same year that the Montgolfiers first sent up their balloon, François Ami Argand, a Swiss, introduced his new lamp. It contained three successive innovations: a hollow wick to provide an enriched or double air supply that burned at a higher temperature and gave a brighter light; a glass cylinder that acted as a chimney and gave the flame a steady brightness; and a mechanism for raising and lowering the wick, thereby regulating the intensity of the light. Although still essentially a traditional oil lamp, the Argand burner would become an inseparable part of the nineteenth-century household, as essential as the electric light became to the twentieth century.

Ironically, the best original documentation of Argand's invention comes not from the man himself but from his imitators and plagiarizers. In 1783, the year of his invention, Argand went to England to promote his new lamp. By 1785, at least two Frenchmen, A. B. Lange and A. A. Quinquet, were passing off the new lamp as their own invention. Thus, in the 1785 edition of the Paris journal *Bibliothèque Physico-Économique* is a nine-page article by Lange titled "Description de la Lampe physiocopneumatique à cylindre, inventée par M. L'Ange." There is also one plate showing several aspects of the lamp. The Library of Congress has a full run of this journal. In the end, Argand fared badly in the courts and died in poverty in 1787. While Argand had thought of his lamp primarily in terms of domestic lighting, it would prove to be most dramatically useful for lighthouses. Prior to the adoption of the Argand system, a lighthouse could consume up to 300 tons of coal a year to keep its fire burning. Large Argand lamps with as many as ten concentric wicks soon came to provide efficient, smokeless bright light for the next century.

Within a generation, the modern era in the history of illumination would start with the beginnings of gaslight. What shocked and then puzzled most people upon first seeing the dazzling brightness of gas illumination was the absence of a wick. This seemed somehow against both nature and their common sense. But the combustible properties of certain gases had been known since the late-sixteenth-century discovery by Johannes Van Helmont of "gas pingue" or inflammable gases, and if most people were taken by surprise, many men of science were not. The first detailed description of the phenomenon of producing coal gas by distillation and then lighting it to produce illumination was published in *Philosophical Transactions of the Royal Society* in 1739, which is in the Library's collections. Originally written by an amateur chemist, John Clayton, in 1691 as a letter to Robert Boyle, the account tells how he "observed that it catched flame, and continued burning at the end of the pipe, though you could not discern what fed the flame." This statement shows that not

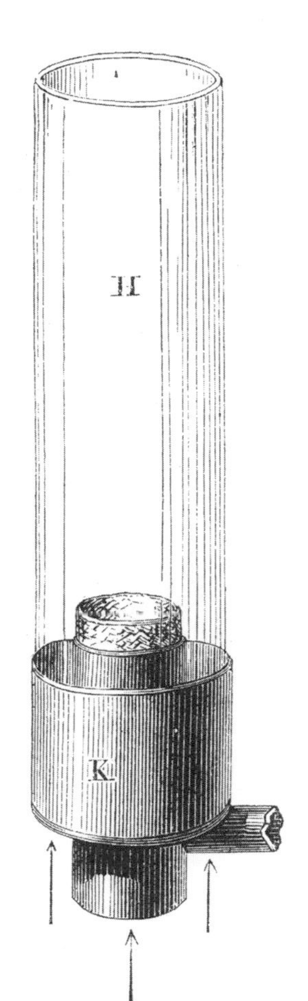

During most of the eighteenth century, interior lighting by candles and lamps was very poor, often smoky, and always in need of attention. The first significant advance toward improving interior illumination was this simple design by Ami Argand in 1784. Argand devised a tubular wick that could be raised or lowered, and which was essentially a hollow cylinder enclosed in a glass chimney. Because the wick was hollow, the inside of the flame was given oxygen, resulting in a tenfold increase in light from a single wick. It also did away with most of the smoke, smell, and flickering. Argand's burner was used in lighthouses, where it provided a dramatic increase in illumination. *Les Merveilles de la science*. vol. 4, 1867–69. Louis Figuier. LC-USZ62-110394.

even Clayton knew what this new "inflammable air" or "spirit" was all about. Nor did he or anyone else at the time know how to use it for practical purposes.

By 1800, however, both chemistry and technology had advanced to the point where the foundations for gas lighting were fully developed. In October 1801, the first public exhibition of gas for lighting and heating took place in a Paris hotel called the Seignelay. One of Philippe Lebon's "thermolamps" warmed and lit the interior, and another illuminated the outdoor garden. Although Lebon proved to be an imaginative individual who anticipated nearly all the applications of gas that were to follow, he was unsuccessful in gaining backers and died at thirty-four, stabbed to death by robbers on the Champs-Elysées.

In England, the situation was more favorable, and William Murdock, who worked for the flourishing engineering firm of Boulton and Watt, began to systematically investigate all aspects of the problem of lighting by gas. After learning of Lebon's work, Murdock was able to make important advances and, in 1802, he lit up the Boulton and Watt Soho Foundry at night. In doing this, Murdock created the prototype of all gas-based systems to come, for he was able to use piping to connect the gas production site with the place where the gas was to be consumed. Four years later, he lit a textile factory in Salford, England, and, in 1807, street lighting by gas was introduced in London when Pall Mall was lit by a commercial company that eventually became the Gas Light and Coke Company. In 1808, Murdock wrote of his accomplishment, and the Library of Congress has his nine-page article which appeared in the *Philosophical Transactions of the Royal Society* under the title "An Account of the Application of the Gas from Coal to Economical Purposes."

The unfortunate Lebon left little documentation behind, and none that the Library has. But he is mentioned in two later works that credit the priority of his accomplishments. One is *The Theory and Practice of Gas-Lighting*, by Thomas S. Peckston, regarded as the first modern textbook on gas, which first appeared in London in 1819. The Library of Congress has Peckston's second edition, published in London in 1823. Lebon also turns up in what is considered to be the first history of gas lighting, William Matthews's *Historical Sketch of the Origin, Progress, and Present State of Gas-Lighting*. The Library has the first edition of Matthews's book (London, 1827).

Another lighting achievement during this time saw its application entirely underground. Once English coal miners exhausted the surface seam deposits and had to sink shafts as deep as six hundred feet, they began to encounter an increasing volume of dangerous methane gas. Explosions caused by the open flames of their candles or lamps caused many a deadly accident, and in 1813 a particularly disastrous explosion led to a general appeal to scientists for help. One of those approached was a young, flamboyant professor, Humphry Davy. In a model of the experimental method, Davy visited the mines, talked to people, and began to study the gas itself. After learning the conditions under which it explodes and its general nature and behavior, he experimented with a variety of physical lamp

The modern gas industry began with William Murdock, who was the first to market successfully a gas lighting system. Shown here is the apparatus he used to produce the coal gas with which he lit the entire Boulton and Watt factory at Soho. Coal gas was produced by heating coal to decomposition in a retort (M). The gas produced was eventually led into an outlet pipe (N) from which it was directed by piping to the site to be illuminated. Lighting a building during darkness meant night shifts for factories whose machinery could be run continuously. *Les Merveilles de la science,* vol. 4, 1867–69. Louis Figuier. LC-USZ62-110396.

barriers, and arrived at a metallic gauze cylinder that would surround the naked flame. It worked by allowing oxygen in to feed the flame while the metal dissipated its dangerous heat. He had created the modern safety lamp in less than three months. The Library of Congress has Davy's initial publications on the safety lamp, both of which appeared in the *Philosophical Transactions of the Royal Society*. The first, "On the Fire-damp of Coal Mines, and On Methods of Lighting the Mines So As to Prevent Its Explosion," appeared as a twenty-two-page article in 1816. The second, "An Account of an Invention for Giving Light in Explosive Mixtures of Fire-damp in Coal Mines, by Consuming the Fire-damp," appeared in 1816.

It was also during the beginnings of industrialization that the age of electricity began. Even before the oil lamp was perfected by Argand and well before gas lighting first was used, the nature of the electrical phenomenon was being investigated and slowly coming to be better understood. First it was Benjamin Franklin who tested, via his famous kite experiment, his hypothesis that atmospheric lightning was an electrical phenomenon similar to the spark produced by the frictional machines of his day. The Library of Congress has Franklin's *Experiments and Observations on Electricity Made at Philadelphia in America*, first published between

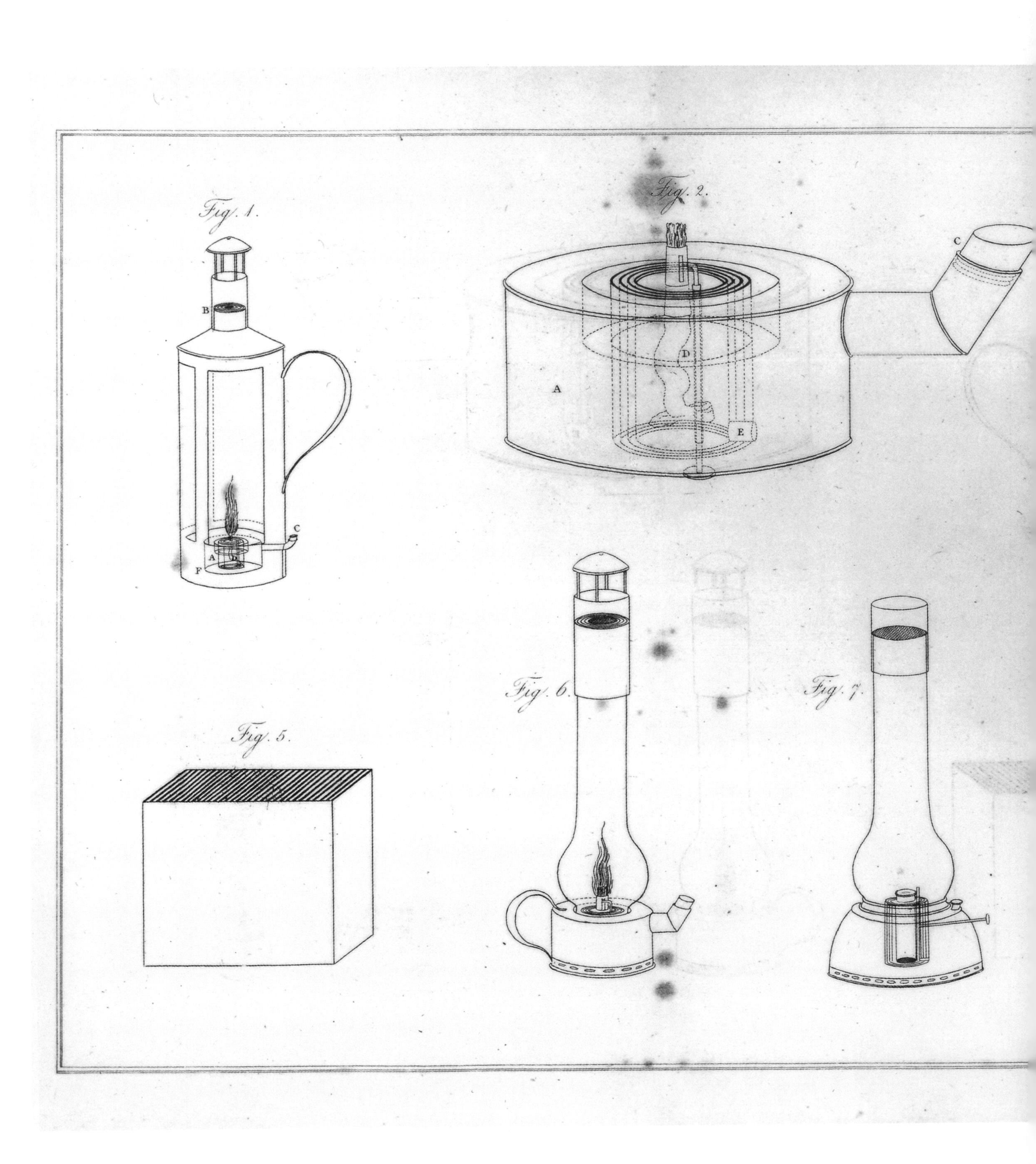
Fig. 1.
B
C
A
D
F
Fig. 2.
C
D
A
E
Fig. 5.
Fig. 6.
Fig. 7.

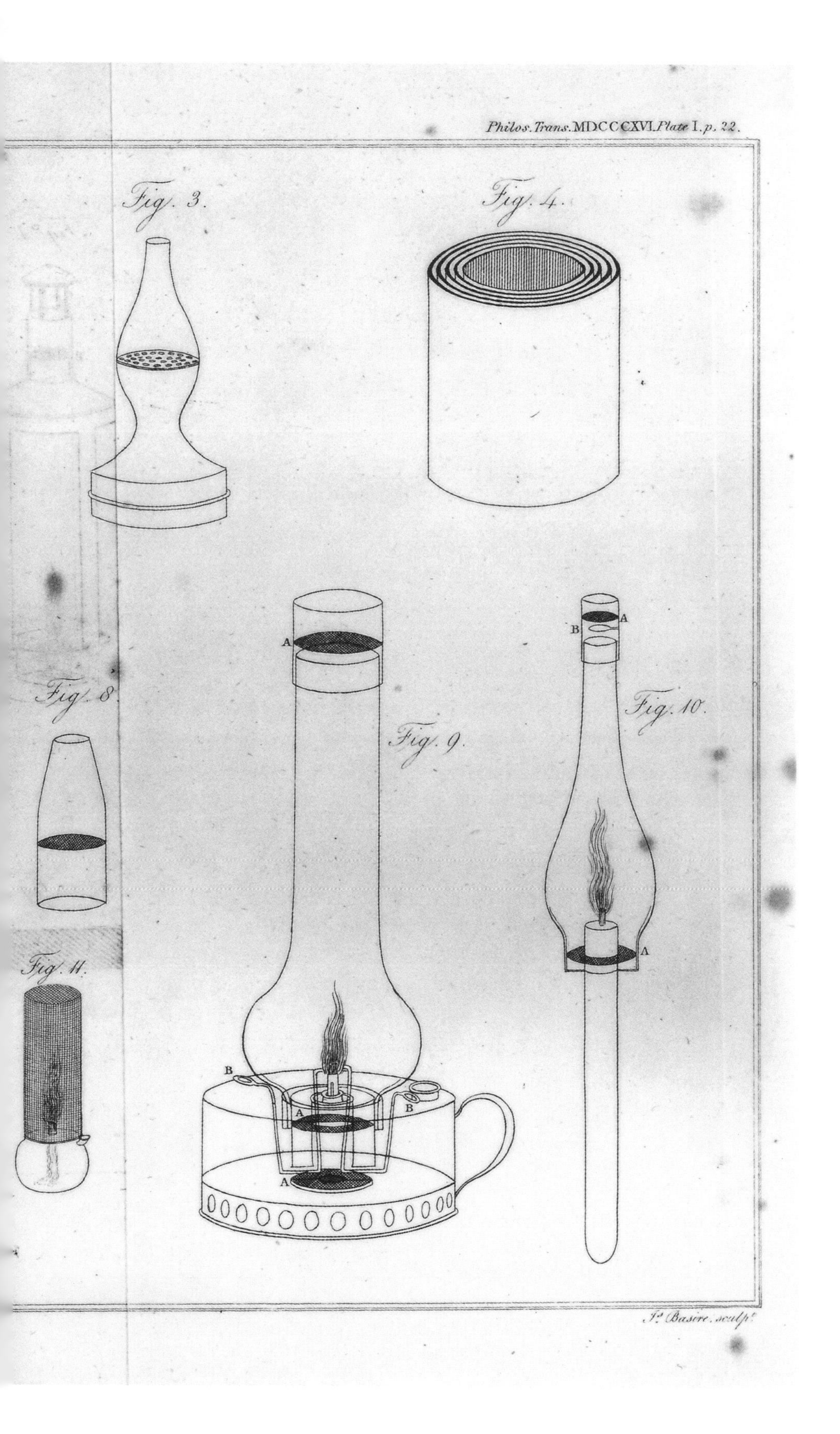

Humphry Davy's invention of the safety lamp is an example of science spurring technology. The most common cause of death among miners was the explosion of underground gas by the naked flames of their lamps. In 1815, the well-known Humphry was asked to find a solution, and within three months he produced several versions of a safety lamp, as seen here. Davy quickly ascertained that it was undetectable methane gas that was being ignited by the high temperature of the miners' lamps. He then sought to produce a lamp whose flame temperature was lower than the ignition point of methane. Davy did this by modifying the standard oil lamp so as to disperse its heat. He surrounded the flame with a fine metal gauze screen through which oxygen could pass but whose surface temperature remained fairly low. Not only was his safety lamp a success at preventing explosions, it had the added advantage of acting as an early warning system, since its flame turned blue in the presence of any quantity of methane. *Philosophical Transactions of the Royal Society*, 1816. Humphry Davy. LC-USZ62-110397.

1751 and 1753 in London. This work contains not only the results of his experiments as communicated to his London friend Peter Collinson but also Franklin's major theory of general electrical action.

Research in electricity soon became a scientific pursuit as popular and ubiquitous as Franklin's international reputation. Thus in 1791, when Luigi Galvani noted the convulsive movement made by a frog's leg he was skinning when it was touched by a scalpel, he first made a correct guess—that electricity was involved—and then misinterpreted its meaning, believing that it came from within the animal itself. Galvani's Italian countryman, Alessandro Volta, then took up Galvani's ideas and subjected them to rigorous testing. Within two years, Volta was able to demonstrate that the twitch of the frog's leg was caused by the contact of two different metals in a moist environment. The animal, argued Volta, had nothing to do with the production of electricity. Rather, he had discovered that electricity can be produced at will by chemical action alone. In 1800, the Royal Society published an account of Volta's discovery in its *Philosophical Transactions* which showed how he was able to produce and sustain an electric current by means of a "crown of cups." This was Volta's version of the first electric battery—a pile of alternating discs of different metals (zinc and silver) separated by saline-soaked leather or felt pads. The Library of Congress has the issue containing Volta's article.

The long period covered in this chapter—1725 to 1825–encompasses much of what is commonly known as the Industrial Revolution, and it can be argued that it is one of the eras that truly lives up to its name. There may have been no other time period when so many aspects of life and so many attitudes were so radically altered. It was during this time that the machine would begin to insinuate itself as a permanent factor in society. The development of new sources of power and the concomitant industrialization that followed transformed every aspect of people's lives. New tools begat new techniques, which in turn begat whole new technical systems, which again changed things, making for a kind of unstoppable technological momentum.

At the beginning of this amazing era, most people lived a difficult but regular and unhurried life dominated by the traditional ways and means of centuries. By around 1825, the arbitrary close of this chapter, European and American citizens had witnessed the appearance of the steamboat, railroad, and balloon as well as iron bridges, automated factories, and street lighting. The social consequences in terms of mobility alone are obvious, and politically, it does not seem surprising that during this time there should have been three great wars and two epoch-making revolutions.

Less dramatically, this was also the time when the long-separated dominions of science and technology began to have at least a rapprochement if not a real unity. Previously, when the particulars of a certain craft were kept deliberately difficult and secret within a small community of craftsmen, now the trend was toward openness and broad dissemination of information. This permeation and eventual transformation of the old attitude of technology to the relative openness of science is certainly one of

Much of Benjamin Franklin's fame rests on this often-depicted kite experiment but in fact, it was the least of his scientific and technical contributions. Although the 1752 experiment demonstrated that lightning is a form of static electricity and formed the basis of his lightning rod (which diverted lightning from buildings), Franklin's work in the field already had earned him a deserved international reputation as the "Newton of Electricity." An American original, Franklin's interests and abilities led him to accomplishments in meteorology and other aspects of physics besides electricity, and his ever-practical side led to such useful inventions as bifocals and a stove which served as a room heater. At the time of this experiment, Franklin was actually a robust and vigorous forty-six years old, and not the old man seen here. He was also accompanied by his twenty-year-old son, who is portrayed in this nineteenth-century print as much younger. Prints and Photographs Division, Library of Congress, LC-USZ62-30750.

the unheralded and most significant changes of the Industrial Revolution. This does not mean that technology suddenly became the application of science. Rather, it simply describes the change in attitude by practitioners of both, in which men of science became more interested in the practical side of an idea and craftsmen took up a new appreciation and interest in the scientific aspects of their work.

This change did not come about by chance, but rather was the slow result of the efforts of many enlightened individuals who formed organized societies. Their purpose was to bring scientific attitudes and methods to the technicians, and their primary weapon in furthering this goal was publication. Consequently, this is also the age of the technical encyclopedia and the technical journal. The first general technical encyclopedia in any language was John Harris's *Lexicon technicum: or, an Universal English Dictionary of Arts and Sciences Explaining Not Only Terms of Art, But the Arts Themselves*. The Library of Congress has this large, two-volume work as it was first published in London between 1704 and 1710. Harris was the secretary of the Royal Society and his publication broke new ground by making the simple distinction between a subject-oriented book (an encyclopedia) and a word-oriented book (a dictionary). Together, his two volumes of technical information cover 10,100 terms or subjects, and among the contributors, the most famous is Isaac Newton.

The next major encyclopedic work is also in English, also two volumes, and is represented by its 1741 edition in the collections of the Library of Congress. Ephraim Chambers began his major work, *Cyclopaedia, or an Universal Dictionary of Arts and Sciences*, as an updating of Harris's volumes, and it first was published in London in 1728. One of his major contributions was the introduction of another simple device—cross references. Chambers included these in the text so that "a chain may be carried on from one end of an art to the other." The influence of his work on subsequent encyclopedias cannot be overemphasized, and it is known that Diderot's *Encyclopédie* was begun originally as a translation of it.

Chronologically, the next major encyclopedic effort in the field of technology was *Machines et inventions approuvées par l'Académie royale des sciences*, first published by the Académie des Sciences in Paris between 1735 and 1777. Although more a patent digest than a real encyclopedia, its seven volumes contain some five hundred plates on all aspects of the technology of its day. The Library of Congress has a complete set of this work in first edition.

One of the most influential works of any kind published in the eighteenth century was the *Encyclopédie*, the first volume of which appeared in 1751 under the joint editorship of Denis Diderot and J. L. R. D'Alembert. Published in Paris from 1751 to 1765 as seventeen folio volumes of text, and eleven volumes of plates (between 1762 and 1772), the full title of this monumental work is *Encyclopédie ou Dictionnaire raisonné des sciences, des arts, et des métiers*. This controversial centerpiece of the French Enlightenment was begun as a French translation of Chambers's *Cyclopaedia*, but it soon took on the larger ambitions of Diderot and D'Alembert, the latter stating that he regarded "the art of instructing and enlightening men" as "the noblest portion and gift within human reach." Judged solely from the standpoint of the history of technology and not as a work of political propaganda or anything else, this monumental work contains excellent descriptions and illustrations of innumerable technical processes, devices, inventions, and techniques. Organized alphabetically, it contains over three thousand elegantly precise engravings that graphically and accurately illustrate the technology of its day. The Library of Congress has the complete first edition of this great work.

One of the more extraordinary examples of the encyclopedic impulse as applied to technology is the *Descriptions des arts et métiers*, published by the Académie des Sciences of Paris from 1761 to 1788. Its goal was to present a scientific picture of all of the industrial processes employed in France during the eighteenth century. For nearly two decades, the academy published yearly from four to six monographs, each describing accurately and clearly the methods of operation of a particular handicraft or industry. This effort was both inspired and calculated, for it was attempting to inform French industry of the best current technological practices available, to help stimulate its growth and advancement. Typically, an individual monograph consists of three sections, one explaining the varieties of raw materials employed, another describing the tools and equipment used, and a final one showing in detail the actual steps required to produce the intended product. A series of engravings usually picture both the tools and the workmen in the shops.

From the time of its publication until now, there has been a wide variety of opinion as to what constitutes a "volume" in the series, and it has always been a cataloger's nightmare. It is known for certain, however, that only two full sets are found in the United States—one in the Smithsonian

Opposite page:
The inclusion of artificial limbs in this Académie des Sciences publication illustrates the breadth and range of its technical subjects. The science of prosthetics dates back at least two centuries to the sixteenth-century French surgeon Ambroise Paré (1510–1590), whose book included many clever mechanical limbs. It is doubtful, however, whether any such artificial arms ever proved useful or even were built. *Machines et inventions approuvées par l'Académie royales des sciences*, vols. 5–6 (figures), 1735–77. LC-USZ62-110398.

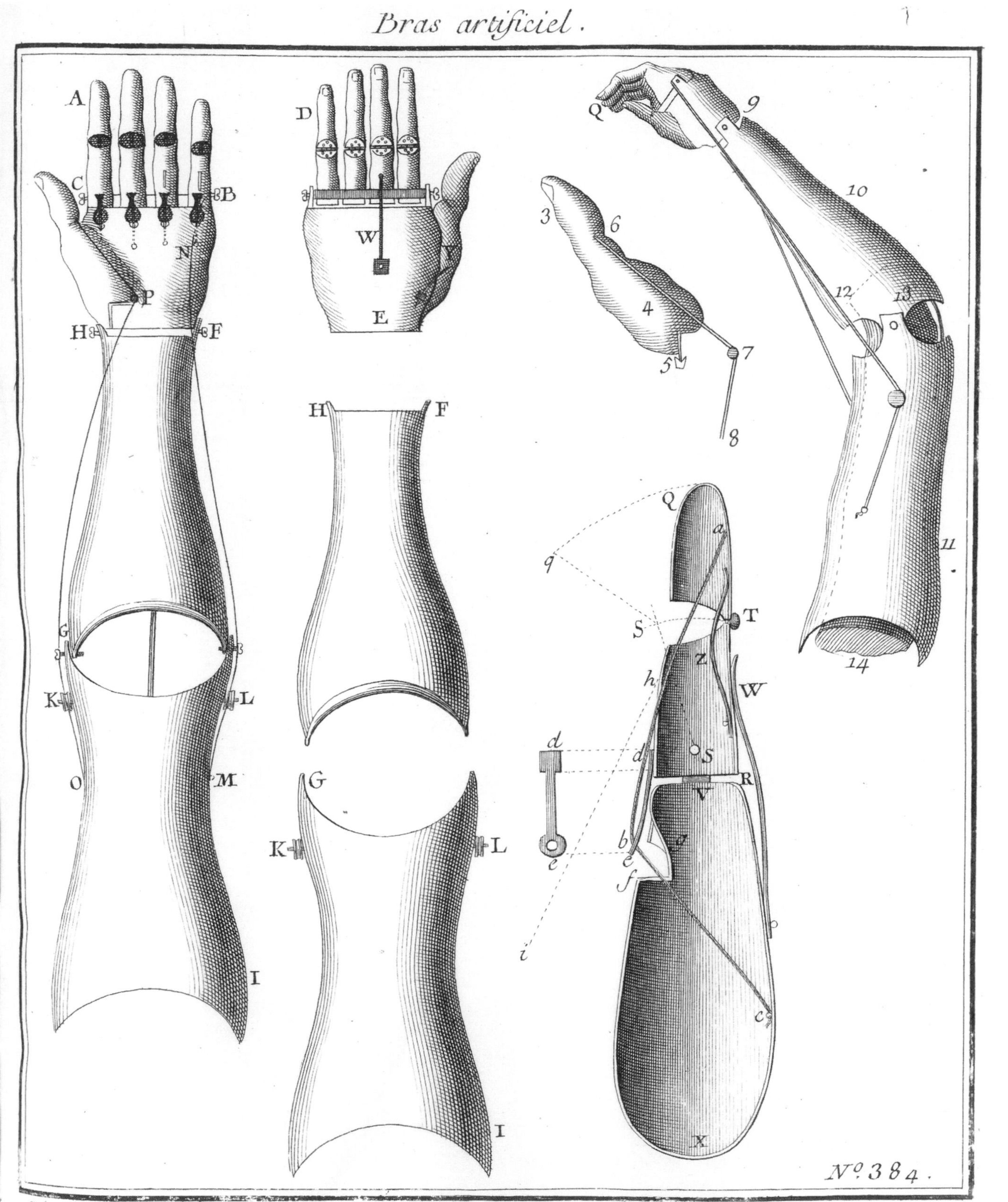
Bras artificiel.
N°. 384.
Dheulland Sculp.

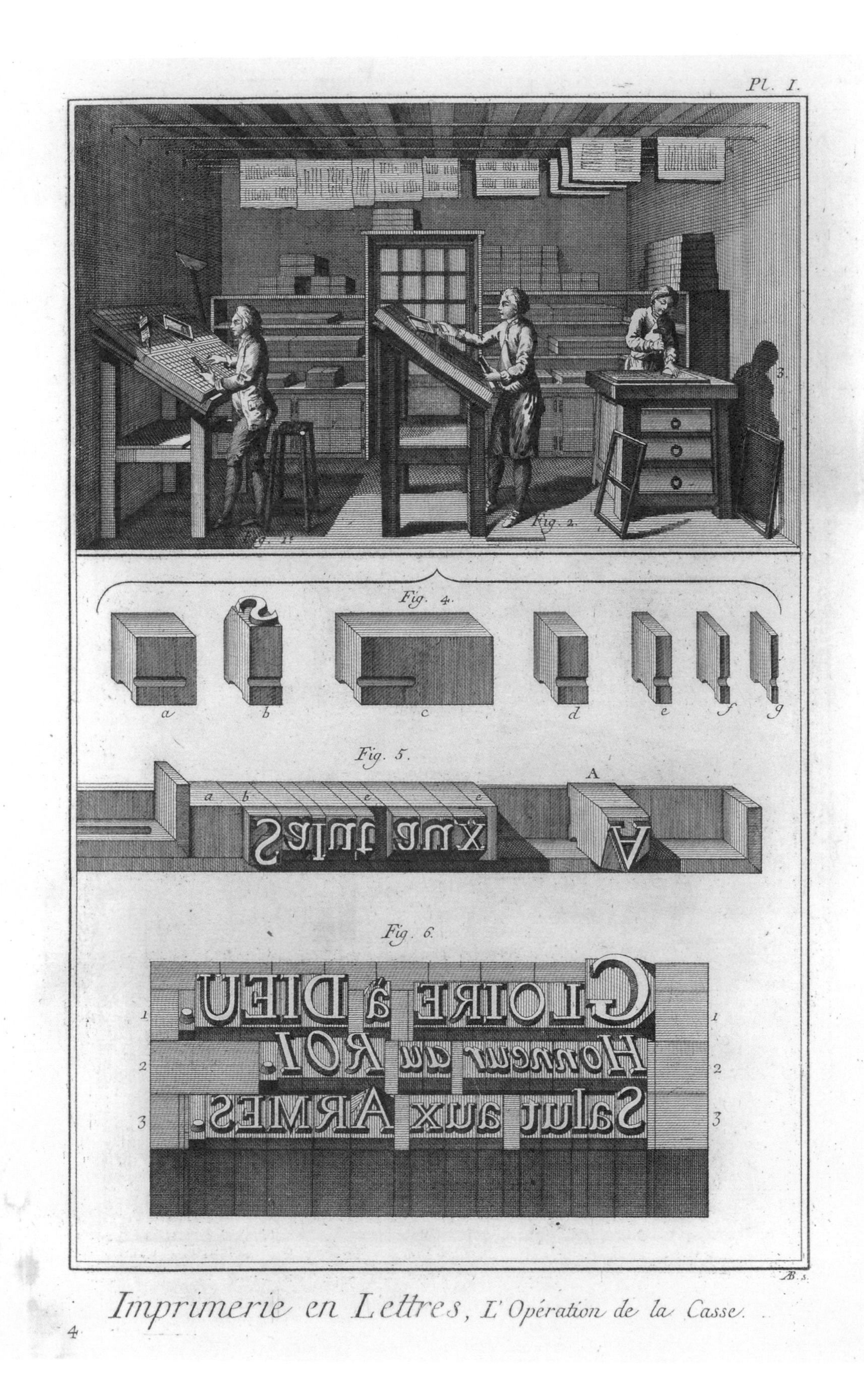
Pl. I.
Fig. 1.
Fig. 2.
3.
Fig. 4.
a
b
c
d
e
f
g
Fig. 5.
A
Fig. 6.
1
2
3
Imprimerie en Lettres, L'Opération de la Casse.
4

Institution in Washington, D.C., and the other at the Metropolitan Museum in New York. Also, seven known sets of varying degrees of completeness are in other American libraries. The Library of Congress has a total of eleven individual titles, detailing among them the art of the blacksmith, of paper-making, and of bookbinding. Perhaps the most significant title in the Library's collection is *L'Art du facteur d'orgues* (*The Art of the Organ-Builder*), published in 1776–78. This work was recognized immediately as one of the most complete and fundamental works of its kind, and was reprinted and translated many times. It is perhaps the best

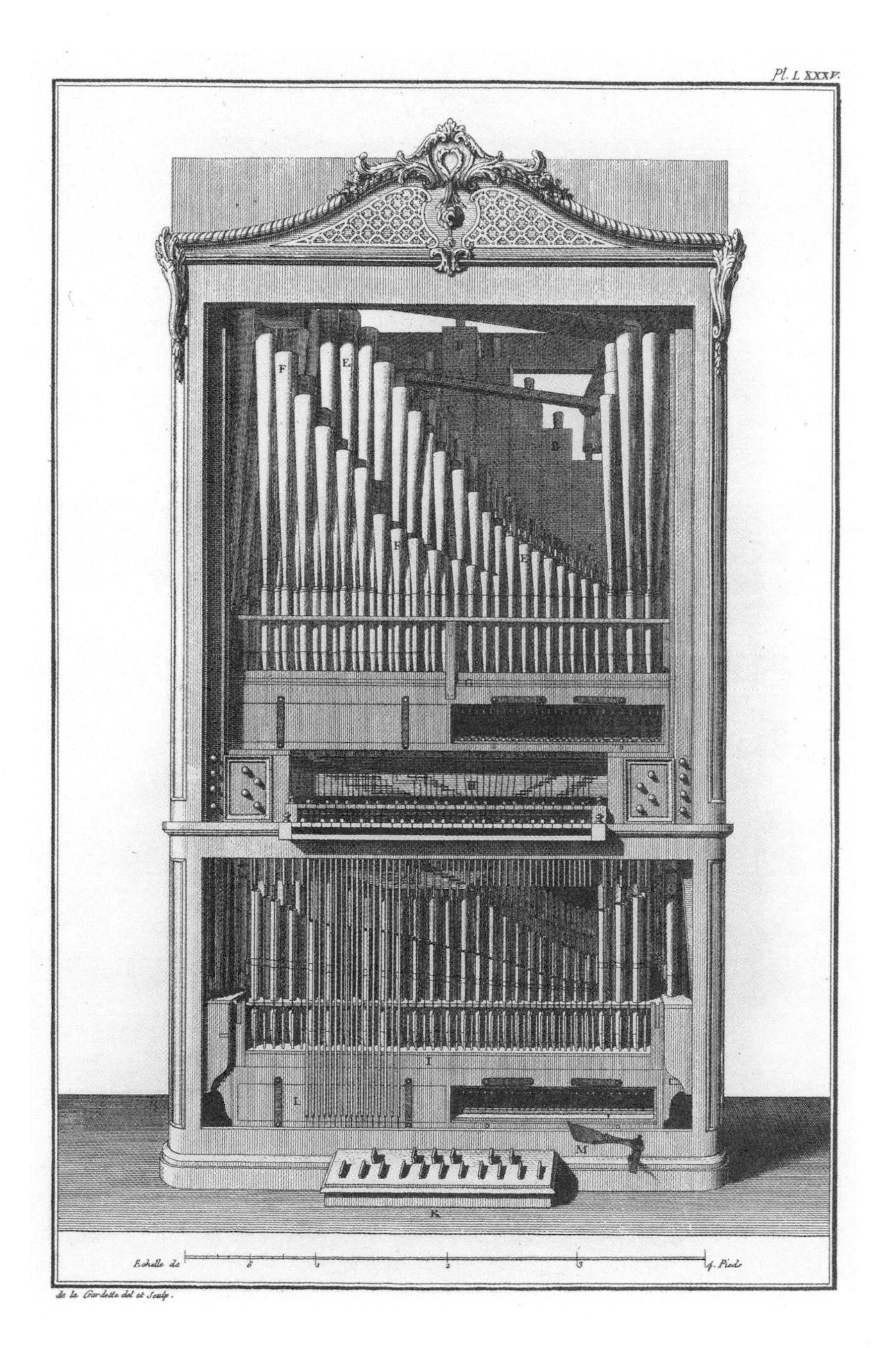

Although Germany led the world in organ building from 1500 to 1800, one of the most complete and fundamental books on organ building was written by a French Benedictine monk, François Bedos de Celles. His book offers a valuable description of classical French organ building and its extensive plates, charts, and diagrams touch on all aspects of organ design and construction. It also describes a square piano combined with an organ, a claviorganum, and a "vielle" or hurdy-gurdy. The authority of this work explains why it became part of the ambitious encyclopedia of French industrial arts and crafts, *Description des arts et métiers: L'Art du facteur d'orgues,* 1776–78. François Bedos de Celles.

Opposite page:
This plate is typical of those in Diderot's *Encyclopédie,* in that it has two parts, a picture above, usually a large-scale scene with people in it, and a detail below, showing specific parts or processes. Here, the typesetting part of a mid-eighteenth-century printing shop is shown. Figure 6 at the bottom shows Canon type in roman and italics. Its size was between forty-four and forty-eight points, and it was the largest type having a specific name. *Encyclopédie ou Dictionnaire raisonné des sciences, des arts, et des métiers,* vol. 7 (plate 1), 1751–72. LC-USZ62-110399.

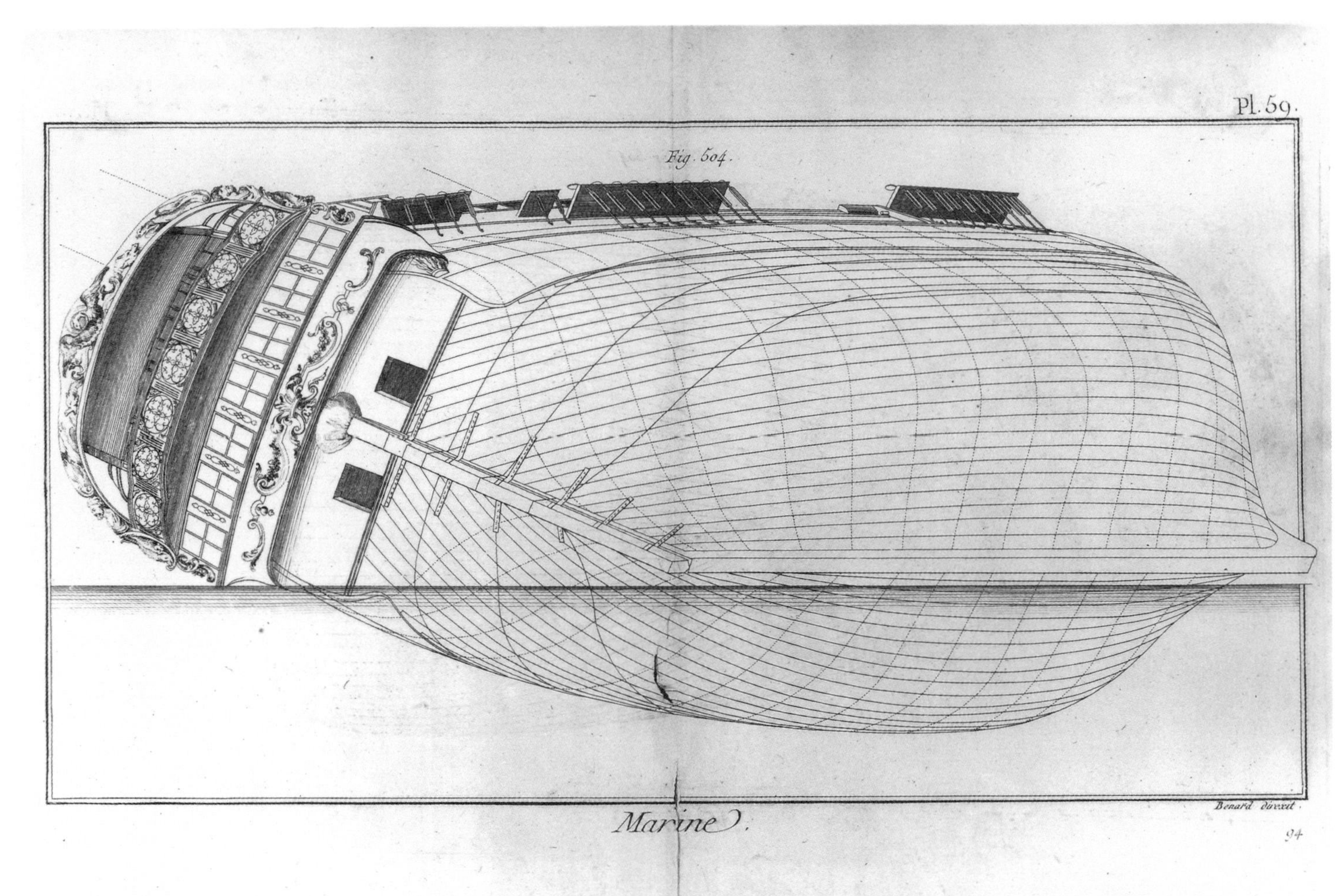

Besides the inherent beauty and artistry of this eighteenth-century image and the precision, knowledge, and skill demanded of the draftsman who made it, what most strikes the eye of a near-twenty-first-century observer is how closely it resembles today's computer-generated graphics. That observer might also conclude that both the eighteenth-century person who drafted it and those who would use it might be completely comfortable if placed in front of a computer terminal. *Encyclopédie méthodique,* vol. 166, 1782–1830. LC-USZ62-110400.

example of the consistently high quality of the material contained in the entire series. The series ended in 1788 and was eclipsed by the greater scope and renown of Diderot's *Encyclopédie*.

Another French effort that was both an outgrowth and a borrower from the *Encyclopédie* and which is often confused with it, is the *Encyclopédie méthodique, ou par ordre des matières*. This enormous and seemingly endless series was published in Paris for nearly fifty years, from 1782 to 1830. The Library of Congress has a nearly complete collection totaling 202 volumes (about one-fourth of which are plates). Unlike the *Descriptions des arts et métiers*, its range includes all the sciences as well. Thus, roughly only one-sixth of the series deals with technology, which includes artillery, architecture, and agriculture as well as the typical "manufactures, arts et métiers."

The Country Builder's Assistant has the distinction of being the first architectural work written by an American. It was written by Asher Benjamin and was published in Greenfield, Massachusetts, in 1797. Besides these unique credentials, what makes it especially exemplary is that Benjamin wrote it to fill a peculiarly American need. The available builders guides—all written and printed in England—were nearly useless, he felt. These emphasized large, ornamented country houses which were mostly inappropriate, both practically and esthetically, to American wants and needs. As a designer and builder, the eminently practical Benjamin knew exactly what he and people like him wanted in a guide, and the first edition of this very popular book has, besides text, thirty plates of instructions showing the basics of laying out and building a simple house.

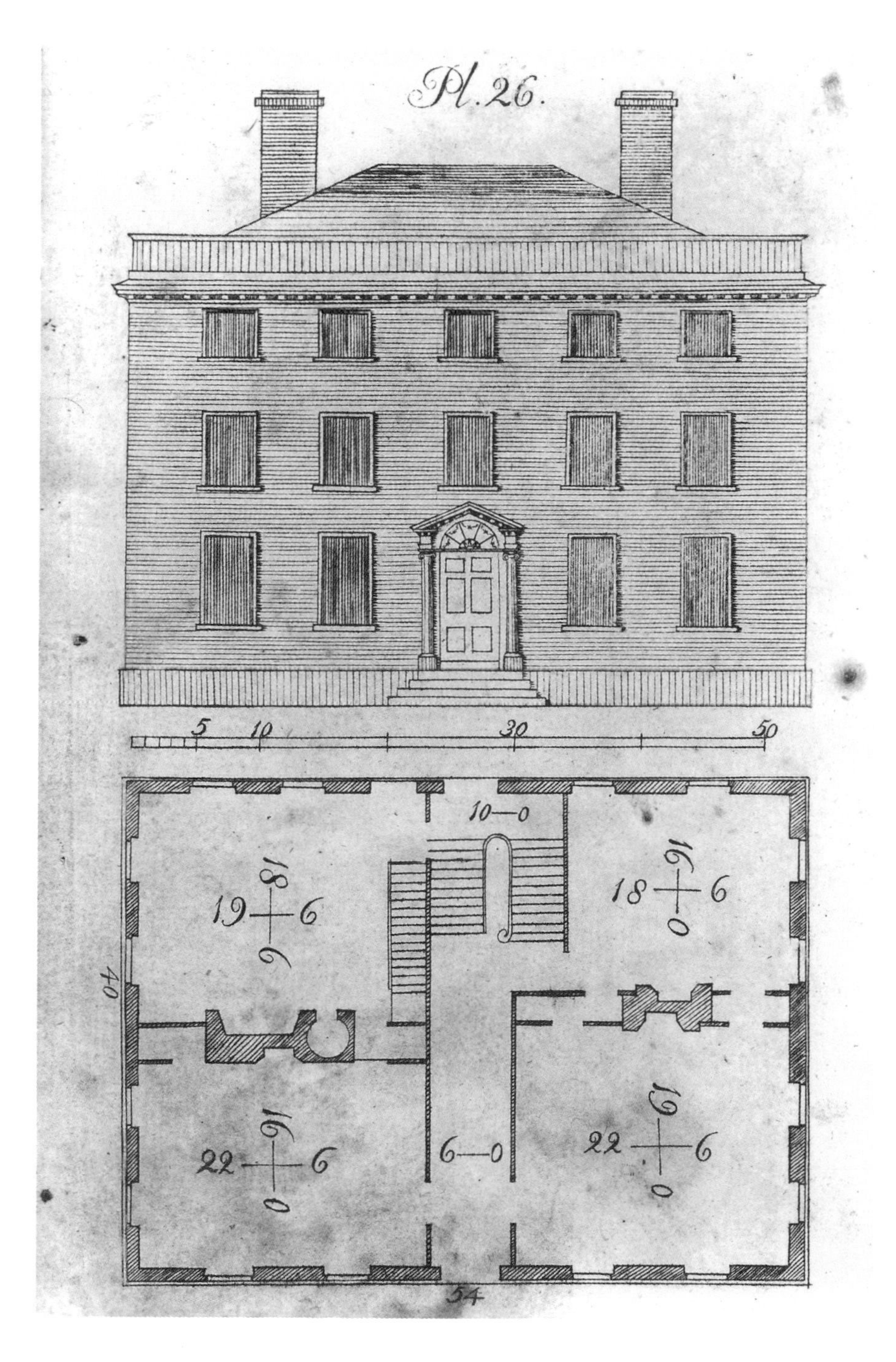

In writing what was the first architectural book by an American, Asher Benjamin (1773–1845) wanted only to produce a work that would speak directly to the emerging American building style. As a practical man, he also wanted to write a book that would consider what materials American builders used and had available, what their abilities were, and especially what purpose a building would serve. Benjamin was reacting to the fact that nearly all the builder's guides then available were imported from England and were geared to erecting large houses with ornate decorative furnishings. His books became so popular that the practical influence of this Connecticut native was felt in many corners of the new nation well into the nineteenth century. *The Country's Builder's Assistant,* 1797. Asher Benjamin. LC-USZ62-110401.

Agriculture, Labourage

The larger lesson found in his popular book is that good technology is not something that stands on its own and forces its users to comply to its standards and methods. Rather, technologies that become accepted and useful are those that meet real people's real needs by adapting in any and all ways. Flexibility and adaptability made the Benjamin book useful and popular, and resulted in this modest, eighteenth-century how-to builder's book joining the company of some of the greatest technical books in the collections of the Library of Congress.

Appropriately, this section concludes with an account of an institution that began over two hundred years ago and still flourishes today. Spurred by the practical and enlightened spirit of the middle of the eighteenth century, the Royal Society of Arts was formed in London early in 1755 for the purpose of creating a "Society for the Encouragement of Arts, Manufactures, and Commerce in Great Britain." Its founding members thought of the society not as a learned group reading and publishing memoirs but rather as an active, utilitarian institution which would stimulate and even reward discovery, invention, or improvements that were socially valuable. By 1783, it came to realize the importance of publication and dissemination of information, and initiated its *Transactions* series in that year. The Library of Congress has a complete set of this regular and still on-going journal from when it first appeared in 1783. Over the years, it has published in the broadest range of technical areas, from saving life at sea, producing coal gas, improving agricultural tools and techniques, introducing the Dutch system of curing herrings, to the use of machines instead of boys as chimney sweeps.

During this beginning industrial age technology insinuated itself into society as a permanent factor. The new power sources alone, and their numerous applications in factories and transportation, gave technology a dominating and heady new presence in everyday life. Equally important was its increasing rapprochement with science and its eventual full embrace of the values and methods of science—specifically the value science placed on openness and the wide dissemination of information. This went against the old technological tradition of secretiveness and gave a newer, braver tradition to technology. This closing of the gap between science and technology meant in real terms that men of science were becoming more aware of and more interested in the practical side of an idea, and that craftsmen similarly assumed a new appreciation for the scientific aspects of their work.

Opposite page:
Some details of mid-eighteenth-century farming in France are provided in this unidealized, rustic portrait. At the bottom, an ordinary plough (figure 2) is contrasted with the English plough of Jethro Tull (figure 3). A man opens a furrow in figure 1, and a woman pushes a sowing machine in figure 4. At the top, from left to right, a man leads a horse pulling a clod-breaking roller, which smooths and levels the ground; another scatters seed by hand; and a third cracks the whip at a horse pulling a harrow, which pulverizes the soil. Agricultural change during the Industrial Revolution did not come to the farm quickly, however, and often was embraced by the farmer very reluctantly. *Encyclopédie ou Dictionnaire raisonné des sciences, des arts, et des métiers*, vol. 1, plate 1, 1751–72. LC-USZ62-110402.

6. *Technology in the Nineteenth Century*

The nineteenth century can be viewed as the time when the modern world was born. Nearly all the material goods and evils that benefit or plague us today had their beginnings during that time. Despite its substantial and continuing influence, it is an age that is often overlooked or, when considered, is thought of as a quaint, premodern era important only in terms of what succeeded it. Consequently, this formative period has been given no single catchy, distinguishing name by historians to set it distinctly apart from other times. Yet an appreciation of what really occurred during this century leads to the realization that it competes admirably with other eras considered revolutionary, like the Scientific Revolution or the Industrial Revolution.

During the nineteenth century, some truly revolutionary things happened that really did drastically change the world. The dominating reality of the time, and what sets it apart from all that went before, is the intrusion of the machine into virtually every aspect of life. Technology and the machine came to pervade, influence, and even define people's lives, not just in the factory or on the job, but in their homes and private lives as well. It was impossible to avoid the ever-present machine and its counterpart, mechanization. At some point during this dynamic process, technology ceased its subordinate, supportive role and took center stage as a dominant player. Technology eventually came to exert a social or cultural influence simply by its existence. It was then that we witnessed the creation of a material civilization based on technology.

Politically, the technology-based triumvirate of capitalism, nationalism, and materialism led to the inevitable domination of much of the world by its exponents—the nations of western Europe and the United States. In many cases, it was straightforward technological superiority that gave the West the edge, whether in the competition of the marketplace or the battlefield. But politics alone was not responsible for the steady push for material progress during this time, for it became the common credo of the average citizen to embrace change unquestioningly. Once social acceptance or societal approval was added to the existing forces of politics and economics that favored technology's advance, technology's insinuation into all aspects of life was complete. For if the man or woman in the street perceived the machine as good and embraced it unhesitatingly, little or no selling job was required by anyone.

The machine entered and altered our culture so easily and in so many ways. New power sources were discovered, harnessed, and then rapidly applied. The nature of energy was finally understood and its practical

Opposite page:
As the nation celebrated its first centennial with an international exhibition at Philadelphia in 1876, the dynamo became the symbol of that century's innocent belief in progress. Here, the centennial is officially opened as President Grant and Brazilian Emperor Dom Pedro prepare to turn two silver-plated cranks that start the 1500-horse-power Corliss steam engine towering in the background. Built with conventional technology but on a monumental scale, the Corliss engine became the dominant motif for the highly successful exhibition. The centennial opened the world's eyes to America which, after 1876, was seen as the land of material progress. America had embraced the machine and was about to become the world's largest industrial and economic power. Prints and Photographs Division, Library of Congress, LC-USZ62-96109.

conversion became easier. The availability of cheap, unlimited energy made the process of mechanization easier and more rapid. Mechanization in turn both created and satisfied the need for new processes, techniques, tools, and instruments. New and stronger materials were developed which again acted as a spur to further development. Iron and steel were coming into their own and were used to create the infrastructure of the emerging new world. Transportation was transformed by the harnessing of new energy forms like steam, internal combustion engines, and electricity, causing a very real sense of the shortening of time and space. People were no longer limited to animal power for transportation. As they could move about more rapidly, so too could they communicate quickly and more effectively at a distance. Again, time and distance begin to shrink with the telephone and telegraph, and large numbers of people could be reached at once via film and radio. Finally, the entire rapidly growing system and population boom were supported by major advances in agricultural productivity. All aspects of the production and distribution of food were affected by technological breakthroughs and improvements.

It could be said with little exaggeration that each of these factors is characteristic of the world as we know it today, just shy of the twenty-first century. Despite our sometimes startling advances and discoveries, all we have really done is to improve upon and continue the trends started in the nineteenth century. The modern world really began in the nineteenth century.

A survey of its major technological accomplishments will barely do justice to the long litany of nineteenth-century achievement. Nearly all the tangible things we associate with progress and modernity came into being at that time. It was during this age that iron bridges of all types spanned great spaces, railroads crossed continents, the automobile and the subway were invented, rockets, dirigibles, assembly lines, streetcars, electric lights, photographs and cinema, telegraph and telephone, radio and gramophone all came to be at a dizzying pace. The list seems endless—calculating machines, cotton gins, six-guns, X rays, elevators, and sewing machines.

The social and political costs of such rapid and probably helter-skelter change can only be imagined, for while these new devices and inventions certainly made things quicker, more efficient, and sometimes easier, they must have also engendered a considerable number of negative factors. It is probably not accidental that this century of unprecedented, constant, and suddenly pervasive technological change culminated in a horribly mechanized, seemingly endless, and almost pointless total war.

As with other historical eras characterized by opportunity, fertility, and rapid change, the nineteenth century contains many intriguing opposites—innocence and cynicism, ennui and exhilaration, angst and hope, charm and decadence. The single thing that unites them all is how inextricably each is bound with the central reality of the machine or, in Henry Adams's astute analogy, the "dynamo." To Adams, who was born in 1838 and who saw how rapidly and irrevocably technology was altering

the traditions of centuries, the dynamo's anarchic energy was replacing the certainties of past life that he referred to as the "virgin"—his symbol of coherence and unity. Said another way, the order and unity provided by God was being replaced by a new God, the machine, whose only constant was change. Technology had made modern life unpredictable, even chaotic to Adams, who admitted he could not come to grips with it.

Despite his anti-technology attitude and his longing for medieval quietude and absolutes, Adams was one of the few individuals who was able to get some distance on the age in which he lived and who was perceptive enough to try to place it in a larger historical context. Whatever his opinion of technology, he seemed to know in his bones that the nineteenth century was an age after which nothing would ever be the same. In this he was correct.

A survey of the published works central to the development of technology in the nineteenth century that also are held in the collections of the Library of Congress will, of necessity, reflect the nature and style of that time. In other words, it will be more of a rapid-fire review of invention and technical accomplishment than a tightly linked and easily woven essay. Simply too much that was too important happened in too short a period of time.

Although the nineteenth century would eventually come to discover the link between light, electricity, and magnetism and would achieve a classic synthesis of these three main fields of physics, at its beginning it had no more understanding of the nature of energy than did the century before. Science—understanding the why as well as the how—would prove its worth to technology in this century, especially in the field of power and the conversion of energy.

While steam engines of all sorts were proliferating by the first quarter of the century and were beginning to drive factory output to new levels of productivity, their construction and operation was pragmatic, with no theoretical understanding of the nature and relationship of heat and mechanical power. In 1824, however, a young French physicist published his only work, *Reflexions sur la puissance motrice du feu*, and effectively

The opportunistic American inventor Samuel Colt displayed his newest revolver at the 1876 centennial exhibition. Although the principle of a revolving cylinder was known and tried long before Colt, a practical weapon was not produced until 1835–36 when Colt obtained his patent. As in all subsequent revolvers, a revolving chamber loaded cartridges successively into a single barrel cylinder, allowing multishot firing. Colt's weapon did not become popular until it had proved its usefulness in the Mexican War in 1846–48. Following that success, Colt exhibited his multishot handgun at the Crystal Palace in London, and during the second half of the century, he was mass-producing revolvers using the new manufacturing system of fully interchangeable parts. *Four Thousand Years of the World's Progress,* 1878. Samuel J. Burr. LC-USZ62-110403.

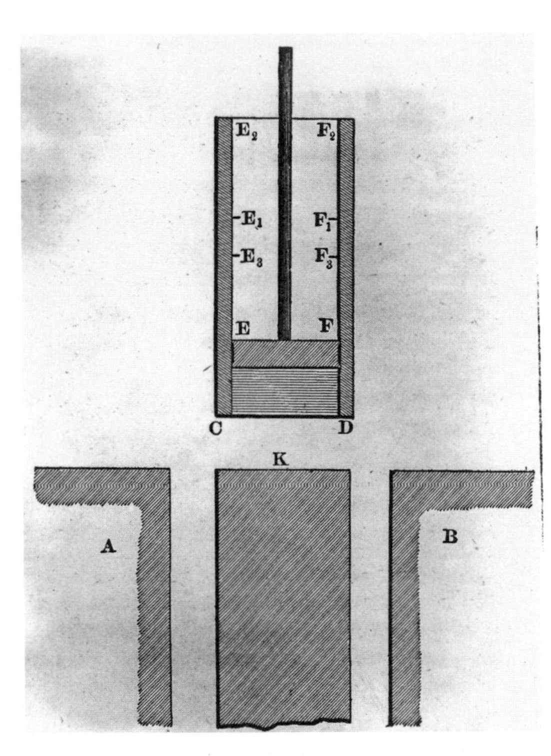

For his theoretical efforts at converting heat into mechanical energy, the young Sadi Carnot is considered the founder of thermodynamics. Here he illustrates a theoretical power cycle of maximum efficiency by demonstrating that its efficiency depends on the temperature difference in an engine. In this picture, A is a much higher temperature than B. Carnot's four-step cycle not only provided a measurable rationale that was better than any previous model for how engines worked but also offered a general or theoretical understanding that led science into the new domain of thermodynamics. Thus the origins of the high science of thermodynamics are seen emerging from the practical and even limited investigations of technology. Carnot died of cholera at thirty-six, and his work was unknown for an entire generation. *Reflections on the Motive Power of Heat,* 1890. N. L. S. Carnot. LC-USZ62-110404.

founded the science of thermodynamics (or the science of heat movement). The twenty-eight-year-old-genius was Nicolas-Leonard Sadi Carnot, and it was he who determined the general conditions under which mechanical power can be obtained from heat.

Carnot was the first to analyze quantitatively the manner in which heat and work interconnect, and he pointed out that every heat engine needs a hot body, or heat source, and a cold body, or condenser. An engine works, he pointed out, because heat passes from the hotter body to the colder one. This seemingly ordinary statement enabled him to construct a very serviceable cyclic theory of a "universal" or ideal motor. It was from his general understanding or theory that the science of thermodynamics later came to be. In fact, Carnot had formulated its first law—the conservation of energy—in his unpublished notebooks. While the Library of Congress does not have Carnot's 1824 publication, it does have the work as it first appeared in English in 1890, published in New York and edited by R. H. Thurston. The full English title is *Reflections on the Motive Power of Heat and on Machines Fitted to Develop That Power*.

In 1834, a French engineer, Benoit P. Clapeyron, wrote a detailed exposition of Carnot's *Reflexions* in which he transformed Carnot's verbal analysis into the symbolism of calculus. This, along with his graphic representations of the Carnot cycle, made Carnot's ideas understandable and familiar to engineers. Clapeyron's article was first published in *Journal de l'École polytechnique*. The Library has it, however, as it appeared in Richard Taylor's English translation, *Scientific Memoirs, Selected from the Transactions of Foreign Academies of Science and Learned Societies and from Foreign Journals*, published in London in 1837. Most now agree that Carnot's work led to the eventual formulation of both original laws of thermodynamics. In the 1840s, James P. Joule offered the experimental proof that energy can be neither created nor destroyed. This first law of conservation was enunciated formally by Hermann L. F. von Helmholtz in 1847. The second law of entropy (that a closed system can only lose but never gain energy) was stated by Rudolf J. E. Clausius, via Clapeyron, in 1850, and by Lord Kelvin in 1851. Carnot was unable to pursue his findings, dying at the age of thirty-six during a cholera epidemic.

This new scientific understanding of thermodynamics was applied practically to the design of steam engines by William J. M. Rankine, a Scottish engineer. Despite his obviously lofty talents (he had read Isaac Newton's *Principia* in Latin when he was only fourteen years old), Rankine was able to translate the abstract theory of thermodynamics for the average engineer. His *Manual of the Steam Engine and Other Prime Movers*, first published in London in 1859, became a standard work and influenced later practitioners to develop both the internal combustion engine and the steam turbine. More importantly, the fundamental principles of the conversion of energy through engines finally were established by the end of the century. The Library of Congress has Rankine's book in its third edition, published in 1866 in London.

Unlike steam power, which developed largely as the result of a good deal of hit-and-miss experimenting and only later benefited from scien-

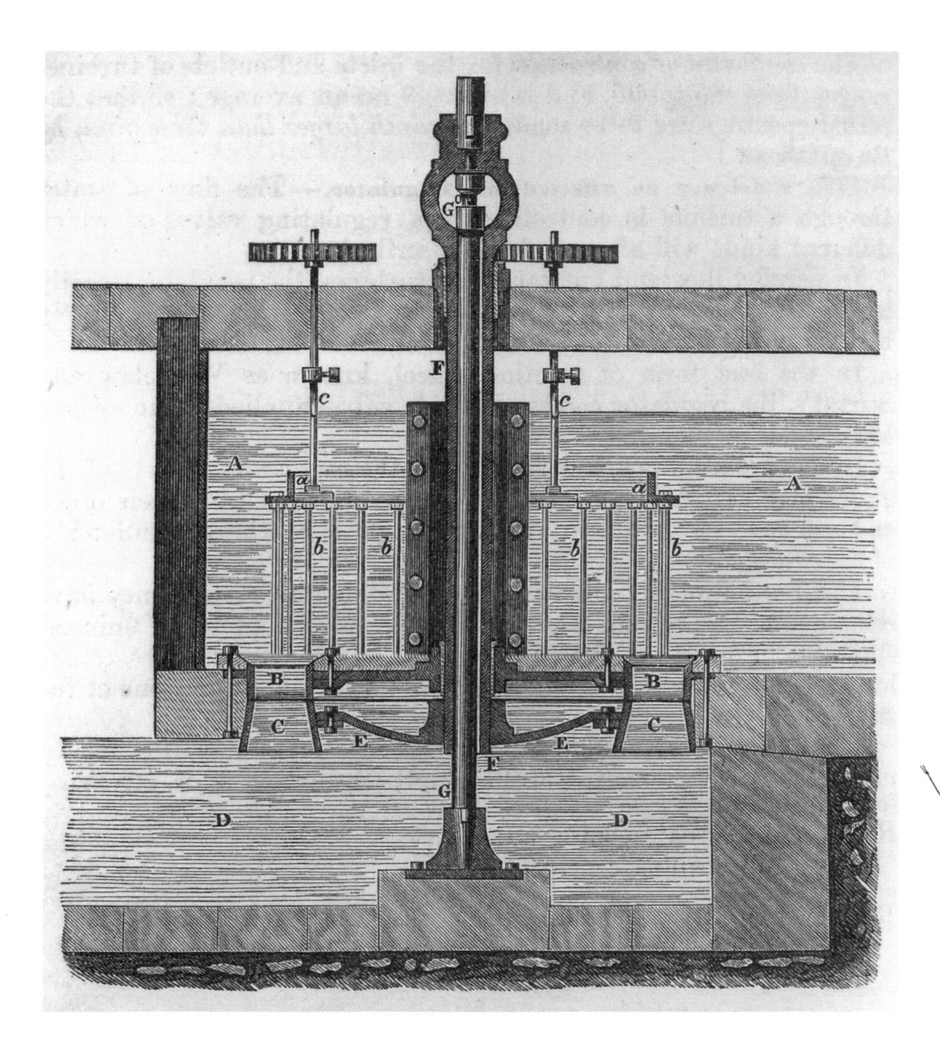

The highly versatile William J. M. Rankine wrote in 1859 the first textbook on the new thermodynamics of Carnot. Although a linguist, classicist, musician, and poet, Rankine was also trained in physics and eventually became an engineer. His classic treatment of this new field gave engineers a systematic understanding of how an engine works and contributed to the growing dominance of the machine. Here in his chapter on turbines, he analyzes a parallel-flow turbine invented by Fontaine-Baron and details how it operates most efficiently. *A Manuel of the Steam Engine and Other Prime Movers,* 1866. William John Macquorn Rankine. LC-USZ62-110405.

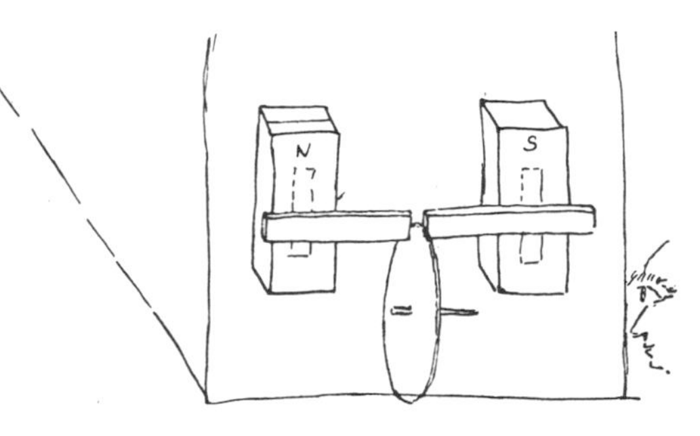

The large-scale use in industry of electricity did not come into being until the latter half of the nineteenth century. It was made possible by the landmark work of Michael Faraday who, in 1831, invented the electric generator. This rough drawing in his diary shows Faraday's breakthrough experiment in which he turned a copper wheel in such a way that its edge passed between the poles of a permanent magnet, generating an electric current in the wheel that continued as long as the wheel turned. Faraday demonstrated that the kinetic energy of motion could be turned into electricity. With greater understanding of this phenomenon and larger magnets, the first really efficient, large dynamos would be constructed within a generation. *Faraday's Diary,* vol. 1, 1932. Thomas Martin, editor. LC-USZ62-110406.

tific understanding, the creation of the electric industry in the nineteenth century was almost the direct result of scientific research. Once Michael Faraday demonstrated in 1831 how to use electric and magnetic fields to produce mechanical motion, the crucial point of departure was reached. This discovery showed both how to generate electricity and how to use it productively. The Library of Congress has Faraday's account of his breakthrough experiment as it appeared first in the Royal Society's *Philosophical Transactions* (1831), and as it was later published, along with his other works, in *Experimental Researches in Electricity*. The Library has all three volumes of the latter in first edition, each of which appeared in a different year in London—1839, 1844, and 1855.

Following Faraday's direction, electric generators and electric motors were simply a matter of time. Once the dynamo made large-scale generation of electricity commercially feasible, markets necessarily had to be found. One of the obvious markets was domestic lighting, and the inventive and opportunistic genius of Thomas Alva Edison, who produced the carbon-filament lamp in 1879, cornered this market. Besides illumination, another major use of electricity was to operate electric motors, and the lessons learned in the development of an electric railway system even-

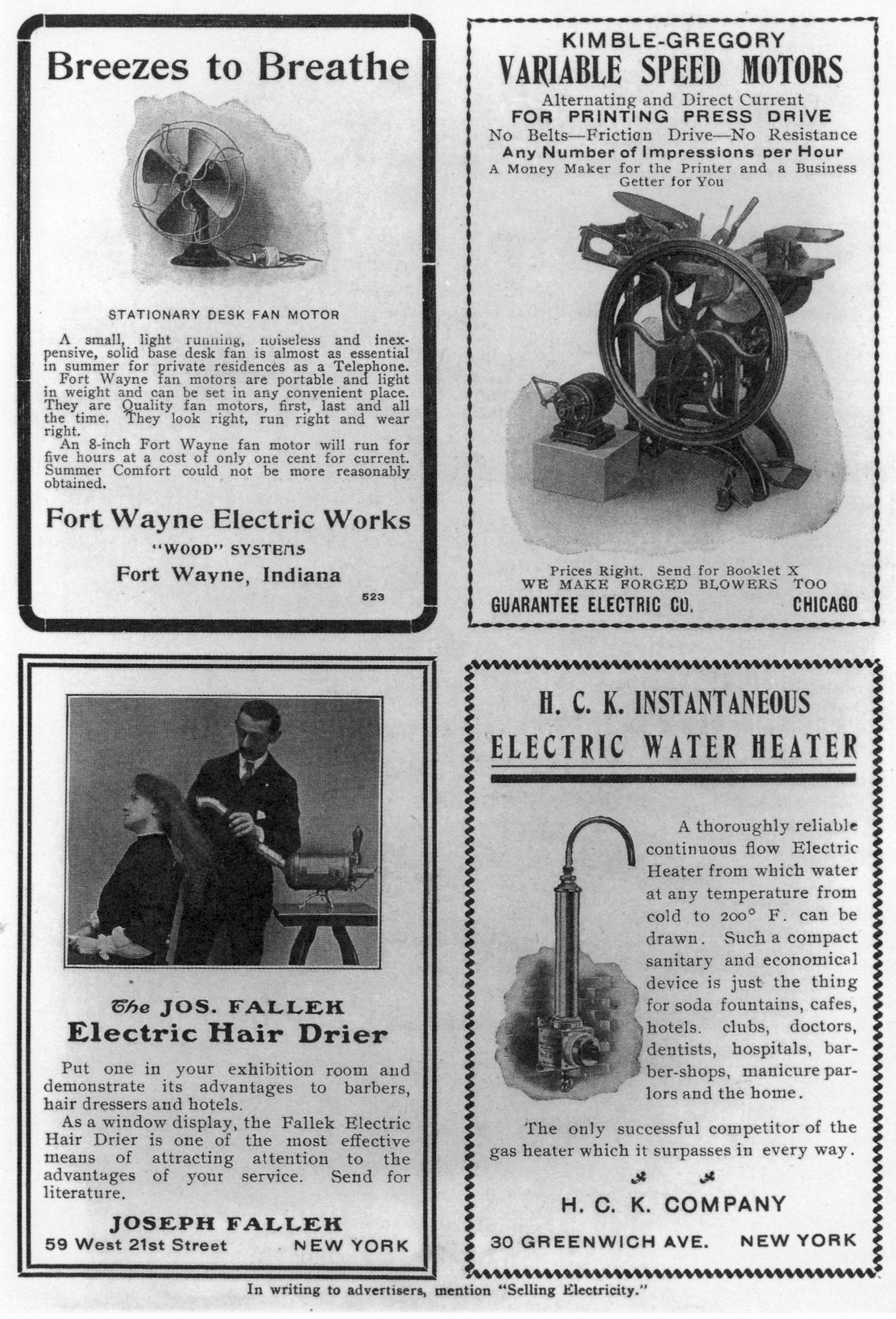
SELLING ELECTRICITY ADVERTISERS.
Breezes to Breathe
STATIONARY DESK FAN MOTOR
A small, light running, noiseless and inexpensive, solid base desk fan is almost as essential in summer for private residences as a Telephone.
Fort Wayne fan motors are portable and light in weight and can be set in any convenient place. They are Quality fan motors, first, last and all the time. They look right, run right and wear right.
An 8-inch Fort Wayne fan motor will run for five hours at a cost of only one cent for current. Summer Comfort could not be more reasonably obtained.
Fort Wayne Electric Works
"WOOD" SYSTEMS
Fort Wayne, Indiana
523
KIMBLE-GREGORY
VARIABLE SPEED MOTORS
Alternating and Direct Current
FOR PRINTING PRESS DRIVE
No Belts—Friction Drive—No Resistance
Any Number of Impressions per Hour
A Money Maker for the Printer and a Business Getter for You
Prices Right. Send for Booklet X
WE MAKE FORGED BLOWERS TOO
GUARANTEE ELECTRIC CO. CHICAGO
The JOS. FALLEK
Electric Hair Drier
Put one in your exhibition room and demonstrate its advantages to barbers, hair dressers and hotels.
As a window display, the Fallek Electric Hair Drier is one of the most effective means of attracting attention to the advantages of your service. Send for literature.
JOSEPH FALLEK
59 West 21st Street NEW YORK
H. C. K. INSTANTANEOUS
ELECTRIC WATER HEATER
A thoroughly reliable continuous flow Electric Heater from which water at any temperature from cold to 200° F. can be drawn. Such a compact sanitary and economical device is just the thing for soda fountains, cafes, hotels, clubs, doctors, dentists, hospitals, barber-shops, manicure parlors and the home.
The only successful competitor of the gas heater which it surpasses in every way.
H. C. K. COMPANY
30 GREENWICH AVE. NEW YORK
In writing to advertisers, mention "Selling Electricity."

tually allowed electric motors to power the engines of the major manufacturing industries. The pioneer in this field, called electric traction, was Frank J. Sprague, who in 1888 built the Richmond Line, the first complete electric street railway system. A onetime employee of Edison's, Sprague attended the Crystal Palace Electrical Exhibition in London in 1882 as a representative of the U.S. Navy, and his report to the secretary includes a detailed discussion of electrical machinery that led, indirectly, to Sprague taking on the Richmond project. The Library of Congress has Sprague's report, titled *Report of the Exhibits at the Crystal Palace Electrical Exhibition, 1882*, published in the General Information Series, no. 2, in 1883.

After the incandescent lamp and the electric motor, the third major application of electricity was in the field of communications. Although it might be difficult to imagine more intrusive and dramatic technical innovations than the electric light bulb and the urban railway, the communications revolution begun by the telegraph, telephone, and radio serves as an equally potent innovative phenomenon.

Most credit the enterprising American, Samuel F. B. Morse, with the invention of the telegraph, but it was really two Englishmen, William F. Cooke and Charles Wheatstone, who first patented the idea of an electric wire telegraph in 1837. Interestingly, both Wheatstone and Morse benefited from the freely given help of the American physicist, Joseph Henry. By 1831, Henry had already devised and then demonstrated that he could move back and forth a small iron bar a mile away by turning on and off the current running through a connecting wire to an electromagnet. This is essentially the practical principle behind the telegraph, and it is noteworthy that both Wheatstone and Morse made their respective breakthroughs only after talking directly to Henry.

Henry should be considered one of the saints of science and technology, in that he always acted out his belief that the benefits of his discoveries should be shared with everyone and should not be used primarily to enrich his reputation or his pocketbook. Because of this, Henry sometimes surrendered priority to others because of circumstance (as to Faraday, whose discovery of electromagnetic induction was preceded by Henry's unpublished work), and other times never fought for what was rightly his (as with Morse's refusal to acknowledge Henry's assistance). In addition to his major contributions in physics and electricity, Henry worked in astronomy and meteorology, served his country as the first secretary of the newly formed Smithsonian Institution, and was also one of the founders of the National Academy of Sciences. The Library of Congress has the first edition of the two-volume *Scientific Writings of Joseph Henry*, published by the Smithsonian Institution in 1886.

Although Morse did labor long and hard and built his own telegraph apparatus after Cooke and Wheatstone had already done the same in England, his primary contribution was the development of a universally accepted alphabet code made up of long and short pulses of electricity. The British system directly recorded each letter sent, using a needle to point to the letter, but this was slow and far from satisfactory. Morse's new system of dots and dashes was based on the efficient idea that the most

Opposite page:
The availability of affordable electricity increased substantially by the end of the nineteenth century, and electricity was put to imaginative uses far beyond electric trolleys and home lighting. In this four-part advertising page, electricity is touted as the preferred, economical, and decidedly modern way of powering portable fans and hair driers, running a printing press, and heating water. *Selling Electricity,* vol. 1, no. 5 (May 1907). LC-USZ62-110407.

PROCEEDINGS OF THE REGENTS. 105

I did not refer exclusively to the needle telegraph when, in my paper, I stated that the *magnetic* action of a current from a trough is at least not sensibly diminished by passing through a long wire. This is evident from the fact that the immediate experiment from which this deduction was made was by means of an electro-magnet and not by means of a needle galvanometer.

At the conclusion of the series of experiments which I described in Silliman's Journal, there were two applications of the electro-magnet in my mind: one the production of a machine to be moved by electro-magnetism, and the other the transmission of or calling into action power at a distance. The first was carried into execution in the construction of the machine described in Silliman's Journal, vol. 20, 1831, and for the purpose of experimenting in regard to the second, I arranged around one of the upper rooms in the Albany Academy a wire of more than a mile in length, through which I was enabled to make signals by sounding a bell, (fig. 7.) The mechanical arrangement for affecting this object was simply a steel bar, permanently magnetized, of about ten inches in length, supported on a pivot and placed with its north end between the two arms of a horse-shoe magnet. When the latter was excited by the current, the end of the bar thus placed was attracted by one arm of the horse-shoe, and repelled by the other, and was thus caused to move in a horizontal plane and its further extremity to strike a bell suitably adjusted.

Fig. 7.

This arrangement is that which is alluded to in Professor Hall's letter* as having been exhibited to him in 1832 It was not, however, at that time connected with the long wire above mentioned, but with a shorter one put up around the room for exhibition.

At the time of giving my testimony, I was uncertain as to when I had first exhibited this contrivance, but have since definitely settled the fact by the testimony of Hall and others that it was before I left Albany, and abundant evidence can be brought to show that previous to my going to Princeton in November, 1832, my mind was much occupied with the subject of the telegraph, and that I introduced it in my course of instruction to the senior class in the Academy. I should

* See the Report of the Committee, page 96, and Proceedings of the Albany Institute, anuary, 1858.

Although Joseph Henry is among the giants of American science and technology, his name is less recognized than those of many individuals of lesser talent and accomplishment. As early as 1831, Henry had built the first electromagnetic telegraph by ringing a bell through a mile of wire. Although he anticipated Morse by several years, he did not patent this invention. Here in miniature is his device, which uses an electromagnet for "the transmission of or calling into action power at a distance." Henry also preceded Faraday in his discovery of the principle of induction, but he delayed publication and again lost priority in this major breakthrough. Despite his later position as first head of the Smithsonian Institution and being one of the founders of the National Academy of Sciences, Henry never received the popular recognition he deserved. *Annual Report of the Board of Regents of the Smithsonian Institution . . . for the Year 1857*, p. 105. 1858. LC-USZ62-110408.

commonly used letters should be represented by the shortest of codes. His research revealed what newspaper typesetters knew from experience—that the most heavily used letters were *A*, *E*, and *T*. Naturally, the three least-used letters were given the longest codes—*Q*, *V*, and *Z*.

After successfully performing a private demonstration of his system in 1837, Morse then petitioned the U.S. Congress for financial assistance. On February 21, 1838, he demonstrated his telegraph before Congress and the president, and a bill was passed in 1843 giving thirty thousand dollars to build an experimental telegraph line between Washington, D.C., and Baltimore, Maryland. On May 24, 1844, Morse sent his famous message, "What hath God wrought?" In the Morse Collection in the Library of Congress, which totals approximately ten thousand items (consisting of family and general correspondence, diaries, notebooks, scrapbooks, clippings, and photos), there is a tape of this historic first telegraphic message.

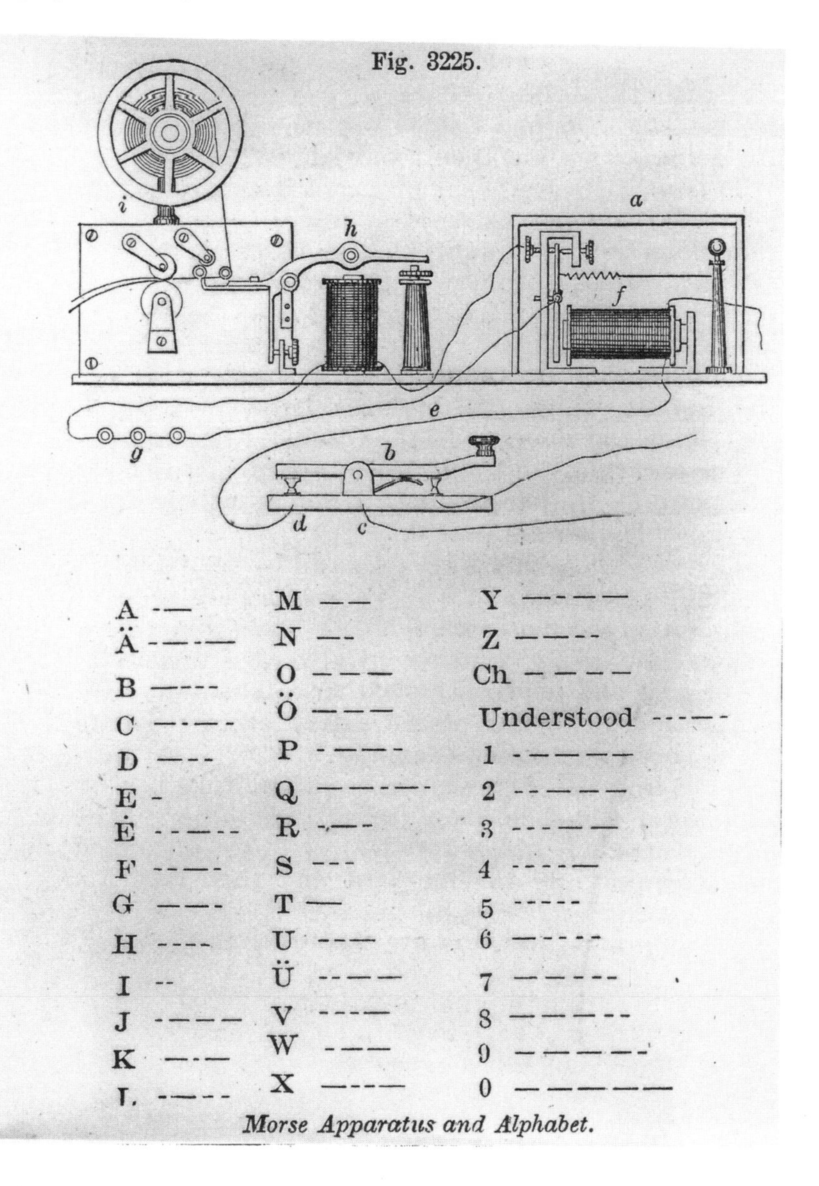

Morse Apparatus and Alphabet.

The first successful electrical telegraph was built by a financially unsuccessful artist who was a minister's son and a graduate of Yale. In the 1830s, electrical experimentation was the vogue in America, and Samuel Morse enthusiastically embraced the notion of sending messages by wire. Although he lacked scientific understanding as well as any real practical savvy, he compensated for his deficiencies by obtaining Joseph Henry's scientific and technical advice. Even the final code that bears his name was the product of another, being the work of his employee, Albert Vail. Nonetheless, the telegraph became a viable communication system in America as early as it did because of Morse's resolute belief in his idea and his unrelenting drive and determination to make it successful. Here, Morse's coded alphabet and the basics of his apparatus are seen in a nineteenth-century American technical dictionary. *Knight's American Mechanical Dictionary,* vol. 2. 1877. LC-USZ62-110409.

It was with this model of the "gallows" telephone that Bell and Watson discovered the principle of the electric speaking telephone in 1875. The telephone differed from the telegraph in that the latter transmitted sharply defined, on-off pulses of electric current, each pulse having the same intensity but varying only in length (dot or dash). A telephone, however, required a continuous current whose intensity varied exactly as sound waves did in the medium of the air. The gallows telephone is essentially a mechanical and electrical model of the human ear. It uses a vibrating medium (a circular, parchment membrane) and a magnet suspended above the membrane to convert the continuously varying sound waves into continuously varying electric currents. Prints and Photographs Division, Library of Congress, LC-G9-Z4-68812-T.

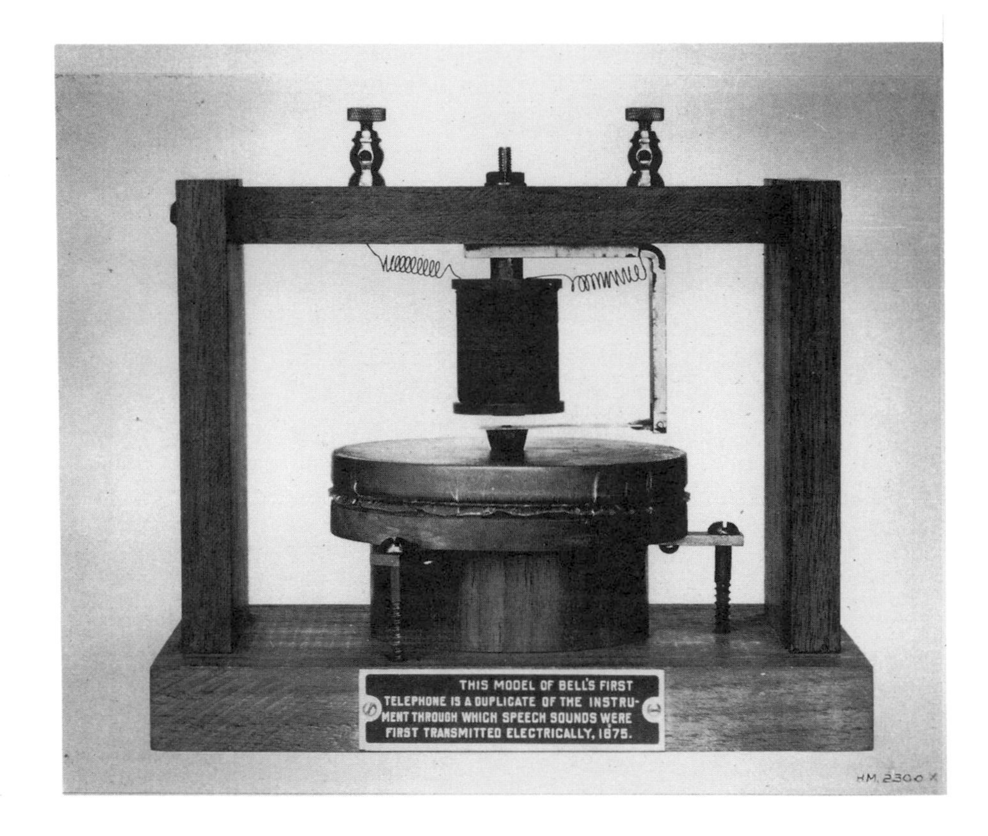

Opposite page:
Alexander Graham Bell and his family first visited Baddeck, Nova Scotia, in 1885 looking for a summer retreat. The thirty-eight-year-old Bell was already rich and famous, and it was among the Scot settlers there that Bell set up a summer laboratory and continued to work. His investigations included a wide and varied range of subjects from sheep-breeding and hydrofoils to powered flight and experimental kite-building. Here Bell is seen at the center behind his grandson, Melville Grosvenor, helping to fly one of his tetrahedral kites. Prints and Photographs Division, Library of Congress, LC-G9-Z3-116,837-AB.

The second great communications breakthrough which resulted from a better understanding of electromagnetism was the telephone. Alexander Graham Bell, a Scotsman who became an American citizen at twenty-seven, used the knowledge given by Oersted and Faraday, as well as Helmholtz's theory of sound, and conceived of the idea of transmitting speech electronically. Both Bell's father and grandfather had studied the mechanics of sound, and his father was a pioneer teacher of speech for the deaf. The young Bell was a professor of vocal physiology at Boston University, and, in his research to develop an efficient hearing aid, explored every connection between electricity and acoustics. He soon realized that if sound wave vibrations could be transformed into a fluctuating electric current, the current could be carried across wires and reconverted into identical sound waves at the other end of the circuit.

Bell's real technical breakthrough was in the use of what he called an "undulating" current—one in which the sound waves in the air could be made to vary the strength of an electric current in a wire. Having seen a new device at the Massachusetts Institute of Technology which he recognized as a mechanical ear—a vibrating membrane that reproduced the original sound effect—he built a microphone in which the membrane would vary an electric current. Then he built a receiver that would reconstitute these variations into audible frequencies. It is said that the first words transmitted by telephone occurred while Bell was working on his new invention, when he spilled some battery acid on his trousers and automatically called for his assistant, saying, "Watson, please come here.

I want you." Although Watson was on another floor, he was at the other end of the instrument circuit, and he heard Bell's voice—the first words transmitted by telephone. Bell patented his invention on March 7, 1876, and in May of that year presented his findings in a paper titled "Researches in Telephony." The Library of Congress has this paper as it was first published in the *American Academy of Arts and Sciences, Proceedings*.

The Library of Congress also has the Bell Family Papers, totaling approximately 140,000 items. These include Bell's correspondence, diaries and other writings, and several hundred volumes of laboratory notebooks which record his work from 1865 to 1922. While much of the material relates to Bell's invention of the telephone, there is also a good deal that reflects his broad interests in the education of the deaf, eugenics, marine engineering, and even aeronautics. The collection is also a family archive, encompassing material from various members of the Bell, Hubbard, Fairchild, and Grosvenor families. Finally, the Library of Congress also has a substantial collection of family photographs, including some early daguerreotypes, which document Bell's professional and private life.

The third breakthrough in communications during the nineteenth century came as a direct result of James Clerk Maxwell's proposal in 1873 that "light is an electromagnetic disturbance propagated according to electromagnetic laws." Maxwell's classic synthesis of three main fields of physics—electricity, magnetism, and light—remained only a hypothesis, however, until fifteen years later when a German physicist, Heinrich R. Hertz, proved it experimentally. Hertz produced a device which demonstrated that energy could be transmitted from place to place in the form of electromagnetic radiation. His experiment showed that a flow of current in one electrical circuit could result in a corresponding flow in a similarly "tuned" circuit not directly connected with the first. In an ingenious way, he was actually able to see that the waves had traveled some distance when a spark appeared at a small gap in the receiving circuit. In this way, Hertz verified Maxwell's hypothesis and proved that electric and light waves were essentially similar, the difference being in the length of their waves. Hertz's experimental skill thus laid the foundation for the later theory of wireless telegraphy, or radio. This experiment and others were reported from 1887 onward, and were collected and published in Leipzig in 1892 as *Untersuchungen uber die Ausbreitung der elektrischen Kraft*, which the Library of Congress has in first edition.

The practical potential of what came to be called "Hertzian waves" appeared endless to a twenty-year-old physics student, Guglielmo Marconi. Born in Bologna, Italy, Marconi had no university degree but rather was being tutored by well-known professors. When he came across an account in 1894 of Hertz's experiment, he began an intense period of experimentation, hoping to produce what he called "wireless telegraphy," transmitting the electric signal of the telegraph without the connecting wires. Concentrating mainly on the receiving end, he devised a "coherer" which, when hit by radio waves sent by his coil and spark gap generator, conducted a measurable surge of electric current. Failing to gain any financial support in Italy, Marconi went to England; his mother was Irish and

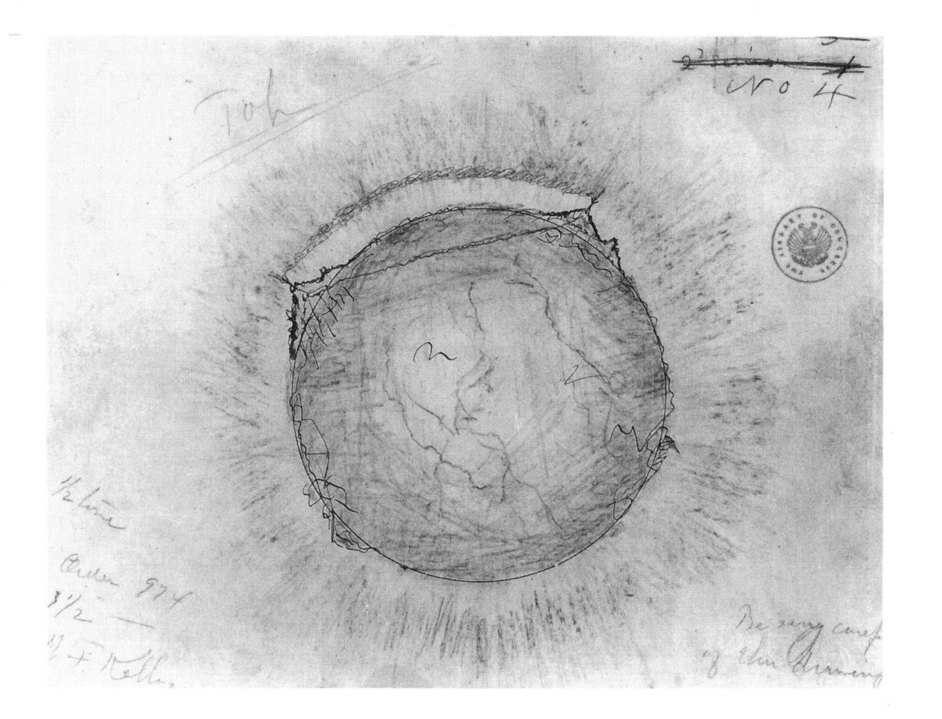

Sketch showing how wireless telegraphy might link California and Japan. Mahlon Loomis Papers, Manuscript Division, Library of Congress. See p. 110.

so his English was excellent. By 1897 he was sending messages over a distance of nine miles, and two years later, across the English Channel.

Marconi's real technical achievements were complemented by his correct conviction that radio waves would follow the curve of the earth and not radiate straight out into space, as most scientists then believed. Acting upon this belief, which he could not prove, he used balloons to lift antennae as high as possible on December 12, 1901, when he sent a radio signal across the Atlantic Ocean. This wireless, Morse code transmission from the southwest tip of England to Newfoundland can probably best serve as the date of the invention of the radio. Marconi was well known even before his spectacular success, having written and traveled considerably to publicize his emerging invention. The Library of Congress has one of his earliest articles, "Wireless Telegraphy," first published in London in the 1899 *Proceedings of the Institution of Electrical Engineers*. In that same year, Marconi visited the United States and his article was reprinted in the *Smithsonian Institution Annual Report, 1901*, which the Library also has in its collections.

As unbelievable and astounding to nineteenth-century sensibilities such dramatic inventions as the telephone and the radio must have been, it may also have seemed that capturing an image on paper of something real and tangible almost smacked of sorcery. Transmitting a human voice must have been a bizarre first-time experience, but fixing the exact likeness of a person's face on paper or stopping time by freezing one particular moment must surely have seemed a bit spooky. Thus photography joined those other "magical" technologies—like the telephone sending a human voice through a wire and then radio ultimately dispensing with the wire—that emerged in the nineteenth century.

As with many inventions and discoveries, in the story of photography there is contention over who first discovered what. In fact, significant parts of the entire photographic process were known before the nineteenth century. The earliest of these is the camera obscura. This Latin name, meaning "dark room," dates back to antiquity and was certainly known to the eleventh-century Arab physicist Alhazen. In simple terms, it consists of a small, darkened room (or a hand-held box) that admits light only through one tiny hole. This results in an inverted image being cast on the opposite wall, which is usually whitened. For centuries, portable versions of the camera obscura were used to view eclipses of the sun without damaging one's eyes, and by the Renaissance, they were used as an aid to drawing.

Given that an image could be obtained by this age-old process, the next step was to try to somehow capture it on a surface. This was done ingeniously by coating a surface with a substance that was sensitive to or would react to light. Again, the piecemeal knowledge lay in waiting, for as early as 1727, the chemical effect of sunlight on silver salts was already known. Much later, the first successful but impermanent attempt to produce an image by this light-sensitive means was reported by Thomas Wedgwood and Humphry Davy in an 1802 paper in the *Journal of the Royal Institution of Great Britain*.

Much of this came together when a wealthy, inventive Frenchman, Joseph-Nicéphore Nièpce, frustrated by his inability to make a realistic lithograph on his own, decided sometime after 1813 to attempt to capture an image mechanically or automatically. After beginning with pewter and coating it with various light-sensitive substances, he progressed to silver chloride on paper and then to a light-sensitive form of bitumen on glass—a petroleum substance or a kind of asphalt that hardens when exposed to light. By 1826, he had produced what he called heliographs on paper sensitized with silver chloride, but could not keep the image permanent. Soon after, he was joined by Louis J. M. Daguerre, a scene painter for the Paris opera who had become famous for his realistic dioramas and who also had been experimenting on how to fix an image.

The two men formed a partnership in 1829, and when Nièpce died in 1833, Daguerre continued to experiment alone. Finally, in 1837, he produced a real photograph taken of his own studio. Remarkably detailed and faithful to the actual scene, it was made after Daguerre discovered that a silver-plated sheet of copper coated with a thin film of silver iodide would react to light. The image was then developed by exposure to mercury vapor and fixed or made permanent by removing the unaffected silver iodide with what was called hyposulphite of soda (sodium thiosulfate). The result was named a daguerreotype and was announced in 1839.

The Library of Congress has the first disclosure of Daguerre's process, as published in the form of an 1839 report by François Arago, a member of the French Chamber of Deputies. Titled "Rapport de M. Arago sur le Daguerréotype," it was published in Paris by the Académie des Sciences because Daguerre and Nièpce's son had sold full rights to the daguerreotype to the French government in return for annuities for life. The Li-

These daguerreotype portraits of the newly elected congressman from Illinois, Abraham Lincoln, and his wife, Mary Todd Lincoln, were taken in 1846. Daguerreotypes first became available in the United States in 1840, and by the beginning of the 1850s no fewer than ten thousand individuals were estimated to be earning a living as daguerreotypists. To this impressive figure must be added the thousands of people who worked in the allied trades, supplying chemicals, plates, and apparatus. Prints and Photographs Division, Library of Congress, LC-USZ62-12457 (Abraham Lincoln) and LC-USP6-2707A (Mary Todd Lincoln). See p. 111.

brary also has Daguerre's own book in first edition, *Historique et description des procédés du Daguerréotype*, which he published in Paris in 1839.

During the years of Daguerre's experiments, an Englishman was also laboring across the Channel, perfecting what would be called the talbotype or later, improved, the calotype. In January 1839, when news of Daguerre's accomplishments reached England, William Henry Fox Talbot, a physicist, rushed into publication a report of his own process called "photogenic drawing." His methods were analogous to Daguerre's, with the major improvement being his invention of the photographic negative from which any number of positive prints could be made. In 1844 he published his lyrically titled book, *The Pencil of Nature*, in London, which the Library of Congress has in first edition. It contains not only a full account of his process but twenty-four photographs as well, making this the first book to be illustrated with photographs.

Priority of discovery is still a matter of dispute, and many regard Fox Talbot as the real inventor of the first practical method of photography. Unquestionably, his development of negatives was a step which went far beyond Daguerre. The two pioneering processes did have several differences: daguerreotypes were on metal and each was unique, but they had a remarkable degree of detail. Talbot's negatives were on paper and could

The calotype or Talbotype was the first major improvement over Daguerre's process. With its use of a photographic negative from which any number of positive prints could be made on paper, it laid the foundation of modern photography. Early calotypes like this one taken by William H. F. Talbot at his family's estate at Lacock Abbey were not very popular since they lacked the clarity and precise detail of a daguerreotype. But Talbot eventually improved the image quality and drastically reduced exposure time, so that sitters no longer had to be clamped in rigid, unnatural positions. Prints and Photographs Division, Library of Congress, LC-USZ62-83363.

be endlessly reproduced, but had a fuzzy or diffused effect. Fox Talbot also was a man of vision who foresaw many applications of this new technology, specifically what he called "living pictures," later known as moving pictures. In distinction to Daguerre, who made his fortune and retired, Fox Talbot continued his pioneering work in photography and distinguished himself in the field of mathematics as well. Further, he was one of the first to decipher the cuneiform inscriptions discovered at the ancient Assyrian city of Nineveh during the 1850s.

Once Daguerre had revealed his process in 1839, photography immediately spread throughout the world, capturing the interest and imagination of everyone who saw its startling products. Portrait studios opened in all the major cities and technical advances followed one upon another. One of the most important was Fox Talbot's discovery of how to drastically reduce exposure time. By 1877, a photograph of a moving horse taken in less than a thousandth of a second was published by Eadweard Muybridge. Earlier, Muybridge had been commissioned by the wealthy California railroad magnate, Leland Stanford, to settle the ancient question of whether all four legs of a horse ever leave the ground simultaneously. His ingenious battery of twelve (and later twenty-four) cameras activated by electromagnets fired by trip wires (the cameras' shutters were released by the breaking threads as the horse dashed by) produced his famous series of photographs demonstrating that all four legs were indeed off the ground during brief periods of trotting.

Muybridge was an unusual character with a somewhat checkered history. Born in England as Edward James Muggeridge, he emigrated to the

Following Talbot's pioneering work at stopping action with the aid of an electric flash, there developed both popular and scientific interest in photographing objects and phenomena that were not ordinarily visible to the naked eye. One of the most celebrated of these photographic experiments was Eadweard Muybridge's series of racehorse images demonstrating that all four of the animal's feet actually do leave the ground at one time. Prints and Photographs Division, Library of Congress, LC-USZ62-52703.

United States and changed his name. He set himself up as a photographer in San Francisco, using the $2,500 in damages he received after being thrown from a coach. During his career, he was able to survive the scandal of having killed his wife's lover, an act for which he was acquitted.

Despite the notoriety of his private life, Muybridge built on his experience photographing the horse and went on to become world-famous for his studies of human and animal locomotion, publishing some twenty thousand photographs before his death in 1904. The Library of Congress has the sixteen-volume first edition of his major study, *Animal Locomotion. An Electro-Photographic Investigation of Consecutive Phases of Animal Movements* (Philadelphia, 1887). It also has the single-volume edition of his book, *Animals in Motion* (London, 1907). The multivolume study was assured success despite its high cost when buyers noticed that the largest category

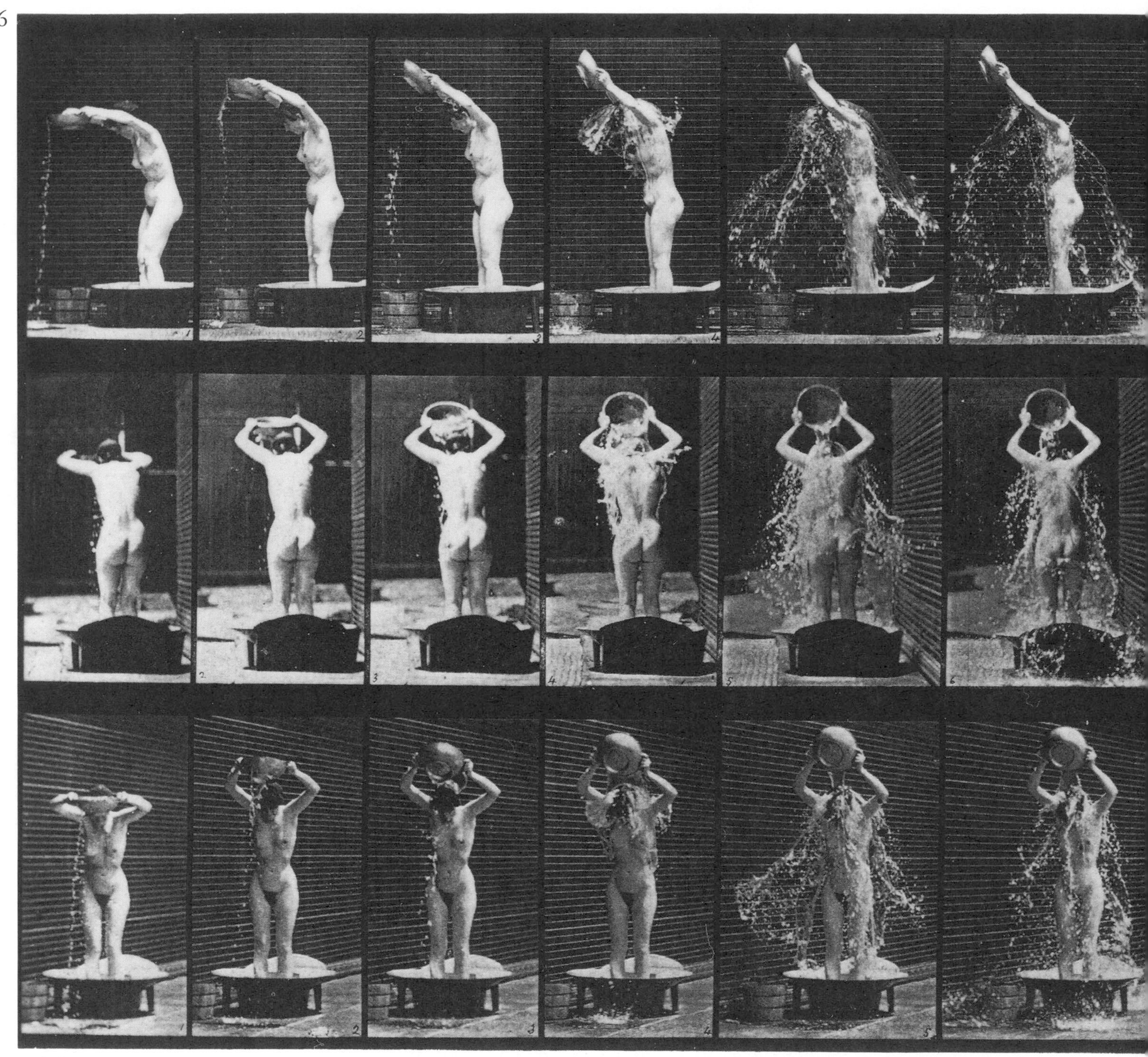

In 1884 Eadweard Muybridge began a serious photographic study of human and animal movement that was sponsored by the University of Pennsylvania. Photographing men, women, and zoo animals in a wide range of actions, he eventually produced a body of images that was published in eleven folio volumes in 1887 and sold for $600. Muybridge later made his magnum opus available to a larger audience with a selected, more affordable edition. *Animals in Motion,* 1907. Eadweard Muybridge. LC-USZ62-110410.

Autochrome photograph by Arnold Genthe. Prints and Photographs Division, Library of Congress. See p. 112.

From *The Park and the Crystal Palace: A Series of Seven Picturesque Views,* 1851. Philip Brannon. See p. 113.

of pictures was of nude women. The Library also has some of his early single prints.

Muybridge also contributed in a way to motion pictures, for he often projected his images of the running horse on a screen in rapid succession, producing the illusion of movement. Cinematography, or the projection upon a screen of motion pictures, was perfected by Louis and Auguste Lumière. Following upon George Eastman's introduction of photographic celluloid film and Thomas Edison's "Kinetoscope," the Lumière brothers solved all the technical problems of capturing and reproducing realistic action and patented a machine which could photograph, develop, and project the motion picture. In Paris in 1895, the brothers pioneered commercial cinema by conducting the first motion picture show before a paying audience. The Library of Congress has the brothers' trade catalog, *Notice sur le cinématographe*, first published in Lyons in 1897. This was the first public announcement of their invention. The Lumières are also known for their invention of the Autochrome process for color photography. Patented in 1904, their process covered photographic plates with tiny color filters (actually grains of potato starch dyed violet, orange, and green) and made positive glass transparencies with remarkably soft and realistic colors. The Library of Congress has, in addition to several of the Lumières' own plates, 384 Autochrome plates made by Arnold Genthe, one of America's most notable photographers of the early twentieth century.

In addition to the vast and broad collection of individual photographs and special collections that the Library has amassed over the years, it has one of the best and most comprehensive collections of books illustrated with photographs. Beginning with Talbot's *The Pencil of Nature*, the first book to be illustrated with photographs, the Library's collection is highlighted by the *Exhibition of the Works of Industry of All Nations*. Issued in 1852 in London in only 130 presentation copies, this rare work documents the great Crystal Palace Exhibition of 1851, named for the iron and glass fantasy constructed in Hyde Park by Joseph Parton to house the first of the great technological exhibitions. This four-volume work contains 155 calotypes of the major technological objects on display. It has significance not only to the history of photography, but also to the overall history of technology, documenting as it does the technology of its day.

The first cameras were often used as documenting devices, and during the Crimean War, between March and July 1855, a camera was first used to cover a war. An English photographer, Roger Fenton, made photographs of the actual battlefields and camp conditions. Interestingly, Fenton's photographs, though of superb quality, are not very warlike, for they display none of the actual horrors or terrors of real war. The Fenton Collection in the Library of Congress consists of 265 original "salted paper" photoprints.

Better known as a war photographer for his grittily realistic work is the American Mathew B. Brady. Brady had made his name as a photographer of prominent Americans well before the Civil War brought him posthumous fame. When the war broke out in 1861, Brady decided to make as complete a record of the conflict as possible and hired a staff of

When the daguerreotype was replaced by the quicker wet plate or wet collodion method, photography was ready to capture military conflicts for the first time. The American Civil War was the first major conflict to be thoroughly photographed, and the fact that so much documentation exists can be attributed primarily to the vision and determination of Mathew Brady. When the war began, Brady left a lucrative portrait business and used his political connections to gain access to battlefields. He soon hired a cadre of twenty photographers who documented everything except actual battle scenes, which would have been blurred, since exposure time was still measured in seconds. Throughout the war, graphic scenes of death and dying were brought to the Northern home front in such weeklies as *Harper's* and *Leslie's,* whose artists transformed photographs into true-to-life wood engravings. This photograph taken by Alexander Gardner, one of Brady's cadre who later became a rival, shows the mammoth Union mortar called the "Dictator" during the siege of Petersburg in October 1864. The nearly nine-ton, thirteen-inch mortar was transported by rail. Prints and Photographs Division, Library of Congress, LC-B8184-4794.

twenty photographers, whom he dispatched to the war zones. Brady himself photographed the battlefields of Bull Run, Antietam, and Gettysburg. By war's end, Brady had invested $100,000 in the mistaken belief that the government would buy his war collection. Following government uninterest and the later financial panic of 1873, Brady was unable to pay even the storage bill for his negatives and was forced to sell his plates at public auction. They were purchased by the War Department for $2,840. Although Brady eventually received $25,000 from Congress through the intervention of friends, he never regained his financial footing and died alone and unremembered in a hospital charity ward in 1896.

Brady's visual legacy eventually found its way to the Library of Congress, which has his magnificent collection of more than ten thousand original glass plate negatives and prints taken during the Civil War by him and his colleagues. It also has approximately five hundred of his original daguerreotypes, the majority of which were taken in Brady's studio during his stint as a portrait photographer in the 1840s and 1850s.

Nineteenth-century breakthroughs in communications, such as the telegraph, telephone, and radio and the later sound recordings on Thomas Edison's wax cylinders and Emile Berliner's phonograph records, made for an entirely new and potentially revolutionary phenomenon that might be called the mobility of ideas. These new communications technologies would mark the beginning of the modern cliché that "it's a small world." Indeed, by the end of the century an individual's voice (and therefore his ideas and influence) could be transmitted almost instantaneously around the world or captured on a disc for later listening. The listener could also know what the speaker or singer looked like from a photograph in a newspaper or magazine. These new media were quickly taken advantage of by those with commercial and political astuteness.

The motion picture made its commercial debut in America in 1896. On the evening of April 23, 1896, at Koster & Bial's Music Hall on Herald Square in New York City, screen projection of a motion picture was first shown to an amazed audience. The projector, called a Vitascope, was invented by an American, Thomas Armat, who had made a previous arrangement with Thomas A. Edison whereby Edison would manufacture Armat's projector and introduce it under his own, more famous name. Among the early Edison film subjects were dancing girls, seen pictured in this lively advertisement, chase scenes, and recreations of Broadway plays—one of which included the first close-up screen kiss by John Rice and May Irwin. In the motion picture collections of the Library of Congress—which holds the nation's most important and comprehensive film collection—are deposited the "incunabula" of motion pictures produced by the Edison, Biograph, Méliès, Lubin, and other early companies. For the next half-century, audiences for motion pictures would continue to grow at a staggering rate, reaching their peak in the late 1940s. During those pretelevision years, the weekly movie attendance was well over ninety million people. Prints and Photographs Division, Library of Congress, LC-USZ62-33505.

As ideas, trends, notions, and similarly intangible cultural and political forces became more mobile, so did the people who generated them. During the nineteenth century, man began moving real distances without assistance from animal power, using bicycles, trains, steamships, motorcycles, automobiles, and dirigibles. The technology that most affected the largest number of people was, of course, the internal combustion engine which made the automobile possible.

An engine works when the heat generated by a contained explosion (internal combustion) enters a cylinder and moves a piston. The first of these engines was probably built in 1680 by Christiaan Huygens. In this experiment, Huygens found that the gas generated by a gunpowder explosion drove a piston up a cylinder. But working with gunpowder was dangerous and impractical, and the creative and multitalented Huygens went on to other things in different fields.

Despite attempts by others, the first real practical success was not achieved until 1859 by the self-taught, Belgian-French inventor Jean Joseph Étienne Lenoir. Using the steam engines of his time as a model, Lenoir built a piston engine in which a mixture of coal gas and air was ignited by a spark from an induction coil. The recent availability of this cheap and relatively safe gas made it a logical combustible for Lenoir to use. Lenoir's new engine, although wasteful of fuel, was much smaller than a steam engine and quicker and easier to start, since unlike a steam engine which had to boil water, it needed only a spark. What engines Lenoir did sell were bought because they were a convenient and compact source of industrial power.

The Library of Congress has a rare, twenty-page booklet published in 1866 and titled *The Lenoir Gas-Engine*. Put out by the Lenoir Gas Engine Company of New York, it is essentially a catalog that gives prospective buyers operating details about this new invention. It contains two engravings and one line drawing of the engine and not only tells how it works but describes all its basic systems. Among the many uses it proposes, the only one that relates in any way to transportation is a reference

This three-horse engine built in 1860 by Jean Etienne Lenoir was the first gas engine made available for public use. The engine design was not strikingly novel, nor was it anything more than an ordinary high-pressure steam engine with valves arranged to admit gas and air and discharge exhaust. But it was an internal combustion engine that ran on coal gas, and it did work. Despite advertising statements to the contrary, his engine was less economical than conventional steam engines, primarily because Lenoir was unaware of the need to compress the gas mixture. *The Gas and Oil Engine,* 1896. Dugald Clerk. LC-USZ62-110411.

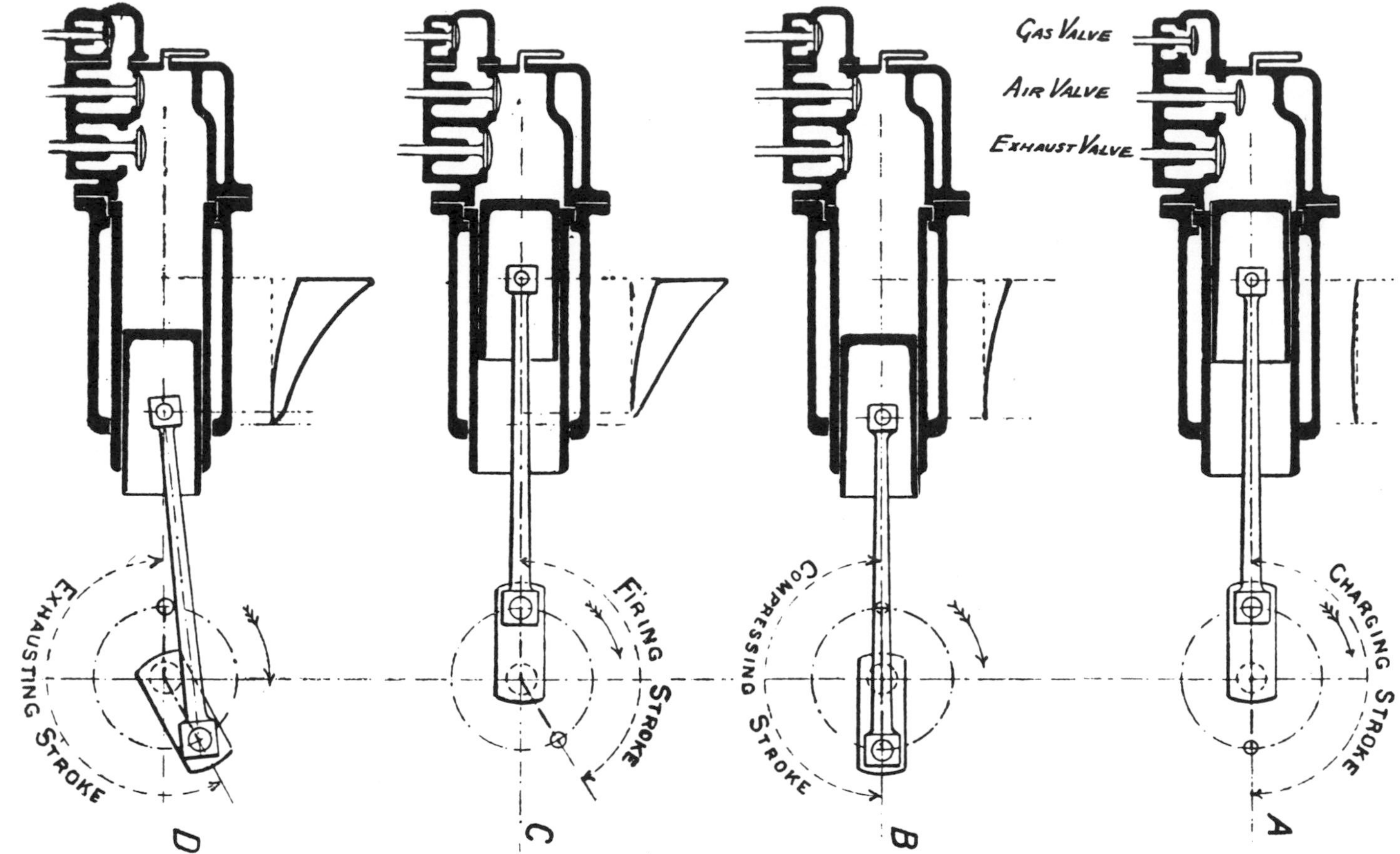

The Otto Silent Engine of 1876 was the forerunner of the modern automobile and aircraft engine. Seeking to make the one-cylinder engine he had built as an improvement on the Lenoir version both quieter and more powerful, Nikolaus A. Otto arrived at the notion of a four-stroke cycle in the cylinder. Here his engine's four cycles (charging, compressing, firing, and exhausting) are illustrated top to bottom. The key improvement is Otto's compression of the gas-and-air mixture. *A Practical Treatise on the "Otto" Cycle Gas Engine,* 1896. William Norris. LC-USZ62-110412.

to "propelling city rail cars." Although Lenoir did mount his engine on a horse cart and drive about in 1862, his engine needed a great deal of improvement before it would become practical and useful for either industry or transportation.

It was Nikolaus Otto who would provide the necessary improvements to Lenoir's engine and who ultimately would create an engine that made both the automobile and the airplane possible. Beginning by attempting to recreate Lenoir's 1860 engine, about which he had read in the newspaper, Otto built an improved version and went into business in 1867 with a shrewd capitalist, Eugen Langen. Ten years later, he produced what is known as the "four-stroke" engine—the precursor of the modern gasoline engine.

Although earlier theoretical studies had shown that four strokes or cycles constituted the most efficient internal movements within such an engine, Otto was ignorant of any such theories, being first a traveling salesman, despite his obvious mechanical ability. Discovering the advantages of four cycles himself, he built the Otto Silent Engine in 1876, which delivered three horsepower at 180 revolutions per minute. Its four strokes or cycles accomplish induction, when the descending piston draws in gas and air; compression, when the rising piston squeezes or compresses the mixture which ignites from a spark; power, when the spark flares and the charge bursts into flame, the expanding gases forcefully driving the piston down; and exhaust, when the rising piston expels the hot gases out the exhaust valve which has just opened.

Otto's new single-cylinder engines were such an improvement over the gas and steam engines of the time, and his partner Langen marketed them so well, that 200,000 were sold, mainly for industrial uses, by the end of the century. The details of his engines are best documented not by anything Otto wrote but by an 1886 book by an Englishman, Dugald Clerk, called *The Gas and Oil Engine*. The Library of Congress has the version of this technical work published in New York in 1896.

Where Otto cared little about using his engines for anything more than industrial purposes, one of his assistants, Gottlieb Daimler, envisioned them powering some sort of vehicle. After finally leaving Otto in 1883 and taking a brilliant engineer, Wilhelm Maybach, with him, Daimler developed the first light, high-speed internal combustion engine and later, a carburetor that made it possible to use gasoline as fuel. In 1885, he installed one of his new engines on a bicycle, and two years later engines powered a boat as well as a four-wheeled carriage. By 1889, he had the engine running a four-wheeled vehicle that was not just an adapted carriage but a vehicle designed originally as an automobile.

At about the same time in Germany, Carl Benz was working toward the same goal as Daimler, each ignorant of the other's efforts. Like Daimler, Benz planned and built an automobile—not a converted horse-drawn carriage—powered by an internal combustion engine (although not a high-speed one like Daimler's). His three-wheeler first ran in 1885 and was patented the next year. The automobile was born.

Since neither man wrote very much, the best contemporary documentation, besides the original patents, is a book by an interested and competent third party. William W. Beaumont's *Motor Vehicles and Motors*, first published in 1900, provides a wealth of detail and illustrations about the work of both Benz and Daimler. The Library of Congress has this book in its larger, revised edition of 1902, published in Westminster and Philadelphia.

Another version of the internal combustion engine—the diesel engine—also came to be at about the same time. Its inventor, Rudolf Diesel, was the brilliant product of the Technische Hochschule in Munich, where he became aware of how significant the long-neglected Carnot heat cycle was to the design of an efficient engine. In this regard, Diesel sought to improve upon the inefficient internal combustion engine by devising an engine that would work on the heat generated by compressing the fuel-air mixture. His extremely high compression ratio heats the air so much that the fuel ignites spontaneously upon injection. This not only does away with the need for an electric spark but allows for the use of cheaper, safer fuel like kerosene.

Diesel made his discovery known in his 1893 paper "Theorie und Konstruction eines rationellen Wärmemotors zum Ersatz der Dampfmaschinen und der heute bekannten Verbrennungs-motoren." Although the Library of Congress does not have this paper, it does have Diesel's lengthy and very technical 1897 lecture as it was translated into English and pub-

This three-wheeled vehicle was the first practical automobile powered by an internal combustion engine. Built in 1885 by the German mechanical engineer Carl Benz, the vehicle was from the beginning designed around the engine, making it something entirely new rather than a converted carriage. Benz's design was not an immediate commercial success, but with improvements and a four-wheel version, he prospered, and in 1926 the Benz Company merged with Daimler-Motoren-Gesselschaft to form Daimler-Benz. This company went on to produce the Mercedes-Benz cars, although Benz himself left the firm around 1906. This image is from the Library's Daimler-Benz Collection, whose fifty-five volumes contain a total of about twenty-five hundred mounted original photoprints with letterpress captions in German. Daimler-Benz Collection, Prints and Photographs Division, Library of Congress. LC-USZ62-48300.

lished in the gas industry journal, *Progressive Age*. Appearing under the title "Diesel's Rational Heat Motor," it was published in four parts in that journal between December 1, 1897, and January 15, 1898. In addition, the February 15, 1898, issue contains seven pages of black-and-white photos of Diesel's engines.

Although Diesel's engines were simply too large and heavy for most early automobiles, they proved ideal for heavy trucks and, of course, for locomotives and ships. The immensely wealthy Diesel died under very suspicious circumstances while crossing the English Channel by boat in 1913, drowning after going overboard. Many believe that the neurotic and depressed Diesel committed suicide. Despite his sad end, his name lives on today in his invention and he is recognized as one of the major contributors to the new age of power.

Mechanical power is obviously what made the airplane an inevitable invention, yet it is ironic that the individuals who contributed most to the achievement of powered flight were those who had the patience, insight, and skill to attempt to understand first the intricacies of gliding, or unpowered flight. The previous chapter discussed George Cayley in England, who might be considered the precursor of all of the remarkable contributors to the development of aviation. As the century progressed, there were many others who tried to advance what they considered the new science of flight both by theorizing and by experimenting. Although few left behind a written record of any kind, one nineteenth-century man decided to set the record straight by trying to document these efforts as much as possible. This was the eminent American civil engineer, Octave Chanute.

In 1894, Chanute published *Progress in Flying Machines*, which became the bible of flying. In it, Chanute presents to the interested reader or would-be experimenter the accomplishments in aeronautics up to that time. Not simply a history, however, his book treats the subject of "flying machines" as a serious, scientific endeavor and contains line drawings as well as tables and formulas. Chanute went on to design and build his own biplane and triplane gliders, and ultimately was an early friend, confidante, and adviser of the Wright brothers. The Library of Congress has a first edition copy of this most important book, published in New York in 1894.

Perhaps of even greater significance is the Library's collection of correspondence between Chanute and the Wright brothers. This series of several hundred letters dates from May 1900 to May 1910 and touches on practically every phase and stage of the development of the airplane. It is a complete collection since Chanute kept virtually every scrap of paper he received from the Wrights, and these were donated to the Library by his heirs. Together with the Wright papers in the Library of Congress, which

In this first authoritative account of aviation history, the American civil engineer Octave Chanute summarized the entire progress of flight up to his time, and in doing so, he produced what the Wright brothers regarded as their Bible. Here, Chanute details the 1876 design of Alphonse Penaud whose all-wing, twin-propeller monoplane was highly advanced but never built. Chanute later became a source of encouragement and a confidante of the Wrights. *Progress in Flying Machines*, 1894. Octave Chanute. LC-USZ62-110413.

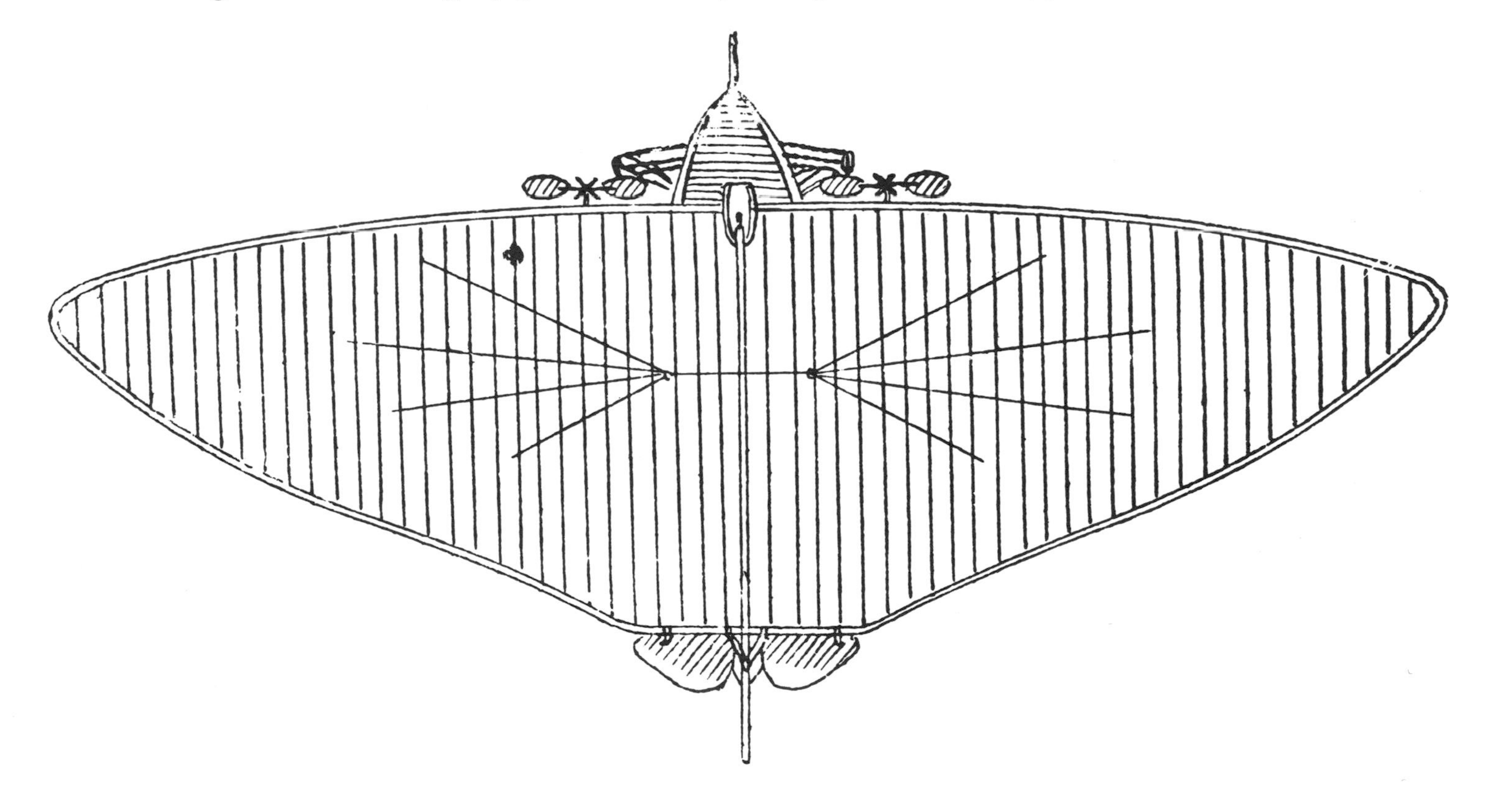

Otto Lilienthal is seen from below flying the two-surface glider in which he made some of his longest and highest glides. Octave Chanute called Lilienthal "The Flying Man" in his 1894 book, and Wilbur Wright later marked the beginning of his real interest in aviation at 1896 when he heard the shocking news that Lilienthal had died as a result of one of his crashes. Before the Wrights ever even began to consider powered flight, they mastered the subtleties of gliding flight much as their brave idol, Lilienthal, had done. Prints and Photographs Division, Library of Congress, LC-USZ62-99633.

include letters Chanute had sent to the brothers, the complete Wright-Chanute correspondence is one of the most important sources in aeronautical history. In this unique collection, the events leading up to and flowing from one of the more significant technological achievements of modern time—the first powered, sustained, and controlled flight—is recorded, documented, discussed, and even elucidated by the participants themselves.

Both the Wright brothers and Octave Chanute readily admitted that it was the example of one man that inspired them to work so hard and faithfully on what seemed a dream. The desire to fly was a passion that consumed the great German engineer, Otto Lilienthal, and at twenty-nine years of age he built his first experimental model with curved wings, learning of their superiority to flat surfaces. The Franco-Prussian War interrupted his work, to which he finally returned in 1886. Three years later he published in Berlin what would become a classic in the history of aviation, his *Der Vogelflug als Grundlage der Fliegekunst*. In this work he examines in detail the types and structures of bird wings, the method and aerodynamics of bird flight, and the application of the data he had obtained by experiment to the problem of human flight. The Library of Congress has Lilienthal's book in both its first German edition and its first English version, translated as *Bird Flight as the Basis of Aviation* (London, 1911).

By 1891, Lilienthal had built himself a type of hang-glider and was ready to put his theorizing into practice. In that year, he launched himself successfully on his first glide, and by 1896, he had conducted over two thousand glides, all the while documenting his experiments and accomplishments. He was also making a name for himself and was known as a skilled glider pilot. At one time, he soared for more than a quarter of a mile. Photographs of his flights were published around the world in technical and popular journals, and Octave Chanute and the Wrights became inspired to follow his lead. His shocking death on August 10, 1896, apparently did not deter the Wrights but actually pushed them forward. Having already demonstrated to the world that flying was a possibility, Lilienthal died the day after he stalled in a glide, sideslipped, and crashed, breaking his spine. His final words—"Sacrifices must be made"—were not made up by an admiring biographer. Later, Wilbur Wright would say, "My own active interest in aeronautical problems dates to the death of Lilienthal." The accomplishments of the Wrights and how they are documented in the Library of Congress collections will be discussed in the following chapter.

Although the Wright brothers, Lilienthal, and others worked with traditional, if not ancient, materials like wood and fabric to build their experimental gliders, the age they labored in was unquestionably the age of iron and steel. Steel—the result of smelting certain types of iron ore in a certain way—was no new discovery and was virtually as old as iron. Although known to have the strength of cast iron and the malleability of wrought iron, steel's availability and use were limited by the prohibitive cost of the extremely slow batch process involved in making it.

How to make large quantities of steel more cheaply was the problem that Henry Bessemer addressed in his 1856 presentation to the British Association for the Advancement of Science. Despite its wonderfully misleading title—"On the Manufacture of Malleable Iron and Steel without Fuel"—Bessemer's speech elaborated on his new process of making steel without any additional fuel, and therefore at a fraction of the cost. In steel-making, carbon has to be burned off. Bessemer avoided most of the cost for this step by adding not more expensive fuel, but oxygen directly to the molten stage by a blast of air. He unknowingly had been anticipated in this by an American ironmaster, William Kelly, who had built seven experimental converters to develop what he called his "air boiling" process. Kelly never applied for a patent, however, until he learned in 1857 of Bessemer's discovery the prior year. Kelly did receive a U.S. patent superseding Bessemer's, but his business fortunes soon foundered and, unlike his rival who became a millionaire, Kelly went bankrupt and had to settle on Bessemer's terms.

Bessemer's seminal paper of 1856 is represented in the collections of the Library of Congress as it appeared in the *Journal of the Franklin Institute* for 1859 (volume 68). His seven-page article is titled "On the Manufacture of Malleable Iron and Steel." Although Kelly wrote little, he lived long enough to contribute an account of his discovery to a book published in his lifetime. In James M. Swank's *History of the Manufacture of Iron in All Ages*, the author offers three pages of what he says is "a complete account of Mr. Kelly's invention prepared for these pages by Mr. Kelly himself." The Library of Congress has this work as it was first published in Philadelphia in 1884.

Perhaps the most dramatic nineteenth-century use of iron and the new steel was in building bridges. Although the first iron bridge—the famous

Burnt gases spew in flame from the top of this Bessemer converter as the result of a high-pressure blast of air from below. This stage of the Bessemer process of making steel, now obsolete, was always the most spectacular part. In trying to make a better and less expensive wrought iron from cast iron, Bessemer used the cheapest form of oxygen there is—ordinary air, which is about 21 percent oxygen—and blasted it onto the molten cast iron. When this spectacular volcano subsided, he was left with what he called "malleable iron" or steel. It took another twenty years from the time of his discovery for his process to be modified and improved, but once refined, it spread throughout the industrial world. Steelmaking both stimulated and was stimulated by the late eighteenth-century boom in railroad building as well as the emergence of steel ships and skyscrapers. Prints and Photographs Division, Library of Congress, LC-USZ6-1720.

Coalbrookdale bridge over the Severn in England, designed by Pritchard and built by Darby and Wilkinson—was completed in 1779, the first iron bridge to be scientifically designed was built by an American engineer, Squire Whipple. In 1840, Whipple's bridge over the canal at Utica, New York, ushered in a whole new era in bridge building. In his book, *A Work on Bridge Building*, privately published in 1847, Whipple rationalized his work and scientifically analyzed the strains on a bridge truss. In it he provided the theoretical understanding of bridge construction to a profession that until then had gotten by with strictly empirical knowledge. The Library of Congress has an enlarged edition of Whipple's book, published in Albany in 1869. The Library also has photographs and design details of the Whipple Bowstring Truss Bridge in Albany, New York. The information on this bridge, one of seven known surviving Whipple bridges, is deposited in the Library as part of the Historic American Engineering Record (HAER) collection.

It was in America also that the first major steel bridge was built. In the 1850s and 1860s, the railroads of an exuberant and expanding nation had made the city of Chicago an economic rival of the river-oriented city of St. Louis. The latter felt a growing need to bridge the great Mississippi River, to connect directly with the country east of the river. It was at this point that James Buchanan Eads stepped in and offered his vision of a steel bridge spanning the wide and powerful river. Eads turned to steel—despite the fact that no steel structures of any kind existed in America—because iron simply would not do. His plan called for three enormous arches to span the river, and he knew that iron was not strong enough.

The monumental bridge he completed in 1874, then called the St. Louis Bridge, and now called the Eads Bridge, still is used and is a structural and aesthetic marvel. It was ahead of its time in many ways besides

The first rational or scientifically designed metal bridge was built by an obscure, mostly self-taught engineer with the uniquely nineteenth-century American name of Squire Whipple. This 1969 photograph by the Historic American Engineering Record shows one of the few remaining Whipple Bowstring Truss bridges still standing. It spans a ravine north of Normanskill, within the city limits of Albany, New York. Its semicircular shape is similar to a bow, with its curved, top-chord compression member held together by a bottom-chord tension member. Once the merits of his new design were proven over time—his bridges were inexpensive, lightweight, and strong—Americans began using Whipple's ideas as the basis for many new bridges. After the publication of his 1847 book, they were also able for the first time to understand the principles underlying their work. For Whipple had analyzed the strains on a bridge truss and stated systematically and scientifically what had previously been only empirical knowledge. Prints and Photographs Division, Library of Congress, HAER NY, 1-ALB, 19-4.

Opposite page:
James Buchanan Eads was described by Andrew Carnegie as "an original genius minus the scientific knowledge." Eads's formal education ceased when he was thirteen, and he spent his formative years as a purser on a Mississippi steamboat. By the age of twenty-two, he had designed a diving bell and began a profitable business as a salvager. It was this intimate and hard-won knowledge of the bottom of the Mississippi River that gained him a hearing for his idea to span his beloved river with a bridge of steel. When he was charged with its construction in March 1867, it was primarily because he alone could best determine how to construct the all-important permanent piers on the treacherous, silt-covered bottom. Eads built them on bedrock (located, for the east pier, at 103 feet below the mud) using pneumatic caissons. After four years of difficult and dangerous work finishing the piers, his next task was to build the connecting steel arches without blocking passage on the river. This was accomplished by cantilevering the two halves of each arch out from the masonry pier. Nearly everything about the Eads bridge was without precedent, from the materials to the construction methods, and today this engineering tour de force still carries highway traffic. Prints and Photographs Division, Library of Congress, LC-USZ62-69754.

the first extensive use of steel in bridge construction. It was also the first to use pneumatic caissons in the founding of the huge bridge piers and the first to use hollow tubular chord members. The bridge has two decks—a lower deck carrying a railroad line and an upper deck for a roadway. Eads gave his health to the river he loved, having worked on and under the Mississippi since he was a teenager, and his genius and spirit are legend today. The Library of Congress has four unique documents all signed by Eads. The longest of these is the seventy-seven-page *Report of the Engineer-in-Chief of the Illinois and St. Louis Bridge Company* (May 1868), the first report the maverick Eads had to make to the company's directors, who distrusted him and his methods throughout. Perhaps the most interesting individual item is his eight-page *Improvement of the Mouth of the Mississippi River*, published in 1874. The title page is inscribed: "President Grant with the compliments of Jas. B. Eads."

Despite the grandeur and success of Eads's arched bridge, his contemporary, John A. Roebling, thought the arch was in no way superior to the principle of suspension—something he specialized in, having spanned the Niagara Gorge in 1851 with a suspension bridge. His Brooklyn Bridge, or the Great East River Bridge, begun in 1869 and completed in 1883, became the definitive American bridge. The design for this behemoth, then the largest suspension bridge in the world, was solely Roebling's, and the story of its construction is replete with enough drama, tragedy, and tension to warrant its entrance into the mythology of America. Roebling himself did not live to see the ground broken, as he died two weeks after contracting tetanus in an accident which crushed his toes. It was his talented son, Washington Augustus Roebling, himself a trained and experienced engineer, who took up his father's plans and actually directed the building of the bridge.

The three major elements of this unique bridge—the underwater piers or foundations, the stone towers, and the steel suspension cables—all work together to make it truly a unique achievement. Recognized in 1964 as a National Historic Landmark, the bridge is best documented in the collections of the Library of Congress by an 1872 report written by Washington A. Roebling which gives a detailed account of the work on the foundation and towers. This ninety-two-page document, containing several fold-out drawings, is titled *Pneumatic Tower Foundations of the East River Suspension Bridge* (New York, 1873). It also contains John A. Roebling's detailed original designs as he described them in his preliminary report of 1867.

The Brooklyn Bridge is actually the least of many examples that support the notion that the modern world was born in the 1800s. Nineteenth-century techniques and technologies abound and can be linked to characteristically twentieth-century phenomena. The capability for individual violence increased dramatically when Samuel Colt offered America his mass-produced six-shot revolver in 1848. Domestic independence similarly increased in 1854 when Isaac Singer made his sewing machine available for home use. And urban expansion received a literal boost when Elisha Graves Otis invented the first safety elevator, thus eliminating the

Like Eads, John A. Roebling believed that steel was "the metal of the future," and he was the first to advocate the use of steel wire instead of iron for the cables of suspension bridges. Having planned since the 1850s how to span the East River to link two boroughs of New York City, Roebling finally got his chance after the Civil War. His suspension bridges in Cincinnati and Niagara Falls must have been sufficient to certify his genius, for Roebling was named chief engineer in 1867 without any competitive bidding. After his accidental death two years later, the actual building of the bridge was accomplished by his son, Washington Augustus Roebling. Built of stone and steel, this monumental bridge spans two eras as it links the two great commercial centers of Manhattan and Brooklyn. Since its completion in 1883, the Brooklyn Bridge has entered a special realm of American popular culture, surpassing even its status as a major achievement of structural engineering. Even to the casual observer, an image such as this one is a powerful and moving sight. Prints and Photographs Division, Library of Congress, LC-USZ62-74616.

major obstacle to an architect's desire to use the new steel to go vertical with his buildings.

Steam power was the precursor of other power technologies, and the spectacular results achieved by James Nasmyth's 1839 steam hammer foreshadowed today's almost frightening ability to literally reshape the earth. The relentless tunneling methods devised by Marc I. Brunel during his spectacular triumph building the Thames Tunnel (1826–42) presaged a time when human transportation would literally brook no obstacle. In this same context of manipulating the natural environment, William Perkin's spectacular discovery in 1856 of how to make artificial dye created the synthetic dye industry and stimulated the development of synthetic organic chemistry—foreshadowing today's commonplace accomplishments of remaking and often outdoing nature itself.

Finally, possibly the most characteristically modern technology of all—manipulating information in all its aspects—can be seen very early on in Charles Babbage's "analytical engine." As the first real calculating machine, this 1822 machine was a network of wires, gears, and levers that nonetheless anticipated the modern computer and its ability to manipulate information. The introduction in 1876 of the Remington no. 1 typewriter would also prove an invaluable tool in this regard.

Many of these seminal inventions and discoveries are documented in some form in the collections of the Library of Congress. The Library has the 1884 New York edition of *James Nasmyth, an Autobiography*, in which he details much about his technical accomplishments. Included, for example, is a copy of his first drawings of the steam hammer in 1839. Marc I. Brunel offers much of the same concerning his varied engineering career in his *Memoir of the Life of Sir Marc Isambard Brunel* (London, 1862), of which the Library has a copy.

Finally, the great but failed goal of Charles Babbage—a calculating machine that was directed by an external program and could carry out any mathematical operation—is documented in the Library of Congress by the writings of contemporaries who took up his cause, since Babbage himself wrote very little. First, the Library has what is usually described as the definitive contemporary account of Babbage's "difference engine," the predecessor to the analytical engine. This was written by Dionysius Lardner and appeared in the July 1834 *Edinburgh Review*. Babbage's work on the analytical engine was publicized first by a military engineer, L. F. Menabrea, whose twenty-four-page article "Notions sur la machine analytique de M. Charles Babbage" appeared in the November 1842 edition of *Bibliothèque Universelle de Genève*. The Library has this journal as well as the article's subsequent translation into English the next year by Ada Lovelace. The Countess of Lovelace not only translated the article but, more importantly, made Menabrea's article understandable by her addition of an extensive set of notes documenting exactly how the machine would function. This article appeared in volume 3 of *Scientific Memoirs Selected from the Transactions of Foreign Academies of Science and Learned Societies*, edited by Richard and John Taylor. The Library has this 1843 publication in first edition.

Opposite page:
This great steam hammer was as tall as three men and was built by James Nasmyth to forge a paddle shaft for the huge steamship *Great Britain.* A steam cylinder at the top of its massive shoulders drove the hammer block downward in guides like a guillotine and was eventually made double-acting. In 1845, Nasmyth adapted the same principle for a steam-powered pile-driver. One of his favorite tricks when demonstrating how precisely he could regulate the speed and frequency of his hammer's strokes was to crack an eggshell placed in a wineglass without damaging the glass. *Cyclopaedia of Useful Arts and Manufactures,* 1854. Charles Tomlinson. LC-USZ62-110414.

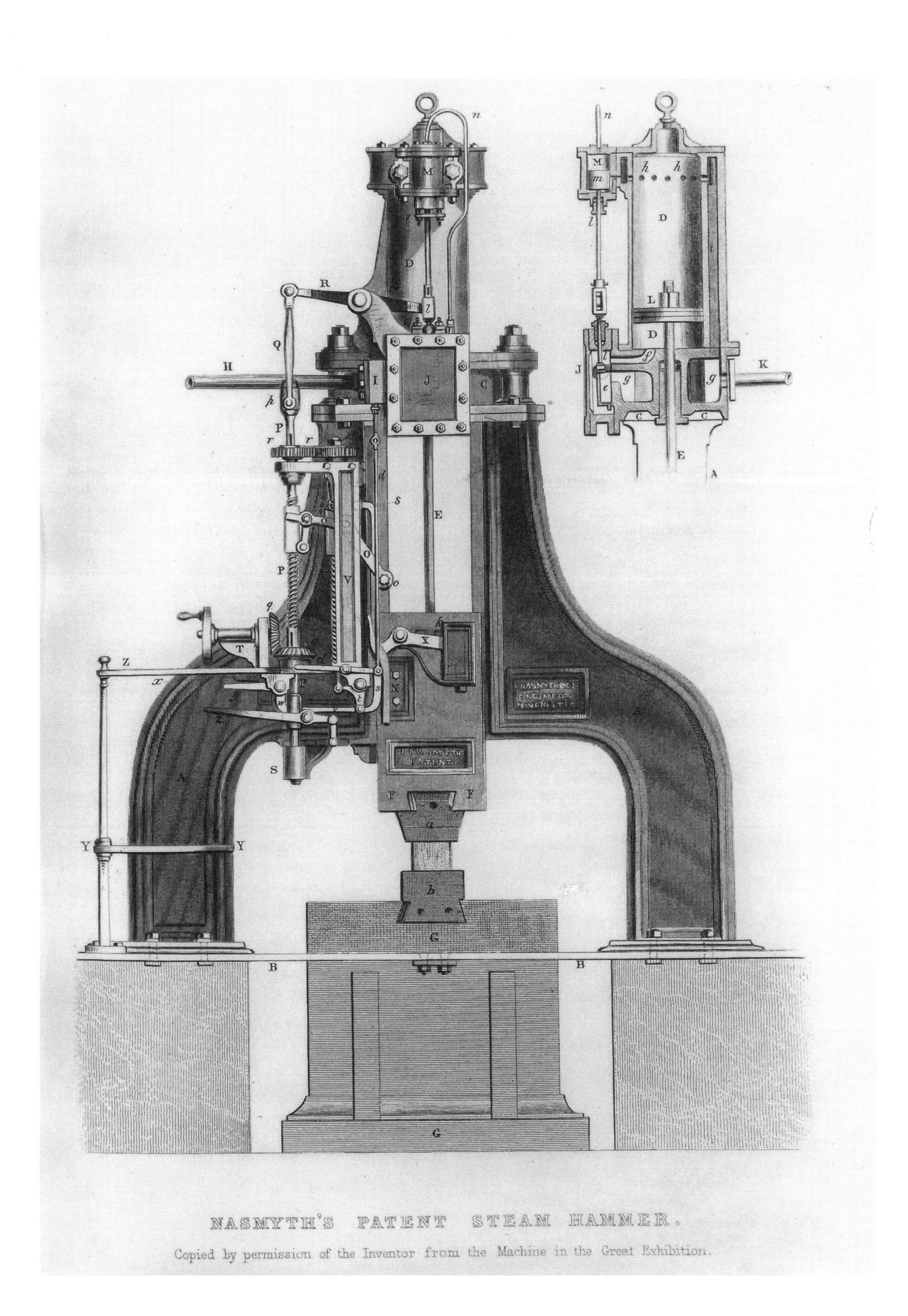

NASMYTH'S PATENT STEAM HAMMER.

Copied by permission of the Inventor from the Machine in the Great Exhibition.

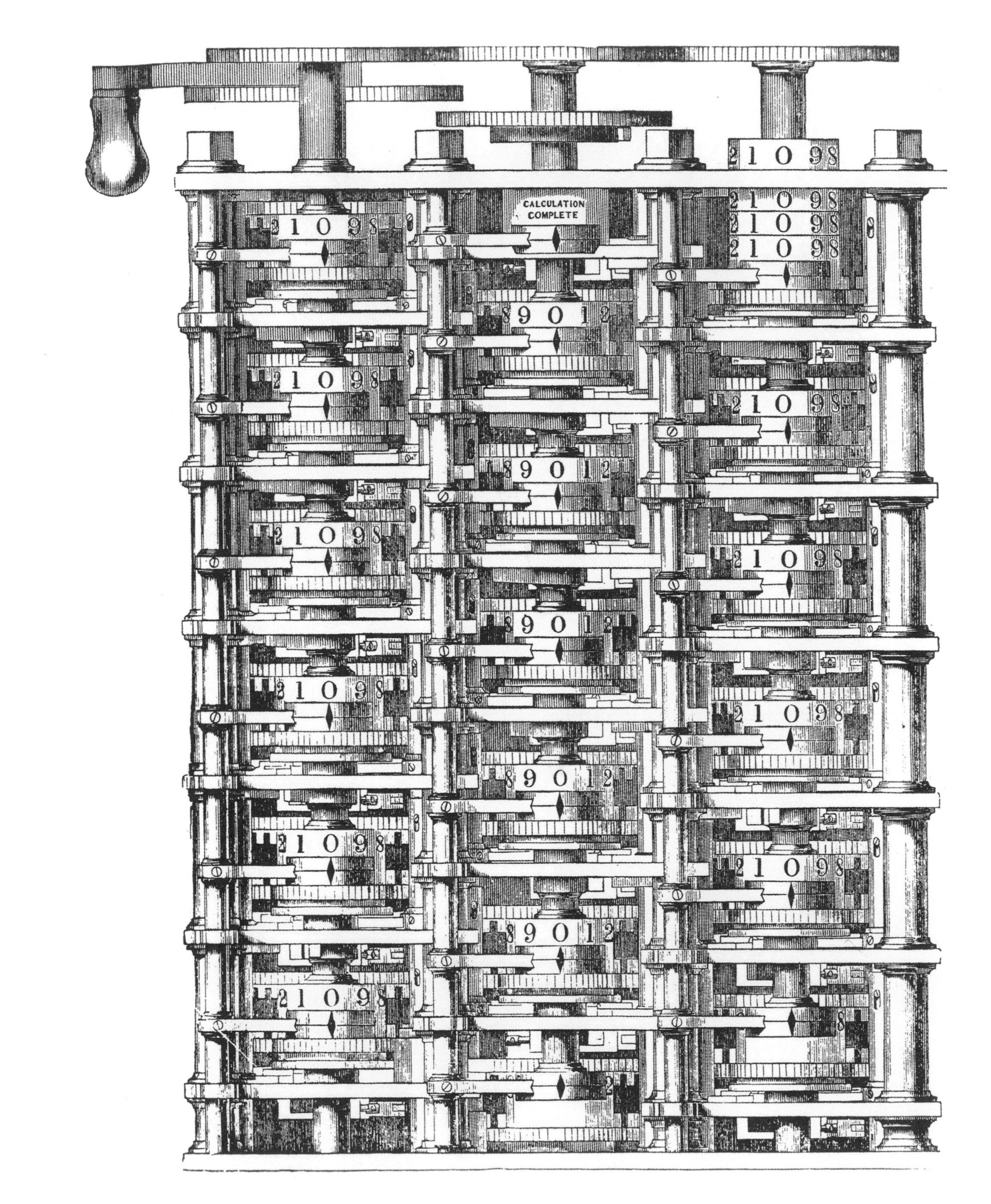

THE TRADITION OF TECHNOLOGY

If technology began to insinuate itself into society during the previous period, it became entrenched in the nineteenth century. The tradition of technology had evolved from one that existed only on the fringes of life and society to one that was beginning to dominate the culture which spawned it. Technology had by now ceased playing a secondary, supportive role and had seized center stage as a dominant actor. As a phenomenon that was almost impossible to avoid, it soon came to define the West as a material civilization based on technology.

It was also during the nineteenth century that the modern world really began. By the turn of the century, many of the tangible things we associate with modernity had already come to be, and it is not an exaggeration to say that we have merely built upon or continued trends initiated a century before.

The rapid development of so many major technological accomplishments made for a heady but also confusing time. Technology and its deep and broad effects were mostly responsible for the intriguing opposites so characteristic of the nineteenth century—angst and hope, innocence and cynicism, ennui and exhilaration. The nineteenth century fully embraced technology and learned that, once so committed, it could never turn away.

Opposite page:
In this woodcut of one small part of Charles Babbage's mechanical calculating machine, the modern term "number crunching" takes on a very real meaning. In his vision—which was far more ambitious than that of Pascal or Leibniz—Babbage saw a machine made of gears, levers, and wires that would not only perform mathematical computation but could store partial answers for later use as well. He also sought to print out the results. He spent most of his life and money (as well as a large amount of British government money) on this mostly failed goal. Although the eccentric, tactless, but brilliant Babbage thought out the principles that guide modern electronic computing, he was necessarily restricted to the limited technology of his time and suffered the fate of those born too early. *Babbage's Calculating Engines,* 1889. Henry Prevost Babbage. LC-USZ62-110415.

7. *Technology in the Twentieth Century*

Surely the most spectacular, if not defining, technological event in this century was the building, testing, detonation, and use of an atomic bomb in 1945. This mid-century event was—besides the first weapon of mass destruction—in many ways both very typical and radically different from any technology that had gone before it.

First, it was unlike any traditional power-generation technology in that it involved the complex and almost magical transformation of matter. This new physics that began with Einstein's profound discovery that mass or matter was not a fixed, inert thing but could be transformed into energy (since both were but different aspects of the same phenomenon) finally achieved the centuries-long studied goal of dreamers and alchemists. Mankind had come to learn one of nature's secrets—that matter itself evolves—and was able to produce an event similar to the transmutation of matter that powers the sun.

Second, no matter how unconventional or revolutionary this new technology was or would become, it still shared the essential nature of any tool or technology whose value is judged by how it is used. In fact, nuclear power could be said to embody in the fullest sense this equal tendency for good or ill inherent in all technologies, since atomic energy is intrinsically dangerous to life whether used for peaceful or aggressive purposes. Thus an overdose of X rays or accidental exposure in a nuclear power plant can be just as deadly to an individual as a nuclear bomb.

It was this first bomb, however, whose sheer power and destructive force forever shattered mankind's pose of technological innocence, that will mark this century as different from any that has gone before. This bomb was a deliberate construction, built frantically and almost desperately during a global conflict for the express purpose of doing exactly what it did—obliterating in one instant a major city of the enemy. The political and moral aspects of building and using such a weapon notwithstanding, the technological side itself tells a fascinating tale. From its theoretical beginnings implicit in Albert Einstein's $E = MC^2$, to the successful detonation of an experimental atomic bomb in a desert area near Alamogordo, New Mexico, on July 16, 1945, there is little doubt that never before or since has so much brainpower and money been devoted to solving a technological problem of immense complexity in such a short time. Only three years before, the secret wartime Manhattan Project had achieved the first self-sustaining nuclear reaction in a laboratory at the University of Chicago on December 2, 1942.

Opposite page:
The signature mushroom cloud of a nuclear explosion became an immediately familiar image for the twentieth century. This bomb was dropped on the Japanese city of Nagasaki on August 9, 1945, by the United States, destroying about half the city's area and bringing World War II to an end. The making and use of the bomb represents at once both a technological dream come true and a technological nightmare. On one hand, the successful conversion of mass into energy symbolized mankind's technological coming-of-age—scientists not only understood the forces that power the sun but were able to recreate them at will. This was surely near the ultimate in technological accomplishment. On the other hand, the use of this massive force for destructive purposes symbolized mankind's final loss of technological innocence. No longer could we believe or even try to convince ourselves that we could unquestioningly embrace any and all of the products of "progress." Seemingly every day since 1945, we have been reminded of this dual nature of things technological—that every gain has a cost, every advantage a risk, every benefit a loss. So it is that twentieth-century technology shares the condition of this modern age with its characteristic equivocation, ambivalence, and most of all, uncertainty. Prints and Photographs Division, Library of Congress, LC-USZ62-36452.

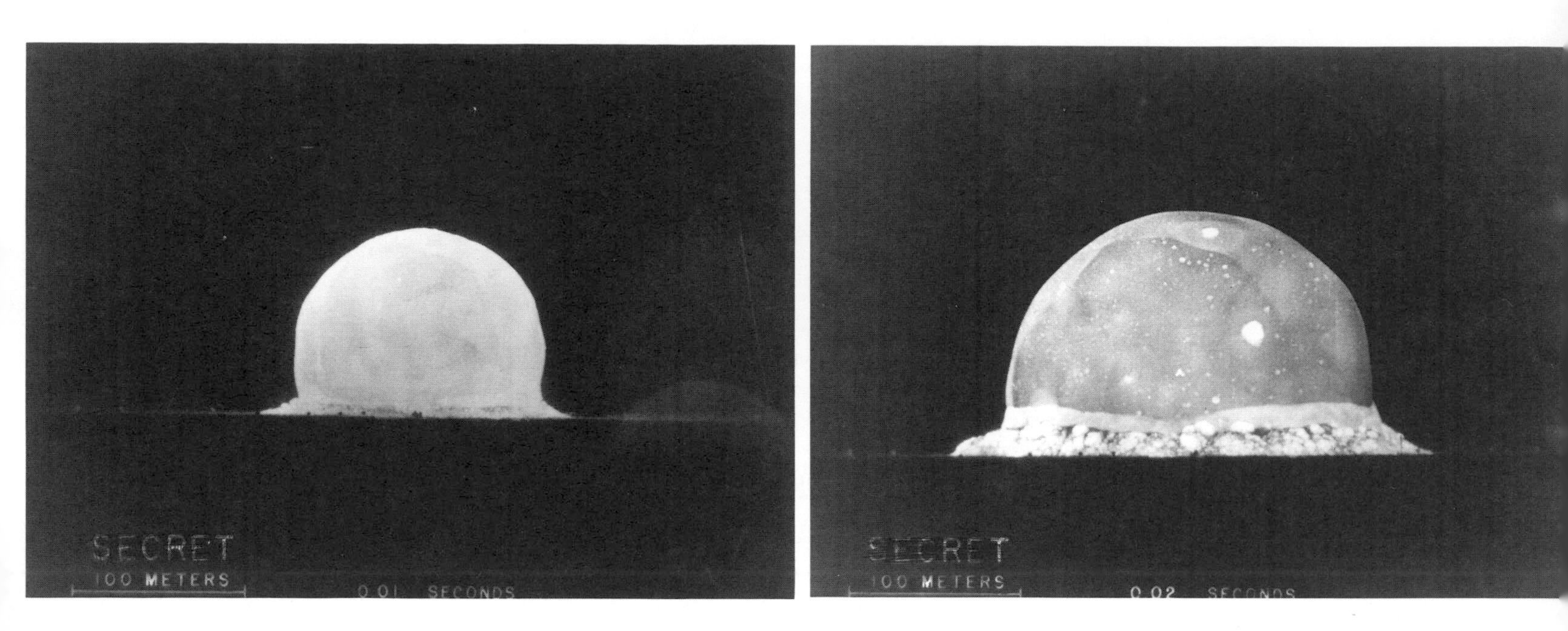

In these sequence photos of the first nuclear bomb explosion at the Trinity test site in New Mexico on July 16, 1945, the first image was recorded at one-one hundredth of a second and the sixth at one-tenth of a second after detonation. Even from six miles away, the sight of the blinding fireball awed all those who witnessed it. The physicist I. I. Rabi described it as the brightest light he had ever seen, one that, "blasted . . . pounced . . . [and] bored its way right through you. It was a vision which was seen with more than the eye." Almost every witness later expressed an essential ambivalence composed both of pride in their technical achievement and fear of its destructive implications. Harvard physicist Kenneth T. Bainbridge called it, "a foul and awesome display," and Los Alamos director J. Robert Oppenheimer later recounted that he had remembered a line from the Hindu scripture, *Bhagavad-Gita,* which said, "Now I am become Death, the Destroyer of worlds." The sensitive, philosophical Oppenheimer had learned Sanskrit to be able to read this dialogue called "The Lord's Song," and as the overworked, 116-pound director watched, he held tightly to a post to keep himself from falling down. The physicist Edward Teller wore dark glasses under his welder's shield, his face smeared with sunburn lotion, and gloves on his hands. Chemist George Kistiakowsky was knocked down by the shock wave, while physicist Enrico Fermi attempted an on-the-spot calculation of the bomb's

The scientists and engineers of the Manhattan Project were working at the very edges of knowledge about nuclear fission and did not fully grasp the totality of the force they would unleash in their desert test in the dawn of July 16, 1945. Theoretical estimates of the energy about to be released ranged from the equivalent of one thousand tons of TNT to the most optimistic guess of five thousand tons. The explosion that morning produced an energy, or yield, equivalent to twenty thousand tons of TNT.

The paper trail leading up to this spectacular, if terrifying, accomplishment begins in the unclassified literature. Following the discovery of artificial radioactivity in the early 1930s, an Italian physicist, Enrico Fermi, performed a series of experiments in which he bombarded uranium with neutrons. His objective was to form a new element, above uranium (element 92) on the periodic table. Although he believed that what he obtained was transuranic, it was later demonstrated by the German scientists Otto Hahn and Fritz Strassmann that this "new" element was actually barium (element 56).

The explanation for this startling phenomenon was taken up by the Austrian physicist Lise Meitner, who was spending the Christmas 1938 holiday with her nephew, Otto R. Frisch, also an Austrian physicist. Their laboratory work quickly revealed that when a neutron is sent into a uranium nucleus, violent internal motions occur that cause the uranium to split into two elements of more-or-less equal weight—for example, barium and krypton equal the atomic number of uranium on the periodic table. To Frisch, this splitting process was so similar to the division of a biological cell that he suggested the name "fission" for the new phenomenon. The final revelation in this new and surprising discovery was impossible to miss, for the by-product of this splitting was an enormous quantity of released energy, calculated by Meitner and Frisch at 200 million electron volts.

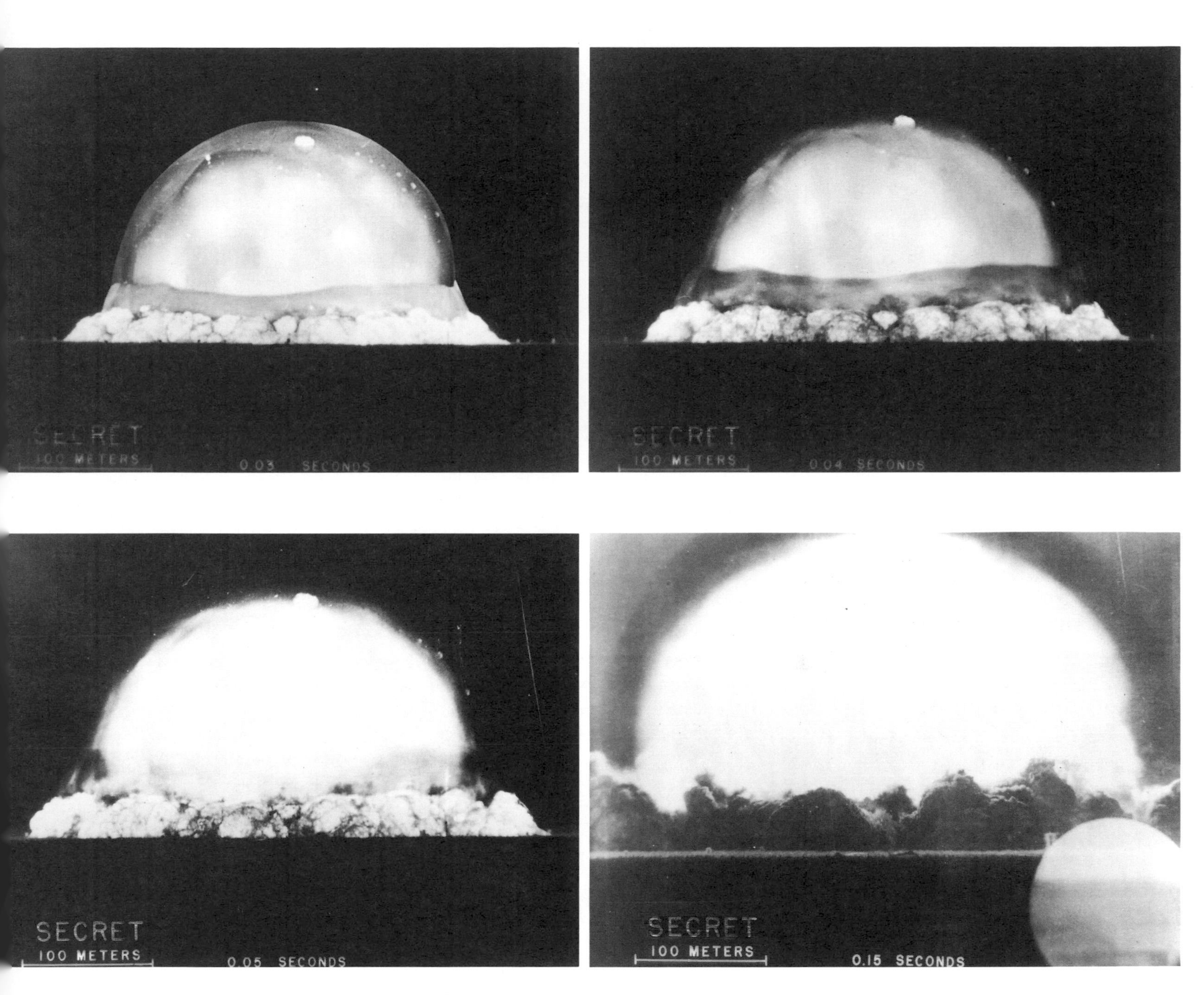

yield. He was able to estimate roughly the bomb's strength by dropping six small pieces of paper before, during, and after the passage of this shock wave. His estimate of ten thousand tons of TNT was short by at least half. Technically, the Trinity blast proved that the highly complex detonation method by implosion actually worked. It also demonstrated that a highly radioactive dust skirt encircled the blast (first seen in the second sequence photo). Since this had created a smothering effect actually weakening the bomb's power, Trinity planners realized that had the "gadget" been detonated at a much higher level, the radioactivity would not only have been much less but the destructive range of the bomb would have been far greater. The army followed this advice and exploded the Hiroshima bomb at a very high altitude, with the expected results. J. Robert Oppenheimer Papers, Manuscripts Division, Library of Congress.

Meitner and Frisch exploded their discovery on the world in two letters to the British journal *Nature*, published in February 1939. Much of the serious physics world already knew of their results, since they had communicated with the Danish physicist Neils Bohr (whose laboratory they had used), and he had already described their work to a meeting of the American Physical Society in Washington, D.C. The Library of Congress has the issue of *Nature* which contains these highly important letters to the editor. The first was written by Meitner and Frisch and was titled "Disintegration of Uranium by Neutrons: A New Type of Nuclear Reaction." The second letter was written by Frisch alone and is titled "Physical Evidence for the Division of Heavy Nuclei under Neutron Bombardment." The Library of Congress has a complete run of *Nature* from its beginnings in 1869. The important work of Hahn and Strassmann was first published in *Die Naturwissenschaften*. Their article, "Über den Nachweis und das Verhalten der bei der Bestrahlung des Urans mittels Neutronen entstehenden Erdalkalimetalle," appeared in the January 6, 1939, issue. The Library of Congress has this journal from its inception in 1913. The earlier work of Fermi can be found in the two-volume set of his *Collected Papers* published by the University of Chicago between 1962 and 1965. The Library of Congress has this set in its first edition. The first volume covers Fermi's work in Italy between 1921 and 1938, and the second volume, his work in the United States between 1939 and his death in 1954.

The theory of fission was developed very rapidly from this point, and once it was realized that besides the fission or division process, the release of free or secondary neutrons also occurred, the implication of a self-sustaining chain reaction was obvious. These secondary neutrons freed by the fission process could, in turn, further bombard and cause more fissions, which could cause yet more, and so on. The theoretical possibility of a self-perpetuating chain reaction was established in Paris at the College de France by Frederic Joliot-Curie and his collaborators, H. von Halban and L. Kowarski. Together they sent a letter to *Nature*, published on March 18, 1939, with the title "Liberation of Neutrons in the Nuclear Explosion of Uranium." Interestingly, Joliot-Curie had added his wife's name to his. Having married French physicist Irene Curie, daughter of Marie and Pierre, Joliot chose this way to perpetuate the Curie name and assure it would not be eclipsed by his. He and his wife won a Nobel Prize in 1935, as the Curies themselves had in 1903.

Finally, the possibility that a new sort of weapon of fantastic power might be feasible was offered by Niels Bohr and his former student, an American physicist, John A. Wheeler. In what is now recognized as the classic analysis of the fission phenomenon, they postulated that fission was more likely to occur in the light isotope of uranium (called uranium-235) than in regular or natural uranium (which is uranium-238), and clarified the special conditions under which a very fast chain reaction might produce a very large release of energy—in effect, a nuclear bomb. Bohr and Wheeler published their findings in a paper titled "The Mechanism of Nuclear Fission," which appeared in the September 1, 1939, is-

Opposite page:
It was in this short letter to the British journal *Nature* that the term *fission* first was used to describe the splitting apart of the uranium nucleus and the release of energy that accompanied it. Meitner and her nephew Frisch had developed a theory as well as a name for Hahn's discovery, and their explanation of fission would prove to be the critical science needed to drive a new technology. For in theory, the energy or few neutrons released by the first fission could cause other fissions which could cause still more, forming a chain reaction. This would prove the essential link in the development of the atomic bomb. *Nature,* February 11, 1939 (vol. 143, no. 3615, pp. 239–40). LC-USZ62-110422, LC-USZ62-110423.

Letters to the Editor

The Editor does not hold himself responsible for opinions expressed by his correspondents. He cannot undertake to return, or to correspond with the writers of, rejected manuscripts intended for this or any other part of NATURE. *No notice is taken of anonymous communications.*

NOTES ON POINTS IN SOME OF THIS WEEK'S LETTERS APPEAR ON P. 247.

CORRESPONDENTS ARE INVITED TO ATTACH SIMILAR SUMMARIES TO THEIR COMMUNICATIONS.

Disintegration of Uranium by Neutrons: a New Type of Nuclear Reaction

ON bombarding uranium with neutrons, Fermi and collaborators[1] found that at least four radioactive substances were produced, to two of which atomic numbers larger than 92 were ascribed. Further investigations[2] demonstrated the existence of at least nine radioactive periods, six of which were assigned to elements beyond uranium, and nuclear isomerism had to be assumed in order to account for their chemical behaviour together with their genetic relations.

In making chemical assignments, it was always assumed that these radioactive bodies had atomic numbers near that of the element bombarded, since only particles with one or two charges were known to be emitted from nuclei. A body, for example, with similar properties to those of osmium was assumed to be eka-osmium ($Z = 94$) rather than osmium ($Z = 76$) or ruthenium ($Z = 44$).

Following up an observation of Curie and Savitch[3], Hahn and Strassmann[4] found that a group of at least three radioactive bodies, formed from uranium under neutron bombardment, were chemically similar to barium and, therefore, presumably isotopic with radium. Further investigation[5], however, showed that it was impossible to separate these bodies from barium (although mesothorium, an isotope of radium, was readily separated in the same experiment), so that Hahn and Strassmann were forced to conclude that *isotopes of barium* ($Z = 56$) *are formed as a consequence of the bombardment of uranium* ($Z = 92$) *with neutrons.*

At first sight, this result seems very hard to understand. The formation of elements much below uranium has been considered before, but was always rejected for physical reasons, so long as the chemical evidence was not entirely clear cut. The emission, within a short time, of a large number of charged particles may be regarded as excluded by the small penetrability of the 'Coulomb barrier', indicated by Gamov's theory of alpha decay.

On the basis, however, of present ideas about the behaviour of heavy nuclei[6], an entirely different and essentially classical picture of these new disintegration processes suggests itself. On account of their close packing and strong energy exchange, the particles in a heavy nucleus would be expected to move in a collective way which has some resemblance to the movement of a liquid drop. If the movement is made sufficiently violent by adding energy, such a drop may divide itself into two smaller drops.

In the discussion of the energies involved in the deformation of nuclei, the concept of surface tension of nuclear matter has been used[7] and its value has been estimated from simple considerations regarding nuclear forces. It must be remembered, however, that the surface tension of a charged droplet is diminished by its charge, and a rough estimate shows that the surface tension of nuclei, decreasing with increasing nuclear charge, may become zero for atomic numbers of the order of 100.

It seems therefore possible that the uranium nucleus has only small stability of form, and may, after neutron capture, divide itself into two nuclei of roughly equal size (the precise ratio of sizes depending on finer structural features and perhaps partly on chance). These two nuclei will repel each other and should gain a total kinetic energy of *c.* 200 Mev., as calculated from nuclear radius and charge. This amount of energy may actually be expected to be available from the difference in packing fraction between uranium and the elements in the middle of the periodic system. The whole 'fission' process can thus be described in an essentially classical way, without having to consider quantum-mechanical 'tunnel effects', which would actually be extremely small, on account of the large masses involved.

After division, the high neutron/proton ratio of uranium will tend to readjust itself by beta decay to the lower value suitable for lighter elements. Probably each part will thus give rise to a chain of disintegrations. If one of the parts is an isotope of barium[5], the other will be krypton ($Z = 92 - 56$), which might decay through rubidium, strontium and yttrium to zirconium. Perhaps one or two of the supposed barium-lanthanum-cerium chains are then actually strontium-yttrium-zirconium chains.

It is possible[5], and seems to us rather probable, that the periods which have been ascribed to elements beyond uranium are also due to light elements. From the chemical evidence, the two short periods (10 sec. and 40 sec.) so far ascribed to ^{239}U might be masurium isotopes ($Z = 43$) decaying through ruthenium, rhodium, palladium and silver into cadmium.

In all these cases it might not be necessary to assume nuclear isomerism; but the different radioactive periods belonging to the same chemical element may then be attributed to different isotopes of this element, since varying proportions of neutrons may be given to the two parts of the uranium nucleus.

By bombarding thorium with neutrons, activities are obtained which have been ascribed to radium and actinium isotopes[8]. Some of these periods are approximately equal to periods of barium and lanthanum isotopes[5] resulting from the bombardment of uranium. We should therefore like to suggest that these periods are due to a 'fission' of thorium which is like that of uranium and results partly in the same products. Of course, it would be especially interesting if one could obtain one of these products from a light element, for example, by means of neutron capture.

It might be mentioned that the body with half-life 24 min.[2] which was chemically identified with uranium is probably really ^{239}U, and goes over into an eka-rhenium which appears inactive but may decay slowly, probably with emission of alpha particles. (From inspection of the natural radioactive elements, ^{239}U cannot be expected to give more than one or two beta decays; the long chain of observed decays has always puzzled us.) The formation of this body is a typical resonance process[9]; the compound state must have a life-time a million times longer than the time it would take the nucleus to divide itself. Perhaps this state corresponds to some highly symmetrical type of motion of nuclear matter which does not favour 'fission' of the nucleus.

LISE MEITNER.

Physical Institute,
Academy of Sciences,
Stockholm.

O. R. FRISCH.

Institute of Theoretical Physics,
University,
Copenhagen.

Jan. 16.

[1] Fermi, E., Amaldi, F., d'Agostino, O., Rasetti, F., and Segrè, E. *Proc. Roy. Soc.*, A, **146**, 483 (1934).

[2] See Meitner, L., Hahn, O., and Strassmann, F., *Z. Phys.*, **106**, 249 (1937).

[3] Curie, I., and Savitch, P., *C.R.*, **206**, 906, 1643 (1938).

[4] Hahn, O., and Strassmann, F., *Naturwiss.*, **26**, 756 (1938).

[5] Hahn, O., and Strassmann, F., *Naturwiss.*, **27**, 11 (1939).

[6] Bohr, N., NATURE, **137**, 344, 351 (1936).

[7] Bohr, N., and Kalckar, F., *Kgl. Danske Vid. Selskab, Math. Phys. Medd.*, **14**, Nr. 10 (1937).

[8] See Meitner, L., Strassmann, F., and Hahn, O., *Z. Phys.*, **109**, 538 (1938).

[9] Bethe, A. H., and Placzek, G., *Phys. Rev.*, **51**, 450 (1937).

A Novel Thermostat

IT is often necessary to maintain an apparatus at a constant temperature. This may be done by immersing it in a circulating liquid maintained at a constant temperature by a thermostat, or by jacketing

TEMPERATURE-CONTROLLED APPARATUS.

it with alternate shells of thermally conducting and insulating materials heated to the selected temperature by means of an internal electric heater. These methods have the disadvantages that the thermostatic system makes the apparatus less accessible, the

In 1939, the gentle Danish physicist and Nobel Prize-winner Niels Bohr came to the United States and collaborated immediately at Princeton with the young American theoretician John A. Wheeler to produce this article. In their theory of the mechanism of nuclear fission—in which they said the nucleus behaves somewhat like a drop of fluid—they postulated correctly that it was the uranium-235 isotope that underwent fission. Since normal uranium contains approximately 0.7 percent of uranium-235, with most of the remainder being uranium-238, the next major technological hurdle to overcome would be how to separate uranium-235 from the other uranium isotopes or to use isotope 238 to produce plutonium. Three different techniques were used—gaseous diffusion, thermal diffusion, and electromagnetic separation—and all were to some degree successful. With the outbreak of World War II occurring two days after this article was published, a fission bomb was certainly on everyone's mind. Shown here is the first of twenty-five pages. *Physical Review,* September 1, 1939 (vol. 56, no. 5, pp. 426–50). LC-USZ62-110424.

Opposite page:
In this memorandum written presumably to Los Alamos director J. Robert Oppenheimer, physicist Edward Teller sets out some of the operating ground rules for "Site Y." This was the code name for the secret bomb-making facility to be built in Los Alamos, New Mexico—a location selected only a month before. The S-1 Project that Teller refers to is the overall bomb project itself, which was administered by Vannevar Bush's Office of Scientific Research and Development (OSRD). Teller's double reference to "the final apparatus" to be built at Site Y was his euphemism for the fission bomb that later became known to Los Alamos workers as "the gadget." Teller's emphasis on secrecy and isolation at Site Y reflected the criteria established by Manhattan Project head Gen. Leslie R. Groves, who had selected the site. This is the first page of Teller's lengthy memorandum. J. Robert Oppenheimer Papers, Manuscript Division, Library of Congress.

SEPTEMBER 1, 1939 PHYSICAL REVIEW VOLUME 56

The Mechanism of Nuclear Fission

NIELS BOHR
University of Copenhagen, Copenhagen, Denmark, and The Institute for Advanced Study, Princeton, New Jersey

AND

JOHN ARCHIBALD WHEELER
Princeton University, Princeton, New Jersey

(Received June 28, 1939)

On the basis of the liquid drop model of atomic nuclei, an account is given of the mechanism of nuclear fission. In particular, conclusions are drawn regarding the variation from nucleus to nucleus of the critical energy required for fission, and regarding the dependence of fission cross section for a given nucleus on energy of the exciting agency. A detailed discussion of the observations is presented on the basis of the theoretical considerations. Theory and experiment fit together in a reasonable way to give a satisfactory picture of nuclear fission.

INTRODUCTION

THE discovery by Fermi and his collaborators that neutrons can be captured by heavy nuclei to form new radioactive isotopes led especially in the case of uranium to the interesting finding of nuclei of higher mass and charge number than hitherto known. The pursuit of these investigations, particularly through the work of Meitner, Hahn, and Strassmann as well as Curie and Savitch, brought to light a number of unsuspected and startling results and finally led Hahn and Strassmann[1] to the discovery that from uranium elements of much smaller atomic weight and charge are also formed.

The new type of nuclear reaction thus discovered was given the name "fission" by Meitner and Frisch,[2] who on the basis of the liquid drop model of nuclei emphasized the analogy of the process concerned with the division of a fluid sphere into two smaller droplets as the result of a deformation caused by an external disturbance. In this connection they also drew attention to the fact that just for the heaviest nuclei the mutual repulsion of the electrical charges will to a large extent annul the effect of the short range nuclear forces, analogous to that of surface tension, in opposing a change of shape of the nucleus. To produce a critical deformation will therefore require only a comparatively small energy, and by the subsequent division of the nucleus a very large amount of energy

Just the enormous energy release in the fission process has, as is well known, made it possible to observe these processes directly, partly by the great ionizing power of the nuclear fragments, first observed by Frisch[3] and shortly afterwards independently by a number of others, partly by the penetrating power of these fragments which allows in the most efficient way the separation from the uranium of the new nuclei formed by the fission.[4] These products are above all characterized by their specific beta-ray activities which allow their chemical and spectrographic identification. In addition, however, it has been found that the fission process is accompanied by an emission of neutrons, some of which seem to be directly associated with the fission, others associated with the subsequent beta-ray transformations of the nuclear fragments.

In accordance with the general picture of nuclear reactions developed in the course of the last few years, we must assume that any nuclear transformation initiated by collisions or irradiation takes place in two steps, of which the first is the formation of a highly excited compound nucleus with a comparatively long lifetime, while

[1] O. Hahn and F.

[3] O. R. Frisch, Nature **143**, 276 (1939); G. K. Green and Luis W. Alvarez, Phys. Rev. **55**, 417 (1939); R. D. Fowler and R. W. Dodson, Phys. Rev. **55**, 418 (1939); R. B. Roberts, R. C. Meyer and L. R. Hafstad, Phys. Rev. **55**, 417 (1939); W. Jentschke and F. Prankl, Naturwiss. **27**, 134 (1939); H. L. Anderson, E. T. Booth, J. R. Dunning, E. Fermi, G. N. Glasoe and F. G. Slack, Phys. Rev. **55**, 511 (1939).

sue of *Physical Review*. This was two days before war broke out in Europe.

Until now, these spectacular advances in nuclear physics were available for all to read, but the coming war would soon change the prevailing spirit of openness and internationalism into one of competition and secretiveness. In fact, even before Germany invaded Poland and England declared war, two Hungarian physicists living in the United States became fearful of the military implications of nuclear fission and took action. In the summer of 1939, Leo Szilard and Eugene P. Wigner drafted a letter to President Franklin D. Roosevelt and persuaded Albert Einstein to sign it. An intermediary took the letter, which warned that nuclear weapons

142264

This document contains information affecting the national defense of the United States within the meaning of the Espionage Act, U. S. C. 50; 31 and 32. Its transmission or the revelation of its contents in any manner to an unauthorized person is prohibited by law.

CANCELLED

E. Teller
December 31, 1942

C380 SECRET CANCELLED

SUMMARY OF PLANS FOR SITE Y

Proposed Work at Site.

The purpose of the work at Site Y is to find out the best possible uses of the materials produced by the different branches of the S-1 Project. Some of this work will have to be postponed until sufficient quantities of the materials will be available. But two branches of research can be started immediately. First, experimental investigation of fast neutron reactions and nuclear experiments on auxiliary materials. Second, considerations on the theory of the final apparatus which is going to be built. Some progress along both of these lines has been already made at various places. The plan is to centralize all this work at Site Y. The magnitude of the undertaking makes it necessary to expand considerably the personnel working on these questions. In later stages of the work experiments on the products received from the other project will have to be undertaken. Most of the construction work on the final apparatus will have to be done at Site Y.

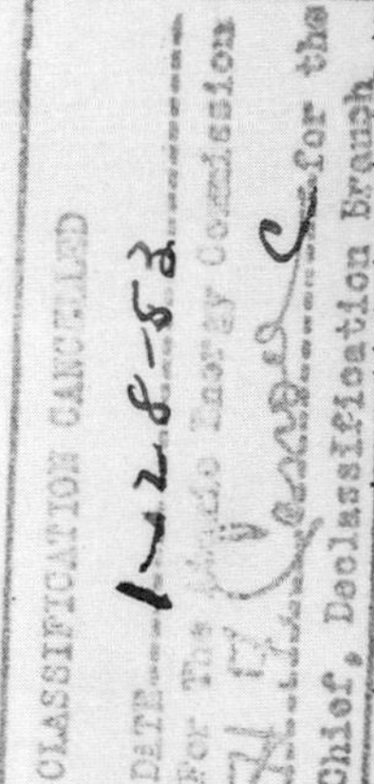

Personnel.

It is planned at present to have at Y a little less than 100 scientists. In addition there will have to be some highly trained personnel for work in the shop, and a relatively small number of persons whose work will be to facilitate living conditions at Site Y. It is planned that families will come along. All together one may expect that there will be approximately 300 people at Site Y. Generally it is planned that secretarial jobs and other jobs making living conditions better at Site Y, will be open to the wives of the scientists.

Military Restrictions.

In the later stages of the work at Site Y, secrecy requirements will be higher than they have been heretofore. This makes more strict military control necessary. Site Y is going to be a restricted area. The commandant in whom will rest the military authority will be a Colonel. In the restricted area he will have authority over civilians as well as over any member of the Armed Forces in the area. The military restrictions will include restrictions on travel and censorship of communications. However, the travel restrictions will not operate at the beginning, and at least until September 1, 1943 people who have joined the project will be at liberty to quit the work if they feel that they have weighty reasons to do so. In the later stages of the work it will not be possible to release anybody from his duties.

Even after full travel restrictions have been introduced, travelling will be permitted for three main reasons:

1. Scientific contacts.
2. Cases where the health of a person is definitely in danger.
3. Cases where family circumstances make travel absolutely necessary.

could be developed by Nazi Germany, to the president. As a result, in October the president appointed an Advisory Committee on Uranium, and the groundwork for what would become the Manhattan Project was set in place. The United States had decided to try to build the first nuclear bomb.

From this point on, the record becomes less public, and by June 1942 the U.S. War Department had reorganized the project and given it the secret code name of Manhattan Engineer District Office. By then, an Office of Scientific Research and Development had been established in the United States under the leadership of an American scientist, Vannevar Bush, and refugee scientists Otto R. Frisch and Rudolf E. Peierls, living in England, had written their now-famous memorandum for top British officials. This memorandum was published for the first time in 1964 in Margaret M. Gowing's history, *Britain and Atomic Energy: 1939–1945*; the Library has a copy of her book. In three large-format pages, titled "On the Construction of a 'Super Bomb'; Based on a Nuclear Chain Reaction in Uranium," the two men not only declared such a bomb to be feasible, but they detailed its size, suggested how it might be detonated, and recommended what safeguards should be followed. This led to the beginnings of a British program to develop a bomb and to an American proposal for a joint effort. With the British initially cool to this idea, the United States pushed ahead on its own and ultimately dominated the project, code-named the Manhattan Project.

Six days after approximately two-thirds of Hiroshima was destroyed from the air, and three days after the second atomic bomb was dropped on the Japanese city of Nagasaki (Kokura was the original target, and the B-29 bomber spent forty-five minutes over that city trying to locate its target before moving to Nagasaki), one of the more remarkable official publications issued by any government was released. Nicknamed the "Smyth Report," this candid and relatively full recounting tells the administrative history of the research and development work carried out by the group of international scientists under the Manhattan Project. This report, titled *Atomic Energy for Military Purposes*, was written in a context of—if not full disclosure—remarkable openness and candor. Its author was an experimental physicist from Princeton University who worked on the project, Henry DeWolf Smyth, and he wrote not for experts but for a wider, lay audience. Until much later histories were compiled, Smyth's government-sponsored effort was the only technical history available concerning the development of the first nuclear weapon. The Library of Congress has this work in several versions. The earliest in its collections is the 182-page, brown-paper-covered version issued by the government as *A General Account of the Development of Methods of Using Atomic Energy for Military Purposes under the Auspices of the United States Government: 1940–1945*. In August 1945 this sold for one dollar. The Library also has the trade edition published the same year by Princeton University Press, which contains the text of the U.S. War Department news release on the Alamogordo tests as well as an index and photographs. Part of this long news release quotes from General T. F. Farrell, a witness to the first explosion.

Opposite page:
During the year preceding the Trinity blast, Princeton physicist Henry DeWolf Smyth had been preparing a technical history of the bomb's development. Only days after Nagasaki, the U.S. government released his report. Aside from responding in a way to the urging of some of the atomic scientists for more openness now that the war was over, the report served to underscore exactly what atomic information might be made public and what kept secret. In this diagram from the Princeton version of this government publication, the chain-reaction principle behind the bomb is illustrated. *Atomic Energy for Military Purposes: The Official Report on the Development of the Atomic Bomb under the Auspices of the United States Government, 1940–1945*, 1948. Henry DeWolf Smyth. LC-USZ62-110426.

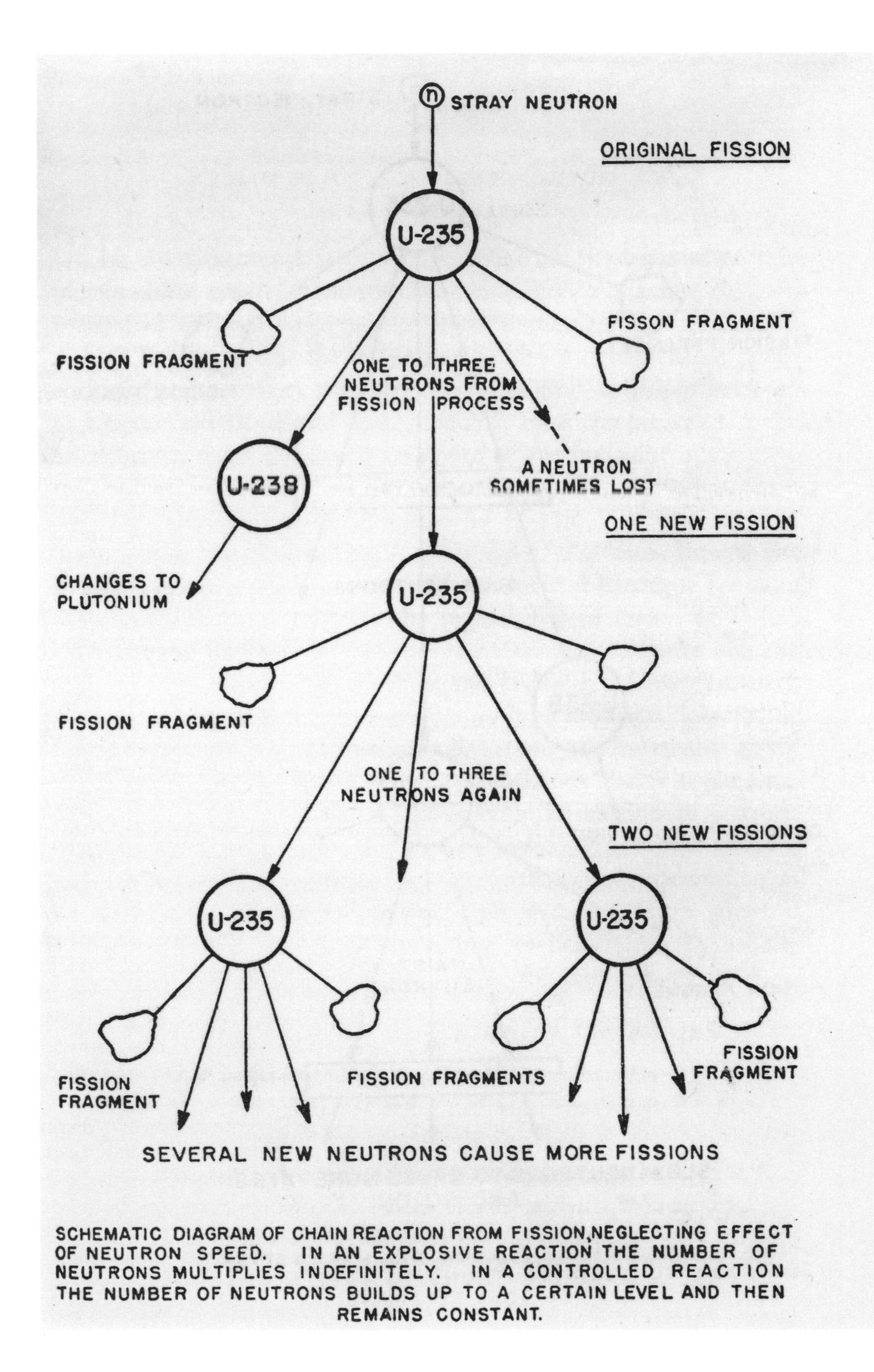

SCHEMATIC DIAGRAM OF CHAIN REACTION FROM FISSION, NEGLECTING EFFECT OF NEUTRON SPEED. IN AN EXPLOSIVE REACTION THE NUMBER OF NEUTRONS MULTIPLIES INDEFINITELY. IN A CONTROLLED REACTION THE NUMBER OF NEUTRONS BUILDS UP TO A CERTAIN LEVEL AND THEN REMAINS CONSTANT.

General Farrell wrote, "The lighting effects beggared description. The whole country was lighted by a searing light with the intensity many times that of the midday sun. It was golden, purple, violet, gray and blue." He concludes with a description of the air blast that followed the light which, he said, "warned of doomsday and made us feel that we puny things were blasphemous to dare tamper with the forces heretofore reserved for the Almighty. Words are inadequate tools for the job of acquainting those not present with the physical, mental, and psychological effects. It had to be witnessed to be realized."

Nowhere in either version is the name Hiroshima or Nagasaki found. The horrific story of the use of nuclear weapons on those cities is told elsewhere. Rather, the Smyth Report concentrates on the technological development alone and culminates with the successful execution of the Trinity Test in the New Mexico desert.

Two collections of papers in the Library of Congress concern America's crash program to develop an atomic bomb. The papers of Vannevar Bush reflect his involvement with the Office of Scientific Research and Development (OSRD). It was Bush's report that resulted in the establishment of the Manhattan Project. The papers of J. Robert Oppenheimer, the director of the Manhattan Project, make up a collection equal in size to Bush's. This significant collection documents not only Oppenheimer's involvement with the development of the bomb, but also his later concern with the political aspects of the control of nuclear energy. A related and very large collection of papers are those of Merle A. Tuve. Tuve was both part of the OSRD and the later creator and director of the Johns Hopkins University Applied Physics Laboratory.

Possibly the only rival to the nuclear bomb as the defining technology of the twentieth century might be television. It could be argued that the bomb itself was an anomoly—a one-time technical spectacular that, in terms of its everyday effect on the lives of most people, might just as well never have happened. Although some in the nuclear energy business might disagree with such a cavalier dismissal, it is difficult to make a strong case that nuclear energy has established itself deeply into the technical fabric of society, despite its use as a safe source of energy. Finally, recent political changes, such as the dissolution of the Soviet Union, are promising to remove even the ultimate military justification for nuclear weapons.

As limited a technology in terms of its usefulness as nuclear energy has been, television has been the reverse. Almost from the beginning, this technology became ubiquitous, seeming to have the seeds for its dispersal within itself. Like a constantly dividing organism, it needed only a host—a viewer or audience—to multiply. Where the atom was complex and terrifying, the television image was understandable and friendly. The former did not seek to communicate but in fact tended to intimidate. The latter not only engaged both eye and ear, it entertained and persuaded, its essence being communication. Where the atom has always been at the far periphery of most people's lives, both threatening and promising, television has become totally embedded in what we are. It is no longer

just a large part of our lives—it has become one with modern life. Late-twentieth-century life without the atom (without its applications, not without its science) seems plausible and even reasonable. Modern life without the communication offered by television is unthinkable.

Like its predecessor technologies, the electric telegraph and the radio, television was a direct medium allowing for immediate communication. From its earliest beginnings, it was always thought of as a medium of instantaneous communication, unlike film which was not a direct medium but was played back at a later time. As early as 1880, the idea of seeing by telegraph was a notion that intrigued many inventors around the world. Representative articles published that year are found in the collections of the Library of Congress. Two such efforts were "Seeing by Electricity," by W. E. Ayrton and J. J. Perry, published in *Nature* (April 22, 1880), and George R. Carey's article with the same title which appeared in *Scientific American* (June 5, 1880).

At that time, however, the level of technological development had not advanced beyond what is best described as the photomechanical (not elec-

This photograph was released to the press by the Jenkins Television Corporation probably at the beginning of 1927. The text accompanying the photo went to great lengths trying to describe to an unknowing public exactly what television was. It described this event as, "the first public demonstration of synchronized sight and sound (talking moving pictures) being transmitted over radio without the use of land wires." The transmissions originated from both Jersey City and Passaic, New Jersey, and were received at this piano studio in Newark. Prints and Photographs Division, Library of Congress, LC-USZ62-90399.

One of the early schemes for television, or "seeing by electricity," was proposed by Bostonian George R. Carey in 1879. In this *Scientific American* article published a year later, Carey's mechanical device takes advantage of two recent discoveries: the 1873 discovery of the light sensitivity of selenium and Bell's 1876 invention of the telephone. Though never actually built, this device laid the principles for practical mechanical television, which was the predecessor of electronic television. *Scientific American,* June 5, 1880. LC-USZ62-110427.

JUNE 5, 1880.]

SEEING BY ELECTRICITY.

The art of transmitting images by means of electric currents is now in about the same state of advancement that the art of transmitting speech by telephone had attained in 1876, and it remains to be seen whether it will develop as rapidly and successfully as the art of telephony. Professor Bell's announcement that he had filed at the Franklin Institute a sealed description of a method of "seeing by telegraph" brings to mind an invention for a similar purpose, submitted

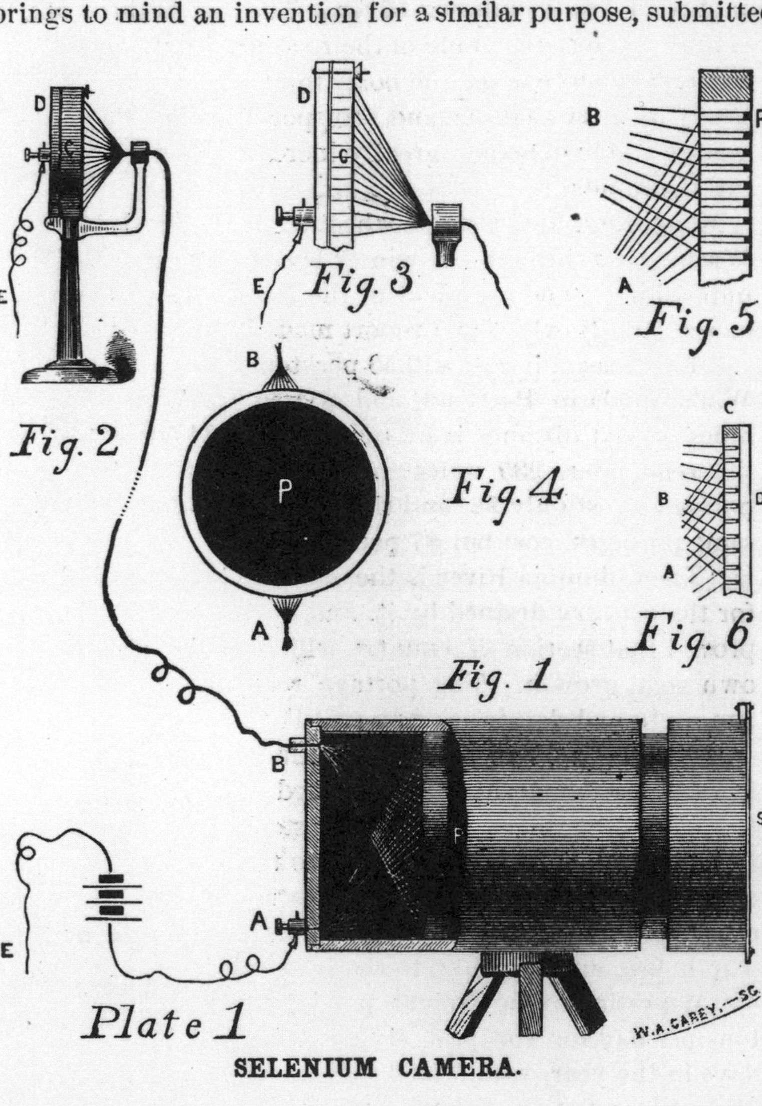

SELENIUM CAMERA.

to us some months since by the inventor, Mr. Geo. R. Carey, of the Surveyor's Office, City Hall, Boston, Mass. By consent of Mr. Carey we present herewith engravings and descriptions of his wonderful instruments.

the electrica
stead of prin
are arrange
Plate P (Fig
connected w
P (Fig. 1), a
The opera
letter, A, up
(Fig. 1), all
letter, A, is,
in the light
wires leadin
relative posi
electricity w
tween C and
upon disk, I
and so trans
that by whi
Figs. 1 an
and recordi
ject that ma
1), by the ca
depends up
duced by th
The clock
and wheel,
N, being fa
volves on i
selenium po
Fig. 3. Th
B, to descri
over every
The selen
current to f
the lights a
plate, T T.
The elect
to the selen
sliding piec
ing screw,
screw, D (
to point, E
paper place

tronic) transmission of an image. It was not until 1897, when a German, K. Ferdinand Braun, perfected the cathode-ray tube, that the technical basis for modern, electronic television was established. Braun produced a cathode-ray tube with a fluorescent screen, that is, a screen that produced visible light when struck by a beam of electrons. The Library of Congress has Braun's important eight-page article as it was first published in the *Annalen der Physik* in February 1897. Everyone considered Braun's cathode-ray tube as a receiver only, and it was not until 1908 that a brilliant suggestion was made by the Scottish engineer A. A. Campbell Swinton.

Responding to an earlier letter in *Nature* which essentially told of how impossible "telegraphic photography and electric vision" was, Campbell Swinton wrote a 275-word response outlining a method that is essentially the basis of modern television. Basically, he made the startling suggestion that cathode-ray tubes be used at both ends of the system—as both transmitter and receiver—making for an all-electric system. It was not until 1911 that Campbell Swinton was able to provide detailed plans on how such a system would work, but even then his ideas and plans were far beyond the available technology. The Library of Congress has both of Campbell Swinton's papers—his landmark letter titled "Distant Electric Vision," as it first appeared in *Nature* (June 18, 1908), and his address in 1911 to the Roentgen Society of London, which appeared in that society's *Journal* in January 1912. This journal became the *British Journal of Radiology* in 1923.

Despite the fact that the basis had been laid for electronic transmission, the first public demonstration of television was accomplished by means of the older, photomechanical method. In 1926, Englishman John L. Baird succeeded in electrically transmitting moving pictures in halftones. These badly flickering, crude images were only a few inches high, and although they were the result of a dead-end and outmoded system, they did much to stimulate further research. Baird had already done the same as early as 1924 and made his work known when his article "An Account of Some Experiments in Television" appeared in the

The first practical successes in image transmission over a distance were obtained using optical-mechanical means. During the 1920s, it was with mechanical systems like that of Englishman John L. Baird that television progressed. Baird devised a photomechanical system in which the picture was scanned by a quickly rotating disk that contained a series of holes arranged in a spiral manner. The light signal from each hole was turned into an electric signal and sent to the receiver via a radio pulse. Baird made a trans-Atlantic broadcast in 1928 and became very successful and famous. His success was short-lived, however, since his photomechanical system would prove to be an evolutionary dead end. *The Wireless World and Radio Review,* May 7, 1924. John L. Baird. LC-USZ62-110428.

May 7, 1924, edition of *Wireless World.* The Library of Congress has this journal in its entirety. Baird continued to make improvements and had some short-lived success, but his mechanical system was still crude and lacked any semblance of sensitivity or definition.

By this time, however, it was apparent to some that fortunes could be made if a high quality electrical system could be perfected. Thus the large companies, with their sizable laboratories, jumped in. Vladimir K. Zworykin, a Russian electrical engineer, had already joined the Pittsburgh-based Westinghouse Electric and Manufacturing Company to help them catch up with the newly formed Radio Corporation of America (RCA). Zworykin eventually worked for RCA as well, but while he was at Westinghouse he filed his first patent for the iconoscope camera tube. This was to become the first effective television camera and would eventually prove the soundness of Campbell Swinton's theoretical ideas.

At about the same time, an independent American inventor, Philo T. Farnsworth, developed a television system based on what he called an "image dissector tube." This was a novel system in that it was not based on the ideas of Campbell Swinton. Farnsworth's camera worked by transforming the picture into photoelectrons which were moved by magnetic coils in front of a fixed electrical probe that did the scanning. In Zworykin's system, the picture to be transmitted was shone into a disc of mica covered with pinpoint islands of photoelectric material. A narrow electron beam scanned this mosaic. While Farnsworth's backers lacked the financial resources of Zworykin (RCA), Farnsworth's patent was a legitimate rival to Zworykin's, and the two companies eventually reached a cross-licensing agreement.

Farnsworth wrote two popular articles documenting and explaining his unique system. The first, "Transmission of Television Images," was written with Harry R. Lubke and appeared in the journal *Radio* in December 1929. This journal eventually came to be called *Audio.* His other article appeared in *Radio Industries* (November 1930) and was titled "An Electrical Scanning System for Television." The Library of Congress has both of these journals as well as *Radio News*, in which Zworykin published his "Television Through a Crystal Globe," in April 1930. This journal later was called *Electronics World.*

By the end of the thirties, electronic television had become a practical possibility, but it was not until after World War II that the television broadcasting industry began to introduce television around the world. Although Electric and Musical Industries (EMI) in Great Britain launched the world's first public, high-definition television service in 1936, and regular television broadcasting began in the United States in 1941, real progress had to wait until the war was over.

The evolution of this new, instantaneous visual technology can be studied in the collections of the Library of Congress. The Library began acquiring programs for its television collection in 1949. Before 1966, it acquired them on a selective and representative basis. Since then, it has sought to be much more comprehensive and will acquire an entire run of a certain series, where before it felt that a single showing or episode was

Opposite page:
When Vladimir Zworykin emigrated to the United States from Russia in 1919 after the revolution, he had already decided that he would try to construct a workable electronic transmitter or television camera. Zworykin had studied in Saint Petersburg under Boris L. Rosing, who experimented with a cathode-ray oscilloscope as a television receiver, and by 1928 Zworykin produced his iconoscope. This first effective television camera was a tube that stored on individual cells the light that had fallen upon them and that would then release this stored information when turned on by a cathode-ray beam. Zworykin wrote this article after he had become director of research for RCA. *Radio News,* April 1930. Vladimir K. Zworykin. LC-USZ62-110429.

TELEVISION *Through a* CRYSTAL GLOBE

New Cone-shaped Tube Reproduces 4 x 5-Inch Picture, Is Quiet in Operation and Does Away With Need of Mechanical Parts in Home Receiver

By V. Zworykin

Reprinted by courtesy of the Institute of Radio Engineers

THE problem of television has interested humanity since early times. One of the first pioneers in this field, P. Nipkow, disclosed a patent application in 1884 describing a scanning of the object and picture, for which purpose the familiar perforated disk was employed and at present the rotating disk is giving excellent results within the mechanical possibilities of our time. The cathode-ray tube, however, presents a number of distinct advantages over all other receiving devices. There is, for example, an absence of moving mechanical parts with consequent noiseless operation, a simplification of synchronization permitting operation even over a single carrier channel, an ample amount of light for plain visibility of the image, and indeed quite a number of other advantages of lesser importance. One very valuable feature of the cathode-ray tube in its application to television is the persistence of fluorescence of the screen, which acts together with persistence of vision of the eye and permits reduction of the number of pictures per second without noticeable flickering. This optical phenomenon allows a greater number of lines and consequently better details of the picture without increasing the width of the frequency band.

This paper will be limited to a description of an apparatus developed in Westinghouse Research Laboratories for transmission by radio of moving pictures using the cathode-ray tube for reception.

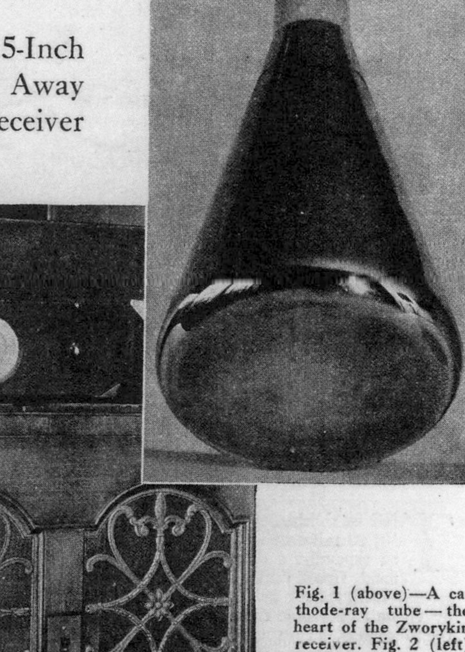

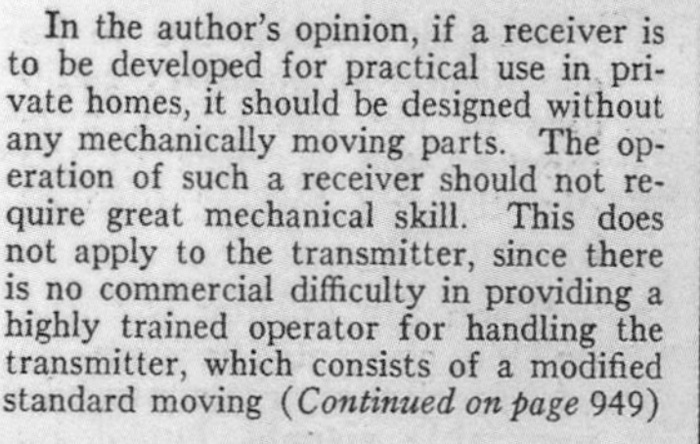

Fig. 1 (above)—A cathode-ray tube—the heart of the Zworykin receiver. Fig. 2 (left)—One type of cabinet receiver housing the Zworykin apparatus

In the author's opinion, if a receiver is to be developed for practical use in private homes, it should be designed without any mechanically moving parts. The operation of such a receiver should not require great mechanical skill. This does not apply to the transmitter, since there is no commercial difficulty in providing a highly trained operator for handling the transmitter, which consists of a modified standard moving (*Continued on page* 949)

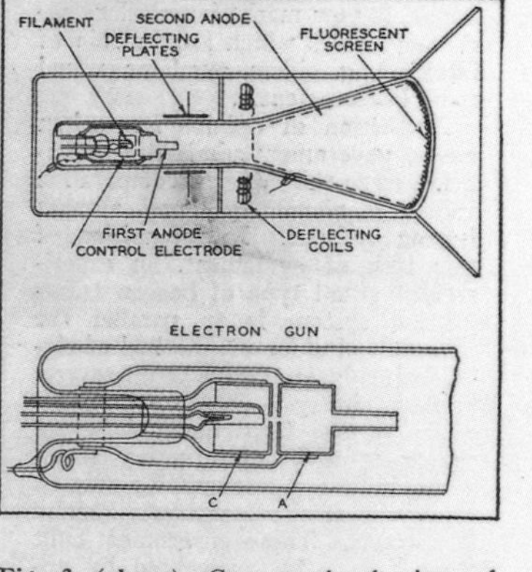

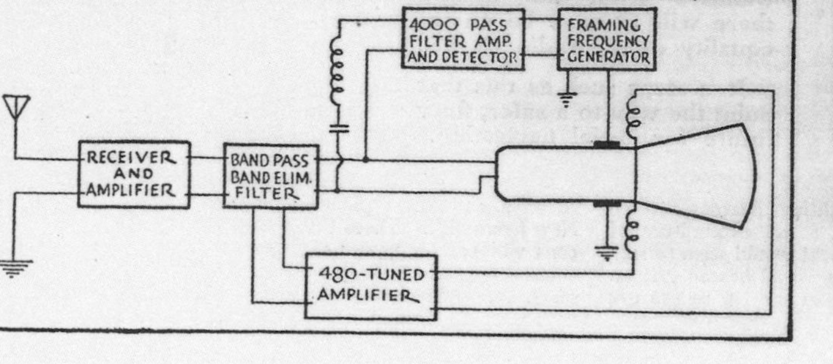

Fig. 3 (above)—Cross-sectional view of cathode-ray tube, including an enlarged drawing of the electron gun. Fig. 4 (left)—Diagram of the band-pass filter which divides the local receiver output into the picture and synchronizing frequencies

At about the same time that Zworykin was doing his pioneering television work, a young American inventor, Philo T. Farnsworth, was achieving similar breakthroughs with very different methods. As a boy of fifteen in Idaho, Farnsworth had drawn diagrams of electronic television systems on a chalkboard for his chemistry teacher and by nineteen, Farnsworth was given modest financial backing by a syndicate. His system was different from Zworykin's in that his "image dissector" or camera was an instantaneous or nonstorage type. Farnsworth and his modest financial backers were at a disadvantage competing against the deep resources of Zworykin's RCA support, but Farnsworth's system was also in fact less sensitive than Zworykin's. It was Zworykin's storage-tube iconoscope that would prove to be the television camera of the future. In this article, which shows a photograph of a Farnsworth television image, the inventor makes the understatement that the transmission of such pictures "can well be considered as having entertainment value." *Radio,* December 1929. Philo T. Farnsworth. LC-USZ62-110430.

Transmission of *Television* Images

Intimations of How It May Be Ultimately Accomplished in Practice

By PHILO T. FARNSWORTH and HARRY R. LUBCKE
Crocker Research Laboratory

Unretouched Picture of Farnsworth Television Image With 20,000 Elements

THE transmission of television images requires the conveyance of information regarding the light intensity of a great number of elemental areas, reproduced in the same relative position they occupied before transmission. Furthermore the process must be repeated with sufficient frequency to give the illusion of motion. In effect, the field of view transmitted must be dissected, an electrical intensity transmitted corresponding to the light intensity on each elemental area, and the electrical intensities converted back to light intensities and placed in their proper relative positions at the receiving terminal, this being accomplished by a system capable of presenting fifteen or more complete pictures to the eye each second.

Thus far man has evolved only one means of electrical communication. His system is a single dimensional time continuum. The commercial wireless, the telegraph, the telephone, and the radio, transmit intelligence that is initially single dimensional, since a dot or dash or the modulations of voice or music are intensity variations occurring in time and time only. In television, however, information must be transmitted regarding space as well; that is, the placement of the various intensities in the area that make up the picture must be conveyed, all of which must be transmitted at a rapid rate if fifteen or more complete pictures are to be presented to the eye each second.

Since the nature of the transmission medium is the same for all cases, it is obvious that as the amount of intelligence transmitted increases, the portion of the medium used must also increase. In terms of radio broadcasting, the sideband width must increase as the amount of information transmitted increases. Thus, a commercial code station requires only 200 cycles as a sideband, and the conventional broadcast station approximately 5000 cycles. Telephoto transmission, the sending of pictures or printed matter, requires 3000 cycles, in which system one picture is transmitted in approximately seven minutes. But for television a sideband of at least 50,000 cycles, and better, 500,000 cycles or more is required, depending upon the detail desired.

A certain minimum amount of detail must be transmitted by a television system in order that the received image may be said to possess "entertainment value." The system must transmit a person's face, for example, with sufficient detail to make the features readily discernible and sufficiently clear to "entertain" the viewer if the person should talk and attempt to convey meanings by facial expressions. This minimum has been specified by some workers in the field as an image consisting of 2500 elementary areas, or elements. The picture shown herewith is made from a photograph of a television image consisting of 20,000 elements.

This image of a lady with her eyes closed was transmitted by the Farnsworth system of electrical scanning and is perhaps the first published American photograph of an actual television image. The original image was approximately 3½ in. square. It can well be considered as having entertainment value. The reticence of those working with a 2500 element picture to publish pictures of an image would seem to indicate that the 2500 element limit was fixed more because of the limitations of the apparatus than because of true entertainment value.

It is felt that real entertainment in television will require an image 8 in. square of some 60,000 elements. An image of this size containing 250 lines per side contains 62,500 elements and can be defined as one of real entertain-

(Continued on Page 85)

Transmitting Equipment Used in Farnsworth System of Electrical Scanning

36 RADIO FOR DECEMBER, 1929

sufficient. While most of its collection was and is being built by copyright acquisition, the Library has two special collections that were given in toto. The NBC Television Collection was acquired in 1986 and is a historic collection of eighteen thousand programs broadcast by NBC from 1948 through 1977. The National Educational Television Programs Collection dates from 1955 to 1969 and totals over ten thousand titles. It is a valuable record of early noncommercial American television.

In terms of social impact, these twentieth-century breakthroughs in communications were matched in many ways by those in transportation. During this century the automobile became a major factor in both modern life and modern economics. Although nearly all the fundamental technical problems associated with the automobile had been solved by the turn of the century, the greatest change in the twentieth century was the switch from custom building to mass production. In 1903, the Ford Motor Company was founded, and five years later it began production of the Model T. In 1913, the assembly line principle was introduced by Ford, and the company would churn out a total of fifteen million Model T's, all without major design changes and all selling relatively cheaply.

It was the American automobile industry that brought about the twentieth-century revolution in production techniques—introducing rationalized mass production, better known as the assembly line. Although Henry Ford was preceded in this by others—Ransom E. Olds produced 5,000 of his "Merry Oldsmobiles" in 1904 and Henry M. Leland's Cadillac Motor Company perfected the notion of standardization and interchangeability of parts—it was Ford who achieved his goal of a "car for the multitude." By providing Americans with a simple, durable, inexpensive car, Ford not only created an industry vital to the American economy but also initiated one of the most significant and pervasive technological agents of cultural change. Here, a row of completed "Tin Lizzies" or Model T's come off the Ford assembly line in 1917. Prints and Photographs Division, Library of Congress, LC-USZ62-63968.

In the same year that the Ford Motor Company was established, essentially founding the modern automobile industry, another industry was having its technical foundations laid. Aeronautics would eventually become another technological staple of the late twentieth century—something we all take for granted. Yet less than a century ago, powered flight was popularly regarded as a pipe dream. Three years after the pioneer experimenter Otto Lilienthal was killed during one of his test glides (1896), the Wright brothers acted upon his inspired example and built their own glider. In their adventures and experiments in unpowered flight—sometimes as many as a hundred soaring flights in one day—the Wrights were able to so improve and perfect their understanding of the phenomenon of flight that they felt capable of attempting powered flight.

After three years of gliding successes and failures, the brothers achieved the first powered, sustained, and controlled flight at Kitty Hawk, North Carolina, on December 17, 1903. Typically, they had designed the ten-horsepower gasoline engine themselves, as well as the entire cloth and wood aircraft. Orville Wright was at the controls when their *Flyer* took off for the first time and flew 120 feet in twelve seconds. On their fourth and last flight that day, Wilbur flew 852 feet in fifty-nine seconds.

Although it seems astonishing today, neither brother had more than a high school education. Yet they were not simply a pair of fortunate opportunists who, through dogged hit-or-miss experimenting, eventually solved a technical problem. They were a collaborative pair of intuitive inventors who nonetheless followed the normal standards and procedures of what was traditionally thought of as the scientific method. Thus they first mastered the extant literature (what there was of it) and focused on a specific, demonstrable goal. They were deliberate and orderly in their three years of gliding experiments, proved willing to change, and were open to new ideas. They sought to understand the "why" of flying in addition to mastering the actual "how."

Here, the Wright Brothers capture the exact moment when another revolution in transportation was born, on December 17, 1903. As Wilbur runs alongside the aircraft, which is already off the ground, Orville pilots the craft from a prone position. The precise and deliberate brothers thought of everything, it seems, as they had prefocused their camera on the spot where the plane would leave its wooden track and instructed a friendly helper to engage its shutter when it reached that spot. This use of photography by the Wrights to record their experiments is consistent with their deliberate scientific methods. Because of this attitude and method, they not only achieved the first free, sustained, and controlled powered flight but were able to provide visual evidence of their spectacular achievement. This is one of 303 glass-plate negatives presented to the Library of Congress by the Wright estate in 1949. Prints and Photographs Division, LC-USZ62-6166A.

It is not surprising, therefore, that when the brothers did finally fly for the first time, they were prepared and able to document their landmark achievement in photographs. The Wrights used photography to document their progress and thus were able to capture the exact moment of take-off of history's first flight. This has become one of the most widely recognized photographs of all time. The Library of Congress has this photograph (and its original glass plate negative) along with three hundred others that the Wrights took and developed themselves. The photographs were part of the Wright Collection given to the Library in 1949 after Orville's death. Totalling over thirty thousand items, the collection's diaries, notebooks, correspondence, scrapbooks, and business papers provide a comprehensive documentary record of the invention of the airplane. Together with the Library's Octave Chanute Collection (see chapter 6), the Library of Congress can offer a unique and completely documented account of the evolution of one of history's great inventions. In 1953 the Library also produced an edited version of the Wright and Chanute collections called *The Papers of Wilbur and Orville Wright*. Edited by Marvin W. McFarland, the two volumes include all papers that were relevant to the evolution of the airplane and the Wrights' discoveries. Altogether,

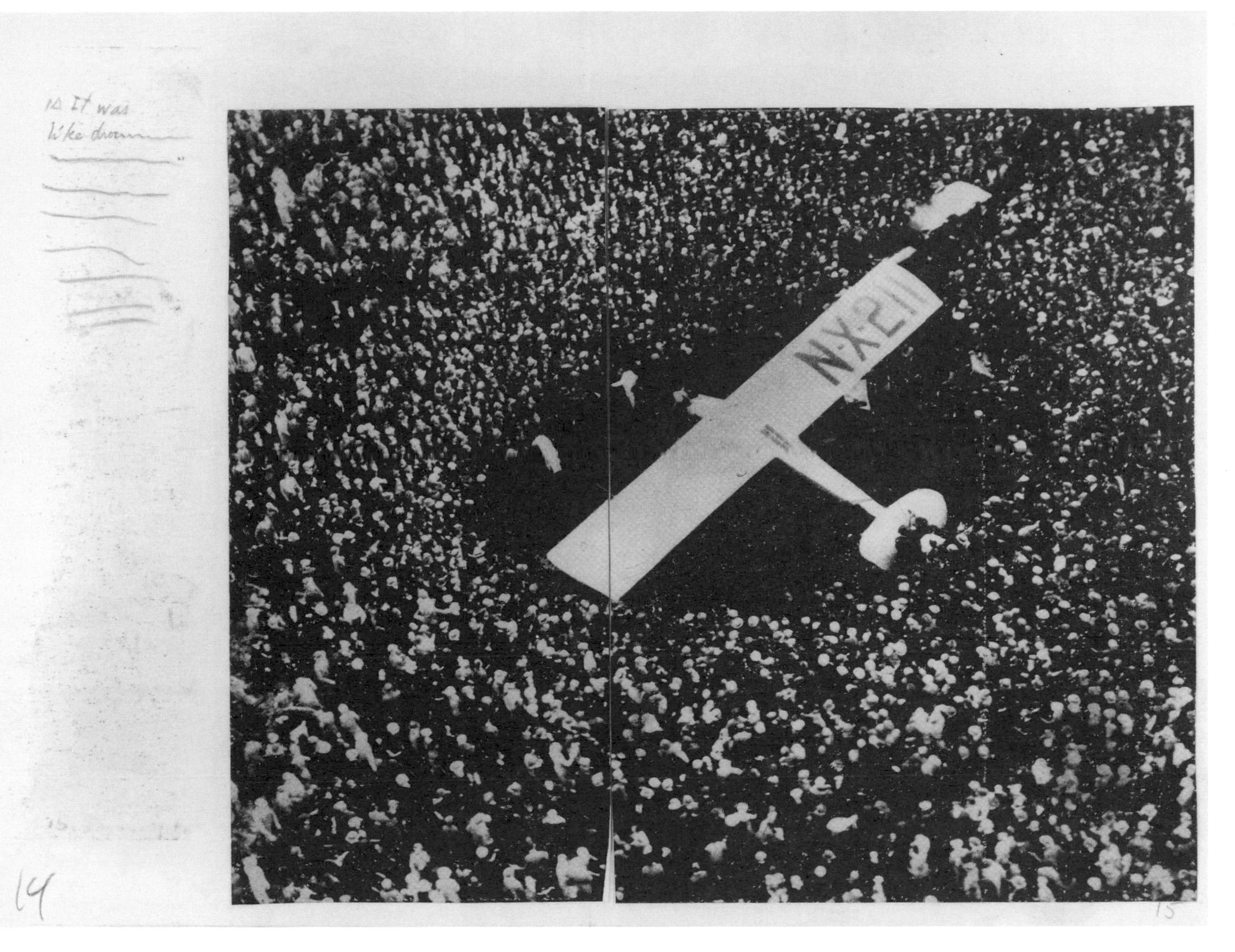

"It was like drowning in a human sea." This is how Charles A. Lindbergh described what occurred as he landed at Le Bourget airport in Paris on May 21, 1927. Here he is engulfed at the Croyden airport, showing that the British were as subject to the post-flight hysteria that surrounded his epic solo journey across the Atlantic as were the French. Lindbergh's thirty-three-hour flight not only captured the imagination of the world and transformed him overnight into an international celebrity, but it created an enthusiasm for aviation that soon was turned into a major industry. This image is included among Lindbergh's papers in the Library of Congress, as it was used as a galley sheet for his 1953 Pulitzer Prize-winning book, *The Spirit of St. Louis.* The handwritten caption at the top left may be his. Charles A. Lindbergh Papers, Manuscript Division, Library of Congress.

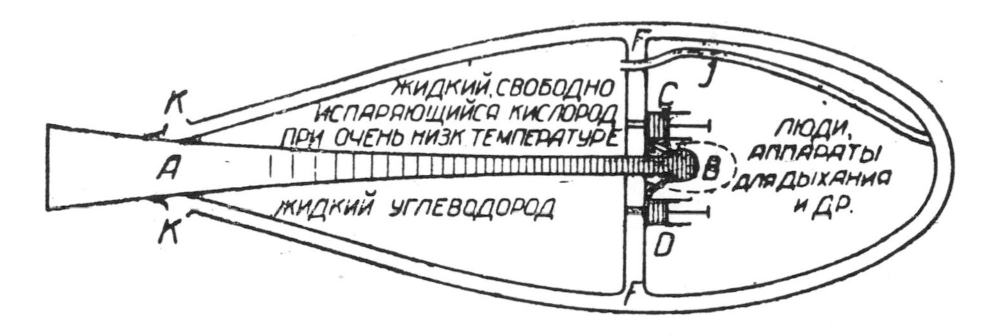

It is difficult today to grasp the deep understanding and vision required to produce this drawing. Not only was its creator postulating flight *beyond* the atmosphere—when mankind had not yet even made its first flight in the atmosphere—but he was nearly deaf and lived in a physically isolated part of Russia. Despite this, Konstantin E. Tsiolkovsky published in 1903 what are essentially the theoretical solutions and basic principles of flight in space based on reaction or rocket engines. This sketch of a rocket reaction motor is probably a 1915 version of his original, and it shows a spacecraft whose nose capsule houses the crew and whose liquid oxygen and liquid hydrogen feed into a combustion chamber. Although he never built any such rocket systems, Tsiolkovsky's theoretical work enabled him to anticipate and solve the practical problems that would be encountered in space flight. *Tsiolkovskii,* 1937. Iakov I. Perelman. LC-USZ62-110432.

the volumes offer over twelve hundred pages of correspondence and diary entries as well as three major appendixes that detail the Wrights' wind tunnel experiments, propellers, airplanes, and motors.

Also in 1903—the year of both the first powered flight and the founding of the American automobile industry—an article appeared in a Moscow journal which discussed the practical problems of flight in space. Written by an obscure, deaf schoolteacher, Konstantin E. Tsiolkovsky, this paper would prove to be as important as any in the history of space research. In it, the self-taught genius, who was intellectually isolated at his native Kaluga in Russia, nonetheless explored the possibility of space travel by means of rockets. In his consideration of the essential physical and mathematical aspects of the problem, Tsiolkovsky proved to be the first to realize and elaborate on the significant difference between an aircraft and a rocket. Travel in space, he concluded, cannot be achieved by the energy of traditional combustion alone, but rather requires a high velocity of exhaust (the reaction principle) to produce a forward motion. He further proposed that, in the vacuum of space, traditional internal combustion engines relying on oxygen would not work, and that a rocket reaction motor using liquid fuel was necessary to produce forward movement. This was a direct application of Newton's third law of motion.

Tsiolkovsky called this visionary paper "Issledovanie mirovykh prostranstv reaktivnymi priborami," translated now as "Investigation of Space by Means of Rockets." Although published for the first time in 1903 in the Moscow journal *Nauchnoe Obozrenie* (*Science Review*), it was actually written and submitted to that same journal in 1898. Thus Tsiolkovsky's landmark investigations in the then-unknown field of astronautics can be said to have predated even the beginnings of modern aeronautics when the Wrights began their work. The Library of Congress does not have this exceedingly rare journal in its collections, nor does it have the equally rare journal *Vestnik Vozdukhoplavania*, which reprinted the 1903 article in its 1911 and 1912 issues. Tsiolkovsky is represented in the Library's collections by a two-volume selection of his works, *Izbrannye Trudy* (Moscow, 1934), which contains the 1903 paper. It is also found in the four-volume collected works, *Sobranie Sochineii* (Moscow, 1951–64). The Library does have many of his very early aeronautical works, most of which are concerned with airships or dirigibles. The earliest of these is his short book on airships, *Aerostat Metallicheskii Upravliaemyi*, published in Moscow in 1892.

Although Robert H. Goddard never sent one of his own rockets into space, he pioneered the technology of rocketry. Having begun his research on rockets in 1906 at the age of twenty-four, he gave up work on solid fuels in the early 1920s, once he realized (as Tsiolkovsky had) that with liquid fuels the rate of consumption can be controlled. Here on a snowy day on March 16, 1926, in Auburn, Massachusetts, Goddard stands next to his A-frame contraption for his wife, who takes this picture. Soon after, he put a blowtorch to its liquid oxygen and gasoline propellants and launched the first successful liquid-fuel rocket. The flight lasted only two and one-half seconds and the rocket accelerated to a height of forty-one feet. Goddard was later forced to move his noisy launches to New Mexico, and although his rockets never reached higher than 1.4 miles up, he had by the start of World War II flown his rockets past the speed of sound and tested a system of gyroscopic stabilization. Goddard and the notion of rockets were basically ignored by the American government during the war, until the German rocket scientists freely admitted to their American captors that it was Goddard's published work that had provided them with the blueprints for their work. Prints and Photographs Division, Library of Congress, LC-USZ62-110433.

Tsiolkovsky remained in near poverty and obscurity until well beyond his sixtieth year, and was only discovered by his countrymen (and the world) after the Russian Revolution. During the 1920s, he continued to work on his new theory of multistage rockets, and it was during that decade that he became known internationally as the father of astronautics.

An American who many argue can contend for that title is rocket pioneer Robert Hutchings Goddard. Like his visionary Russian counterpart, Goddard also worked alone. But unlike his predecessor, he had the benefit of an advanced education, having earned his doctorate in physics. Like Tsiolkovsky, he was intrigued and excited at an early age by the idea of rocketry and space travel. At twenty-four years old in 1906, he began to explore the mathematical practicality of rockets, and by 1912 he was testing reaction engines in his laboratory vacuum.

In 1919, Goddard was seriously pursuing the theory of flight in space and he produced his classic paper titled "A Method of Reaching Extreme Altitudes." Written as a report on his work to date, the state of the art of rocketry, and the future potential of rockets, it first suggests the use of rockets for meteorological research. But after discussing rocket flight within the atmosphere, Goddard obliquely turns the paper to a "calculation of minimum mass required to raise one pound to an 'infinite' altitude." Goddard was finally stating what he had in mind—flight into space. The last page of the article contains the startling suggestion that a rocket could be fired at the moon. The Library of Congress has Goddard's paper as it was first published in 1919 in the Smithsonian Miscellaneous Collections, volume 71, number 2.

After his landmark paper, Goddard continued his research. Unlike Tsiolkovsky, who had neither the interest nor the facilities to be an engineer but who excelled at theory, Goddard put to good use his skills as an expert engineer. Thus in 1923 he was able to construct and test a new type of rocket engine that used liquid fuels instead of solids. As Tsiolkovsky had already theorized, Goddard discovered that the consumption rate of liquid fuels could be controlled. By 1926, Goddard had built a workable rocket, and on March 16 of that year, he launched the world's first liquid-fuel rocket. This first successful rocket launch was as modest as the Wright brothers' first flight, reaching an altitude of only 41 feet, an average speed of 60 miles per hour, and traveling a distance of 184 feet.

Goddard continued to test-launch his rockets, and although he experimented in the open country of his Aunt Effie's farm in Auburn, Massachusetts, neighborhood suspicion and dissatisfaction forced him to stop

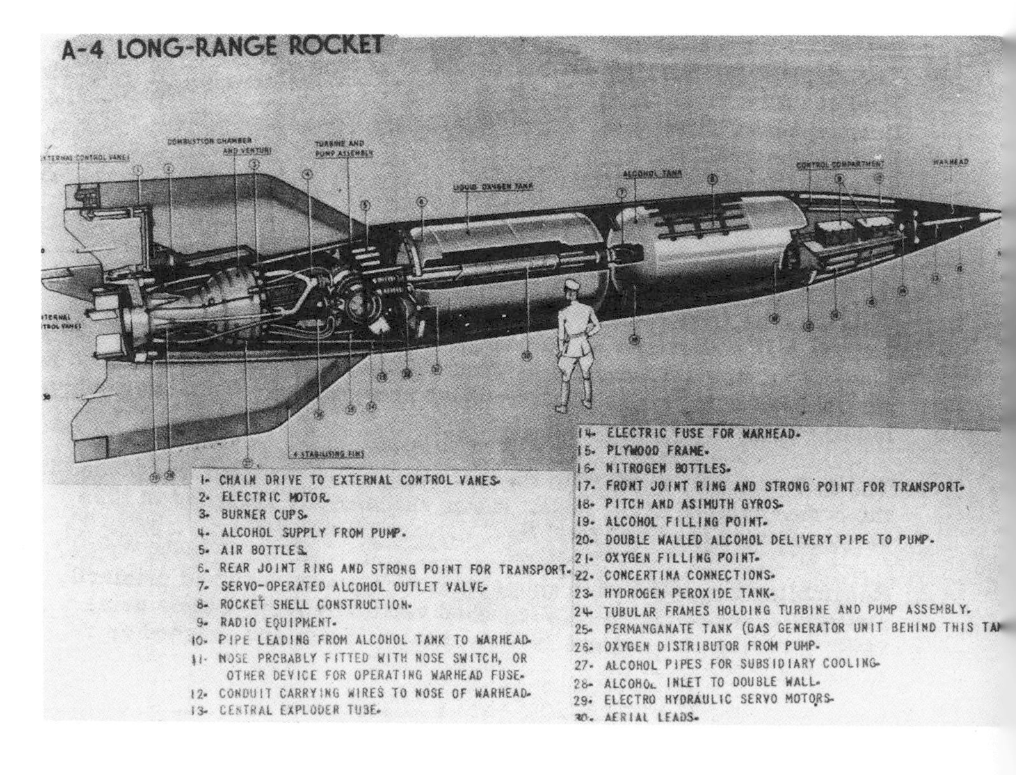

With a man standing in front for a rough feeling of scale, this postwar drawing of a German V-2 rocket gives some specifications of this forty-six-foot long strategic weapon system. When U.S. troops entered the German rocket center at Peenemunde in April 1945, they discovered completely intact assembly lines of V-1 and V-2 rockets. During the following months, they shipped about one hundred V-2 rockets to the desert of New Mexico and eventually brought the key German scientists to America as well. During April 1946, the United States launched the first of these rockets, beginning a program of research and development that culminated in the creation of its own space program in 1958. It can be said that today's Space Shuttle is but a technical extension of these first German rockets. *A Summary of German Guided Missiles,* PB-27795, March 15, 1946. Norman Harlan and Gene McConnell. Science and Technology Division, Library of Congress.

his noisy and dramatic experiments. Fortunately, Charles Lindbergh saw the value of Goddard's work and arranged for him to receive modest financial support from the Guggenheim Foundation as well as the Carnegie and the Smithsonian institutions. Goddard was then immediately off to the desert of New Mexico to resume his launches.

During the 1930s, Goddard succeeded in launching a rocket to a height of one and a half miles, and later developed and received patents for his multistage rocket plans as well as for his pumps, engine cooling system, variable thrust engines, and retrorockets. Although he offered the United States his services during World War II, he was given the task of trying to adapt his rockets to boost aircraft during takeoff. He died in relative obscurity in 1945, and after the war, when the United States brought the German V-2 rocket developers to America, they told their surprised captors that virtually all they knew of rocketry had come from Goddard's published work.

At the beginning of the space age, when the United States sought to compete actively in space with the Soviet Union, Goddard was not only posthumously honored, but his estate received a million dollars for the past use of his patents. It has been said that one could not design, construct, or launch a rocket without infringing on one of Goddard's 214 patents. If the great Tsiolkovsky was the father of the theory of space flight, then Goddard was its first executor. As one of the last of the major individual scientists or inventors, it was Goddard as much as anyone who helped mark the twentieth century as the beginning of the space age.

Yet if this century began the space age as well as the age of nuclear energy and television, it also saw the definitive start of the computer age. Although the earliest expression of the idea of a programmable computer was Charles Babbage's proposed "analytical engine" in the 1830s (mentioned in chapter 6), the first concrete step toward that goal was a modest punch-card tabulator invented by an American to process the data accumulated by the 1890 United States census.

In 1886, statistician Herman Hollerith was working on the census of 1880 and encountered the problems of large-scale calculation and compilation. After studying the card controller of the Jacquard loom, which wove cloth in the early 1800s using a punched card system, and reading of Babbage's plans, Hollerith designed a machine that would use punched cards to represent logical and numerical data. One of his most important innovations was the use of electromechanical sensing equipment to locate the holes in the cards. This was not only faster than mechanical feelers, as in Babbage's machine, but also made the tabulator and sorter much smaller and more compact. Hollerith's new system made it possible to classify and count the data for the United States census of 1890 in one-third of the time it had taken in 1880, and it was an obvious and celebrated success. In 1896 he organized and formed the Tabulating Machine Company which, after several mergers, would evolve into the computer giant IBM (International Business Machine Corporation) in 1924.

In 1972, the Library of Congress acquired the papers of Herman Hollerith as a gift from his heirs. This collection spans the period from 1876

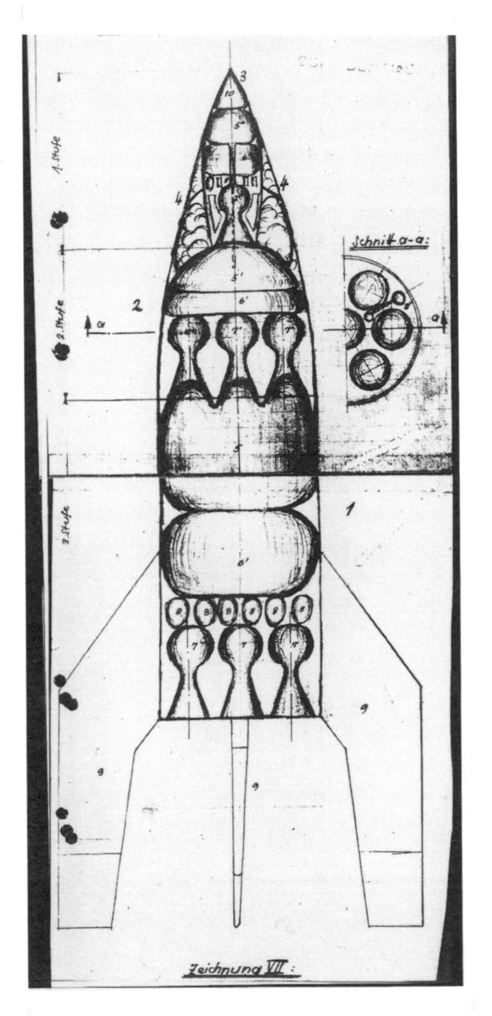

One of the engineers who worked at the German rocket facility under Wernher von Braun was Hermann J. Oberth. With the Allied capture of that facility, the United States became intensely interested in rocketry, and Oberth was among the many scientists and engineers who were debriefed. Here he draws from memory the rough details of a three-stage, long-range rocket that he had designed in 1941. Using liquid fuels, it was intended to carry an explosive payload in its nose (10). In this report, which was translated in late 1945, Oberth also includes a reference to the spaceflight potential of such a rocket, saying that if the upper stage could attain additional velocity, it "would not return to the ground any more, but would rotate around the earth, like the moon, at will, because of the effect of centrifugal force. Such a device, would be of great scientific and practical interest." *The Design of a Long Range Rocket,* PB-22295, October 4, 1945. Hermann Oberth. Science and Technology Division, Library of Congress.

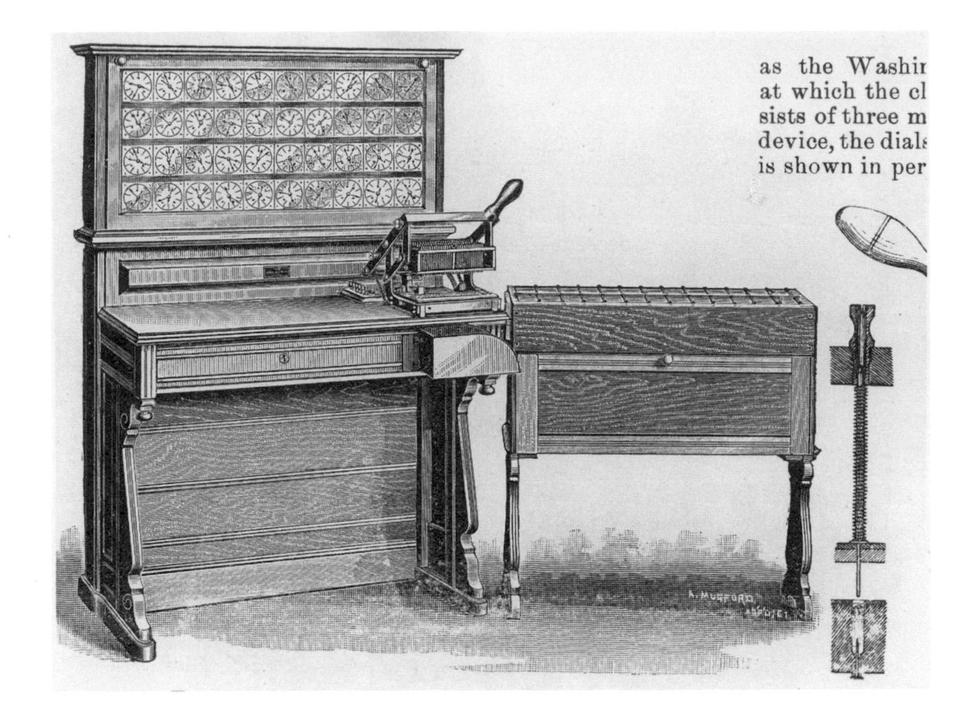

292 A major step in the development of the modern computer was this electromechanical machine that used punched cards. Invented by Herman Hollerith, who borrowed Babbage's punched card system, the machine used electricity to sort and record data compiled for the U.S. census of 1890. The system worked on the idea that if a metal rod could be passed through a series of aligned perforations, an electrical circuit would be completed. The earlier census of 1880 had taken seven and a half years to complete and another year to publish. Hollerith's machines did the job in two and a half years, despite a population increase from 50 to 63 million. In 1896, Hollerith set up the Tabulating Machine Company, which grew through mergers into the International Business Machines Corporation (IBM). Prints and Photographs Division, Library of Congress, LC-USZ62-44851.

to 1929, with the bulk of the material dating from 1910 to 1927. Totalling about ten thousand items, the collection includes business, personal, and special correspondence, printed matter and other material relating to the tabulating business, and scrapbooks with newspaper clippings about the machines and their use in the census. A sampling reveals information relating to the operation of his Tabulating Machine Company, its merger into the Computing-Tabulating-Recording Company in 1911, and his later association with IBM. There is also a chronological file of Hollerith patents as well as blueprints, drawings, and one Hollerith machine punch plate.

From the time of Hollerith's first machine, through World War I and until the mid-1930s, interest in automatic computation was not steadily maintained. In 1937, however, Alan M. Turing, a brilliant British mathematician, published an article entitled "On Computable Numbers" in the *Proceedings of the London Mathematical Society*, in which he explored the idea of a "universal machine" that could perform any calculation or logical operation that a human mind could devise. Although Turing's paper was an exercise in formal logic and he did not attempt to build such a machine, it was a conceptual breakthrough which must have influenced all who could understand it. The Library has the issue of the journal that the article appeared in.

In the same year, Howard Aiken of Harvard University invited International Business Machines to collaborate with him in his attempt to create what was essentially the automatic calculator envisioned by Babbage a century before. By August 1944, Aiken and his IBM colleagues were able to present Harvard University with an imposing machine called the

Automatic Sequence Controlled Calculator (ASCC or Harvard Mark I). Although smaller than Babbage's machine would have been, it was still an enormous machine, fifty-one feet long and eight feet high, weighing fifty-one tons and containing 800,000 parts and 500 miles of wire. Although somewhat slow, it can be considered the first general-purpose digital computer. Despite earlier and similar wartime accomplishments in Germany by Konrad Zuse (1941), as well as John V. Atanasoff's prototype built at Iowa State College in 1942, it is Aiken's design that marks the starting point of American computer development, especially since it entailed the full involvement and commitment of IBM.

Aiken's project and accomplishments are well documented in the collections of the Library of Congress. Besides his paper "The Automatic Sequence Controlled Calculator," written with Grace M. Hopper and published in 1946 in *Electrical Engineering*, the Library has the complete, 561-page version which was excerpted for the article. Called *A Manual of Operation for the Automatic Sequence Controlled Calculator* and published in 1946 by Harvard University, this large report documents as completely as possible the history, technical evolution, and actual workings of the new machine.

As with television, whose future lay not with electromechanical devices but with purely electronic ones, so computers had to make the leap to no moving parts that came with total electronics. The first attempt at this was Howard Aiken's Mark I computer, which was a mechanical calculator that used electricity to power its thermionic tubes (called valves, since they conducted current in only one direction). The massive, fifty-one-foot-long machine is seen here as it appeared in 1946, less than two years after it was first put into operation at Harvard University. Although it was impossibly slow and was programmed by punched-paper tape, this electrically driven automatic calculator is regarded by most as the first general-purpose digital computer. *A Manual of Operation for the Automatic Sequence Controlled Calculator*, 1946. LC-USZ62-110436.

The first truly electronic computer was ENIAC (Electronic Numerical Integrator and Computer). This pretransistor computer used vacuum or radio tubes, and its ancestry is found more in the electronics of radio and telephony than in Aiken's mechanical Mark I computer. Although much faster than the Mark I (it could multiply two ten-decimal numbers in less than one three-thousandth of a second, compared to the three seconds it took the Mark I), it required, as seen here, three walls of a good-sized room. Its utility was proven when, as one of its first tasks, it was given a problem in nuclear physics. ENIAC produced an answer in two hours that would have taken 100 engineers one year to solve with precomputer methods. *Proceedings of a Symposium on Large-Scale Digital Calculating Machinery.* 1948. LC-USZ62-110437.

The next computer generation was also the first large electronic digital computer and was called ENIAC (Electronic Numerical Integrator and Computer). Twice as large as Aiken's Mark I (its U-shape wrapped around three walls), it was also a thousand times faster since it operated with vacuum tubes instead of electromechanical switches. ENIAC was the product of two associates of the Moore School of Engineering at the University of Pennsylvania, John W. Mauchly and J. Presper Eckert. Although the Library of Congress does not have a rare U.S. Army Ordnance Department Report on ENIAC issued in late 1945 (the Army used ENIAC to calculate ballistic tables), it does have the original four volumes of informative and inspiring lectures delivered at the Moore School between July 8 and August 31, 1946, by Eckert, Mauchly, and seventeen other prominent scientists. Published by the University of Pennsylvania under the general title *Theory and Techniques for Design of Electronic Digital Computers*, these volumes contain forty-eight separate lectures, eleven of which were delivered by Eckert and six by Mauchly. These lectures would greatly stimulate postwar competition in the development of stored-program computers and would lead to a succession of "electronic brain" computers such as EDVAC and the well-known UNIVAC.

The notion of a stored program, or of a computer's activity being controlled by a program kept in its internal memory, was the final major step in the development of a general use electronic computer. This need was apparent to all, since a change in ENIAC's program required the unplugging and plugging in of hundreds of wires—a task that could take two days. This problem became the focus of mathematician John von Neumann of the Institute of Advanced Study at Princeton University. Associated with Eckert and Mauchly on the ENIAC project in late 1944, von Neumann wrote his famous 101-page "First Draft of a Report on the EDVAC" in June 1945 for U.S. Army Ordnance. This famous report not only contains a brilliant description of the planned stored memory machine but reveals the reasoning behind the various design decisions. EDVAC (Electronic Discrete Variable Arithmetic Computer) was to be the Army's and von Neumann's successor to ENIAC, but it was not completed until 1951. By this time, the British EDSAC had become the first really practical stored-program computer, and von Neumann had already developed the IAS (Institute for Advanced Studies) computer which broke all existing computational speed records with its new cathode-ray tube memory system.

Despite the tardiness of EDVAC's completion, it is significant for having caused one of the most important papers in the history of the computer sciences to be written. While von Neumann was at Princeton's Institute for Advanced Studies, he wrote a report with Arthur W. Burks and Herman H. Goldstine titled "Preliminary Discussion of the Logical Design of an Electronic Computing Instrument." Published in June 1946, this paper gave a detailed review of the entire field of automatic computation and offered comprehensive designs for a parallel, stored program computer that was substantially different than anything hitherto proposed. It is not exaggerating to say that this paper had a major influence on the form of all subsequent digital computers.

Both the "First Draft" (1945) and the "Preliminary Discussions" (1946) are rare in their original form, as they were generated as part of a contract with the U.S. Army and were issued as reports. They can be found, however, in the collections of the Library of Congress as part of the von Neumann papers. The Library received this collection from the estate during 1974 and 1975, and it comprises approximately 10,400 items. Much of the collection relates directly to von Neumann's involvement with computers. The correspondence files, which date from 1934 to 1956, include an extraordinarily large range of prominent physical scientists and mathematicians. In addition to von Neumann's lecture notes, speeches, draft articles, and other published material, there are very relevant items such as the progress reports from the Institute for Advanced Study projects he conducted, as well as a large number of his unpublished technical reports. The Library also has the five-volume *John von Neumann: Collected Works*, edited by A. H. Taub and first published in New York in 1963.

The final, major breakthrough in computer technology came in 1947 with the invention of the transistor and what might be called the semiconductor revolution. All the early computers were physically huge and very expensive to run. They operated using thermionic valves or vacuum tubes—large, hot, short-lived devices that were also extremely power-hungry. After the war, the Bell Company laboratories began a program headed by William B. Shockley to research solid state devices, and it was out of this effort that the transistor was born. Shockley, along with Bell physicists John Bardeen and Walter H. Brattain, discovered that crystals of semiconductors (materials whose electrical conductivity lies between

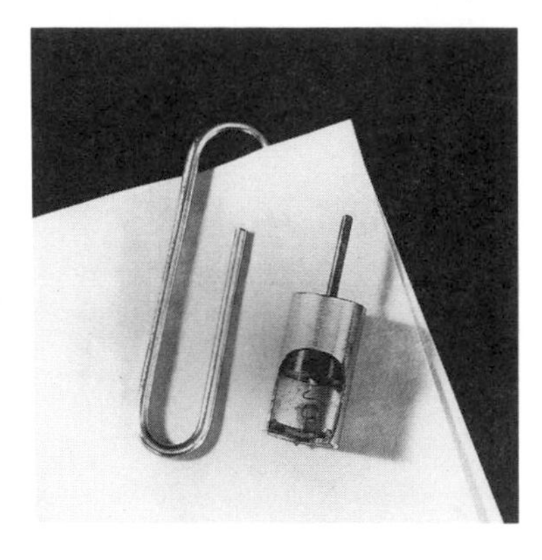

The invention of transistors immediately revolutionized the construction of electronic circuits. Their unique ability to control a large current with a small one, thus producing amplification, is the key to their miniaturization. Once they were demonstrated successfully at the end of 1946, their virtues of small size, light weight, ruggedness and long life, and high efficiency combined with low power requirements seemed almost too good to be true. The versatility and miniaturization possibilities of transistors transformed every technology they touched—from computers to hearing aids—and led directly to today's multibillion dollar silicon chip industry. This 1950 photograph of a phototransistor (one that is light sensitive) shows how small was this first generation transistor. Some predict that fifty years later, a new generation of products based on light (called photonics, since they use photons instead of electrons) may be ready to rival the present dominance of electronics. Prints and Photographs Division, Library of Congress, LC-USZ62-77084.

that of metals and insulators) could be made to perform the exact functions of a vacuum tube but at the same time be smaller, more versatile, and even more reliable. These desirable properties were achieved by the controlled introduction of an impurity into the crystal.

The advantages of miniaturization were quickly realized, and the large and cumbersome vacuum tube was gradually replaced by the small, strong transistor. Although this took several years, the steady improvement of the transistor made possible the construction of more powerful and more compact computers, as well as less expensive ones. Indeed, today's chip technology is but an elaboration of the transistor principle. Finally, transistors were not used solely on computers, but were naturally applied to a wide range of devices requiring electronic circuits. Their small size revolutionized radio and television, and they have been applied to everything from hearing aids to satellites.

The seminal work done by Shockley, Bardeen, and Brattain is documented in a number of talks and papers, the most significant of which may be those published in 1949. Bardeen and Brattain wrote "Physical Principles Involved in Transistor Action," and it was published in two journals the same month: in *Bell System Technical Journal* (April 1949) and in *Physical Review* (April 15, 1949). An article by Shockley, G. L. Pearson, and J. R. Haynes, titled "Investigation of Hole Injection in Transistor Action," first appeared in the February 15, 1949, issue of *Physical Review* and then in the July 1949 issue of *Bell System Technical Journal*, most of which was devoted to what the journal called the practical applications of semiconductors. The Library of Congress has complete sets of both journals.

The commercial use of transistors in computers did not come about until 1959, marking the second generation of computers. The third generation began in the late 1960s, when further improvements in miniaturization were made. These included the fabrication of the integrated circuit (IC), a solid-state device made up of hundreds of transistors, diodes, and resistors on one tiny silicon chip. The fourth generation of the 1980s was characterized by very large scale integration (VLSI), which vastly increased the circuit density of chips. The coming fifth generation of computers promises the creation of artificial intelligence (AI), with machines that are capable of reasoning, recognizing relationships, and learning.

As the computer, the bomb, the automobile, aircraft and space vehicles, and television are synonymous with the twentieth century, so plastics might be called the prototypical material of this century. There is no denying that the characteristic ability of this modern synthetic to change shape without losing cohesion and to retain that shape upon cooling (while staying both tough and light) make it an eminently useful modern material. Although there now seems hardly a facet of modern life that has not been transformed or at least affected by this ubiquitous and useful material, at the century's beginning plastic meant one thing—celluloid.

The first plastic was Parkesine, named in 1862 after its inventor, the English metallurgist Alexander Parkes, who treated cellulose nitrate to achieve a hard, transparent substance. In 1868, an American, John Wesley Hyatt, used camphor on Parkes's nitrocellulose and produced the product called celluloid. It found a wide variety of applications, from combs and toys to automobile windshields and motion picture film.

However, the first completely synthetic plastic was produced from phenol and formaldehyde in 1910 by Leo H. Baekeland. Baekeland was a successful research chemist who had profited from selling his new high speed photographic printing paper to Eastman Kodak, so that he was able to devote himself fully to the challenging goal of creating a synthetic resin—originally, a substitute for shellac and varnish. In February 1909, Baekeland read a paper to the New York Section of the American Chemical Society announcing his results. His talk was titled "The Synthesis, Constitution, and Uses of Bakelite," and it began a new era in industrial chemistry. The Library of Congress has Baekeland's paper as it was first published in the inaugural volume of the *Journal of Industrial and Engineering Chemistry* (1909).

Although Hyatt's celluloid had created an industry of some size, it was small compared to what ultimately was engendered by Baekeland's first commercial product, Bakelite. All manner of new high-heat presses, molds, and extruders were needed, which in turn drove the machine and tool industry to innovate. Following Baekeland's creation of the first real synthetic plastic and compounding the overall effect of his new material, however, was a theoretical discovery of great importance. Ironically, it was from the investigation of a problem in organic chemistry (the nature of natural rubber) that a German chemist, Hermann Staudinger, was able to understand and describe the theoretical nature of polymerization. In polymerization, small molecules combine chemically to form very large, chain-like molecules notable for elasticity and high tensile strength. Staudinger's idea was that polymers were formed by chemical interaction and that they could therefore be synthesized—not a popular idea at the time but one that would provide the theoretical basis for the entire plastics industry to come. The Library of Congress has the seminal paper, in which he defines the "macromolecule," entitled "Uber die Konstitution des Kautschuks," first published in *Berichte der Deutschen chemischen Gesellschaft* in July 1924.

By this time, plastics research was well under way not only in Germany, which was desperate to discover synthetics since it had been deprived of rubber by the victors of World War I, but in the United States as well. One of the more popular products to come out of the laboratory of E.I. du Pont de Nemours & Company was the "superpolyamide" known as nylon. Building on Staudinger's work, Du Pont's Wallace H. Carothers directed the creation of nylon, the world's first synthetic fiber composed of readily available natural elements (carbon, hydrogen, nitrogen, and oxygen). Carothers also developed the first commercially successful, general-purpose synthetic rubber, called neoprene. The Library of Congress has Carother's long article, "Polymerization," as it first appeared in *Chemical Reviews* of June 1931.

More than sixty years after Carothers's landmark paper, the uses of plastics and resins and man-made fibers seem to be limited mostly by our

BAKELITE REVIEW

A PERIODICAL DIGEST OF BAKELITE ACHIEVEMENTS INTERESTING
TO ALL PROGRESSIVE MANUFACTURERS AND MERCHANTS

OCTOBER, 1933
VOL. 5 • NO. 3

COLORS THAT RIVAL PRECIOUS GEMS
(See Page 2)

Bakelite Review,
October 1933.
See p. 114.

Following the 1934 invention of nylon and its full-scale manufacture in 1938, 64 million pairs of nylon stockings were sold in 1939 alone. In this 1942 photograph, however, probably few of the women working at this wartime aircraft plant wore nylons as they cleaned transparent noses for bomber and reconnaissance planes. Eventually marketed under the "Plexiglass" name, this new polymer became available just before the war, and its lightweight transparency made it a perfect substitute for glass in aircraft. Prints and Photographs Division, Library of Congress, LC-USZ62-45642.

imagination. When Leo Baekeland was beginning to think practically about his new Bakelite, he wrote that he could suggest twenty-seven industrial applications of his new product. Today, it has certainly spawned more than that number of actual industries. In fact, there are now very few technologies that do not in some manner use synthetic plastics and fibers.

The tensile strength, economy, and flexibility offered by plastics are virtues that also describe another modern material—concrete. Although concrete has been a constructional material since Roman times and was improved in the nineteenth century, by adding a natural cement or lime composition and then by reinforcing it with imbedded steel bars, it achieved a new status and wider application when prestressed concrete was invented. The development of prestressed concrete is attributed to a French civil engineer, Eugene Freyssinet, who discovered that concrete that was allowed to set around tautly drawn, high-quality steel wires or bars was greatly increased in strength. Freyssinet learned that if the tension is released after the concrete has hardened, the resulting contraction compresses the concrete longitudinally, giving it a very high resistance to tension. The refinement of this process gave twentieth-century architects and builders a strong, new material.

Freyssinet was a master of reinforced concrete, using it to build several bridges in France after the turn of the century, including one with a 300-foot span. After 1928, however, he directed all his time to his new discovery and became a pioneering advocate whose success with prestressed concrete persuaded all nonbelievers. Freyssinet is represented in the collections of the Library of Congress first by a work by Jean Badovici titled *Grandes Constructions par Freyssinet*. Published in Paris in 1927, this book contains drawings, elevations, and photographs of four of his large projects done with reinforced concrete. Among them are his startling dirigible hangars at Orly Airport near Paris and his triple-arch Plougastel Bridge at Brest. Concerning prestressed concrete, the Library has a separate printing of Freyssinet's *Une révolution dans l'art de bâtir: Les constructions précontraintes*, published in Paris in 1941. This mature elaboration of the theory and practice of prestressed concrete set the stage for the rapid spread of its use following World War II.

As a major innovation in structural materials during the twentieth century, prestressed concrete is surpassed only by the use of steel to build a self-supporting internal frame for a building. Once Bessemer steel became perfected and mass-produced, the idea of using steel beams alone to form what is called skeleton construction was probably only a matter of time. The end product of this new capability to build a very tall building without massive foundations and thick, load-bearing walls was the skyscraper, probably one of the most defining technical accomplishments of this century. It could certainly be argued that the skyscraper, with its dominating and at times awful vertical thrust, ranks with such timeless constructions as the Egyptian pyramids or the Great Wall of China: Each in its own way is a massive mark upon the earth, by which man pridefully announces both his presence and his dominion.

The modern skyscraper is not simply a very tall building, however. It marks a total departure from all preceding construction methods. In a traditionally framed building of solid masonry construction, there is a practical height limit that simply cannot be exceeded or the building will not stand. A skyscraper's steel skeleton frame supports itself with a regular grid of vertical columns and horizontal girders and beams that take all the loads and stresses. Combine this steel frame with the availability of a safe, reliable mechanical elevator, and the building heights are virtually unlimited. In 1956, Frank Lloyd Wright actually proposed a mile-high skyscraper of 528 floors.

The first true skyscraper, the Home Insurance Building in Chicago, was of a much more modest size. Designed by an American architect, William Le Baron Jenney, who supposedly took his skeleton frame idea

from Philippine bamboo huts he had seen, the ten-story building was built between 1884 and 1895. Although it looked very weighty and traditional (yet much taller) because of its masonry exterior, this exterior actually hung like a curtain which was itself supported on each floor by the interior metal skeleton framing. By the time this landmark building was demolished in 1931, so many skyscrapers of unprecedented heights were already built in New York that many areas of the city had begun to attain a dark, canyonlike quality.

Jenney's work is sketchily represented in the Library of Congress. There is a large-format early work he wrote with Sanford E. Loring, *Principles and Practice of Architecture*, which contains forty-six folio plates of plans, elevations, and details of churches done by the authors and shows some of Jenney's early efforts with iron. The Library's first edition of this work was published in 1869. There is also his very short article "The Chicago Construction, or Very Tall Buildings on a Compressible Soil," which was published in the *Inland Architect and News Record* for 1891. Here he explicitly details how he was able to reduce the building's piers to a minimum in width and height by letting his exterior columns take the floor loads. Finally, Jenney's skyscraper was the subject of an investigating committee that made daily inspection visits during its demolition

The soaring arches of these massive airship hangars under construction after World War I at Orly Airport near Paris were made possible by the use of prestressed concrete. They were designed by Eugene Freyssinet, the French engineer who first successfully applied the principle of prestressing metals to concrete. Prestressed (and later poststressed) concrete made it possible to use less of both steel and concrete to achieve strength, thereby allowing architects to design graceful, light-looking buildings. Freyssinet's invention, which he used for several bridges as well as the Gare Maritime harbor in France, came into worldwide use beginning in 1938. These hangars were destroyed by bombing during World War II. *L'Architecture vivante,* Copyright Da Capo Press, Inc. 1924. LC-USZ62-110439.

Although the 200-foot-tall, fifteen-story Reliance Building in Chicago is relatively unimpressive by today's height standards, this example of the early Chicago school of architecture can be considered among the first generation of skyscrapers. Its design—by the firm of Burnham and Root—with an interior steel frame, a reticulated structural pattern, and large areas of glass was distinctively American in its functional simplicity. The skyscraper came about not only because of technological advances like the safety elevator and the skeleton frame or cage construction, but also because of a tremendously concentrated upsurge in population and a growing demand for commercial space in America's booming cities. Vertical growth was the perfect solution. HABS ILL, 16-Chig, 30-2, Prints and Photograph Division, Library of Congress.

from October 9 to November 1, 1931. The committee's final report concluded that Jenney's Home Insurance Building was the first building of skeleton construction and should therefore be considered the first true skyscraper. The report appeared in the *Architectural Record* (August 1934), which is in the Library's collections.

Jenney was one of the American architects who made up the famous Chicago School, the group of creative talents who came together after the disastrous Chicago Fire of 1871. Although they designed buildings of all types and functions and worked throughout the United States, they are best known for their creative mastery of steel framing and, thus, for the skyscraper. The work of many of the Chicago School's members can be studied in the Prints and Photographs Division of the Library of Congress. Its most important architectural collection, the Historic American Buildings Survey (HABS), contains measured drawings, photographs, and data sheets documenting many of the major accomplishments of American architecture.

Although skyscrapers may be one of this century's most permanent, tangible, and striking demonstrations of its technological prowess, they share a quality with all other technologies mentioned so far, in that they are separate from mankind. No matter that our computers are essentially interactive with us, or that our televisions need minds to either create programs or watch them, or that our buildings need people to live and work in them to realize their potential, all these and other technologies exist, like any tool, for mankind, but they are not identical with us. This obvious distinction becomes significant when considering the final technology to be discussed here—that of medical technology. For the tools and techniques of medicine—the scalpels, CAT scans, vaccines, and drugs—all impinge directly, even intimately on our bodies. Unlike nearly any of our other technologies, the tools of medicine are used only on its makers. Most of our other tools are for use on something else, usually something inanimate. Medical tools, however, are for use on ourselves. This is a significant distinction and must certainly affect the way we regard this particular technology. We may hate or love the bomb; abhor television or embrace it; detest or revel in the mysteries of the computer. But we seldom feel as ambivalent about a technology as we do about one that is aimed directly at our physical well-being, one that can save our lives or reveal horrible, fatal truths to us.

Medical advances in the late nineteenth century, such as the discovery of a practical anesthesia or Roentgen's X rays, both prepared us and raised our expectations about what was to come in the twentieth century. Altogether, it is safe to say we have not been disappointed. First, this century has witnessed the most revolutionary transformation imaginable in terms of communicable diseases. Tuberculosis, polio, cholera, smallpox, and all the traditional scourges of childhood have been virtually eliminated in countries with a modern health infrastructure. Second, our scientific understanding of the more complicated medical issues continues to improve steadily, with major breakthroughs in such fields as genetics and brain research seemingly on the horizon.

The Woolworth Building in this 1913 photograph rises above Lower Manhattan, its tower soaring fifty-five stories to a height of 760 feet. Designed by architect Cass Gilbert in a vertical style that most describe as Gothic, it became the dominant symbol of commercial America and presaged the great Depression-era surge in skyscraper construction. Still to come were the Chrysler Building, the Empire State Building, and the recent World Trade Center twin towers and the Sears Tower (110 stories, 1,454 feet high). Prints and Photographs Division, Library of Congress, LC-USZ62-106874.

Given this century's spectacular and numerous medical and surgical successes, selecting those medical technologies for consideration here might have been difficult were it not for the obvious simplicity of three easily representative choices. Aspirin, penicillin, and cyclosporin were discovered and made practical respectively at the beginning, in the middle, and toward the end of the twentieth century. Each is a drug with its origins in the natural or organic world, and each bears the promise and stigma of being a "wonder drug." One makes tolerable a painful world; another makes survivable an injurious world; and the last makes possible a renewed world.

Aspirin, or acetylsalicylic acid, is found naturally in some plants and the bark of certain trees and has been known for centuries as having therapeutic properties. Hippocrates found chewing willow leaves (with its small amounts of salicin) to be a pain reducer, and willow bark was written about in England in the mid-eighteenth century as a fever reducer. At the beginning of the twentieth century, the chemical side of medicine

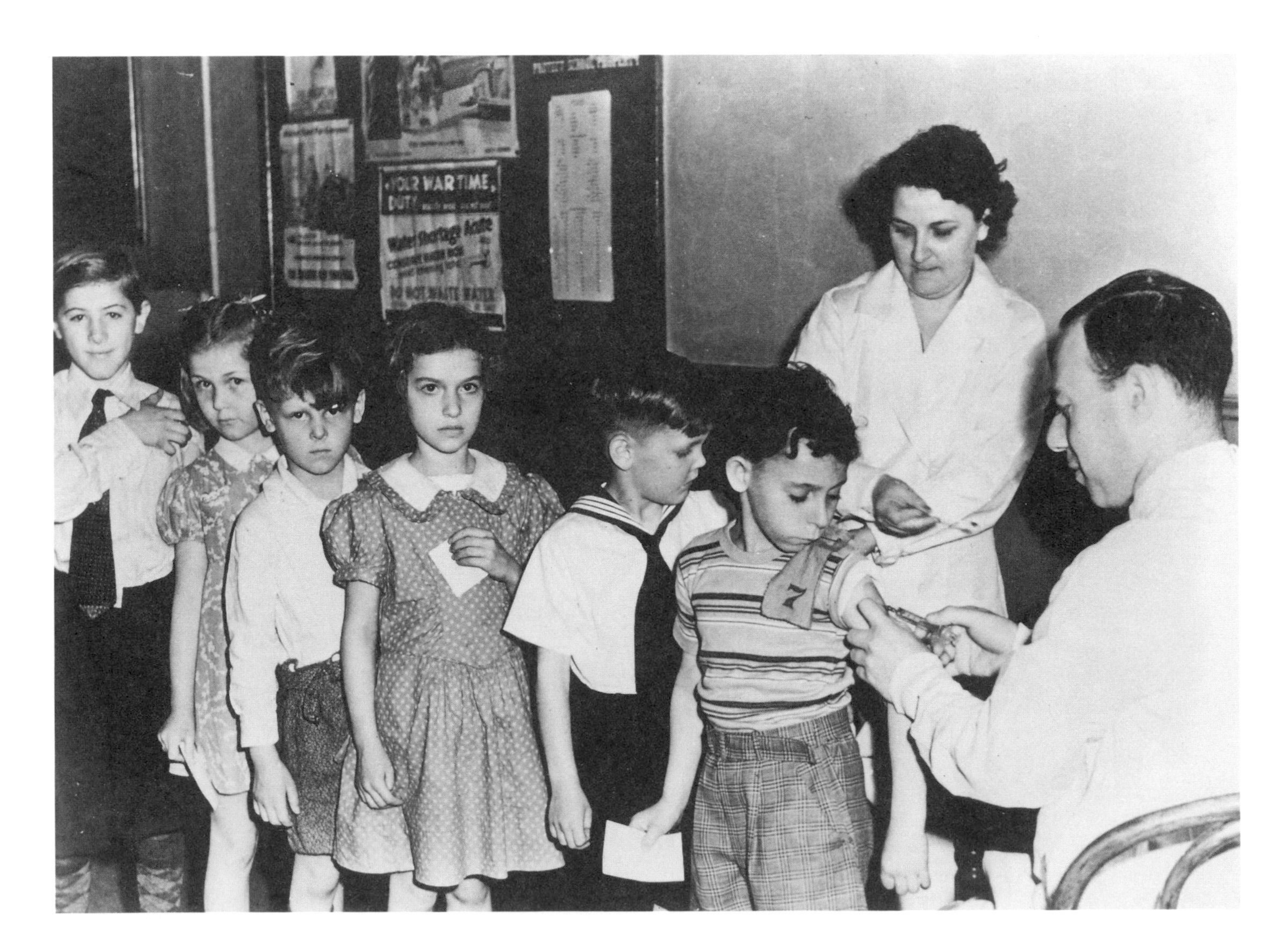

In terms of public health, the twentieth century suffered its greatest pandemic during the First World War. Beginning in late 1917 and ending in early 1919, an influenza epidemic literally swept the globe, killing directly or by secondary bacterial pneumonia an estimated 20 million people. In a matter of a few short weeks, the deadly "flu bug" claimed more lives than the mechanized death machines of the combined armies had extinguished in four years of fighting. Since then, one of the major techniques of preventive medicine has been immunization, whereby the body is stimulated to produce antibodies. Between the wars, large-scale immunization programs like this 1940s inoculation program in New York City were begun in the United States. This scene was repeated again in the

1950s, as every American community mobilized to vaccinate its children with the new Salk polio vaccine. Prints and Photographs Division, Library of Congress, LC-USZ62-110440.

Three years before this ad that was aimed at American druggists appeared in 1920, the German Bayer Company's patent on aspirin expired. Because of the fierce competition that ensued, aspirin advertising exploded in the 1920s and 1930s. Aspirin had sold well since its introduction to America in 1900, but when Bayer discovered how to mass-produce it in tablet form in 1915, the product entered the commercial mainstream. "Take two aspirin and call me in the morning," became the busy doctor's bromide. *National Drug Clerk*, June 1920. LC-USZ62-110441.

had begun to move from the laboratory of nature to the chemical laboratories of pharmaceutical groups formed to search for synthetic substances. By then, salicin had evolved into its more powerful cousin, salicylic acid (SA), which burned both the mouth and throat before it reduced pain and fever. Salicylic acid was the first synthetic drug to play a major role in medicine, but its powerful side effects made it less than fully useful.

In 1893, Felix Hoffman, a research chemist at the German pharmacological laboratory of what is today Bayer AG, was given the task of finding a less irritating antifever drug. Hoffman's father suffered badly from rheumatoid arthritis, and it was to him that Hoffman gave his first synthesis of ASA (acetylsalicylic acid). Hoffman found that ASA had been synthesized over forty years earlier by Charles Frederic von Gerhardt, who had not however discovered its beneficial effects. Hoffman also found that, although bitter-tasting to him and his father, it offered pain relief without the severe side effects of SA. The new drug soon came to the attention of Henrich Dreser, the internationally known head of Bayer's pharmacology, and it was his 1899 article that announced the discovery process and the glowing clinical reports. The Library of Congress has Dreser's thirteen-page article, "Pharmakologisches uber Aspirin (Acetylsalicylsaure)," as it was first published in the journal, *Pflüger's Archiv für die gesamte Physiologie des Menschen und der Thiere* (June 1899).

Since then, aspirin has become the most widely used, cheapest, and easiest-to-administer drug in medical history. In addition to its ability to reduce pain, fever, and inflammation with minimal side effects when taken properly, it is nearly unique in the world of drugs in that it has a strong beneficial effect on several of the body's systems without the usual side effects. Most drugs follow an opposite pattern in which very specific drugs tend to produce more side effects, and the very general ones prove often less than effective. From its beginnings, aspirin has been a wonder drug, with recent studies indicating its potential usefulness to both the body's hormonal and its cardiovascular systems.

As with aspirin, penicillin also has its origins in the natural world. As the first of the modern antibiotics to be discovered, penicillin fulfilled the need for an effective cure for bacterial disease. Once the germ theory of disease was proven valid in the last quarter of the nineteenth century, it remained for medicine to discover a way to kill germs or microorganisms internally the way they could be killed outside the body with antiseptics.

The discoverer of the antibacterial properties of natural molds of the genus *Penicillium* was a Scottish bacteriologist, Alexander Fleming. In 1928 Fleming noticed by chance that, in a dish containing a culture of staphylococcus germs, there was evidence that the bacteria had been killed in spots that had been touched by a certain mold. Although Fleming identified the mold, cultivated and preserved it, and was the first to call attention to its properties, he was unable to isolate or identify the operative substance. This was left for others to do. Fleming did, however, immediately announce his discovery in the *British Journal of Experimental Pathology* (1929) in an article "On the Antibacterial Action of Cultures of a Penicillium, with Special Reference to Their Use in the Isolation of *B. Influenzae*," which the Library of Congress does not have in its collections. The Library does, however, have Fleming's 1946 recapitulation "History and Development of Penicillin," in *Penicillin: Its Practical Applications*, which he edited.

It took ten years before interest in Fleming's discovery was sufficient to motivate others to perform the necessary biological tests on the mold and to isolate it and elucidate its structure. This was achieved by two men of very different backgrounds and skills working together. In 1940, an Australian, Howard W. Florey, and a German biochemist, Ernst B. Chain, were able to announce that they had developed a natural chemotherapeu-

Penicillins are natural by-products from molds of the genus *Penicillium.* The microorganism first identified as a source of penicillin was the mold *Penicillin notatum,* but other species, such as this star-shaped mold *Penicillium chysogenum,* also produce the life-saving drug. By 1939, penicillin was recognized as a chemotherapeutic agent that was virtually nontoxic, but it was not able to be purified and manufactured in large quantity until the United States entered the war in 1941. With the manufacture of penicillin given high government priority, the problems of mass production and quality were solved, as were those involving packaging methods and preservation. Penicillin was eventually produced in large fermentation vats similar to those used in the beer brewing industry. Prints and Photographs Division, Library of Congress, LC-USZ62-92563.

tic drug, now called penicillin, and that it had demonstrated almost miraculous results on mice injected with lethal doses of streptococci. The Library of Congress has their brief report, "Penicillin As a Chemotherapeutic Agent," as it first appeared in the August 24, 1940, issue of *Lancet*. Once the equipment and methodology for mass production of penicillin and other reliable antibiotics had been developed, modern medicine entered its truly "wonder" phase, able to save lives that had been heretofore hopelessly lost. The development of antibiotics must certainly have brought about one of the most radical changes in medical history.

Although chance and luck remain a factor in modern medical discoveries, today's major insights are more often the result of deliberate, dogged, and exhaustive investigation. Such is the story of this century's third wonder drug, cyclosporin, whose immunosuppressive properties have made organ transplants a practical reality. As with aspirin and penicillin, cyclosporin is organic, the by-product of something found originally in nature. In this case, it is an earth fungus found in the soil of Wisconsin and Norway, and its discovery demonstrates how thorough modern methods have become.

Like many drug companies, the Sandoz Corporation in Switzerland requests that its employees who travel abroad return with small samples of foreign soil. These are then tested for microorganisms that may make antibiotics. In 1970, Sandoz researchers set to work on soil from Wisconsin and Norway and discovered that both contained two new strains of fungi—*Tolypocladium inflatum* and *Cylindrocarpon lucidum*. Both new strains also produced a new substance known as Cyclosporin A (CsA). At Sandoz, Jean F. Borel set to work on this new water-insoluble substance to deter-

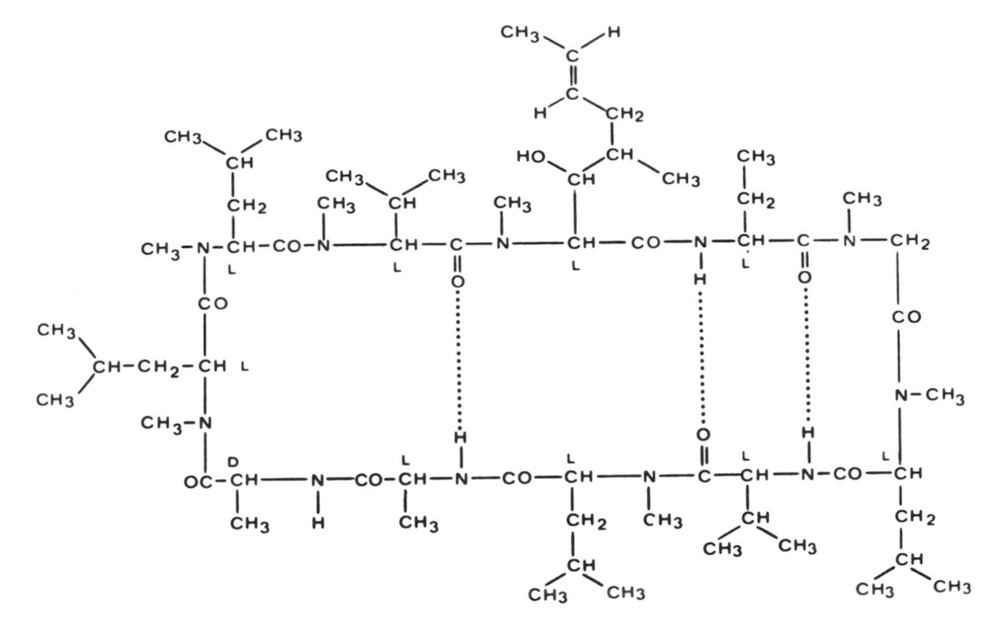

The modern era of organ and tissue transplants began with the development of cyclosporin, the chemical structure of which is shown here. Although not entirely benign—among its side effects are elevated blood pressure and possible kidney damage—cyclosporin revolutionized organ transplantation in the early 1980s. By significantly reducing the need for high dosages of steroids and other antirejection drugs, it allowed technical advances to be made in all aspects of surgical transplantation and organ preservation. Less than two decades ago, achieving the routine transplants of today would have seemed unimaginable, yet medical technology has more than kept pace with its scientific counterpart. With its rapid advances, this technology adds further complexity to the already-daunting moral and ethical questions spawned by these new medical capabilities. "Biological Effects of Cyclosporin A: A New Antilymphocytic Agent," *Agents and Actions,* July 1976. Jean F. Borel, et al. Courtesy *Agents and Actions,* Birkhauser Verlag. LC-USZ62-110442.

mine whether it possessed any anticancer or immunoregulatory properties. Borel discovered that while this new substance seemed to suppress the body's production of T cells (those produced when the body senses a foreign antigen), it did not kill them or any other rapidly dividing cells (such as bone marrow cells) the way most immunosuppressive agents did. Although Sandoz had been planning to discontinue its immunological work, Borel managed to persuade them to continue research on this promising drug, and he worked on purifying and characterizing it. Once its molecular structure was known, the lab was able to completely synthesize it and make relatively plentiful derivatives. In July 1976, Borel published its structure and properties in the journal *Agents and Actions*, and this led in turn to successful clinical trials. In late 1983 it was approved for general use, and today it is routinely used in organ transplantation. The Library of Congress has Borel's paper, "Biological Effects of Cyclosporin A: A New Antilymphocytic Agent," as it first appeared in *Agents and Actions*. Although the drug still has undesirable side effects and no one yet really knows how it works on the molecular level, it has given a real second chance to thousands of dying people, literally restoring them to a normal existence via an organ transplant. Further, it holds out hope that breakthroughs will be made in other areas of immunological research, especially for diseases like diabetes and arthritis.

These three twentieth-century wonder drugs—aspirin, penicillin, and cyclosporin—all were found first in the natural world. The lesson each holds for us would therefore seem obvious were it not apparently ignored nearly every day. As we approach the end of this century of technological marvels, we may be in danger of so altering nature's balance or normal state via our technology that we may lose not only the pleasures of the natural world (many of which have already been lost), but its essential physical diversity as well. Preservation of the earth's remaining stores of biodiversity should become a mainstream goal of science and technology, if only for humankind's own welfare, since it may be from this still-untapped pool of multifarious life forms that future researchers will draw the next cure for the next plague. Who is to say from which fungus or humble toad or familiar tree may come a cure for cancer or for AIDS?

The state of our present medical technology is such that it is rapidly becoming the domain of some of the "big questions" usually reserved for other fields. In the history of science, for example, physics is typically the forum in which issues of fundamental or universal import are appropriately raised and discussed. The answers or hypotheses given in response to these weighty matters often quickly cross over into everyday life and shape the way we perceive our universe and ourselves. So whether we acknowledge it or not, our ideas about the world are often the product of some school of thought—like Descartes's mechanism, Einstein's relativity, or Heisenberg's uncertainty principle.

Today's medical technology has in some ways thrust medicine into the domain of the philosophical. When a major program like the Human Genome Project—a program whose goal is to collect the billions of genetic bits and pieces in the human body that constitute the blueprint of life—is supported by $3 billion and promises to be sustained over the fifteen years it will take to gather, map, sequence, and interpret the DNA data, the implications are more than simply technical. Modern technology now enables us to contemplate a time when we can not only know the makeup of a human being at the chemical or subunit of DNA (deoxyribonucleic acid) level, but can also manipulate or alter that sequence or order for various purposes. Although this knowledge is meant to serve us by explaining chemically the role of genetic factors in a multitude of diseases, it contains nonetheless an unprecedented and dangerous capability for tampering with human life at the most elemental level.

Compared to the lofty and complex goals and methods of the genome project, the discovery of an effective oral contraceptive seems rather simple and straightforward, yet its achievement about three decades ago might be considered the beginnings of mankind's technological intrusion into the inner workings of our own chemical make-up and mechanisms. Understanding of this physical "what and how" of being human, coupled with a more skilled technical capability, can result in the ability to manipulate and control. In the modern and postmodern world especially, scientific enlightenment is seldom unaccompanied by technical application.

It could be argued, then, that if control is one of the major underlying themes of the history of both technology and science, it was never exercised in a more sensitive or controversial domain than that of human biology. When the American biologist Gregory Pincus and his colleagues began their investigation of the antifertility effect of steroid hormones and discovered that they could suppress ovulation, their next step marked the beginning of what might be called this century's biological interventionism. With the perfection of an effective oral contraceptive, modern technology attained a level of subtlety and sophistication heretofore unattained. In a sense, technology itself could be said to have begun its own evolution away from the obvious and the mechanical, toward a more subtle and natural mode of operation. In this case, old medical technology would have sought a mechanical solution to the problem of contraception. New technology, however, seeks its answer in the biological or chemical essence of the physical situation itself and uses its new knowledge to manipulate, alter, or redirect a natural process naturally. Thus Pincus brilliantly altered female physiology via synthetic hormones which basically trick a woman's body by simulating a natural condition (pregnancy).

Pincus's work is well documented in the collections of the Library of Congress. It has the landmark paper he wrote with M. C. Chang and three others (Zarrow, Hafez, and Merrill), titled "Studies of the Biological Activity of Certain 19-Nor Steroids in Female Animals," which first appeared in *Endocrinology* in 1956. This article was followed by more tests during the next few years, and by 1959 an oral contraceptive became available in the United States. The Library also has Pincus's later book, *The Control of Fertility* (New York, 1965), as well as his article in *Science*, "Control of Conception by Hormonal Steroids" (1966).

As technology becomes more effective, more sophisticated, more subtle, and more intrusive, it also becomes more of a threat. With the control it provides come the expected hard choices. Today's world presents decision-makers of all types with options never dreamed of before, and even average people find themselves with the ability to affect and control their

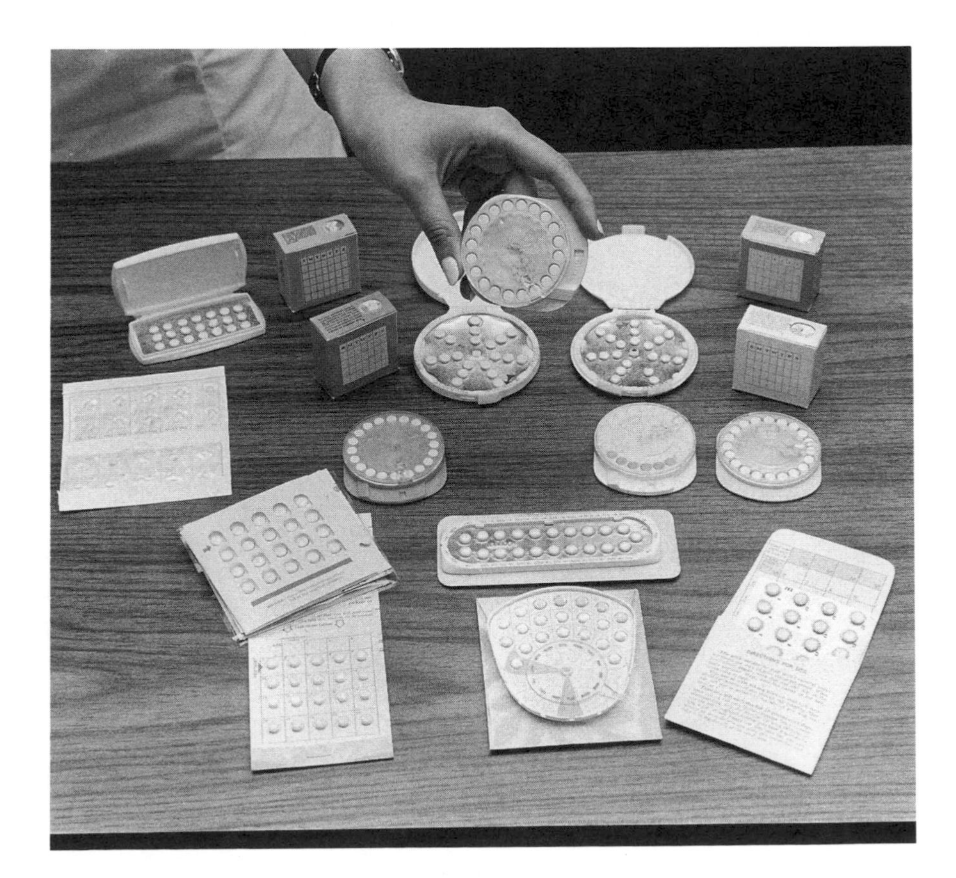

The separation of sex from reproduction was achieved in the second half of this century. With the development of an aesthetically acceptable and highly efficient oral contraceptive—shown here in its various available forms in 1968—pregnancy became an individual, discretionary matter for the woman who conscientiously took her daily pill. Whether the revolution in sexual morals and ethics that has occurred in the past three decades would have happened anyway is a matter of debate. What is more evident, however, is the level of intervention and degree of change which certain technologies can effect in our everyday lives and behavior. *U.S. News and World Report* Collection, Prints and Photographs Division, Library of Congress, LC-U9-19168-23A/24.

lives to an unprecedented degree. The irony of this modern situation is the classic human dilemma writ technologically—with each newly gained freedom comes a concomitant and often extra dimension of responsibility. Technology, no matter how complex or sophisticated, remains a tool whose final worth is judged by the ends it serves.

During the current century, technology and its long-evolving tradition have matured to the point where they appear to us as a force with singular and sometimes frightening autonomy. Following the mid-century technological high point of nuclear fission and its actual use in a desperately destructive manner, mankind's technological prowess seemingly knew no limits. Anything and everything seemed possible.

History has shown us, however, that such technological marvels—as well as those on today's horizon like the recent advances in biotechnology—are usually achieved and sustained at a great and complex cost. As our technology steadily increases in sophistication and intrudes ever more deeply and subtly in our lives, so must the tradition of technology change and broaden to include and take into account many nontechnical factors. If there is one thing certain about the technological tradition of the coming century, it is that it must have broader base values than simply those of efficiency and economy. The technological equation will, from now on, have to include nontechnical factors and values that will guide and inform decision makers. We are being forced by our depleted and spoiled resources to understand that simply because something is technologically achievable, it is not necessarily a better way. The technological roads not taken will be as important as those chosen.

Bibliography

The bibliography furnishes citations to the works discussed in the text, arranged alphabetically in sections corresponding to the chapters. Library of Congress call numbers are given wherever possible. Abbreviations indicate the custodial divisions holding works that are not in the Library's general collections. A key to these abbreviations follows.

Batchelder Coll.	Rare Book and Special Collections Division
Freud Coll.	Rare Book and Special Collections Division
G&M	Geography and Map Division
Jefferson Coll.	Rare Book and Special Collections Division
Law	Law Library
Micro	Microform Reading Room
MSS	Manuscript Division
Rare Bk. Coll.	Rare Book and Special Collections Division
Rosenwald Coll.	Rare Book and Special Collections Division
Thacher Coll.	Rare Book and Special Collections Division
Toner Coll.	Rare Book and Special Collections Division
Vollbehr Coll.	Rare Book and Special Collections Division

A number in parentheses following the designation "Rosenwald Coll." refers to the entry number for that work in the published catalog entitled *The Lessing J. Rosenwald Collection* (Washington: Library of Congress, 1977).

1. Greek and Roman Technology

Alberti, Leone Battista. De re aedificatoria. Florentiae, N. Laurentii, 1485. 204 leaves.

Incun. 1485.A58 Rare Bk. Coll.

———Libri De re aedificatoria decē. Parrhisiis [Impressum opera B. Rembolt & L. Hornken, 1512] clxxiiii leaves.

NA2515.A335 1512 Rosenwald Coll. (951)

Edited by Geoffroy Tory.

———L'architettvra di Leonbatista Alberti. Firenze, Appresso L. Torrentino, 1550. 404 p.

NA2517.A34 1550 Rosenwald Coll. (847)

Apicius. De re coquinaria. Milan, G. Le Signerre, 1498. [42] leaves; first two leaves wanting.
Incun. 1498.A6 Rare Bk. Coll.

Archimedes. Opera, quae quidem extant, omnia. Basileae, I. Heruagius, 1544. 139, 65 p.
QA31.A681 Rare Bk. Coll.

Crescenzi, Pietro de. Ruralia commoda. [Augsburg] J. Schüssler, 1471. [212] leaves, the last three (blank) wanting.
Incun. X.C89 Rare Bk. Coll.

De rebus bellicis. *In* Notitia dignitatum. Basileae [Apvd H. Frobenivm et Episcopivm] 1552. leaves q–r[3].
DG83.5.A1G4 Rare Bk. Coll.

Frontinus, Sextus Julius. De aqvae dvctibvs vrbis Romae. Altonae, Typis Eckstorffii iunioris, sumtibus I. H. Kaven, 1792. 302 p.
TD216.F7 1792 Rare Bk. Coll.

Edited by Georg Christian Adler.

——— ———*In* Cleonides. Harmonicum introductorium. Venice, S. Bevilaqua, 1497. leaves [85]–[94]
Incun. 1497.C65 Rare Bk. Coll.

Hero, *of Alexandria*. Gli artifitiosi et cvriosi moti spiritali di Herrone. Ferrara, V. Baldini, stampator ducale, 1589. 103 p.
QC142.H527 Rare Bk. Coll.

Translated by Giovanni Battista Aleotti.

———De gli avtomati. In Venetia, Appresso G. Porro, 1589. 47 leaves.
TJ215.H4 1589 Rare Bk. Coll.

Translated by Bernardino Baldi.

———Spiritalivm liber. Urbini, 1575. [80] leaves.
QC142.H54 Rosenwald Coll. (875)

Translated by Federico Commandino.

Lipsius, Justus. Poliorceticωn; sive, De machinis, tormentis, telis, libri qvinqve. Ed. 2, corr. & aucta. Antverpiae, Ex officina Plantiniana, apud I. Moretum, 1599. 218 p.
DG89.L5 1599 Rare Bk. Coll.

Palladio, Andrea. I qvattro libri dell'architettvra. Venetia, D. de' Franceschi, 1570. 4 v. in 1.
NA2515.P25 1570 Rosenwald Coll. (873)

———The first book of architecture. London, Printed by J. M. and sold by G. Richards . . . and by S. Miller . . . 1663. [10], 228 (i.e 236) p.
NA2515.P253 1663 Rare Bk. Coll.

English translation by Godfrey Richards of the French translation by Pierre Le Muet.

Piranesi, Giovanni Battista. Le antichita romane. Roma, A. Rotilj, 1756. 4 v.
NA1120.P67 Rare Bk. Coll.

———[Opere. Roma, 1761–99] 20 v.
N6916.P48 Rare Bk. Coll.

Imperfect; v. 7, text after p. lxxxvii wanting.

Pisan, Christine de. [The boke of the fayt of armes and of chyualrye] [Westminister] W. Caxton, 1489. 144 leaves, the last (blank) wanting.
Incun. 1489.P554 Rosenwald Coll. (570)

Translation of *Faits d'armes et de chevalerie*.

Scriptores rei militaris. Romae, E. Silber, 1487. 4 pts. in 1 v.
Incun. 1487.S35 Rare Bk. Coll.
Contents: Aelianus Tacticus. De instruendis aciebus.—Frontinus, Sextus Julius. Strategematicon liber.—Modestus. De uocabulis rei militaris.—Vegetius Renatus, Flavius. De re militari.

———Lutetiae, Apud C. Wechelum, 1532. 279 p.
U101.V3 1532 Rosenwald Coll. (1005)
Contents: Vegetius Renatus, Flavius. De re militari.—Frontinus, Sextus Julius. De strategematis.—Aelianus Tacticus. De instruendis aciebus.—Modestus. De vocabulis rei militaris.

Scriptores rei rusticae. [Scriptores de re rustica] Venetiis, Opera et impensa N. Ienson Gallici, 1472. 60 leaves.
Incun. 1472.P16 Rare Bk. Coll.
Imperfect copy, containing only *De re rustica libri* by Rutilius Taurus Aemilianus Palladius.

———Libri de re rustica, M. Catonis, Marci Terentii Varronis, L. Iunii Moderati Columellae, Palladii Rutilii. Parisiis, Apud I. Parvum [1533] [28], 506 (i.e. 496) p.
PA6139.R8 1533 Rare Bk. Coll.

Valturio, Roberto. De re militari. [Italy, 3d quarter of 15th cent.] [182+] leaves, the last blank.
Rosenwald Coll. ms. no. 14
Imperfect: leaves [106]–[107] and a few leaves at end wanting.

——— ———[North Italy, 3d quarter of 15th cent.] [208] leaves, the last 2 blank.
Rosenwald Coll. ms. no. 13

——— ———[Verona] J. Nicolai de Verona, 1472. [262] leaves; leaves [5], [6], and [175] are blank.
Incun. 1472.V21 Rosenwald Coll. (218)

Vegetius Renatus, Flavius. De re militari. [Cologne] N. G[oetz, ca. 1475] 40 leaves, the first and last (both blank) wanting.
Incun. X.V37 Rosenwald Coll. (61)

——— ———German. [Augsburg, J. Wiener, ca. 1475] 110 leaves.
Incun. X.V4 Rosenwald Coll. (62)

Vitruvius Pollio. De architectura libri dece. [Como, Impressa p̧ G. da Pōte, 1521] clxxxiii leaves.
NA2515.V8 1521 Rosenwald Coll. (804)
Translated by Caesare Caesariano.

2. *Medieval Technology*

Bacon, Roger. Epistolae De secretis operibus artis et naturae, et De nullitate magiae. *In* Theatrum chemicum, praecipuos selectorum auctorum tractatus de chemiae et lapidis philosophici antiquitate, veritate, jure, praestantia, & operationibus, continens. v. 5. Argentorati, Sumptibus heredum E. Zetzneri, 1660. p. 844–868.
QD25.T4, v. 5 Rare Bk. Coll.

Includes John Dee's annotations.

———Epistola De secretis operibus artis & naturae, & De nullitate magiae. *In* Manget, Jean Jacques. Bibliotheca chemica curiosa; seu, Rerum ad alchemiam pertinentium thesaurus instructissimus. t. 1. Genevae, Sumpt. Chouet, G. De Tournes, Cramer, Perachon, Ritter, & S. De Tournes, 1702. p. 616–626.
QD25.M27, v. 1 Rare Bk. Coll.

Eraclius. Here begins the first and metrical book of Eraclius, a very wise man, On the colours and arts of the Romans. Incipit primus et metricus liber Eraclii, sapientissimi viri, De coloribus et artibus Romanorum. *In* Merrifield, Mary P. Original treatises, dating from the XIIth to XVIIIth centuries, of the arts of painting, in oil, miniature, mosaic, and on glass; of gilding, dyeing, and the preparation of colours and artificial gems; preceded by a general introduction; with translations, prefaces, and notes. v. 1. London, J. Murray, 1849. p. 182–257.
MicRR 11938 ND

English and Latin on facing pages.

———Heraclius, Von den Farben und Künsten der Römer. Originaltext und Übersetzung. Mit Einleitung, Noten und Excursen versehen von Albert Ilg. Wien, W. Braumüller, 1873. xxiv, 190 p. (Quellenschriften für Kunstgeschichte und Kunsttechnik des Mittelalters und der Renaissance, 4)
N7420.E7

Grosseteste, Robert, *Bp. of Lincoln*. Commentaria in libros posteriorum Aristotelis cum textu seriatim inserto. Scriptū Gualterii Burlei super eosdem libros posteriorum. [Venice, O. Papiensis, 1497] 38 leaves.
Thacher Coll. (476)

Gundissalinus, Dominicus. De divisione philosophiae. Hrsg. und philosophiegeschichtlich untersucht nebst einer Geschichte der philosophischen Einleitung bis zum Ende der Scholastik von Ludwig Baur. Münster, Druck und Verlag der Aschendorffschen Buchhandlung, 1903. 408 p. (Beiträge zur Geschichte der Philosophie des Mittelalters, Bd. 4, Heft 2/3)
B720.B4, v. 4

Hugo *of Saint Victor*. Didascolicon. *In* Brack, Wenceslaus. Vocabularius rerum. [Basileae, P. Kollicker, 1483] leaves [65]–[97]
Law Incun. X.C33

———Didascali libri VII. De studio legendi. *In his* Opera omnia. t. 3. Mogvntiae, Sumptibus A. Hierat, excudebat I. Volmari, 1617. p. 1–40.
BX890.H75 Rare Bk. Coll.

Joannes *de Sancto Geminiano*. Summa de exemplis ac similitudinibus rerum. Noviter impressa. [Venice, J. et. G. de Gregoriis, 1497] [12], 392 leaves, leaf 12[b] blank.
Incun. 1497.J6 Rare Bk. Coll.

Kilwardby, Robert, *Cardinal, Abp. of Canterbury*. De ortu scientiarum. Edited by Albert G. Judy. [London] British Academy, 1976. lxi, 255 p. (Auctores Britannici Medii Aevi, 4)
Q175.K464 1976

Lull, Ramón. Arbor scientiae. Liber ad omnes scientiae vtilissimus. [Lugduni] 1515. 681, [19] p.
B765.L83A7 1515 Rare Bk. Coll.

Mappae clavicula. Incipit libellus dictus Mappae clavicula. *In* Archaeologia; or, Miscellaneous tracts relating to antiquity, published by the Society of Antiquaries of London. v. 32. London, 1847. p. 187–244.
DA20.A64, v. 32

———Mappae clavicula; a little key to the world of medieval techniques. [Edited and translated by] Cyril Stanley Smith and John G. Hawthorne. Philadelphia, American Philosophical Society, 1974. 128 p. (American Philosophical Society, Philadelphia. Transactions, new ser., v. 64, pt. 4)
T44.M3613 1974
"An annotated translation based on a collation of the Sélestat and Phillips-Corning manuscripts, with reproductions of the two manuscripts."

Muratori, Lodovico Antonio. Compositiones ad tingenda musiva, pelles, & alia, ad deaurandum ferrum, ad mineralia, ad chrysographiam, ad glutina quaedam conficienda, aliaque attium documenta, ante annos nongentos scripta. *In his* Antiquitates italicae medii aevi; sive, Dissertationes. t. 2. Mediolani, Ex typ. Societatis Palatinae in Regia Curia, 1739. columns 365–396.
DG443.M9, v. 2 Rare Bk. Coll.

Neckam, Alexander. The treatise De utensilibus of Alexander Neckam. (Of the twelfth century.) *In* Wright, Thomas, *ed.* A volume of vocabularies, illustrating the condition and manners of our forefathers, as well as the history of the forms of elementary education and of the languages spoken in this island, from the tenth century to the fifteenth. [London] Priv. print., 1857 (A Library of national antiquities, v. 1) p. [96]–119.
PE274.A5W6

Petrus de Sancto Audemaro. Here beginneth the book of Master Peter of St. Audemar On making colours. Incipit liber Magistri Petri de Sancto Audemaro De coloribus faciendis. *In* Merrifield, Mary P. Original treatises, dating from the XIIth to XVIIIth centuries, of the arts of painting, in oil, miniature, mosaic, and on glass; of gilding, dyeing, and the preparation of colours and artificial gems; preceded by a general introduction; with translations, prefaces, and notes. v. 1. London, J. Murray, 1849. p. 116–165.
MicRR 11938 ND
English and Latin on facing pages.

Petrus Peregrinus, *of Maricourt.* Epistola Petri Peregrini de Maricourt ad Sygerum de Foucaucourt militem De magnete. Bullettino di bibliographia e di storia delle scienze matematiche e fisiche, t. 1, mar. 1868: 70–89
QA1.B9, v. 1

———The letter of Petrus Peregrinus On the magnet, A.D. 1269. Translated by Brother Arnold, with introductory notice by Brother Potamian. New York, McGraw Pub. Co., 1904. xix, 41 p.
QC751.P49

Theophilus, *called also* Rugerus. Libri III De diversis artibus; seu, Diversarum artium schedula. Opera et studio R. Hendrie. Londini, J. Murray, 1847. li, 447 p.
N7420.T4 1847
Latin text with English translation.

———De diuersis artibus. The various arts. Translated from the Latin with introduction and notes by C. R. Dodwell. London, New York, T. Nelson [1961] lxxvii, 178 p.
N7420.T4 1961

Villard de Honnecourt. Album de Villard de Honnecourt, architecte du XIIIe siècle. Reproduction des 66 pages et dessins du Manuscrit français 19093 de la Bibliothèque nationale. Paris, Impr. Berthaud frères [1906] 18 p., 35 leaves.
MicRR 32692

———Facsimile of the sketch-book of Wilars de Honecourt, an architect of the thirteenth century; with commentaries and descriptions by M. J. B. A. Lassus and by M. J. Quicherat. Translated and edited, with many additional articles and notes, by Robert Willis. London, J. H. and J. Parker, 1859. 243 p. 73 plates.
MicRR 73154

Vincent *de Beauvais*. Speculum doctrinale. [Strassburg, The R-Printer (Adolph Rusch), not after 1478] [404] leaves, the first and last blank.
Thacher Coll. (12)

3. *Renaissance Technology*

Agnese, Battista. [Atlas of portolan charts, dedicated to Hieronymus Ruffault, abbot of St. Vaast. Venice, 1544] 15 leaves.
Vellum Chart Coll. no. 5 G&M Vault
Manuscript on vellum.

Agricola, Georg. De veteribvs et novis metallis. Libri duo. *In his* [Opera omnia] Basileae [Per H. Frobenivm et N. Episcopivm] 1546. p. [381]–414.
QE25.A34 Rare Bk. Coll.

———De re metallica libri XII. Basileae [Apvd H. Frobenivm et N. Episcopivm] 1556. 538 (i.e. 502), [74] p.
TN617.A25 1556 Rosenwald Coll. (910)

Albertus Magnus, *Saint, Bp. of Ratisbon*. De mineralibus. Pavia, C. de Canibus, 1491. [28] leaves.
Incun. 1491.V47 Rare Bk. Coll.

Apianus, Petrus. Cosmographie; ou, Description des quatre parties du monde. Corrigée & augmentée par Gemma Frison . . . auec plusieurs autres traitez . . . nouuellement traduits en langue françoise. Anuers, I. Bellere, 1581. 333 (i.e. 337) p.
GA6.A63 1581 Rosenwald Coll. (1205)

———Instrument Buch, erst von new beschriben. Ingolstadii, 1533. [107] p.
QB85.A63 1533 Rosenwald Coll. (661)

Arnaldus *de Villanova*. De vinis. [Paris] F. Baligault, for C. Jaumar and T. Julian [ca. 1500] [12] leaves.
Incun. X.A85 Rare Bk. Coll.

Ars moriendi. Ars moriendi. [Germany, 1466] [24] leaves.
Incun. X.A874 Rosenwald Coll. (20)

Bacci, Andrea. De natvrali vinorvm historia de vinis Italiae et de conuiuijs antiquorum libri septem. Romae, Ex officina N. Mutis, 1596. 14, 370 p.
TP548.B2 Rare Bk. Coll.

Bachot, Ambroise. Le govvernail. Lequel conduira le curieux de geometrie en perspectiue dedans l'architecture de fortifications, machine de guerre & plusieurs autres particularitez y contenues. Melun, Imprimé soubz lauteur, 1598. 44 p., 58 leaves.
UG144.B3 Rosenwald Coll.

Baldinucci, Filippo. Vita di Filippo di Ser Brunellesco, architetto fiorentino. Firenze, N. Carli, 1812. 392 p.
NA1123.B8B3

Bartolommeo *da li Sonetti*. Isolario. [Venice, G. Anima Mia, ca. 1485] [56] leaves.
Incun. X.B27 Rosenwald Coll. (245)

Besson, Jacques. Theatrvm instrvmentorvm et machinarum. Lvgdvni, Apud B. Vincentium, 1578. [22] p., 60 plates.
TJ144.B495 Rosenwald Coll. (1079)

Bible. *Manuscripts, Latin.* Biblia latina. April 4, 1452–July 9, 1453. 2 v. (244, 215 leaves) 2 columns, 60 lines.
Rosenwald Coll. ms. no. 28

Bible. *Latin. ca. 1454–55. Mainz. Gutenberg (42 lines).* Biblia latina. [Mainz, Printer of the 42-line Bible (J. Gutenberg) between 1454 and 1455?, not after 1456] [643] leaves (leaves [642] and [643] blank; leaf [642] wanting) bound in 3 v.
Incun. 1454.B5 Rare Bk. Coll.

Bible. *N.T. Revelation. Latin. Selections. ca. 1470.* Apocalypsis Sancti Johannis. [Germany, ca. 1470] [48] leaves.
Incun. 1474.T45 copy 3 Rosenwald Coll. (23)

Biblia pauperum. Biblia pauperum. [Dutch or German, 1465] [40] leaves.
Incun. X.B562 Rosenwald Coll. (19)

Biringucci, Vannuccio. De la pirotechnia. [In Venetia, Per V. Roffinello] 1540. 168 leaves.
TN144.B45 1540 Rare Bk. Coll.

Brunschwig, Hieronymus. Kleines Distillierbuch. Strassburg, J. Grüninger, 1500. [18], ccix leaves.
Incun. 1500.B78 Rosenwald Coll. (202)

Das Buch des edlē Ritters vñ Landtfarers Marcho Polo. *In* Historie von Herzog Leopold und seinem Sohn Wilhelm von Österreich. Augsburg, A. Sorg, 1481. leaves [76[b]]–[135[a]]
Incun. 1481.H5 Rosenwald Coll. (82)

Canestrini, Giovanni. Arte militare meccanica medievale. [Milano] Toninelli [1946] 377 p. (p. [37]–[359] on double leaves) (Le Arti, 1)
U810.C3 Rosenwald Coll.

Chaucer, Geoffrey. The workes, newlye printed, wyth dyuers workes whych were neuer in print before. [London] Printed by W. Bonham, 1542. ccclxxxxii leaves.
PR1850 1542 Rosenwald Coll. (1226)

Colombo, Cristoforo. Epistola de insulis nuper inventis. [Rome, S. Plannck, 1493] 4 leaves.
Incun. 1493.C6 Rare Bk. Coll.

Colonna, Francesco. Hypnerotomachia Poliphili. Venice, A. Manutius, 1499. [234] leaves.
Incun. 1499.C6 Rosenwald Coll. (340)

Digges, Leonard. An arithmeticall militare treatise, named Stratioticos . . . Together with the moderne militare discipline, offices, lawes and dueties in euery wel gouerned campe and armie to be obserued: long since attēpted by Leonard Digges, gentleman, augmented, digested, and lately finished, by Thomas Digges, his sonne. London, Printed by H. Bynneman, 1579. 191 (i.e. 183) p.
U101.D57 Rare Bk. Coll.
Imperfect; title-page and 2 folded diagrams wanting.

———A geometrical practical treatize named Pantometria, diuided into three bookes, longimetra, planimetra, and stereometria . . . First published by Thomas Digges. London, Printed by A. Jeffes, 1591. 195 p.
QA33.D57 Rare Bk. Coll.

Dürer, Albrecht. Etliche Vnderricht, zu Befestigung der Stett, Schloss, vnd Flecken. [Nürenberg, 1527] [52] p. (in part on fold. leaves)
NC760.D84 Rosenwald Coll. (656)

Durantis, Gulielmus, *Bp. of Mende*. Rationale divinorum officiorum. [Mainz] J. Fust and P. Schoeffer, 1459. [160] leaves.
Incun. 1459.D8 Rosenwald Coll. (29)

Dryander, Johannes. Cylindri vsvs et canones. [Marpvrgi, Impressvm per A. Kolbium, 1543] [30] p.
QB214.D7616 Rare Bk. Coll.

Translation of *Usslegung und grundtlicher Verstandt, des nützlichenn Instruments Cylindri*.

Ercker, Lazarus. Beschreibung aller fürnemisten mineralischen Ertzt vnnd Bergkwercks Arten. Franckfurt am Mayn, J. Feyerabendt, 1598. 133 leaves.
TN664.E7 1598 Rare Bk. Coll.

[Estienne, Charles] Praedivm rvsticvm, in quo cuiusuis soli vel culti vel inculti plãtarum vocabula ac descriptiones, earúmque conserẽdarum atque excolendarum instrumenta suo ordine describuntur. Lvtetiae, Apvd C. Stephanum typographũ regium, 1554. 648, [48] p.
SB99.E82 Rare Bk Coll.

Imperfect; p. [45]–[48] wanting.

Fernel, Jean. Cosmotheoria, libros duos complexa. Parisiis, In aedibus S. Colinaei, 1527. [6], 46 leaves.
QB41.F4 copy 2 Rosenwald Coll. (987)

Fine, Oronce. De solaribus horologiis, & quadrantibus, libri quatuor. Parisiis, Apud G. Cauellat, in pingui gallina [1560] 222 p.
QB215.F3 Rare Bk. Coll.

Fontana, Domenico. Della trasportatione dell'obelisco vaticano et delle fabriche di Nostro Signore papa Sisto V. Libro primo. Roma, D. Basa, 1590. 108 leaves.
DT62.O2F6 1590 Rare Bk. Coll.

No more published.
Leaves 106–107 incorrectly bound between leaves 103 and 104.

Gallucci, Giovanni Paolo. Della fabrica et vso di diversi stromenti di astronomia et cosmografia, oue si uede la somma della teorica et pratica di queste due nobilissime scienze. Venetia, R. Meietti, 1597. 228 leaves.
QB85.G28 1597 Rosenwald Coll. (883)

Giorgio, Francesco di. Trattato di architettura civile e militare di Francesco di Giorgio Martini, architetto senese del secolo XV, ora per la prima volta pubblicato per cura del cavaliere Cesare Saluzzo, con dissertazioni e note per servire alla storia militare italiana. Torino, Tip. Chirio e Mina, 1841. 2 v. *and* atlas.
NA2520.G55 folio

Giorgio, Francesco di. Trattati di architettura ingegneria e arte militare. A cura di Corrado Maltese. Trascrizione di Livia Maltese Degrassi. Milano, Il Polifilo, 1967. 2 v. (Trattati di architettura, v. 3)
NA2515.G5

Grapaldi, Francesco Mario. De partibus aedium. [Parma, 1516] [20], 265 leaves.
NA2515.G8 1516 Rare Bk. Coll.

Gregoras, Nicephorus. Insignis philosophi Nicephori Astrolabi. *In* Nicephorus Blemmydes. Logica [with other tracts. Venice, S. Bevilaqua, 1498] leaves 30^{v}–31^{v}.
Incun. 1498.N5 Rare Bk. Coll.

Guasti, Cesare. La cupola di Santa Maria del Fiore, illustrata con i documenti dell'Archivio dell'Opera secolare; saggio di una compiuta illustrazione dell'Opera secolare e del Tempio di Santa Maria del Fiore. Firenze, Barbèra, Bianchi, 1857, 241 p.
NA5621.F7G8

Guasti, Cesare, *ed.* Santa Maria del Fiore; la costruzione della chiesa e del Campanile secondo i documenti tratti dall'Archivio dell'Opera secolare e da quello di stato. Firenze, Tip. di M. Ricci, 1887. cxiv, 321 p.
BX4634.F6A4

Guglielmo da Pastrengo. De originibvs rervm libellvs. [Venetijs, Impressum per N. de Bascarinis, 1547] 131 leaves.
PA8330.G67D4 1547 Rare Bk. Coll.

Hulsius, Levinus. Theoria et praxis quadrantis geometrici &c., Das ist, Beschreibung, Unterricht vnd Gebrauch des gevierdten geometrischen vnd anderer Instrument. Noribergae, Typis Gerlachianis, sumptibus C. de Iudaeis, 1594. 70 p.
TA544 H9 1594 Rosenwald Coll. (728)

Isidorus, *Saint, Bp. of Seville.* Etymologiae. [Augustae Vindelicorum] G. Zainer, 1472. 264 leaves.
Incun. 1472.I81 Rare Bk. Coll.

Leonardo da Vinci. Il Codice Atlantico di Leonardo da Vinci nella Biblioteca Ambrosiana di Milano. Riprodotto e pubblicato dalla Regia Accademia dei Lincei sotto gli auspici e col sussidio del Re e del Governo. Trascrizione diplomaticà e critica di Giovanni Piumati. Milano, U. Hoepli, 1894–1904. 9 v.
ND623.L5A2 folio

Leonardo da Vinci. The Madrid codices. [New York] McGraw-Hill, 1974. 5 v.
Z6616.L58R47 1974
Edited by Ladislao Reti.

Lorini, Buonaiuto. Le fortificationi. Nvovamente ristampate, corrette & ampliate di tutto quello che mancaua per la lor compita perfettione, con l'aggivnta del sesto libro. Venetia, F. Rampazetto, 1609. 303 p.
UG400.L7 1609 Rare Bk. Coll.

Machiavelli, Niccolò. The arte of warre, written in Italian by Nicholas Machiauel, and set foorth in English by Peter VVithorne . . . Nevvly imprinted with other additions. [London, Imprinted by VV. VVilliamson, for I. VVight] 1573. [124] leaves.
U101.M16 1573 Rosenwald Coll. (1241)

Manetti, Antonio. Filippo Brunellesco. Mit Ergänzungen aus Vasari und Anderen, hrsg. von Heinrich Holtzinger. Stuttgart, W. Kohlhammer, 1887. 104 p.
NA1123.B8M3

Mercator, Gerardus. Atlas sive Cosmographicae meditationes de fabrica mvndi et fabricati figvra. Dvisbvrgi Clivorvm [1595] 3 pts. in 1 v.
G1007.A7 1595 Rosenwald Coll. (730)

Münster, Sebastian. Fürmalung vnd künstlich Beschreibung der Horologien, nemlich wie man der Sonnen Vren mit mancherley Weys vnd Form, vnd auff allerley Gattung entwerffen soll an die Mauren, auff die nider vnnd auffgehebte Ebne. Basel, Gedruckt bey H. Peter [1537] clvii p.
QB215.M815 Rare Bk. Coll.
First ed. published in 1531 under title *Compositio horologiorum in plano, muro, truncis, anulo concavo, cylindro et variis quadrantibus.*

———Rvdimenta mathematica, haec in duos digeruntur libros. [Basileae, In officina H. Petri, 1551] 242 p.
QB215.M82 Rare Bk. Coll.

Nigro, Andalo de. Opus astrolabii. [Ferrara] J. Picardus, de Hamell, 1475. 30 leaves, the last blank.
Incun. 1475.N53 Thacher Coll.
Edited by Petrus Bonus Advogarius.

Novus Orbis regionum. Novvs Orbis regionvm ac insvlarvm veteribvs incognitarvm, unà cum tabula cosmographica, & aliquot alijs consimilis argumenti libellis. Basileae, Apvd I. Hervagivm, 1532. [48], 584 (i.e. 586) p.
E141.N93 copy 2 Rosenwald Coll. (898)

Nunes, Pedro. Opera. Basileae, Per S. Henricpetri [1592] [6] leaves, 439 p.
QB41.N8 Rare Bk. Coll.

Ortelius, Abraham. Theatrvm orbis terrarvm. [Antverpiae, Apud A. C. Diesth, 1570] [142] leaves.
Thacher Coll. (85)

Pantheo, Giovanni Agostino. Voarchadvmia contra alchi'miam: ars distincta ab archimi'a, & sophia: cum additionibus: proportionibus: numeris: & figuris opportunis. Venetiis, 1530. 69 leaves.
QD25.P32 Rosenwald Coll. (823)

Platina, Bartolomeo. [De honesta voluptate. Venice, L. de Aquila, 1475] 93 leaves.
Thacher Coll. (278)

Porta, Giovanni Battista della. Magiae natvralis libri XX. Neapoli, Apud H. Salvianum, 1589. 303 p.
Q155.P77 1589 Rare Bk. Coll.

———Natural magick, in twenty books. London, Printed for T. Young and S. Speed, 1658. 409 p.
Q155.P78 1658 Rare Bk. Coll.

Prager, Frank D., *and* Gustina Scaglia. Mariano Taccola and his book De ingeneis. Cambridge, Mass., MIT Press [1972] 230 p.
TJ144.P73

Priscianus. De ponderibus & mẽsuris. *In his* Opera. [Venice, Vindelinus de Spira, 1470] leaves [244]–[249]
Hain 13355 Vollbehr Rare Bk. Coll.

Proclus *Diadochus*. Procli Diadochi Platonici philosophi De fabrica usuq; astrolabi. *In* Nicephorus Blemmydes. Logica [with other tracts. Venice, S. Bevilaqua, 1498] leaves 26ᵛ–30ʳ.
Incun. 1498.N5 Rare Bk. Coll.

Ptolemaeus, Claudius. Cosmographia. Latin. Rome, A. Buckinck, 1478. [124] leaves; leaf [70] blank.
Incun. 1478.P855 Rosenwald Coll. (225)

Ramelli, Agostino. Le diverse et artificiose machine. Parigi, 1588. 338 leaves.
TJ144.R3 Rosenwald Coll. (1086)

[Reisch, Gregor] Margarita philosophica nova. [Ex Argentoraco veteri, I. Grüningerus operis excussor, 1508] [639] p.
AE3.R34 1508 Rosenwald Coll. (609)

Robertus *Anglicus*. De astrolabio canones. De astrolabii compositiõe. [Perugia, ca. 1477] 42 leaves.
Copinger 5134 Rare Bk. Coll.

Sánchez de Arévalo, Rodrigo, *Bp*. Speculum vitae humanae. German. [Augsburg, G. Zainer, 1475–78] [10], clxiiii leaves.
Incun. 1475.S3 Rosenwald Coll. (57)

Scappi, Bartolomeo. Opera [Venetia, M. Tramezzino, 1574?] 369 leaves.
TX711.S4 1574 Rare Bk. Coll.

Schedel, Hartmann. Liber chronicarum. Nuremberg, A. Koberger, 1493. [20], cclxvi, [6], cclxvii–ccxcix, [3] leaves.
Incun. 1493.S3 Rosenwald Coll. (163)

Schopper, Hartmann. Panoplia omnivm illiberalivm mechanicarvm avt sedentariarum artium genera continens. Francofvrti ad Moenvm [Impressvm apud G. Coruinum, impensis S. Feyerabent, 1568] [148] leaves.
GT5780.S4 Rosenwald Coll. (705)

Stöffler, Johann. Elvcidatio fabricae vsvsqve astrolabii. Parisiis, Apud H. de Marnef, & G. Cauellat, 1570. 172 leaves.
QB85.S8 1570 Rare Bk. Coll.

Stuck, Johann Wilhelm. Antiqvitatvm convivialivm libri III. Tigvri, Excvdebat C. Froschovervs, 1582. 397 leaves.
D78.S82 Rare Bk. Coll.

Tartaglia, Niccolò. La noua scientia. [Stampata in Venetia per N. de Bascarini, 1550] 32 leaves.
QC123.T3 1550 Rare Bk. Coll.

———Quesiti et inventioni diverse. [In Venetia per N. de Bascarini] 1554. 128 leaves.
U101.T3 Rare Bk. Coll.

Tortelli, Giovanni. Orthographia. Venice, N. Jenson, 1471. [298] leaves, the first and last (wanting) blank.
Incun. 1471.T6 Rosenwald Coll. (216)

Tory, Geoffroy. Champ flevry. Au quel est contenu lart & science de la deue & vraye proportiõ des lettres attiques. Paris [1529] [8], lxxx leaves.
NK3615.T6 1529 Rare Bk. Coll.

Vasari, Giorgio. Filippo Brvnelleschi scvltore et architetto. *In his* Le vite de piv eccelenti architetti, pittori, et scvltori italiani, da Cimabve insino a' tempi nostri. Firenze, 1550. p. 291–332.
N6922.V2 1550 Rare Bk. Coll.

Verardi, Carlo. Historia Baetica. [Basel, I. Bergmann, de Olpe] 1494. [36] leaves.
Incun. 1494.V47 Rare Bk. Coll.

Vergilius, Polydorus. De inventoribvs rervm libri tres. [Argentorati, In officina M. Schurerij, 1512] lxiiii leaves.
PA8585.V4A6 1512 Rare Bk. Coll.

Von Erkantnus der Berckwerck. *In* Der Vrsprung gemeynñer Berckrecht. [Strassburg, Gedruckt durch J. Knoubloch, 1520?] p. [46]–[87]
Law Holy Roman Empire 4 Mining Law 1520 A

[Waldseemüller, Martin] Cosmographiae introdvctio cvm qvibvsdam geometriae ac astronomiae principiis ad eam rem necessariis. [Deodati, G. Lud, 1507] [103] p.
E125.V6W 1507a Rosenwald Coll. (934)

Whithorne, Peter. Certaine vvayes for the ordering of souldiours in battelray, and setting of battayles, after diuers fashions, with their maner of marching. London, Imprinted by VV. VVilliamson for I. VVight, 1573. [52] leaves.
U101.M16 1573 Rosenwald Coll. (1241)

Zonca, Vittorio. Novo teatro di machine et edificii per uarie et sicure operationi. Padoua, P. Bertelli, 1607. 115 p.
TJ144.Z87 Rare Bk. Coll.

4. *Technology and the Scientific Revolution*

Académie des sciences, *Paris*. Machines et inventions approuvées par l'Académie royale des sciences, depuis son établissement jusqu'à présent; avec leur description. t. 1–7; 1666/1701–1734/54. Paris, 1735–77. 7 v.
TJ71.A6 Rare Bk. Coll.

Accademia del cimento. Saggi di naturali esperienze fatte nell'Accademia del cimento sotto la protezione del serenissimo principe Leopoldo di Toscana e descritte dal segretario di essa accademia. Firenze, G. Cocchini, 1667. cclxix p.
QC17.A3 1667 Rare Bk. Coll.

[An Account of a book] Christiani Hugenii Astroscopia compendaria, tubi optici molimine liberata. Or the description of an aerial telescope. *In* Royal Society of London. Philosophical transactions, v. 14, July 20, 1684: 668–670.
Q41.L8, v. 14 Rare Bk. Coll.

Appier Hanzelet, Jean. La pyrotechnie. Pont a Mousson, Par I. & G. Bernard, 1630. 264 p.
UF144.A7 Rare Bk. Coll.

Biancani, Giuseppe. Sphaera mvndi; sev, Cosmographia demonstratiua, ac facili methodo tradita. Mvtinae, Ex typographia A., & H. Cassiani, 1653. 6 leaves, 232, 24 p.
GA7.B57 Rare Bk. Coll.

Blondel, François. L'art de jetter les bombes. Paris, Chez l'autheur et N. Langlois, 1683. 445 p.
UF144.B55 Rare Bk. Coll.

Böckler, Georg Andreas. Theatrum machinarvm novvm, exhibens aqvarias, alatas, ivmentarias, manuarias. Coloniae Agrippinae, Sumptibus P. Principis, 1662. 55 p., 154 plates.
TJ144.B7 Rare Bk. Coll.

Boone, Nicholas. Military discipline. Boston, Printed for and sold by N. Boone, 1701. 96 p.
U143.B72 Rare Bk. Coll.

Boyle, Robert. Nova experimenta physico-mechanica de vi aëris elastica et eivsdem effectibvs. Genevae, Apud S. de Tovrnes, 1680. 154 p.
QC161.B792 Rare Bk. Coll.
A translation of *New Experiments Physico-Mechanicall* (Oxford, 1660).

———A continuation of New experiments, physico-mechanical, touching the spring and weight of the air, and their effects. Oxford, Printed by H. Hall, for R. Davis, 1669–82. 2 v. in 1.
QC161.B793 Rare Bk. Coll.

Branca, Giovanni. Le machine; volume nuouo et di molto artificio da fare effetti marauigliosi tanto spiritali quanto di animale operatione arichito di bellissime figure conle dichiarationi a ciascuna di esse in lingua uolgare et latina. In Roma, Ad istāza di I. Martuci . . . per I. Mascardi, 1629. 40, 14, 23 leaves.
TJ144.B75 1629 Rare Bk. Coll.

Buonanni, Filippo. Micrographia cvriosa; sive, Rervm minvtissimarvm observationes. Romae, Typis A. de Rubeis, 1703. 106 p.
QH271.B8 Rare Bk. Coll.

Cardano, Girolamo. De rervm varietate libri XVII. Basileae [Per H. Petri] 1557. 6 leaves, 707, [32] p.
Q155.C35 1557a Rare Bk. Coll.

Caus, Solomon de. Les raisons des forces movvantes. Francfort, En la boutique de I. Norton, 1615. 3 v. in 1.
TJ144.C4 Rare Bk. Coll.

Descartes, René. Discours de la methode pour bien conduire sa raison, & chercher la verité dans les sciences. A Leyde, De l'Impr. de I. Maire, 1637. 78, 413, [34] p.
Q155.D43 Rare Bk. Coll.

Fahrenheit, Gabriel Daniel. Experimenta circa gradum caloris liquorum nonnulorum ebullientium instituta. *In* Royal Society of London. Philosophical transactions, v. 33, Jan./Feb. 1724: 1–3.
Q41.L8, v. 33 Rare Bk. Coll.

Félibien, André, *sieur des Avaux et de Javercy*. Des principes de l'architecture, de la sculpture, de la peinture, et des autres arts qui en dependent. Avec vn dictionnaire des termes propres à chacun de ces arts. Paris, J. B. Coignard, 1676, 795 p.
N7420.F4 Rosenwald Coll. (1412)

Flurance, David Rivault, *sieur* de. Les elemens de l'artillerie. Paris, Chez A. Beys, 1605. 16 leaves, 192 p.
4UF 72 Rare Bk. Coll.

Fontana, Francesco. Novae coelestivm, terrestrivm; rerum observationes. Neapoli, Apud Gaffarum, 1646. 151 p.
QB41.F65 Rare Bk. Coll.

Galilei, Galileo. Syderevs nvncivs. *In his* Opere. v. 2. Bononiae, Ex typographia HH. de Ducijs, 1655. p. [1]–41 (1st group)
QB3.G14 1655, v. 2 Rare Bk. Coll.

Gheyn, Jacob de. The exercise of armes for calivres, mvskettes, and pikes, after the ordre of his excellence Maurits, Prince of Orange. Printed at the Hage, 1608. [13] p., 117 plates.
U101.G4213 Rare Bk. Coll.
Imperfect; preface, letterpress, and 39 plates wanting.

———Maniement d'armes, d'arqvebuses, mousqvetz, et piqves. Amsterdam, Imprimé chez R. de Badous, on les vend' ausi a Amsterdam chez H. Laurens, 1607. [11] p., 117 plates.
U101.G43 Rare Bk. Coll.

Grillet, René. Curiositez mathematiques. A Paris, Chez J. B. Coignard, 1673. 12, 60 p.
QA71.G7 Rare Bk. Coll.

Guericke, Otto von. Experimenta nova (ut vocantur) magdeburgica de vacuo spatio. Amstelodami, Apud J. Janssonium à Waesberge, 1672. 244 p.
Q155.G93 Rare Bk. Coll.

Hadley, John. An account of a catadioptrick telescope. With the description of a machine contriv'd by him for the applying it to use. *In* Royal Society of London. Philosophical transactions, v. 33, Jan./Feb. 1724: 1–3.
Q41.L8, v. 33 Rare Bk. Coll.

Hevelius, Johannes. Machinae coelestis. Gedani, Auctoris typis, & sumptibus, imprimebat S. Reiniger, 1668–79. 2 v.
QB41.H423 1673 Rosenwald Coll.

———Selenographia; sive, Lunae descriptio. Gedani, Autoris sumtibus, typis Hünefeldianis, 1647. 563 p.
QB29.H44 Rosenwald Coll. (1321)

Hooke, Robert. An attempt to prove the motion of the earth from observations. London, Printed by T. R. for J. Martyn, 1674. 28 p.
QB633.H6 1674 Rosenwald Coll. (1522)

———Micrographia; or, Some physiological descriptions of minute bodies made by magnifying glasses. London, Printed by J. Martyn and J. Allestry, 1665. 18 leaves, 246, [10] p.
QH271.H79 copy 2 Rosenwald Coll. (1511)

Hoste, Paul. Naval evolutions; or, A system of sea-discipline. London, Printed for W. Johnston, 1762. 90 p.
V167.H825 Rare Bk. Coll.

Translation by Christopher O'Bryen of *L'art des armées navales*.

Huygens, Christiaan. Dioptrica. *In his* Opuscula posthuma. t. 1. Amstelodami, Apud Janssonio-Waesbergios, 1728. p. 1–202.
QC3.H8 Rare Bk. Coll.

———Horologivm oscillatorivm; sive, De motv pendvlorvm ad horologia aptato demonstrationes geometricae. Parisiis, Apud F. Muguet, 1673. [14], 161 p.
TS545.H88 1673 Rare Bk. Coll.

Kepler, Johann. Dioptrice; sev, Demonstratio eorum quae visui & visibilibus propter conspicilla non ita pridem inventa accidunt. Avgvstae Vindelicorvm, Typis D. Franci, 1611. 28, 80 p.
QC353.K383 1611 Rare Bk. Coll.

Kircher, Athanasius. Ars magna lucis et umbrae, in X. libros digesta. Amstelodami, Apud J. Janssonium à Waesberge & haeredes E. Wayerstraet, 1671. 16 leaves, 810 (i.e. 710) p.
Q155.K56 Rare Bk. Coll.

———Mvsvrgia vniversalis; sive, Ars magna consoni et dissoni in x. libros digesta. Romae, Ex typographia haeredum F. Corbelletti, 1650. 2 v.
ML100.A2K5 Mus Case

Leeuwenhoek, Anthony van. Arcana naturae detecta. Delphis Batavorum, Apud H. a Krooneveld, 1695. 568, [14] p.
QH271.L48 Rare Bk. Coll.

———Ontledingen en ontdekkingen . . . Vervat in verscheide brieven. Leiden, C. Boutestein, 1696–1718. 5 v.
QH9.L4 Rare Bk. Coll.

Leibniz, Gottfried Wilhelm, *Freiherr* von. Machina arithmetica in qua non additio tantum et subtractio sed et multiplicatio nullo, divisio vero paene nullo animi labore peragantur. Zeitschrift für Vermessungswesen, 26. Bd., 15 Mai 1897: 301–314.
TA501.Z5, v. 26

Transcription and German translation of a manuscript in the Leibniz Archiv at the Niedersächsische Landesbibliothek in Hannover.

Leupold, Jacob. Theatrum machinarum generale. Schau-Platz des Grundes mecha-
[1] nischer Wissenschafften. Leipzig, Druckts C. Zunkel, 1724. 10 leaves, 240 p.
TJ144.L48 1724 Rare Bk. Coll.

———Theatrum machinarum hydrotechnicarum. Schau-Platz der Wasser-Bau-
[2] Kunst. Leipzig, Druckts C. Zunkel, 1724. 12, 184 p.
TJ144.L48 1724 Rare Bk. Coll.

———Theatri machinarum hydraulicarum; oder, Schau-Platz der Wasser-Künste.
[3] Leipzig, Druckts C. Zunkel, 1724–25. 2 v.
TJ144.L48 1724b Rare Bk. Coll.

———Theatrum machinarium; oder, Schau-Platz der Heb-Zeuge. Leipzig, Ge-
[4] druckt bey C. Zunkel, 1725. 16, 162 (i.e. 164) p.
TJ144.L48 1725 Rare Bk. Coll.

———Theatrum staticum universale. Leipzig, Gedruckt bey C. Zunkel, 1726. 12,
[5] 332 (i.e. 232) p.
TA351.L48 1726 Rare Bk. Coll.

———Theatrum pontificiale; oder, Schau-Platz der Brücken und Brücken-Baues.
[6] Leipzig, Gedruckt bey C. Zunkel, 1726. 16, 153 p.
TJ144.L47 1725 Rare Bk. Coll.

———Theatrum arithmetico-geometricum; das ist, Schau-Platz der Rechen- und
[7] Mess-Kunst. Leipzig, Gedruckt bey C. Zunkel, 1727. 14, 300 (i.e. 200) p.
QA101.L595 1727 Rare Bk. Coll.

Theatrum machinarum molarium; oder, Schau-Platz der Mühlen-Bau-Kunst. Ausge-
[8] fertiget und zusammen getragen von Johann Matthias Beyern und Consorten. Neue verm. Aufl. Dresden, In der Waltherischen Hof-Buchhandlung, 1767–88. 3 v. in 1.
TJ1040.L48 1767 Rare Bk. Coll.

Scheffler, Joachim Ernst. Theatri machinarum supplementum; das ist, Zusatz zum
[9] Schau-Platz der Machinen und Instrumenten. Leipzig, Zu finden bey dem Autore und B. C. Breitkopf, 1739. 12, 100, 96 p.
TJ144.S35 1739 Rare Bk. Coll.

Leurechon, Jean. Mathematical recreations; or, A collection of sundrie excellent problems out of ancient & moderne phylosophers both usefull and recreatiue. [London] Printed for W. Leske, 1653. 19 leaves, 286 p.
Q155.L5 Rare Bk. Coll.

Malpighi, Marcello. Opera omnia. Lugduni Batavorum, Apud P. vander Aa, 1687. 2 v. in 1.
QH9.M2 Rare Bk. Coll.

Morland, *Sir* Samuel, *1st Bart*. The description and use of two arithmetick instruments. London, Printed, and are to be sold by M. Pitt, 1673. [164] p.
QA75.M7 Rare Bk. Coll.

Moxon, Joseph. Mechanic exercises; or, The doctrine of handy-works. London, 1677–[84?] 2 v. in 1.
TT144.M93 1684 Rosenwald Coll.

Napier, John. Mirifici logarithmorum canonis descriptio. Edinbvrgi, Ex officinâ A. Hart, 1614. 57, [91] p.
QA33.N44 Rare Bk. Coll.

———Rabdologiae; sev, Nvmerationis per virgulas libri dvo. Edinbvrgi, Excudebat A. Hart, 1617. 154 p.
QA75.N27 Rare Bk. Coll.

Newton, *Sir* Isaac. A letter . . . containing his new theory about light and colors. *In* Royal Society of London. Philosophical transactions, v. 6, Feb. 19, 16$\frac{71}{72}$: 3075–3087.
Q41.L8, v. 6 Rare Bk. Coll.

———Opticks; or, A treatise of the reflexions, refractions, inflexions and colours of light. London, Printed for S. Smith, and B. Walford, 1704. 144, 211 (i.e. 213) p.
QC353.N556 Rare Bk. Coll.

Nouvelle machine d'arithmetique de l'invention du Sieur Grillet horlogeur. A Paris. Journal des sçavans (Amsterdam), t. 6, 25 apr. 1678: 170–172.
AP25.J7, 1678

Palladio, Andrea. I qvattro libri dell'architettvra. Venetia, D. de' Franceschi, 1570. 4 v. in 1.
NA2515.P25 1570 Rosenwald Coll. (873)

———The architecture of A. Palladio; in four books. Revis'd, design'd, and publish'd by Giacomo Leoni. Translated from the Italian original. The 3d ed. With notes and remarks of Inigo Jones. London, A. Ward, 1742. 2 v.
NA2517.P3 Rare Bk. Coll.

Translated by Nicholas Du Bois.

Papin, Denis. A new digester or engine for softning bones, containing the description of its make and use in these particulars: viz, cookery, voyages at sea, confectionary, making of drinks, chymistry, and dying. London, Printed by J. M. for H. Bonwicke, 1681. 54 p.
TX/49.P27 1681 Rare Bk. Coll.

Bound with his *A Continuation of the New Digester of Bones* (London, Printed by J. Streater, 1687. 123 p.).

———Nova methodus ad vires motrices validissimas levi pretio comparandas. Acta eruditorum, Aug. 1690: 410–414.
Z1007.A1806, 1690

Pascal, Blaise. Œuvres. Nouv. éd. Paris, Lefèvre, 1819. 5 v.
B1900.A2 1819

———Traitez de l'eqvilibre des liqvevrs, et de la pesantevr de la masse de l'air. A Paris, Chez G. Desprez, 1663. 14 leaves, 232, [7] p.
QC143.P3 Rare Bk. Coll.

Persius Flaccus, Aulus. Persio, tradotto in verso sciolto e dichiarato da Francesco Stellvti. Roma, G. Mascardi, 1630. 218 p.
PA6555.A2 1630 Rare Bk. Coll.

Piranesi, Giovanni Battista. Le antichita romane. Roma, A. Rotilj, 1756. 4 v.
NA1120.P67 Rare Bk. Coll.

Rey, Jean. Essais, sur la recherche de la cause pour laquelle l'étain et le plomb augmentent de poids quand on les calcine. Réimpression de l'éd. de 1630, publiée avec préf. par Édouard Grimaux. Paris, G. Masson, 1896. 143 p.
QD27.R4

———The essays. A fascimile [sic] reprint of the original ed. of 1630, with an introd. by Douglas McKie. London, E. Arnold [1951] xliv p., facsim. (143 p.), xlv–lxxxiii p.
QD27.R4 1630a

Santorio, Santorio. Medicina statica; or, Rules of health, in eight sections of aphorisms. English'd by J. D. London, J. Starkey, 1676. 180 p.
RA775.S21 Rare Bk. Coll.

Savery, Thomas. An account of Mr. Tho. Savery's engine for raising water by the help of fire. *In* Royal Society of London. Philosophical transactions, v. 21, June 1699: 228.
Q41.L8, v. 21 Rare Bk. Coll.

———The miners' friend; or, An engine to raise water by fire, described, and of the manner of fixing it in mines, with an account of the several other uses it is applicable unto; and an answer to the objections made against it. *In* Gt. Brit. *Patent Office*. Supplement to the series of letters patent for inventions recorded

in the Great Seal Patent Office, and granted between the 1st March (14 Jac. I.) A.D. 1617, and the 1st October (16 Vict.) A.D. 1852; consisting for the most part of reprints of scarce pamphlets, descriptive of the early patented inventions comprised in that series. Edited by Bennet Woodcroft. v. 1. London, Printed by G. E. Eyre and W. Spottiswoode, printers to the Queen's Most Excellent Majesty, 1858. p. [114]–127.
T257.D5 1617–1852 Suppl. Rare Bk. Coll.

Scheiner, Christoph. Rosa vrsina; sive, Sol. Bracciani, Apud A. Phaeum, 1630. [40], 784 (i.e. 858), [36] p.
QB525.S28 Rosenwald Coll.

Schott, Gaspar. Mechanica hydraulico-pnevmatica. [Francofurti ad M.] Sumptu heredum J. G. Schönwetteri, excudebat H. Pigrin typographus, Herbipoli, 1657. 16 leaves, 488, [14] p.
QC143.S3 Rare Bk. Coll.

———Organum mathematicum libris IX. Opus posthumum. Herbipoli, Sumptibus J. A. Endteri, & Wolfgangi jun. heredum, excudebat J. Hertz, 1668. 18 leaves, 858, [8] p.
Q155.S39 1668 Rare Bk. Coll.

———Technica curiosa; sive, Mirabilia artis, libris XII. comprehensa. Norimbergae, Sumptibus J. A. Endteri, & Wolfgangi junioris haeredum, excudebat J. Hertz, typographus Herbipol., 1664. 20 leaves, 1044, [16] p.
Q155.S4 Rare Bk. Coll.

Starowolski, Szymon. Institvtorvm rei militaris libri VIII. Cracoviae, In officina C. Schedelij, 1639. 523 p.
U101.S79 Rare Bk. Coll.

Strada, Jacobus de, *a Rosberg*. La premiere partie des dessins artificiavlx de tovtes sortes des movlins a vent, a l'eau, a cheual & a la main, auec diuerses sortes des pompes & aultres inuentions pour faire monter l'eau au hault. Impriméz a Francfort svr le Main par P. Iacques, 1617. [5] p., 50 double plates.
TJ144.S89 Rare Bk. Coll.
Imperfect; plates 1 and 37 wanting.

Torricelli, Evangelista. Lezione accademiche. Firenze, Nella stamp. di S. A. R. per J. Guiducci, e S. Franchi, 1715. xlix, 96 p.
Q155.T69 Rare Bk. Coll.

———Opera geometrica. [Florentiae, Typis A. Masse & L. de Landis, 1644] 243, 115 (i.e. 151) p.
QA33.T69 Rare Bk. Coll.

Veranzio, Fausto, *Bp*. Machinae novae. Cvm declaratione latina, italica, hispanica, gallica, et germanica. Venetiis [1615?] 19, 18, 20, 19, 20 p.
TJ144.V4 Rare Bk. Coll.

Wallhausen, Johann Jacobi von. Uchenie i khitrost' ratnagō stroenīīa pekhotnykh līudei. V Moskvīe, 1647. 224 leaves, 35 plates.
U101.W217 Rare Bk. Coll.
Translation of *Kriegskunst zu Fuss*.

[Zeising, Heinrich] Theatrum machinarum. 6. und letzter Theill. [Gedruckt zu Altenburg in Meissen durch J. Meuschken, 1614] 88 p.
ML3805.A2B19 Mus Case
Issued with *Mvsica mathematica* (p. [89]–175), by Abraham Bartolus.

5. *Technology and the Industrial Revolution*

Académie des sciences, *Paris*. Machines et inventions approuvées par l'Académie royale des sciences, depuis son établissement jusqu'à présent; avec leur description. t. 1–7; 1666/1701–1734/54. Paris, 1735–77. 7 v.
TJ71.A6 Rare Bk. Coll.

Adams, George. Astronomical and geographical essays: containing a full and comprehensive view, on a new plan, of the general principles of astronomy; the use of the celestial and terrestrial globes . . . the description and use of the most improved planetarium, tellurian, and lunarium; and also an introduction to practical astronomy. 6th ed., corr. and enl. by William Jones. London, W. & S. Jones, 1812. xvi, 518 p.
QB42.A22

Bedos de Celles, François. L'art du facteur d'orgues. [Paris, Pmpr. de L. F. Delatour] 1776–78. xxxii, 676 p. *and* atlas of cxxxvii plates.
ML555.B4 Mus Case

Benjamin, Asher. The country builder's assistant, containing a collection of new designs of carpentry and architecture which will be particularly useful to country workmen in general. Printed at Greenfield, Mass., By T. Dickman, 1797. [32] p., 30 leaves of plates.
NA2520.B4 1797 Rare Bk. Coll.

Blanchard, Jean Pierre. Journal of my forty-fifth ascension, being the first performed in America, on the ninth of January 1793. Philadelphia, Printed by C. Cist, 1793. 27 p.
TL620.B6A4 45th Rare Bk. Coll.

Translation of his *Journal de ma quarante-cinquieme ascension*.

Blenkinsop, John. [Letter] To the editor. Monthly magazine, v. 37, June 1, 1814: 394–395.
AP4.M8, v. 37

Describes and depicts his steam carriage.

Buchanan, Robertson. Practical essays on mill work and other machinery. The 2d ed. corr., with notes and additional articles, containing new researches on various mechanical subjects, by Thomas Tredgold. London, Printed for J. Taylor, 1823. 2 v.
TJ1040.B9

Bushnell, David. General principles and construction of a sub-marine vessel, communicated in a letter of October, 1787, to Thomas Jefferson, then Minister Plenipotentiary of the United States at Paris. *In* American Philosophical Society, *Philadelphia*. Transactions. v. 4. Philadelphia, Printed and sold by T. Dobson, 1799. p. 303–312.
Q11.P6, v. 4 Rare Bk. Coll.

Cayley, *Sir* George, *Bart*. On aerial navigation. *In* The Aeronautical annual. no. 1; 1895. Edited by James Means. Boston, W. B. Clarke [1894] p. 16–48.
TL501.A3, v. 1 Rare Bk. Coll.

First published in the *Journal of Natural Philosophy, Chemistry, and the Arts*, Nov. 1809 and Feb.–Mar. 1810.

———On the principles of aerial navigation. Mechanics' magazine, v. 38, Apr. 8, 1843: 273–278.
T1.I38, v. 38

———Practical remarks on aerial navigation. Mechanics' magazine, v. 26, Mar. 4, 1837: 417–428.

T1.I38, v. 26

———Retrospect of the progress of aerial navigation, and demonstration of the principles by which it must be governed. Mechanics' magazine, v. 38, Apr. 1, 1843: 263–265.

T1.I38, v. 38

———Sir George Cayley's governable parachutes. Mechanics' magazine, v. 57, Sept. 25, 1852: 241–244.

T1.I38, v. 57

Chambers, Ephraim. Cyclopaedia; or, An universal dictionary of arts and sciences. 5th ed. London, D. Midwinter, 1741–43. 2 v.

AE5.C432 1741a Toner Coll.

Charles, Jacques Alexandre César. Représentation du globe aérostatique qui s'est élevé de dessus l'un des bassins du jardin royal des Thuilleries le 1er. décembre 1783 à 1. heure 40 min.[tes] Avec le récit de son voyage aérién. [Paris, 1783?] xv p.

TL620.C5A4 Rare Bk. Coll.

Clayton, John. An experiment to prove, that water, when agitated by fire, is infinitely more elastic than air in the same circumstances. *In* Royal Society of London. Philosophical transactions, v. 41, July/Oct. 1739: 162–166.

Q41.L8 v. 41 Rare Bk. Coll.

Davy, *Sir* Humphry, *Bart.* An account of an invention for giving light in explosive mixtures of fire-damp in coal mines, by consuming the fire-damp. *In* Royal Society of London. Philosophical transactions, v. 106, pt. 1, 1816: 23–24.

Q41.L8, v. 106

———On the fire-damp of coal mines, and on methods of lighting the mines so as to prevent its explosion. *In* Royal Socety of London. Philosophical transactions, v. 106, pt. 1, 1816: 1–22.

Q41.L8, v. 106

Encyclopédie; ou, Dictionnaire raisonné des sciences, des arts et des métiers, par une société de gens de lettres. Mis en ordre & publié par M. Diderot & quant à la partie mathématique, par M. d'Alembert. Paris, Briasson, 1751–65. 17 v.

AE25.E53 Rare Bk. Coll.

———Supplément. Amsterdam, M. M. Rey, 1776–77. 4 v.

AE25.E53 Suppl. Rare Bk. Coll.

———Recueil de planches, sur les sciences, les arts libéraux, et les arts méchaniques, avec leur explication. Paris, Briasson, 1762–72. 11 v.

AE25.E53 Plates Rare Bk. Coll.

———Suite du Recueil des planches. Paris, Panckoucke, 1777. 22 p.

AE25.E53 Plates Suppl. Rare Bk. Coll.

———Table analytique et raisonnée des matieres contenues dans les XXXIII volumes in-folio du Dictionnaire des sciences, des arts et des métiers et dans son supplément. Paris, Panckoucke, 1780. 2 v.

AE25.E53 Table Rare Bk. Coll.

Encyclopédie methodique; ou, Par ordre de matières: par une société de gens de lettres, de savans et d'artistes. [A Paris, Chez Panckoucke, 1782–1830] 202 v.

AE25.E5

Vol. 1 is in Rare Bk. Coll.

Evans, Oliver. The abortion of The young steam engineer's guide, containing an investigation of the principles, construction and powers of steam engines. Philadelphia, Printed for the author by Fry and Kammerer, 1805. 139 p.
TJ464.E9 Rare Bk. Coll.

———The young mill-wright & miller's guide. In five parts. Philadelphia, 1795. 160, 178, 90, 10, [12] p.
TS2145.E8 Rare Bk. Coll.

Faujas de Saint-Fond, Barthélemy. Description des expériences de la machine aérostatique de MM. de Montgolfier, et de celles auxquelles cette découverte a donné lieu. Suivie de recherches sur la hauteur à laquelle est parvenu le ballon du Champ-de-Mars. Paris, Cuchet, 1783–84. 2 v.
TL617.F3 1783 Rare Bk. Coll.

Fitch, John. The original steam-boat supported; or, A reply to Mr. James Rumsey's pamphlet. Shewing the true priority of John Fitch, and the false datings, &c. of James Rumsey. Philadelphia, Printed by Z. Poulson, junr., 1788. 34 p.
VM619.F54 Rare Bk. Coll.

———[Letter] To the editor of the Columbian magazine. Columbian magazine, v. 1, Dec. 1786: 174
AP2.A2U6, v. 1 Rare Bk. Coll.

Describes "a new invented steam-boat."

Franklin, Benjamin. Experiments and observations on electricity, made at Philadelphia in America. pt. [I]–II. London, Printed and sold by E. Cave, 1751–53, 2 v.
QC516.F85 1751 Rare Bk. Coll.

Fulton, Robert. Torpedo war, and submarine explosions. New-York, Printed by W. Elliot, 1810. 57 p.
V850.F975 Rare Bk. Coll.

———A treatise on the improvement of canal navigation; exhibiting the numerous advantages to be derived from small canals. London, I. & J. Taylor, 1796. xvi, 144 p.
TC744.F97 Rare Bk. Coll.

Gt. Brit. *Laws, statutes, etc., 1702–1714 (Anne).* An act for providing a publick reward for such person or persons as shall discover the longitude at sea. *In its* Anno regni Annae reginae Magnae Britanniae, Franciae, & Hiberniae, duodecimo. London, Printed by J. Baskett, 1714. p. 355–357.
Law Gt. Brit. 1

Hammond, John, *surveyor, comp.* The practical surveyor; shewing, ready and certain methods for measuring, mapping and adorning all sorts of lands and waters, by the several instruments used for this purpose: particularly, of a new theodolite. Extracted from the works of the most experienced artists. London, T. Heath, 1725. 112 p.
TA544.H3 Rare Bk. Coll.

Harris, John. Astronomical dialogues between a gentleman and a lady: wherein the doctrine of the sphere, uses of the globes, and the elements of astronomy and geography are explain'd in a pleasant, easy and familiar way. With a description of the famous instrument called the orrery. London, Printed by T. Wood for B. Cowse, 1719. 184 p.
QB42.H2 Rare Bk. Coll.

———Lexicon technicum; or, An universal English dictionary of arts and sciences: explaining not only the terms of art, but the arts themselves. London, D. Brown, 1704. [926] p.
N33.H3 Rosenwald Coll. (2548)

Harrison, John. The principles of Mr. Harrison's time-keeper; with plates of the same. Published by order of the Commissioners of Longitude. London, Printed by W. Richardson and S. Clark and sold by J. Nourse, 1767. 31 p.
QB107.H3 Rare Bk. Coll.

Hulls, Jonathan. A description and draught of a new-invented machine for carrying vessels or ships out of, or into any harbour, port, or river, against wind and tide, or in a calm. London, Printed for the author, 1737. [London, Reprinted for J. Sheepshanks, 1855] 48 p.
VM600.H91 Rare Bk. Coll.

Jeffries, John. A narrative of the two aerial voyages of Doctor Jeffries with Mons. Blanchard; with meteorological observations and remarks. The first voyage, on the thirtieth of November, 1784, from London into Kent; the second, on the seventh of January, 1785, from England into France. London, Printed for the author and sold by J. Robson, 1786. 60 p.
TL620.B6J4 1786 Rare Bk. Coll.

Lange, Ambroise Bonaventure. Description de la lampe physico-pneumatique à cylindre, inventée par M. L'Ange, distillateur du Roi. *In* Bibliothèque physico-économique, instructive et amusante. 4. année; 1785. A Paris. p. 119–126, 392.
AC20.B5, v. 4

Leybourn, William. The compleat surveyor; or, The whole art of surveying of land, by a new instrument lately invented; as also by the plain table, circumferentor, the theodolite as now improv'd, or by the chain only. 5th ed., in IX books. Every operation both geometrical & arithmetical being examin'd, and an appendix added to the whole, consisting of practical observations in land surveying, by Samuel Cunn. London, Printed for S. Ballard, 1722. 100, 166, 155 p.
TA544.L68 1722 Rare Bk. Coll.

Lunardi, Vicenzo. An account of the first aërial voyage in England, in a series of letters to his guardian, Chevalier Gherardo Compagni, written under the impressions of the various events that affected the undertaking. London, Printed for the author, sold by J. Bell, 1784. 66 p.
TL620.L8A3 Rare Bk. Coll.

McAdam, John Loudon. Remarks on the present system of road making. 7th ed., carefully rev. London, Printed for Longman, Hurst, Rees, Orme, and Brown, 1823. 236 p.
TE243.M11

Matthews, William. An historical sketch of the origin, progress, & present state of gas-lighting. London, R. Hunter, 1827. xxxii, 434 p.
TP751.M45

[Meikleham, Robert] A descriptive history of the steam engine. London, J. Knight and H. Lacey, 1824. 228 p.
TJ461.M46 1824

Murdoch, William. An account of the application of the gas from coal to economical purposes. *In* Royal Society of London. Philosophical transactions, v. 98, pt. 1, 1808: 124–132.
Q41.L8, v. 98

Pearson, William. An introduction to practical astronomy. London, Printed for the author, 1824–29. 2 v.
QB145.P3

Peckston, Thomas Snowdon. The theory and practice of gas lighting; containing much original matter relative to coal-gas, and an entirely new treatise on the

economy of the gases, procured for illuminating purposes from oil, turf, &c. 2d ed., carefully corr. London, T. & G. Underwood, 1823. 444 p.
TP751.P36

Perronet, Jean Rodolphe. Description des projets et de la construction des ponts de Neuilly, de Mantes, d'Orléans & autres; du project du canal de Bourgogne, pour la communication des deux Mers par Dijon; et de celui de la conduits des eaux de l'Yvette et de Bièvre à Paris. A Paris, de l'Impr. royale, 1782–83. 2 v.
4 Rare Books Rare Bk. Coll.

Picard, Jean. Des Herrn-Picard . . . Abhandlung von Wasserwägen, nebst einer Nachricht von etlichen Wasserwägungen, welche von ihm auf königl. Befehl sind verrichtet worden, wie auch einem kurtzen Begriff von dessen Ausmessung der Erd-Kugel; ans Licht gestellt durch den Herrn de La Hire, aus dem Frantzösischen übersetzt . . . durch Daniel Passavant. Berlin, 1749. 11 leaves, 215 p.
QB88.P33 Rare Bk. Coll.
Translation of his *Traité de nivellement*.

Pilâtre de Rozier, Jean François. Premiere expérience de la montgolfiére construite par ordre du roi, lancée en présence de Leurs Majestés, de la famille royale, et de monsieur le comte d'Haga . . . le 23 juin 1784. 2. éd. Paris, De l'Impr. de Monsieur, 1784. 10 p.
TL620.P55A3 1784 Rare Bk. Coll.

Provis, William Alexander. An historical and descriptive account of the suspension bridge constructed over the Menai Strait, in North Wales; with a brief notice of Conway Bridge. From designs by, and under the direction of, Thomas Telford. London, Printed for the author by Ibotson and Palmer, 1828. 105 p.
TG64.M4P75

Rees, Abraham. The cyclopaedia; or, Universal dictionary of arts, sciences, and literature. 1st American ed., rev., corr., enl., and adapted to this country, by several literary and scientific characters. Philadelphia, S. F. Bradford [1810–24] 41, 6 v.
AE5.R33

Robison, John. A system of mechanical philosophy. With notes by David Brewster. Edinburgh, J. Murray, 1822. 4 v. *and* atlas.
Q113.R66

Rumsey, James. A short treatise on the application of steam, whereby is clearly shewn, from actual experiments, that steam may be applied to propel boats or vessels of any burthen against rapid currents with great velocity. Philadelphia, Printed by J. James, 1788. 26 p.
VM619.R88 Rare Bk. Coll.

Smith, Robert. A compleat system of opticks in four books, viz, a popular, a mathematical, a mechanical, and a philosophical treatise. To which are added remarks upon the whole. Cambridge, Printed for the author, 1738. 2 v.
QC353.S65 Rare Bk. Coll.

Telford, Thomas. Life of Thomas Telford, civil engineer, written by himself; containing a descriptive narrative of his professional labours. Edited by John Rickman. London, Payne and Foss, 1838. xxiv, 719 p.
TA140.T3

Thornton, William. Short account of the origin of steam boats, written in 1810, and now committed to the press. Washington, Printed at Elliot's Patent Press, 1814. 20 p.
VM615.T5 1814a Rare Bk. Coll.

Tomlinson, Charles. Cyclopaedia of useful arts, mechanical and chemical, manufactures, mining, and engineering. With an introductory essay on the Great Exhibition of the Works of Industry of All Nations, 1851. London & New York, G. Virtue, 1854. 2 v.
T9.T66

Tredgold, Thomas. A practical treatise on rail-roads and carriages, shewing the principles of estimating their strength, proportions, expense, and annual produce, and the conditions which render them effective, economical, and durable; with the theory, effect, and expense of steam carriages, stationary engines, and gas machines. London, J. Taylor, 1825. 184 p.
TF144.T77

Trevithick, Francis. Life of Richard Trevithick, with an account of his inventions. London, New York, E. & F. N. Spen, 1872. 2 v. in 1.
TJ140.T8T8

Wood, Nicholas. A practical treatise on rail-roads, and interior communication in general; with original experiments, and tables of the comparative value of canals and rail-roads. London, Knight and Lacey, 1825. 314 p.
TF144.W7 1825

6. *Technology in the Nineteenth Century*

Arago, Dominique François Jean. Le Daguerréotype. *In* Académie des sciences, *Paris*. Comptes rendus, t. 9, 19 août 1839: 250–267.
Q46.A14, v. 9

Beamish, Richard. Memoir of the life of Sir Marc Isambard Brunel, civil engineer, vice-president of the Royal Society, corresponding member of the Institute of France, &c. London, Longman, Green, Longman, and Roberts, 1862. xvi, 359 p.
TA140.B753B3

Beaumont, William Worby. Motor vehicles and motors, their design, construction and working by steam, oil and electricity. 2d ed. rev. Westminster, A. Constable; Philadelphia, J. B. Lippincott Co., 1902. xvi, 636 p.
TL145.B4

Bessemer, Henry. On the manufacture of malleable iron and steel. *In* Franklin Institute, *Philadelphia*. Journal, v. 68, Dec. 1859: 390–396.
T1.F8, v. 68

Brannon, Philip. The park and the Crystal Palace; presenting a complete series of exterior views, displaying the beautiful combinations of that magnificent structure with the surrounding scenery. London, Ackermann, 1851. [2] p., 7 plates.
DA687.C9B8 P&P Case

Carnot, Sadi Nicolas Léonard. Reflections on the motive power of heat and on machines fitted to develop that power. Edited by R. H. Thurston. New York, J. Wiley, 1890. 260 p.
QC311.C29

Translation of his *Réflexions sur la puissance motrice du feu*.

Chanute, Octave. Progress in flying machines. New York, American Engineer and Railroad Journal [1894] 308 p.
TL670.5.C5 1894 Rare Bk. Coll.

Clapeyron, Benoît Paul Émile. Mémoire sur la puissance motrice de la chaleur. *In* Paris. École polytechnique. Journal. t. 14, 23. cahier. A Paris, De l'Impr. royale, 1834. p. 153–190.
QA1.P19, v. 14

———Memoir on the motive power of heat. *In* Scientific memoirs, selected from the transactions of foreign academies of science and learned societies, and from foreign journals. Edited by Richard Taylor. v. 1. London, Printed by R. and J. E. Taylor, 1837. p. 347–376.
Q111.S3, v. 1

Clerk, *Sir* Dugald. The gas and oil engine. 7th ed., rev. and enl. New York, J. Wiley [1896] 558 p.
TJ755.C63

Daguerre, Louis Jacques Mandé. Historique et description des procédés du daguerréotype et du diorama. Nouv. éd., corr., et augm. du portrait de l'auteur. Paris, A. Grioux, 1839. 76 p.
TR365.D14

———An historical and descriptive account of the various processes of the daguerréotype and the diorama. London, McLean, 1839. 86 p.
TR365.D13

Diesel, Rudolf. Diesels rationeller Wärmemotor, *In* Verein Deutscher Ingenieure. Zeitschrift, Bd. 41, 10.–17. Juli 1897: 785–791, 817–821.
TA3.V5, v. 41, Heft 2

———Diesel's rational heat motor. Progressive age: gas, electricity, water, v. 15, Dec. 1–15, 1897: 575–578, 602–607; v. 16, Jan. 1–15, 1898: 5–6, 30–35.
TP700.G14, v. 15–16

Eads, James Buchanan. Improvement of the mouth of the Mississippi River. [St. Louis, 1874] 8 p.
TC425.M65E16

Faraday, Michael. Experimental researches in electricity. London, R. and J. E. Taylor, 1839–55. 3 v.
QC503.F21 Rare Bk. Coll.

Henry, Joseph. Scientific writings. Washington, Smithsonian Institution, 1886. 2 v. in 1. (Smithsonian miscellaneous collections, v. 30)
Q11.S7, v. 30

Hertz, Heinrich Rudolph. Untersuchungen ueber die Ausbreitung der elektrischen Kraft. Leipzig, J. A. Barth, 1892. 295 p.
QC661.H59 Rare Bk. Coll.

Illinois and St. Louis Bridge Company. Report of the engineer-in-chief [James B. Eads] St. Louis, Missouri Democrat Book and Job Print. House, 1868. 77 p.
TG25.S15 1868 Rare Bk. Coll.

[Lardner, Dionysius] [Babbage's calculating engine] Edinburgh review, v. 59, July 1834: 263–327.
MicRR 32977

Lenoir Gas Engine Company, *New York*. The Lenoir gas-engine. [New York] 1866. 20 p.
TJ776.L57

Lilienthal, Otto. Der Vogelflug als Grundlage der Fliegekunst. Ein Beitrag zur Systematik der Flugtechnik. Auf Grund zahlreicher von O. und G. Lilienthal ausgeführter Versuche. Berlin, R. Gaertner, 1889. 187 p.
TL570.L5 Rare Bk. Coll.

———Birdflight as the basis of aviation; a contribution towards a system of aviation, compiled from the results of numerous experiments made by O. and G. Lilienthal. With a biographical introduction and addendum by Gustav Lilienthal. Translated from the 2d ed. by A. W. Isenthal. London, New York, Longmans, Green, 1911. xxiv, 142 p.
TL570.L6 Rare Bk. Coll.

London. Great Exhibition of the Works of Industry of All Nations, 1851. Reports by the juries on the subjects in the thirty classes into which the exhibition was divided. By authority of the Royal Commission. London, Spicer Bros., Wholesale Stationers; W. Clowes, Printers, Contractors to the Royal Commission, 1852. 4 v.
T690.D8 P&P Case Z
"Presented by Her Majesty's Commissioners for the Exhibition of M.D.CCC.LI to the Government of the United States."

Lumière, Auguste, *and* Louis Jean Lumière. Notice sur le cinématographe. Lyon, Impr. L. Decléris, 1897. 29 p.
TR880.L8 Rare Bk. Coll.

Marconi, Guglielmo, *marchese*. Wireless telegraphy. *In* Smithsonian Institution. Annual Report of the Board of Regents, showing the operations, expenditures, and condition of the Institution for the year ending June 30, 1901. Washington, Govt. Print. Off., 1902. p. 287–298.
Q11.S66, 1901

Menabrea, Luigi Federico, *marchese* di Valdora. Notions sur la machine analytique de M. Charles Babbage. Bibliothèque universelle de Genève, nouv. sér., 7. année, oct. 1842: 352–376.
AP24.B5, Oct. 1842

———Sketch of the analytical engine invented by Charles Babbage Esq. *In* Scientific memoirs, selected from the transactions of foreign academies of science and learned societies, and from foreign journals. Edited by Richard Taylor. v. 3. London, Printed by R. and J. E. Taylor, 1843. p. 666–731.
Q111.S3, v. 3
Translated and extensively annotated by Ada King, Countess of Lovelace.

Mr. Babbage's invention. Copies of the correspondence between the Lords Commissioners of His Majesty's Treasury and the President and Council of the Royal Society, relative to an invention of Mr. Babbage. Ordered, by the House of Commons, to be printed, 22 May 1823, [London] 1823. 6 p. (Gt. Brit. Parliament. House of Commons. Sessional papers, 1823. no. 370)
J301.K6, 1823, v. 15
Includes text of letters from Charles Babbage and George Harrison to Sir Humphry Davy in his capacity as president of the Royal Society, and the Society's brief report on Babbage's letter.

Muybridge, Eadweard. Animal locomotion. An electro-photographic investigation of consecutive phases of animal movements. 1872–1885. Published under the auspices of the University of Pennsylvania. Plates. Philadelphia, 1887. 16 v.
QP301.M8 Plates Rare Bk. Coll.
Plates 128, 137, and 531 in duplicate; plates 47, 115, 128 (2), 131, 137 (1), 150, 195, and 531 (1) mutilated.

——— ———Prospectus and catalogue of plates. Philadelphia, J. B. Lippincott Co., 1887. 18, xxxii p.
QP301.M8 Prospectus Rare Bk. Coll.

———Animals in motion. An electro-photographic investigation of consecutive phases of muscular actions. Commenced 1872. Completed 1885. London, Chapman & Hall, 1907. 264 p.
QP301.M83

Nasmyth, James Hall. James Nasmyth engineer; an autobiography, edited by Samuel Smiles. New ed., rev. and corr. for American readers. New York, Harper, 1884. xvii, 461 p.
TJ140.N25A3

Rankine, William John Macquorn. A manual of the steam engine and other prime movers. 3d ed., rev. London, C. Griffin, 1866. xxxii, 575 p.
TJ250.R21

Roebling, Washington Augustus. Pneumatic tower foundations of the East River suspension bridge. New York, Averell & Peckett, 1873. 92 p.
TG25.N53R7

Sprague, Frank Julian. Report on the exhibits at the Crystal Palace electrical exhibition, 1882. Washington, Govt. Print. Off., 1883. 169 p. (U.S. Office of Naval Intelligence. Information from abroad. General information series, no. II)
V1.U5

Swank, James Moore. History of the manufacture of iron in all ages, and particularly in the United States for three hundred years, from 1585 to 1885. Philadelphia, The Author, 1884. 428 p.
TN703.S97

Talbot, William Henry Fox. The pencil of nature. London, Longman, Brown, Green, & Longmans, 1844. [80] p.
TR144.T3 Rare Bk. Coll.

Whipple, Squire. Bridge-building; being the author's original work, published in 1847, with an appendix, containing corrections, additions, & explanations, suggested by subsequent experience: to which is annexed an original article on the doctrine of central forces. Albany, N.Y., 1869. 250, 15 p.
TG145.W57

7. *Technology in the Twentieth Century*

Aiken, Howard Hathaway, *and* Grace Murray Hopper. The automatic sequence controlled calculator. Electrical engineering, v. 65, Aug./Sept.–Nov. 1946: 384–391, 449–454, 522–528.
TK1.A61, v. 65

Ayrton, William Edward, *and* John Perry. Seeing by electricity. Nature, v. 21, Apr. 22, 1880: 589.
Q1.N2, v. 21

Badovici, Jean. Grandes constructions, realisées par E. Freyssinet. [Paris] Éditions A. Morancé [1927?] 17 p., 25 plates.
TA683.B2 1927a

Baekeland, Leo Hendrik. The synthesis, constitution, and uses of Bakelite. Journal of industrial and engineering chemistry, v. 1, Mar. 1909: 149–161.
TP1.I6, v. 1

Baird, John Logie. An account of some experiments in television. Wireless world and radio review, v. 14, May 7, 1924: 153–155.
TK5700.W55, v. 14

Bardeen, John, *and* Walter Houser Brittain. Physical principles involved in transistor action. Bell System technical journal, v. 28, Apr. 1949: 239–277.
TK1.B425, v. 28

——— ———Physical review, 2d ser., v. 75, Apr. 15, 1949: 1208–1225.
QC1.P4, s. 2, v. 75

Bohr, Niels, *and* John Archibald Wheeler. The mechanism of nuclear fission. Physical review, 2d ser., v. 56, Sept. 1, 1939: 426–450.
QC1.P4, s. 2, v. 56

Borel, Jean F., Camille Feurer, H. U. Gubler, *and* H. Stähelin. Biological effects of cyclosporin A: a new antilymphocytic agent. Agents and actions, v. 6, July 1976: 468–475.
RM1.A3, v. 6

Braun, Ferdinand. Ueber ein Verfahren zur Demonstration und zum Studium des zeitlichen Verlaufes variabler Ströme. Annalen der Physik und Chemie, n.F., Bd. 60, No. 3, 1897: 552–559.
QC1.A6, n.F., v. 60

Carothers, Wallace Hume. Polymerization. Chemical reviews, v. 8, June 1931: 353–426.
QD1.A5635, v. 8

Chain, *Sir* Ernst Boris, Howard Walter Florey, *Baron* Florey, Arthur Duncan Gardner, Norman George Heatley, Margaret Augusta Jennings, Jean Orr-Ewing, *and* Arthur Gordon Sanders. Penicillin as a chemotherapeutic agent. Lancet, v. 239, Aug. 24, 1940: 226–228.
R31.L3, v. 239

Dreser, Heinrich. Pharmakologisches über Aspirin (Acetylsalicylsäure). Archiv für die gesammte Physiologie des Menschen und der Thiere, 76. Bd., 5./6. Heft, 1899: 306–318.
QP1.A63, v. 76

Farnsworth, Philo Taylor, *and* Harry Raymond Lubcke. Transmission of television images. Radio, v. 11, Dec. 1929: 36, 85–86.
TK6540.R17, v. 11

Farnsworth, Philo Taylor. An electrical scanning system for television. Radio industries, v. 5, Nov. 1930: 386–389.
TK6540.R52, v. 5

Fermi, Enrico. Collected papers (Note e memorie). [Chicago] University of Chicago Press [1962–65] 2 v.
QC3.F39

Fleming, *Sir* Alexander. History and development of penicillin. *In* Fleming, *Sir* Alexander, *ed.* Penicillin, its practical application. London, Butterworth, 1946. p. 1–23.
RM666.P35F5 1946

Freyssinet, Eugène. Une révolution dans l'art de bâtir; les constructions précontraintes. Paris, Librairie de l'enseignement technique, L. Eyrolles, 1941. 27 p.
TA683.F83

Frisch, Otto Robert. Physical evidence for the division of heavy nuclei under neutron bombardment. Nature, v. 143, Feb. 18, 1939: 276.
Q1.N2, v. 143

Goddard, Robert Hutchings. A method of reaching extreme altitudes. City of Washington, Smithsonian Institution, 1919. 69 p. (Smithsonian miscellaneous collections, v. 71, no. 2)
Q11.S7, v. 71

Gowing, Margaret M. Britain and atomic energy, 1939–1945. With an introductory chapter by Kenneth Jay. [New York] St. Martin's Press [1964] xvi, 464 p.
QC773.A1G6 1964

Hahn, Otto, *and* Fritz Strassmann. Über den Nachweis und das Verhalten der bei der Bestrahlung des Urans mittels Neutronen entstehenden Erdalkalimetalle. Die Naturwissenschaften, 27. Jahrg., 6. Jan. 1939: 11–15.
Q3.N7, v. 27

Halban, Hans von, Frédéric Joliot, *and* Lew Kowarski. Liberation of neutrons in the nuclear explosion of uranium. Nature, v. 143, Mar. 18, 1939: 470–471.
Q1.N2, v. 143

Harvard University. *Computation Laboratory*. A manual of operation for the automatic sequences controlled calculator. With a foreword by James Bryant Conant. Cambridge, Mass., Harvard University Press, 1946. 561 p.
QA75.H325 Rare Bk. Coll.

General editor: Lt. Grace M. Hopper.

Haynes, James Richard, *and* William Bradford Shockley. Investigation of hole injection in transistor action. Physical review, 2d ser., v. 75, Feb. 15, 1949: 691.
QC1.P4, s. 2, v. 75

Jenney, William Le Baron. The Chicago construction; or, Tall buildings on a compressible soil. Inland architect and news record, v. 18, Nov. 1891: 41.
NA1.I5, v. 18

Loring, Sanford E., *and* William Le Baron Jenney. Principles and practice of architecture. Chicago, Cobb, Pritchard, 1869. 62 p., 45 plates.
NA2610.L7 folio

Meitner, Lise, *and* Otto Robert Frisch. Disintegration of uranium by neutrons: a new type of nuclear reaction. Nature, v. 143, Feb. 11, 1939: 239–240.
Q1.N2, v. 143

Pennsylvania. University. *Moore School of Electrical Engineering*. Theory and techniques for design of electronic digital computers. Lectures given at the Moore School 8 July 1946–31 August 1946. Philadelphia, 1947–48. 4 v.
QA75.P4 Rare Bk. Coll.

Pincus, Gregory, Min Chueh Chang, Meyer X. Zarrow, Elsayed Saad Eldin Hafez, *and* Anne Merrill. Studies of the biological activity of certain 19-nor steroids in female animals. Endocrinology, v. 59, Dec. 1956: 695–707.
QP187.A25, v. 59

Pincus, Gregory. Control of conception by hormonal steroids. Science, v. 153, July 29, 1966: 493–500.
Q1.S35, v. 153

———The control of fertility. New York, Academic Press, 1965. xvii, 360 p.
RG136.P55

Seeing by electricity. Scientific American, new ser., v. 42, June 5, 1880: 355.
T1.S5, n.s., v. 42 folio

About the inventions of George R. Carey.

Smyth, Henry De Wolf. Atomic energy for military purposes; the official report on the development of the atomic bomb under the auspices of the United States Government, 1940–1945. Princeton, Princeton University Press, 1945. 264 (i.e. 266) p.
QC173.S4735 1945a Rare Bk. Coll.

———A general account of the development of methods of using atomic energy for military purposes under the auspices of the United States Government, 1940–1945. Washington, D.C. [U.S. Govt. Print. Off.] 1945. 182 p.
QC173.S474 1945 Rare Bk. Coll.

Staudinger, Hermann. Uber die Konstitution des Kautschuks (6. Mitteilung). *In* Deutsche Chemische Gesellschaft, *Berlin*. Berichte, 57. Jahrg., Nr. 7, 1924: 1203–1208.
QD1.D4, v. 57

Swinton, Alan Archibald Campbell. Distant electric vision. Nature, v. 78, June 18, 1908: 151.
Q1.N2, v. 78

———Presidential address. *In* Röntgen Society, *London*. Journal, v. 8, Jan. 1912: 1–13.
QC1.B7, v. 8

T͡Siolkovskiĭ, Konstantin Ėduardovich. Aėrostat metallicheskiĭ upravli͡aemyĭ. Izdanīe S. E. Chertkova. Moskva, Tip. M. G. Bolchaninova, 1892. 83 p.
TL654.T8A3

———Izbrannye trudy K. Ė. T͡Siolkovskogo; pod obshcheĭ redakt͡sieĭ inzh.-mekh. E. V. Latynina, s biograficheskim ocherkom prof. N. D. Moiseeva. [Moskva] ONTI NKTP SSSR, Gosmashmetizdat, 1934. 2 v.
TL507.T7

———Sobranie sochineniĭ. Moskva, Izd-vo Akademii nauk SSSR, 1951–64. 4 v.
TL507.T67

Turing, Alan Mathison. On computable numbers, with an application to the Entscheidungsproblem. *In* London Mathematical Society. Proceedings, ser. 2, v. 42, pt. 3–4, 1936: 230–265.
QA1.L5, s. 2, v. 42

Von Neumann, John. Design of computers, theory of automata and numerical analysis. Oxford, New York, Pergamon Press, 1963. 782 p. (*His* Collected works, v. 5)
QA3.V5942, v. 5

Was the Home Insurance Building in Chicago the first skyscraper of skeleton construction? Architectural record, v. 76, Aug. 1934: 113–118.
NA1.A6, V. 76

Includes text of the report submitted by a committee of architects appointed by the Marshall Field Estate to investigate and decide upon the question.

Wright, Wilbur, *and* Orville Wright. The papers of Wilbur and Orville Wright, including the Chanute-Wright letters and other papers of Octave Chanute. Marvin W. McFarland, editor. New York, McGraw-Hill Book Co. [1953] 2 v.
TL540.W7A4

Zworykin, Vladimir Kosma. Television through a crystal globe. Radio news, v. 11, Apr. 1930: 905, 949, 954.
TK6540.R668, v. 11

Index

Numbers in **boldface type** indicate pages where illustrations are found.